ELECTRONICS TECHNOLOGY

DEVICES AND CIRCUITS

William E. Dugger, Jr.
Professor and Administrative Leader
Virginia Tech

Howard H. Gerrish

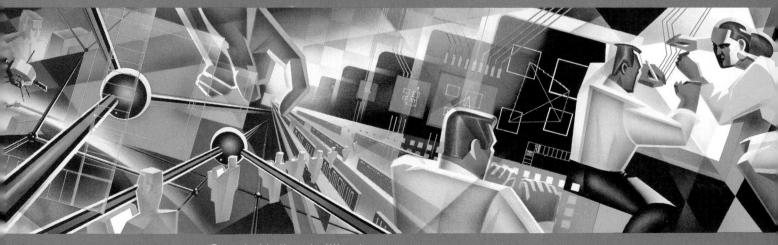

South Holland, Illinois
The Goodheart-Willcox Company, Inc.
Publishers

Copyright 1994

by

THE GOODHEART-WILLCOX COMPANY, INC.

All rights reserved. No part of this book may be reproduced, stored in a retrieval system, or transmitted in any form or by any means, electronic, mechanical, photocopying, recording, or otherwise, without the prior written permission of The Goodheart-Willcox Company, Inc. Manufactured in the United States of America.

Library of Congress Catalog Card Number 93-25857
International Standard Book Number 0-87006-085-6

1 2 3 4 5 6 7 8 9 10   94   98 97 96 95 94

Library of Congress Cataloging in Publication Data

Dugger, William.
  Electronics technology : devices and circuits / by William E. Dugger, Jr. and Howard H. Gerrish.

    p.   cm.
  Includes index.
  ISBN 0-87006-085-6
  1. Electronics.   I. Gerrish, Howard H.   II. Title.
TK7816.D77   1993
621.381--dc20   93-25857
                                          CIP

Artwork on cover is courtesy of **Fujitsu Microelectronics, Inc.**, San Jose, CA.
Artwork was created by Transphere International, San Francisco, CA.

Introduction

Electronics Technology: Devices and Circuits is about the field of electricity and electronics. This important part of technology is the basis for the *Information Age*—the period of time in which we now live. The advent of this era began with the development of the first electronic computer, which happened around the time the transistor was invented. Its invention made possible the invention of the integrated circuit, or IC, a decade later. The IC has been the primary factor behind the evolution of the modern computer. These advancements are most important to study, as well as their impacts on society.

This text is intended to instruct students seeking background information on electricity and electronics fundamentals, components, circuits, and applications. It explores the use of electronic components in fields of communication, automation and control, and computer and space technology. It shows how solid-state electronics has made possible the miniaturization of electronic equipment and improved circuit reliability to a remarkable degree. Its depth of coverage will provide you with a comprehensive background in this exciting field.

Chapters 1 through 11 herein present a study of direct current and alternating current electricity. This information is essential to the understanding of semiconductor devices and circuits.

Chapters 12 through 21 present theory and applications that include discrete diodes, transistors, and other solid-state devices. Integrated circuits and linear and digital applications are also presented. Chapter 21 is devoted exclusively to computers and microprocessors.

Important careers in electricity and electronics are presented in Chapter 22. This information is very useful to the secondary-school, community-college, or university student planning a career in this field. Also, the adult who is either working in this area or is planning a career change to this ever-expanding field can benefit from this career information.

This text also includes many projects and experiments. Each completed project will aid you in understanding the theory presented and help develop your manipulative skill in electronic circuit construction. In addition, an Activity Manual that is correlated to this textbook is available for the student or experimenter. The Activity Manual contains numerous experiments that help clarify the theory presented in the text. It also contains review questions, to be used as a study guide. These questions assist in mastering the textbook information on a chapter-by-chapter basis.

William E. Dugger
Howard H. Gerrish

Contents

Section III

Section IV

Activities and Projects

There are a number of simple activities presented in this text that demonstrate various concepts of electricity. These activities can be performed by the student or instructor if so desired. In addition, the following projects are included in this text:

Acknowledgments

The preparation of a text of this nature is necessarily the result of many people in the field of electronics sharing their knowledge and experience. To those who have assisted us, we express sincere appreciation and hope that together we have made some significant contribution toward better education in schools. We wish to thank these industries or associations for generously supplying illustrative material and technical information.

AEMC Corp.
Allen-Bradley Co.
AP Products
Apple Computer, Inc.
ASCO
AT&T
Baldor
Beckman Instruments, Inc.
B&K Precision
Brown and Sharpe
Casio, Inc.
Centralab
Cerwin-Vega
Charles Shuler, California Univ.
  of Pennsylvania, California, PA
Cobra Div., Dynascan Corp.
Commonwealth Edison
Dale Electronics
Delco Products Div., General
  Motors Corp.
Delco Remy Div., General Motors
  Corp.
Denning Mobile Robotics Inc.
E.F. Johnson
ESB, Inc. Automotive Div.
Eveready
Fujitsu Microelectronics, Inc.
Furnas Electric Co.
General Motors Corp.
Graymark Electronics
Hewlett-Packard
Hickok Electrical Instruments, Inc.
Hitachi Ltd.
Hughes Aircraft Co.
IBM Corp.
IDESCO Corp.
Intel Corp.
International Rectifier Corp.
IRC, Inc.
IRC, Inc., Semiconductor Div.
Japan Railways Group

J.W. Miller
Lab-Volt Div., Buck Engineering Co.
LeCroy Corp.
Littelfuse
L.S. Starrett Co.
Magnecraft
Marantz
Martin Marietta
Maxell Corp.
Microtran Company, Inc.
Miller Electric Mfg. Co.
Monolithic Memories Inc.
Motorola, Inc.
National Semiconductor Corp.
NCR Corp.
NEC Electronics, Inc.
NEC Telephones, Inc.
Ohmite Mfg. Co.
Omega Engineering, Inc.
Parker Hannifin Corp.
Pioneer
Potter & Brumfield
Rexroth Corp.
Ricoh Corp.
Sargent-Welch Scientific Co.
Sharp
Siemens
Sonin, Inc.
Sprague Electric Co.
Texas Crystals
Triad
Triplett Corp.
Uniden Corp. of America
Union Carbide Co.
United Transformer Co.
U.S. Air Force
U.S. Department of Energy
Waterloo Manufacturing Software
Westinghouse
Weston Instruments, Inc.
W.W. Grainger, Inc.

SAFETY PRECAUTIONS FOR THE
ELECTRICITY-ELECTRONICS LABORATORY

There is always an element of danger when working with electricity. Observe all safety rules that concern each project and be particularly careful not to contact any live wire or terminal, even if it is connected to low voltage. Projects do not specify dangerous voltage levels. However, keep in mind at all times that it is possible to experience a surprising electric shock under certain circumstances. Even a normal healthy person can be injured or seriously hurt by the shock or what happens as a result of it. In addition, tools used in the electronics laboratory can cause injury if used carelessly. The importance of following safe practices and procedures cannot be overemphasized. There is no such thing as a 100% safe electronics laboratory. Develop sound safety habits!

Section I

ELECTRICAL ENERGY AND POWER

1 Theory of Energy
2 Sources of Electricity
3 Insulators, Semiconductors, and Conductors
4 Resistive Circuits

Some say that the field of electricity and electronics is the prime mover of the Information Age in which we now live. The organized field of electricity and electronics is only two centuries old, hardly a pebble on the beach of time. The number of inventions and innovations in electricity has grown dramatically in the last fifty years.

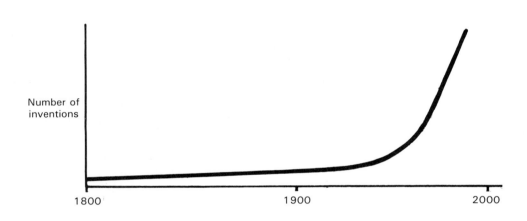

Radio is about a century old, while television is about 50 years old. The transistor was invented in 1948, and, a decade later, the integrated circuit was invented. Personal computers were commercially developed and sold for the first time in 1975. Superconductors were refined for a potential market in the late 1980s.

The first four chapters of this text present the basics of dc electricity. The purpose of the section is to provide a solid foundation for learning.

Chapter 1 discusses the atomic theory underlying the basis of the electron and the essence of electricity. Static electricity is discussed along with the law of charges.

Some common sources of electricity are presented in Chapter 2. Electric current and voltage are explained in detail. The conversion of chemical, heat, light, and mechanical forms of energy to electrical energy is explained.

Chapter 3 gives an in-depth explanation of insulators, semiconductors, and conductors. The field of superconductors is presented. Resistance, the opposition to electricity, is discussed in this chapter.

In Chapter 4, the fundamental *Ohm's law* is presented. Basic series, parallel, and combination circuits are explored. Electrical power is also presented.

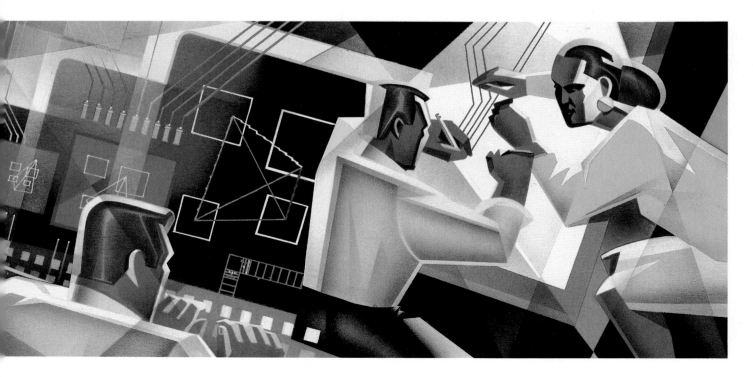

Applications of Electronics Technology:
This pocket-sized NICAD battery charger has a retractable wall plug that plugs
into wall outlets. Its computerized technology performs with integrated circuitry. (SAFT)

Chapter 1   THEORY OF ENERGY

The fundamentals of electricity and electronics are founded in the atomic building blocks that are common to all matter.

After studying this chapter, you will be able to:
☐ *Discuss the nature and atomic structure of matter.*
☐ *Explain the energy shells of the atom.*
☐ *Talk about electrostatic fields and the law of charges.*
☐ *Explain Coulomb's law.*
☐ *Demonstrate how an electroscope can be used to detect electrostatic charges.*

INTRODUCTION

Since the beginning of time, people have searched for ways to convert energy into useful work. Only in recent history have we emerged from the dark ages of mystery and superstition and made intelligent investigations into the sources and uses of energy. A great expanse of time extends from the crude inventions of the past to the now sophisticated age of orbiting satellites, nuclear power plants, and computerized machinery. It is said that more has been learned in the field of science during the last one hundred years than in all the previous years human beings have existed on Earth.

Astronauts who traveled to the Moon have disclosed it to be a cold and barren outpost. Can you image the life, if any, on Earth if it were not for the Sun and its continuous supply of light and heat energy? Energy is our life. It grows our food and supplies our water. It is harnessed to manufacture our homes, our clothes, and many conveniences. Its conversion to transportation has provided us with ships and trains, automobiles, and planes. Energy conversion has joined remote world locations within a world community by communications.

What does conversion of energy have to do with electronics? Electronics is a study of energy conversion and control. It is a study of the discovery and the development of an electrical potential and its transfer and conversion into work. At the core of this study is *matter*.

THE STRUCTURE OF MATTER

Matter may be defined as anything that occupies space and has mass. It may be composed of any number of elementary substances found in nature. Matter may take the form of a solid, a liquid, or a gas. It may be an *element*, a *compound*, or a *mixture*.

The **element** is the basic building block of nature. It is the basic or purest form of matter. It was originally defined as a substance that could not be divided or decomposed into simpler substances. However, through the development, construction, and use of accelerators, or atom smashers, scientists have divided some of the heavier elements. In doing so, they have found the source of nuclear energy.

Familiar examples of elements that you use daily, along with their chemical symbols, are iron (Fe), copper (Cu), aluminum (Al), carbon (C), gold (Au), and silver (Ag). There are over 100 of these elements that have been identified in nature and in the laboratory. Figure 1-1, the **Periodic Table of the Elements**, arranges all of the elements into columns and rows according to similarities in their properties.

As mentioned, matter may also be a compound or a mixture. A **compound** is a substance that is made of two or more *chemically combined* elements. The elements of a compound always occur in the same proportion. These elements are only separated by chemical reaction. A **mixture** is also a combination of two or more elements. However, these elements are not chemically combined, proportions of the elements may vary, and they retain their own properties.

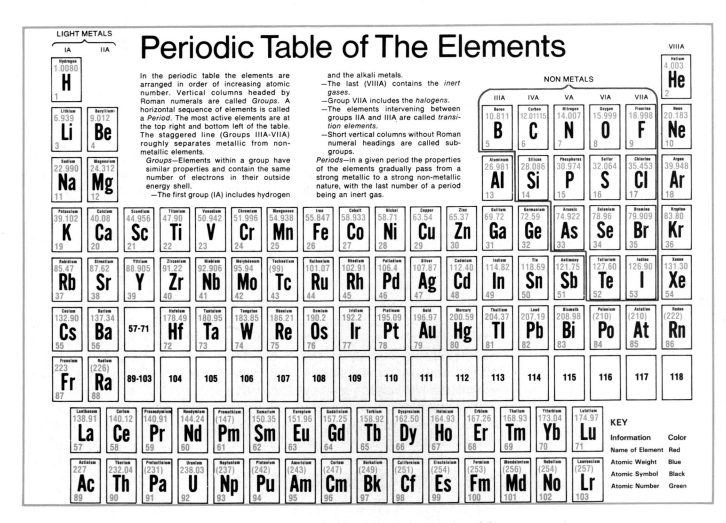

Figure 1-1.   *The Periodic Table of Elements classifies all known elements by atomic weight and atomic number.   (U.S. Air Force)*

ATOMS AND MOLECULES

The element is the basic building block of nature. However, the element itself is composed of small particles called *atoms*. An **atom** is defined as the smallest particle of an element that retains the properties of that element. The atom is the underlying structure of all matter. Each atom is a source of energy.

The idea of atoms was first suggested over 2000 years ago. However, it was not until the late 1700s that a theory was developed—a historic step in the understanding of chemical behavior. This theory was the inspiration of an English chemist named John Dalton. It is known as **Dalton's atomic theory**. Among ideas of his theory are:

- All elements are composed of tiny particles called atoms.
- Atoms of the same element are identical. The atoms of any one element are different from those of any other element.

Dalton first formulated a rudimentary table of atomic weights. He devised a system of classification of chemical compounds based on the number of atoms involved. His observations were crude, but his atomic theory and concept of atomic weights were fundamental.

Atoms can combine with other atoms. The result is a *molecule*. A **molecule** is a *neutral* group of atoms, either like or unlike, that are bound together. Molecules constitute matter.

Many compounds, *but not all*, are bound together in molecules. Compounds that are bound together in molecules are called **molecular compounds**. Examples of these compounds are carbon monoxide (CO), water (H_2O), and hydrogen gas (H_2). The smallest particle of a molecular compound that retains all of the chemical properties of the compound is a molecule. Further, all of the molecules of any given molecular compound are identical. Molecules of one compound, however, differ from any other compound. (Compounds that are not molecular are **ionic compounds**. Ionic compounds are composed of positively and negatively charged atoms called **ions**. Table salt, or sodium chloride [NaCl], is an example.)

If a molecular compound is divided, it will retain the properties of the compound up to a point. If further subdivision is made, the molecule will not be the same substance. For example, water is a chemical combination of two hydrogen atoms (H_2) and one oxygen atom (O). If you divided a drop of water again and again, you would still have water—two hydrogen atoms bound to an oxygen atom. However, if the smallest division, or molecule, were chemically divided, it would no longer be water.

Atomic structure

Atoms of all elements are composed of minute particles of electrical charges. Elements differ only in the number of particles and their arrangement. An English physicist named Sir Joseph John Thomson received a Nobel Prize for investigating the transmission of electricity through gases. His discoveries are the foundation for the present day theory of *electron flow*, which defines *current*, the movement of charge, in terms of electron flow.

Thomson proposed a theory that an atom was a small ball containing an equal amount of positive and negative charges of electricity. Its overall charge was said to be neutral (neither positive nor negative). Another British physicist, named Ernest Rutherford, further refined the picture of the atom. He proposed the theory that an atom consisted of a compact, positive nucleus surrounded by negatively charged electrons in motion. Thus emerged our present concept of the structure of an atom.

According to the classic **Bohr model**, atoms have a planetary type of structure that consists of a central **nucleus** surrounded by orbiting particles called *electrons*. Concentrated at the nucleus are *positively charged* particles called **protons** and *neutrally charged* particles called **neutrons**. The **electrons** revolve in specific orbits and are *negatively charged*.

In the electrically neutral atom, the number of protons equals the number of electrons. In the positively charged atom, or **cation**, the number of protons is greater than the number of electrons. In the negatively charged atom, or **anion**, the number of electrons is greater than the number of protons.

The nuclei of the atoms of any given element must all contain the same number of protons, but the number of neutrons may vary. The number of neutrons is always equal to or greater than the number of protons (except for hydrogen). Atoms of the same element that vary in the number of neutrons are called **isotopes**. Atomic structures of hydrogen, carbon, silicon, and germanium are shown in Figure 1-2.

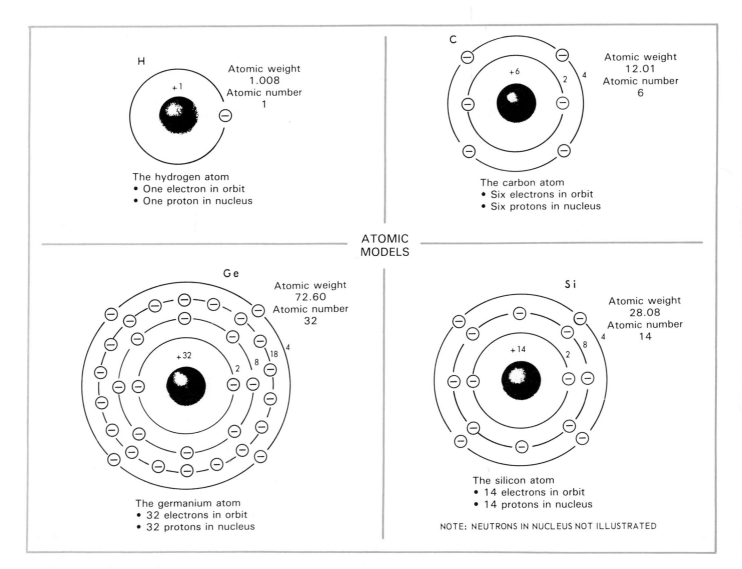

Figure 1-2. The atomic structure of hydrogen, carbon, silicon, and germanium.

All elements are classified in the Periodic Table of Elements by *atomic number* and by similar properties. The **atomic number** designates the number of protons in the nucleus of the atom. Referring again to Figure 1-2, the atomic number of hydrogen is 1, carbon is 6, silicon is 14, and germanium is 32.

Atomic weight is also given in the periodic table. The **atomic weight** is the relative mass of an atom. It is nearly equal to the sum of protons and neutrons in the nucleus of the atom. Units of atomic weight are **atomic mass units (amu)**, where 1 amu, by definition, is equal to 1/12 the mass of a carbon-12 atom. (Carbon-12 is a carbon atom with 6 protons and 6 neutrons. It has an atomic weight of 12 amu.) In practical terms, 1 amu is equal to the weight of a single proton or a single neutron.

The mass of a carbon-12 atom, determined by an instrument called a *mass spectrometer*, is 1.99268×10^{-23} gram. In grams, therefore, 1 amu is equal to 1/12 of this mass, or 1.66057×10^{-24} gram. It is important to remember that the mass of any atom is made up of the *neutrons* and *protons*. The mass of an electron is only 1/1845 that of a proton and, therefore, does not contribute significantly to the total mass.

Atomic energy levels

As stated, electrons revolve in orbits around the nucleus. It is hard to believe that electrons are in motion in an apparently solid material such as a piece of copper. Yet, it is true. The question arises about how far from the nucleus the electrons orbit. The orbiting electron is held at one of a number of discrete distances from the nucleus. In one of these discrete orbital paths, *centripetal force* is nearly in balance with the outward force from the motion of the electron. (**Centripetal force** is the inward, *radial* force [directed toward the center] required to keep a particle or object moving in a circular path. Centripetal force on the orbiting electron is provided by the *electrostatic force of attraction* between it and the nucleus.)

Two different forms of energy determine the energy level of an electron. These are:

- **Potential energy (PE).** This is the energy of position (or configuration). At the atomic level, the position is the distance of the electron from the nucleus. The further the electron is from the nucleus, the greater the potential energy. Potential energy is a stored energy with the *potential* to do work — thus the name. As an example of potential energy (on a larger scale): There is considerable energy stored in a mountain lake because of elevation. If an outlet is provided, the energy is exhibited in water rushing down the mountainside.
- **Kinetic energy (KE).** This is the energy of mass in motion. The kinetic energy of the electron increases with the square of its velocity. With the energy of motion, a moving object tends to stay in motion. As an example of kinetic energy: As the water of the mountain lake rushes down the mountainside, energy of position is transformed to energy of motion. At the base of the mountain, this energy could be transferred to turbine blades and transformed to electricity.

The total energy of an electron is the sum of the potential and kinetic energies. The orbital path used by the electron is based on its total energy. Electrons with less energy are found orbiting closer to the nucleus. Electrons with greater energy are found further out.

Energy shells. The concept of energy levels in an atom is most important in the study of solid-state electronics. This concept suggests that electrons orbit in specific **orbital shells**, or layers, around the nucleus. Each shell corresponds to a certain energy level. The maximum number of electrons in each shell is given by the formula, $2n^2$, where *n* is the number (called **quantum number**) of the shell. See Figure 1-3.

Each major shell is divided into **subshells**, which represent a level of energy. These subshells are illustrated in Figure 1-4 with letter designations of *s*, *p*, *d*, and *f*. The number of electrons in each subshell is limited to: *s* shell, 2; *p* shell, 6; *d* shell, 10; *f* shell, 14.

Excitation, photons, and phonons. The interesting study of energy shells of an atom would not be complete without mention of *excitation*. **Excitation** is the addition of energy to an atom by means of heat, light, or an electric field. Excitation causes electrons to jump outward into the next energy shell. Every atom has any number of possible electron configurations. That associated with the lowest energy level is called the **ground state**. Any other configuration is an **excited state**.

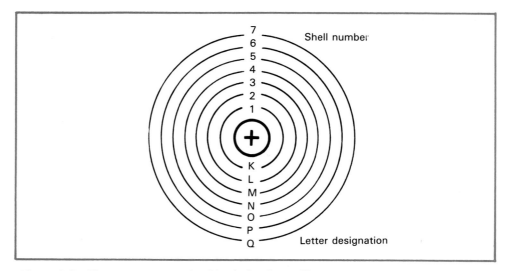

Figure 1-3. Electrons are contained in shells of specific energy levels around the nucleus. Each shell is made up of subshells. Shell letter and number designations are shown.

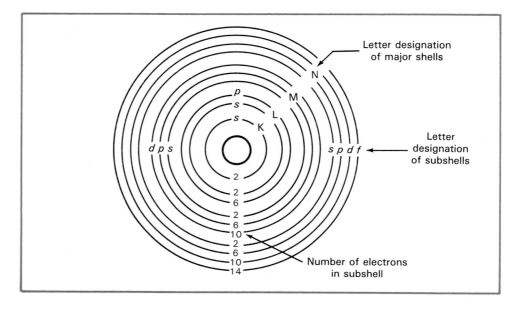

Figure 1-4. Each shell is divided into subshells. Each subshell represents a specific energy level. Subshell letter designations are shown with electrons in each shell.

Photons and *phonons* cause excitation. A **photon** is a discrete quantity of light energy. A **phonon** is a discrete quantity of heat energy caused by thermal vibration. (Light energy and heat energy have come to be considered as kinetic or potential energy at the atomic or molecular level. The same is true for other forms of energy—nuclear, chemical, and electrical.) When these discrete quantities of energy, or **quantums**, are accepted by the atom, the electrons move outward from their subshell to the next higher energy subshell. Addition of energy causes electrons to move outward. Conversely, an electron moving inward to a lower energy subshell gives off energy. This could be compared to the heating of an object. As temperature is applied, it picks up heat energy. As the object cools, it gives off heat energy.

As mentioned, the influence of an electric field can also cause electrons to move to outer orbits. This will be studied later in this chapter.

Valence electrons

Valence electrons are the electrons in the *s* and *p* subshells of the *outer* orbital shell (**valence shell**) of an atom. For example, the electron configuration of boron is $1s^22s^22p^1$; the numbers preceding the subshells are the quantum numbers, and each raised number is the number of electrons in that subshell. From the configuration, we see that there are three valence electrons in the outer shell ($2s^22p^1$).

On the periodic table, elements are grouped vertically by number of valence electrons. This number is at the top of the table. For example, boron is in the IIIA column since it has three valence electrons. Elements of the IIIA group are called **trivalents**. Silicon is in the IVA group. It has four valence electrons. Elements of this group are called **tetravalents**. Nitrogen, in the VA column, has five valence electrons. Elements of this group are called **pentavalents**.

The number of electrons in the valence shell of an atom is an indication of its ability to gain or lose electrons, and it determines the electrical and chemical properties of the atom. Due to the nature of things, an atom will try to seek the stablest level possible. This level is achieved when the valence shell is full. An atom that has almost its full complement of electrons in the valence shell will easily gain electrons to complete its shell. However, a relatively large amount of energy is required to free any of its electrons. Conversely, an atom that has only a small number of electrons in the valence shell, compared to its permitted amount, will lose these electrons quite easily.

To give this a little more meaning, a stable atom has eight electrons in its valence shell. It is a good insulator. A good conductor has one valence electron. An atom with four valence electrons is neither a good conductor nor a good insulator. Atoms with four valence electrons are called **semiconductors**.

Ionization. We know that neutral atoms have an equal number of protons and electrons; cations have more protons; anions have more electrons. To make a positive or negative ion out of a neutral atom, external energy can be added to upset the balance of neutrality—a process called **ionization**. A neutral atom becomes a cation by giving up electrons; an anion by accepting electrons.

To remove an electron from an atom, energy levels must be raised. As stated earlier, this may be accomplished by heat, light, or an electric field. The energy required to do this is called the **ionization potential**.

Cations and anions have opposite charges. Opposite charges attract. (This will be discussed in more detail later in the chapter.) The forces of attraction that bind positive and negative charges are called **ionic bonds**. Ionic compounds are electrically neutral.

Covalent bonding. Most solid substances are crystalline in structure. (Others, like rubber, are **amorphous**, or lacking an internal structure.) In a crystal, the atoms, ions, or molecules are arranged in an orderly, repeating, three-dimensional pattern. This pattern is called a **crystal lattice**. See Figure 1-5.

From Figure 1-5, you can see that every crystal is like its neighbor and arranged in a precise manner. Note that silicon has four valence electrons. To complete the valence shell, eight electrons are required. Since no free electrons are available, each atom shares its valence electrons. The atoms are firmly attached in a crystal lattice by **covalent bonding**. A crystal of this type will not conduct electricity because there are no *free electrons* to transfer the energy.

The law of charges

Existing in space around a charged body is an invisible and static force field called an **electrostatic field**, Figure 1-6. Both positively and negatively charged bodies support an electrostatic field. By convention, a negative electrostatic

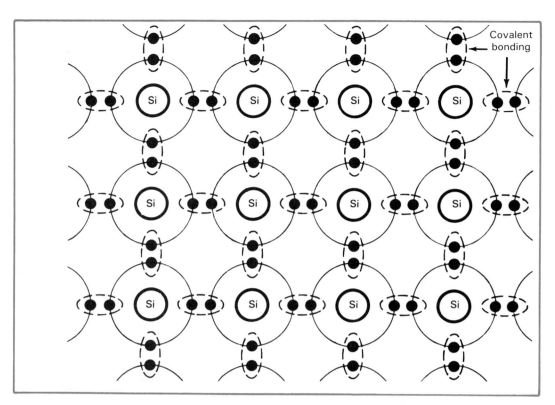

Figure 1-5. The crystal lattice structure of silicon. Only valence electrons are shown for simplification.

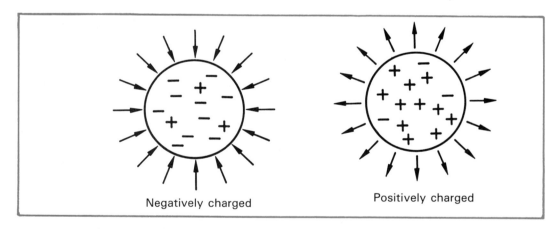

Figure 1-6. The electrostatic field of force existing around a negatively and positively charged body.

field is illustrated by arrows pointing toward the mass. A positive field is denoted by outwardly pointing arrows.

If two bodies of unlike charge are brought close together so that their fields interact, they will be brought together by an attractive force. If two bodies possess the same charge, either positive or negative, a repulsive force will tend to hold them apart. See Figure 1-7. This is an important law in the study of electricity — the **law of charges**. It states:

Like charges repel each other. Unlike charges attract each other.

This law will be applied to explain the theory and operation of many components, circuits, and devices.

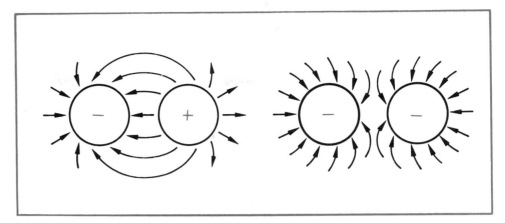

Figure 1-7. The law of charges. Unlike charges attract. Like charges repel.

The coulomb. The atom is an infinitesimal particle of matter. In fact, more than several billion atoms make up the head of a pin alone. The electron is even smaller. Electron charge is a measurable quantity—a very small measure. A quantity of electricity, or **electrical charge (Q)**, is given in units of **coulombs (C)**. It is, by definition, the amount of electricity delivered when current of 1 ampere is maintained for 1 second (1 coulomb = 1 ampere-second). A single electron is measured as having a charge of 1.602×10^{-19} C. The reciprocal of this gives the number of electrons in 1 C: 6.24×10^{18} electrons. This is 6,240,000,000,000,000,000 electrons!

Charles Coulomb was a French physicist. He conducted extensive research in the measurement of small forces. His experiments verified the laws of electrostatics. His work made possible the definition of quantity of electric charge, which is named in his honor. Through his experiments, came **Coulomb's law**, which states:

The force between two charges is directly proportional to the product of the charges and inversely proportional to the square of the distance between the charges. $(F = kQ_1Q_2/r^2)$

The electroscope. An instrument used in the laboratory to detect a charge and determine its polarity is called an **electroscope**, Figure 1-8. It consists of a glass jar or chamber with a center rod and ball. Two gold leaves hang on the rod at the center of the chamber. If they become charged alike, either negatively or positively, they will spread apart. This will happen because the like charges on both leaves will repel each other. This demonstrates the phenomenon of **static electricity**—electricity in motionless charges.

To use the electroscope, vigorously rub a vulcanite rod with a piece of wool. This action puts a negative charge on the rod by friction. Now, bring the rod close to the ball on top of the electroscope but do not touch the ball. The gold leaves will spread apart. See Figure 1-9.

Here is what happened: The negative electrostatic field around the rod repelled the electrons from the ball down to the gold leaves. Since both leaves now have a negative charge, they repel each other.

While the rod is in position and the leaves are apart, touch the ball with your finger. Remove your finger, then remove the rod. The leaves remain apart. What happened? When you touched the ball, electrons escaped through your finger and into your body. When your finger and the rod were removed, the electroscope had no chance to regain its lost electrons. It, therefore, became

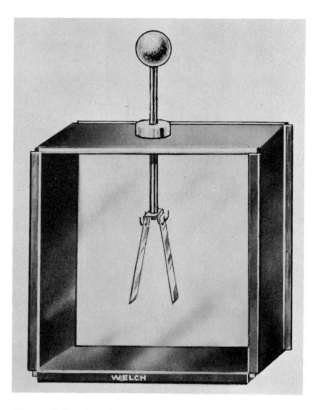

Figure 1-8. *An electroscope will detect a charge and reveal its polarity.*
(Sargent-Welch Scientific Co.)

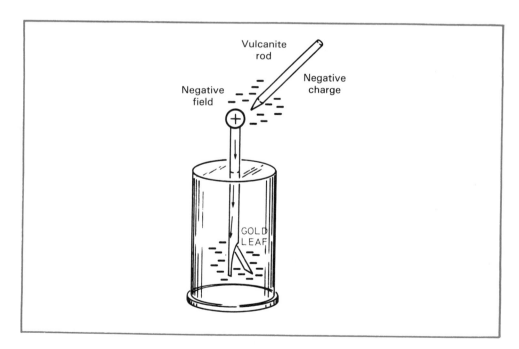

Figure 1-9. *Negative electrostatic field around the rod drives electrons down to the leaves*
of the electroscope and causes them to spread apart.

charged positively, and the leaves remained apart. Touch the ball of the electroscope. It regains its lost electrons and becomes neutral. The leaves collapse.

So far, the electroscope has been charged by *induction*, since no contact

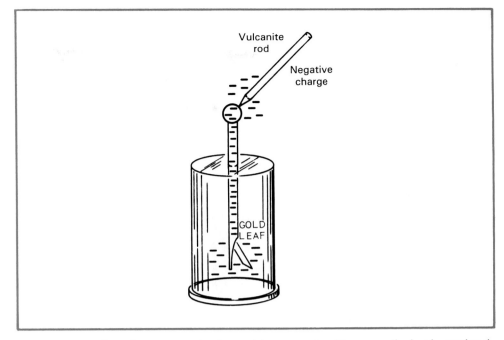

Vulcanite
rod

Negative
charge

GOLD
LEAF

Figure 1-10. The electroscope is charged by contact with a negatively charged rod.

was made between the rod and the electroscope. Rub the vulcanite rod again, but this time *touch* the ball of the electroscope with the rod. Remove the rod and the electroscope remains charged—*negatively*. The rod shared its electrons with the electroscope. This is charging by *conduction*. Refer to Figure 1-10. Touch the ball to discharge it.

These experiments may be repeated using a glass rod and a piece of silk. Rubbing will put a positive charge on the glass rod. A positive rod will charge the electroscope positively by induction and positively by conduction.

LESSON IN SAFETY: As you work in your electronics lab, you will learn how to use tools correctly and safely and develop good working habits. Study the job at hand. Think through your procedures and your methods in the application of tools, instruments, and machines. Use the right tool for the job at hand. Improper tool selection may cause damage to equipment and personal injury.

ELECTROSCOPE PROJECT

The electroscope shown in Figure 1-11 is made with an Erlenmeyer flask. Any glass jar, such as a canning jar, would be suitable, if you made an insulated cover for it.

Secure a piece of 3/8 in. brass rod and file one end until it has a flat side. Round the opposite end for a finished appearance. Attach a sheet of gold leaf with just a drop of cement at one end only. Assemble the electroscope as shown in Figure 1-11 and try it out.

LESSON IN SAFETY: Is an electric shock fatal? The effects of an electric current can be generally predicted, Figure 1-12. The current does the damage! Notice that current as small as 100 milliamperes (0.1 amperes) can be fatal. Currents below 100 milliamperes can be serious. These, too, can be fatal.

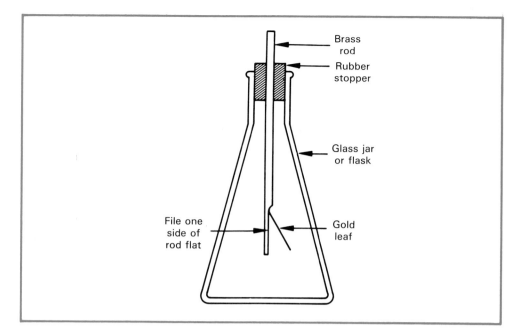

Figure 1-11. A plan for an electroscope.

READINGS		EFFECTS
Safe Current Values	1 mA or less 1 mA to 8mA	Causes no sensation—not felt. Sensation of shock, not painful; individual can let go at will since muscular control is not lost.
Unsafe Current Values	8 to 15 mA	Painful shock; individual can let go at will since muscular control is not lost.
	15 to 20 mA	Painful shock; control of adjacent muscles lost; victim cannot let go.
	20 to 50 mA	Painful, severe muscular contractions; breathing difficult.
	50 to 100 mA	Ventricular fibrillation, a heart condition indicated by un-coordinated contractions of the heart, is possible; where-upon DEATH LIKELY.
	100 to 200 mA	Ventricular fibrillation occurs. DEATH LIKELY.
	200 mA and over	Severe burns and tissue damage, severe muscular con-tractions—so severe that chest muscles clamp the heart and stop it for the duration of the shock. (This prevents ventricular fibrillation.) DEATH POSSIBLE.

Figure 1-12. Physiological effects of electric currents.

SUMMARY

- Matter is anything that occupies space or has mass.
- Elements are basic or pure forms of matter.
- Compounds are combinations of one or more elements.
- Atoms are the simplest form of an element, still having the unique characteristics of that element.
- Molecules are the simplest form of a molecular compound, still having the unique characteristics of that compound.
- Potential energy is energy resulting because of position, and sometimes, it is called stored energy.
- Kinetic energy is *released* energy in motion.
- Atoms have energy shells. Electrons occupy the shell according to their energy level.

- The negatively charged particle of an atom is the electron; the positively charged particle is the proton.
- Like charges repel each other, while unlike charges attract each other.
- The coulomb is a quantity of electrons (6.24×10^{18} electrons).

KEY TERMS _____

Each of the following terms has been used in this chapter. Do you know their meanings?

amorphous	electrostatic field	nucleus
anion	element	orbital shell
atom	excitation	pentavalent
atomic mass unit (amu)	excited state	Periodic Table of the Elements
atomic number	ground state	phonon
atomic weight	ion	photon
Bohr model	ionic bonds	potential energy (PE)
cation	ionic compound	proton
centripetal force	ionization	quantum
compound	ionization potential	quantum number
coulomb (C)	isotopes	semiconductor
Coulomb's law	kinetic energy (KE)	static electricity
covalent bonding	law of charges	subshell
crystal lattice	matter	tetravalent
Dalton's atomic theory	mixture	trivalent
electrical charge (Q)	molecular compound	valence electron
electron	molecule	valence shell
electroscope	neutron	

TEST YOUR KNOWLEDGE _____

Please do not write in this text. Place your answers on a separate sheet of paper.

1. Anything that occupies space and has mass is called _____.
2. Elements cannot be further subdivided. True or False?
3. The smallest particle of an element is a molecule. True or False?
4. Table salt is an example of a molecular compound. True or False?
5. Electrons make up the largest percentage of the mass of an atom. True or False?
6. The atomic weight of an atom is governed by:
   a. The number of neutrons, protons, and electrons in the atom.
   b. The number of neutrons and protons contained in its nucleus.
   c. The number of orbital electrons in the atom.
7. Cations are attracted to each other. True or False?
8. Considering the valence of an atom, which would be a better conductor — an atom with one valence electron or an atom with five valence electrons?
9. Potential energy of an atom is the energy of:
   a. Mass.
   b. Motion.
   c. Position.
10. Kinetic energy of an atom is the energy of:
    a. Mass.
    b. Motion.
    c. Position.

11. The weight of an atom is referenced to the weight of _____, which is 12 amu.
12. A positive electrostatic field is denoted by outwardly pointing arrows. True or False?
13. Existing in space around a charged body is an invisible _____ field.
14. What is the letter symbol for charge?
    a. C.
    b. K.
    c. Q.
15. What is the letter symbol for coulombs?
    a. C.
    b. K.
    c. Q.
16. Electroscopes demonstrate static electricity. True or False?
17. How does the distance between two charged bodies affect the forces between them?
18. Electric currents as small as _____ milliamperes can be fatal.

Chapter 2

SOURCES OF ELECTRICITY

Where does electricity come from, or where can we find a source of electricity? We know energy cannot be created or destroyed. However, it can be converted from one form to another.

After studying this chapter, you will be able to:

☐ *List common sources of electrical energy.*
☐ *Discuss how potential difference provides an electromotive force.*
☐ *Explain the difference between electron flow and conventional current theory.*
☐ *Name common units of measurement used in electricity.*
☐ *Explain how chemical cells work in providing a source of electricity.*
☐ *Discuss various types of cells.*
☐ *Describe the effect of connecting cells in series and in parallel.*
☐ *Explain how electricity can be created by light, heat, and pressure.*

INTRODUCTION

To produce any electrical or electronic action, a source of energy must always be provided. Consider how static electricity is produced by rubbing a vulcanite rod with a piece of wool. Here, the source of energy comes from friction. Friction is a heat energy, which, according to atomic theory, is kinetic energy of rapidly moving molecules.

A *source* is not necessarily a reservoir of energy that is ready for use. More correctly, a source is a means for converting other forms of energy into electrical energy. A flashlight battery stores no electrical energy. It is a means of converting chemical energy into electrical energy. A generator is only a means of converting mechanical energy into electrical energy. The many circuits, components, and devices used in electricity and electronics, then, convert electrical energy from the source into other forms of energy to perform useful work.

POTENTIAL DIFFERENCE

Through ionization, a body may become charged either positively or negatively, respectively, by losing or gaining electrons. When two bodies or terminals have a difference of charge or, more specifically, a difference in **potential** (defined as electric *potential energy* per unit *charge*), a **potential difference**, exists between them. Only potential *differences*, not *potentials*, can be physically measured.

In a completed electrical circuit, a potential difference between two terminals will cause a movement, or transfer, of energy. This moving energy has the ability to do *work*. (The amount of work done in the circuit is the product of the potential difference and the quantity of charge moved. In general, **work** *[W]* is the amount of force applied to a body times the displacement of the body.) In the circuit, work will continue as long as there is a potential difference.

In a battery, there is a potential difference across the positive and negative terminals. The voltage across these terminals, when no current flows to an external circuit, is referred to as **electromotive force (emf)**. In Figure 2-1, a battery is connected to a lamp. Follow the energy conversion: the battery changes chemical energy to electrical energy, which is converted to heat and light energy by the lamp. The lamp will become dimmer as the chemical energy of the battery is used up and the potential difference between its terminals approaches zero.

Figure 2-1. A battery converts chemical energy to electrical energy. The lamp converts electrical energy to heat and light energy.

Potential difference (or emf) is commonly referred to as **voltage**. The unit of measurement of voltage is the **volt (V)**, named in honor of the Italian physicist, Count Alessandro Volta. The symbols used in this text to represent a *quantity* of voltage are the letters *V* and *E*. *V* will be used when dealing with electronic circuits, and *E* will be used with circuits of basic electricity. However, *units* of volts, when abbreviated, will *always* be represented by the letter V.

The value of 1 V can be understood when you realize that a certain potential difference is required to move a quantity of electric charge through a circuit. When this movement occurs, work is done. It was stated that work in an electrical circuit is the product of potential difference and electrical charge. Potential difference, then, is work per unit of charge, where work is in units of **joules (J)**, and charge is in coulombs. (Note that 1 joule is roughly 0.738 foot-pounds, or 1 newton-meter of work.)

One volt is equal to the amount of potential difference between two points for which 1 coulomb of charge will do 1 joule of work in moving from one point to the other. Therefore, 1 volt is equal to 1 joule/coulomb. Voltage may be expressed by the equation:

$$E = \frac{W}{Q} = \frac{\text{work in joules}}{\text{charge in coulombs}}$$

The fixed value of 1 volt has been established by the International Electrical Congress and by law in the United States.

ELECTRIC CURRENT

Current *(I)* is the movement of electric charge. Most often, this movement is by way of a conductive pathway (**conductor**). In an electrical circuit, voltage does the work in moving electrons; in turn, current imparts energy to accomplish the task at hand. So, while voltage is the driving force, it is current that actually lights a light, heats a toaster, drives a solenoid, etc. In a physiological sense, voltage might be thought of as the heart of the electrical circuit, and current, the lifeblood.

In this text, we consider current to be **electron flow** unless otherwise stated. Direction of electron flow is from negative to positive, in the *external* circuit.

In contrast, some texts consider current to be the movement of *positive charges*. This is called **conventional current**. It is opposite of electron flow. As an example, the current in a battery (*internal* circuit) is a result of an *ionization current* of *positive ions*. This current is from negative to positive and is opposite of electron flow (which means that inside the battery electrons flow from *positive* to *negative*). Also, you will learn later that *internal* current in some semiconductors is the movement of positive charges called *holes*. Hole current is opposite that of electron current. The actual mechanics of current is described in detail in Chapter 3.

The unit of measurement of electric current is the **ampere (A)**, or, commonly, the **amp**. Subunits of amps are milliamps (mA) and microamps (μA). The word *ampere* comes from the French physicist, Andre Ampere. Ampere's work involved identifying the cause and nature of electric currents. He identified electrochemical and electromagnetic properties and has been called the father of the theory of electromagnetism.

One ampere is equal to the amount of current in a conductor when 1 coulomb of charge (6.24×10^{18} electrons) moves past a given point in 1 second. Current may be expressed by the following equation:

$$I = \frac{Q}{t} = \frac{\text{charge in coulombs}}{\text{time in seconds}}$$

In Chapter 1, however, coulombs was defined in terms of amperes. Since it is very difficult to measure charge, to find coulombs, you need to know amperes. Technically, 1 ampere is defined as *that current moving in each of two long parallel conductors 1 meter apart which results in a force of exactly* 2×10^{-7} *newtons (SI metric unit of force) per meter of length of each conductor*. It is far easier to measure force between the two conductors than it is to measure the quantity of charge. Therefore, while 1 ampere is, in fact, 1 coulomb per second, the measurement standard is based on units of force.

ELECTRICITY FROM CHEMICAL ENERGY

At the close of the 18th century, an Italian physiologist by the name of Luigi Galvani experimented with dissected frog legs. He proposed a theory of animal electricity that became the basis of **galvanism**—producing a direct current by chemical action. Galvani conducted his experiments by fastening a dissected frog leg to a bench with copper wire; then touching the leg with a steel scalpel. Each time it was touched by the scalpel, the leg would twitch. Galvani reasoned that the frog contained animal electricity, which caused its muscles to contract when touched with the knife.

Galvani's assumption was proved to be incorrect by Volta, the Italian physicist. Volta showed that what was thought to be "animal" electricity was, in truth, "metallic" electricity. He did this when he produced electricity by immersing two dissimilar metals in a fluid that chemically reacted with one of the metals. Through his research, Volta developed the voltaic cell that bears his name.

We must give a great deal of credit to these scientists. The cells and batteries that we use so frequently to power electronic circuits originated through the inventiveness and genius of the early experimenters. It should be interesting to perform upcoming experiments, since they will lead to a better understanding of this valuable source of electrical energy.

CELLS

A familiar source of electromotive force, or voltage, is the chemical *cell*. A **cell** is a single unit that produces electrical energy from chemical energy. The symbol for one is shown in Figure 2-2. Cells produce a current that is steady and unidirectional. A current that is unidirectional is called a **direct current (dc)**. Cells, then, are a dc voltage source. (It may be redundant, but we use "dc voltage" or "dc current," which means "direct voltage" or "direct current.") Experiments follow that provide examples of a simple cell.

* Secure a dime, a penny, and a small square of blotting paper. Soak the blotting paper in a strong salt/water solution. Now, assemble the cell shown in Figure 2-2. A voltmeter connected to this simple cell will indicate an emf.
* Insert a piece of copper and a piece of zinc in a lemon. Refer to Figure 2-3. A voltmeter connected as shown will indicate an emf.

Notice that in both Figure 2-2 and Figure 2-3, plus (+) and minus (−) signs were placed by the terminals to indicate *positive* and *negative*. It is important to observe the correct polarity when measuring a dc current or voltage with an analog (needle indicator) meter. Correct polarity is positive to positive and negative to negative. Notice that the copper has a negative polarity in one instance and a positive polarity in the other. This will depend upon the tendency of one electrode to give up electrons relative to the other. The one giving up electrons more readily will take on a negative charge with respect to the other.

Why was an emf indicated in these experiments? A brief explanation is in

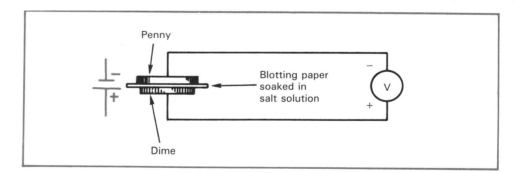

Figure 2-2. *A voltmeter connected to this cell will show a potential difference. The symbol for a cell is pictured left.*

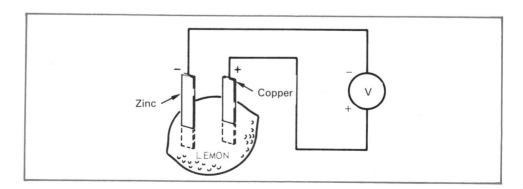

Figure 2-3. *Zinc and copper electrodes inserted in a lemon will produce a potential difference.*

order. In the experimental *voltaic cell*, depicted in Figure 2-4, a jar contains a mild *electrolytic solution* of acid and water. An **electrolytic solution** is *electrically conducting*. (Not all solutions will conduct electricity.) This solution is also referred to as an **electrolyte**. A conductor used to establish electrical contact with the nonmetallic medium, called an **electrode**, is first placed in the solution. The electrode, in this case, is made of zinc (Zn). Small bubbles of hydrogen start to appear around the electrode, showing that the acid and zinc are reacting. In the solution, the terminal of the zinc electrode is found to have a negative charge. Next, a carbon rod is placed in the solution. Although the acid and the carbon do not chemically react, the carbon terminal becomes positively charged.

In Figure 2-5, a chemical explanation is given using an electrolytic solution of sulfuric acid (H_2SO_4) and water. The acid **dissociates** (separates into ions) in the water to form positive, hydrogen ions ($2H^+$) and negative, sulfate ions (SO_4^{2-}). The sulfate ions move toward the zinc and combine with it to form zinc sulfate ($ZnSO_4$). This action causes the zinc to become negatively charged, since the neutral zinc must give up some positive ions to form the zinc sulfate. The $2H^+$ ions move toward the carbon rod and form a blanket of hydrogen bubbles, or positive ions, around the rod. It becomes positively charged. Refer to Figures 2-5 and 2-6.

LESSON IN SAFETY: In general, if mixing acid and water, always pour acid into water. Do not pour water into acid. Acid will burn your hands and your clothing. Should your skin come in contact with acid, wash the affected area at once with clear water. Acid may be neutralized with baking soda.

Primary and secondary cells

Notice in the voltaic cell, the zinc is eaten away, or used up, as the cell is used. This cell cannot be recharged. The chemical action cannot be reversed. A cell that cannot be recharged is classified as a **primary cell**. The familiar carbon-zinc *dry cell* is an example of a primary cell.

Some cells can be recharged. They are called **secondary cells**. Secondary cells are also called **storage cells**. The word *storage*, however, is misleading since the cell does not *store* electrical energy, but *converts* it from chemical energy. The nickel-cadmium cell, to be discussed, is an example of a secondary cell.

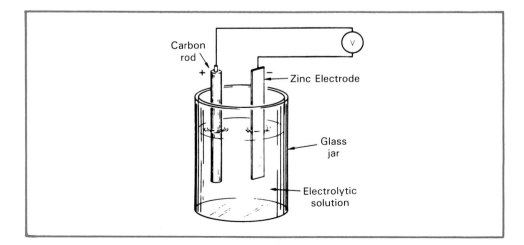

Figure 2-4. Carbon and zinc electrodes in acid electrolyte will produce a potential difference.

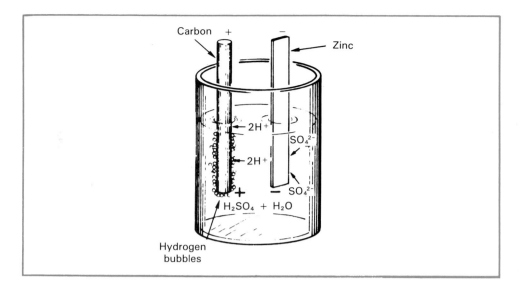

Figure 2-5. The chemical reaction in a simple voltaic cell.

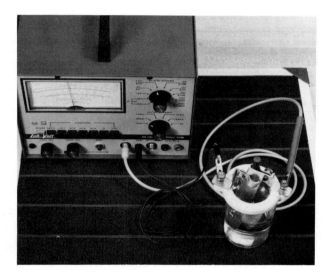

Figure 2-6. An experimental setup of a voltaic cell. (Lab-Volt Div., Buck Engineering Co.)

Carbon-zinc cells

The **carbon-zinc cell** is perhaps the most common type of dry cell. (A **dry cell** is a voltage-generating cell having a paste-like electrolyte, or conducting medium.) In Figure 2-7, it is easy to identify the zinc electrode, or can, of the dry cell. The can is negative. The carbon rod through the center is the positive electrode. It is immediately surrounded by a paste of manganese dioxide (MnO_2) and carbon black. The electrolyte is in a paste form, consisting mostly of ammonium chloride (NH_4Cl). The chemical equation may be represented as:

$$Zn + 2NH_4Cl + 2MnO_2 \rightarrow Mn_2O_3 + H_2O + 2NH_3 + ZnCl_2$$

The left side of the equation shows the approximate chemical content of a new dry cell. As current is drawn from the cell, the zinc, ammonium chloride, and manganese dioxide combine to produce manganese *sesqui*oxide (Mn_2O_3), water, ammonia (NH_3), and zinc chloride ($ZnCl_2$). The carbon is inert and does not appear in the reaction.

Note that manganese dioxide is used as a *depolarizer*. During the discharge of a cell, hydrogen bubbles collect around the carbon rod and form an effective insulating blanket. This is called **polarization**. A **depolarizer** is used in some primary cells to prevent formation of hydrogen bubbles at the electrode. The depolarizer combines with the hydrogen and other products of the action to form water and Mn_2O_3. The depolarizer adds substantially to the life and output of the cell.

Mercury cells

A widely used type of dry cell is the **mercury cell**. Mercury cells come in several shapes and sizes. A cutaway of a miniature button-type cell is shown in Figure 2-8. The negative electrode seen here is a gelled mixture of zinc powder and electrolyte. The electrolyte is potassium hydroxide (KOH). The positive, depolarizing plate is compressed mercuric oxide (HgO). A porous barrier separates the positive and negative plates.

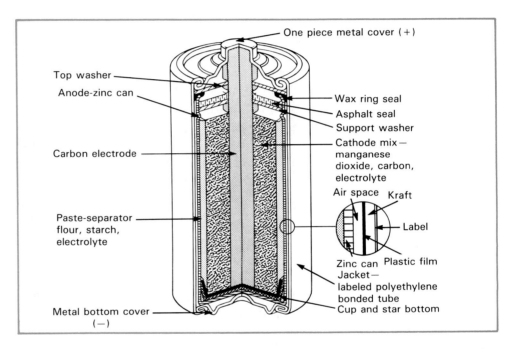

Figure 2-7. Carbon-zinc cell with cutaway view. This type of cell is also called a Leclanche cell. (Eveready)

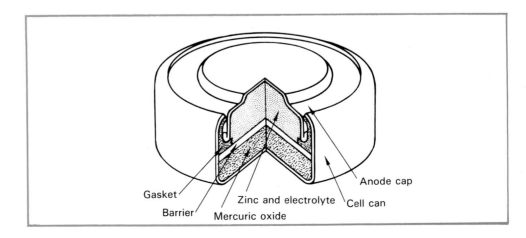

Figure 2-8. A cutaway view of a mercury cell. (Eveready)

Much of the miniaturized electronic gear in use today would not have been possible without the development of the mercury cell as an energy source. Mercury cells have a few distinct advantages over carbon-zinc cells. They have a high density of electrical energy in a very small package and a reliable terminal voltage. They also enjoy a long shelf life without deterioration. One drawback is their relatively high cost compared to the carbon-zinc cell.

Alkaline cells

A primary cell that is similar to the carbon-zinc cell is the **alkaline cell**. The main difference is the alkaline cell uses potassium hydroxide, a caustic alkali, for the electrolyte. Also, this cell has about 50 to 100 percent more capacity and has a better shelf life. This cell is also known as an *alkaline-manganese cell.*

Nickel-cadmium cells

Another type of cell that uses an alkaline electrolyte is the **nickel-cadmium cell**, or **NICAD cell.** Since they are rechargeable, nickel-cadmium cells are secondary cells. They are popular in cordless home appliances such as electric toothbrushes and carving knives. They are used everyday in such equipment as tape recorders, razors, portable TVs, and lawn trimmers. These cells have contributed to the development and availability of a whole new family of devices for home and family use.

Nickel-cadmium cells have a very low internal resistance and can deliver high currents with very little loss of terminal voltage. They are small, yet have the ability to be recharged, usually over 1000 times. A nickel-cadmium cell will hold a charge for a relatively long period of time. Other features are that they have a long shelf life and can be recharged for quick use.

Figure 2-9 shows a NICAD cell. It consists of a positive plate, a separator, a negative plate, an electrolyte, and a cell container. The plates are made by sintering powdered nickel into a wire screen, also of nickel, which makes a strong and flexible plate. One plate is impregnated with a nickel salt solution, making this plate positive. The other is impregnated with a cadmium salt solution, making it negative. The separator is made of an absorbent, insulating material. The electrolyte is a solution of potassium hydroxide. The container is a steel can, which is sealed after the cell is placed in it. The open circuit terminal voltage of this type of cell is 1.33 V at room temperature.

A major disadvantage of this battery is in its use of cadmium. This is because cadmium is highly toxic and is considered hazardous waste. Alternate technologies include *nickel-metal-hydride* and *lithium* batteries. These substitutes are less of an environmental risk. They are not considered hazardous waste.

Lithium cells

Lithium cells have gained popularity in the past few years as high quality, lightweight, sources of electricity that give extremely long life. Lithium, which is used as the negative electrode in these cells, has the highest potential of all metals for giving off electrons by chemical action. A number of materials are used in commercial cells for the positive electrode. See Figure 2-10.

Silver oxide cells

Silver oxide cells have been developed fairly recently. Typically, they are used in hearing aids and in electronic watches. These cells possess the highest level of electrical capacity and leakage resistance of any chemical cell of their size. They have a very stable discharge voltage of 1.5 V. They are small, having a button-type construction.

BATTERIES

Up to this point, only the *cell* has been discussed. When two or more cells are connected either in *series* or in *parallel*, the combination is called a **battery**. Do not be confused by this term. Many people call a flashlight *cell* a flashlight *battery*. This is technically incorrect.

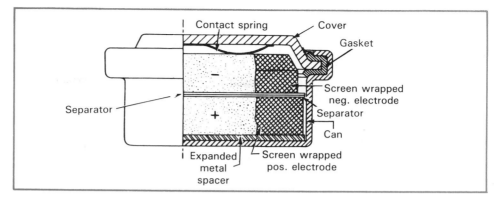

Figure 2-9. A cutaway view of a nickel-cadmium cell.   (Union Carbide Co.)

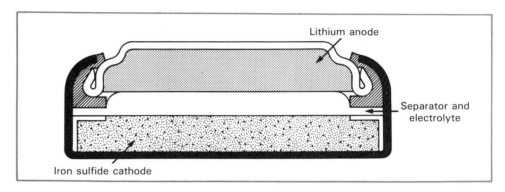

Figure 2-10. A cutaway view of a lithium cell.   (Eveready)

In Figure 2-11, cells connected in *series* are connected end to end—*plus to minus*. Note that both pictorial and schematic representations are shown. Cells connected in this manner are said to be *series-aided*. The result is an increase in voltage of the battery above a single cell. The battery output voltage (E_{out}) is the sum of all cell voltages. If all cells have the same voltage, output voltage may be determined by the formula:

$$E_{out} = n \times E_c$$

where *n* is the number of cells and E_c is the voltage of one cell.

In Figure 2-12, the four cells are connected in *parallel*—all positive terminals are connected together and all negative terminals are connected together. Connected this way, the voltage output does not increase, but the **capacity**, or the current-output capability over a given time period, is substantially increased. (Bad cells should not be connected in parallel with good cells. The good cells will supply more current and may be overloaded. Also, a cell with lower output voltage acts as a load resistance. This will drain current from cells of higher output voltage.)

To increase voltage *and* capacity, cells may be connected in some combination of series and parallel. This may be as series groups, with the groups connected in parallel; or as parallel groups, connected in series. These groupings are shown schematically in Figure 2-13.

Which arrangement is best for a given condition? Generally speaking, if the load applied to the battery is a high resistance load, it is better to use the series connection. For a low resistance load, use the parallel connection to produce a larger current. However, as load current increases, so does the

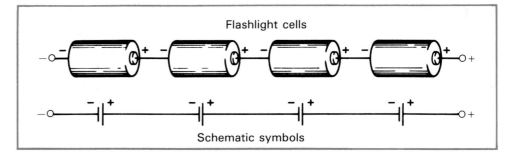

Figure 2-11. A battery of four series cells.

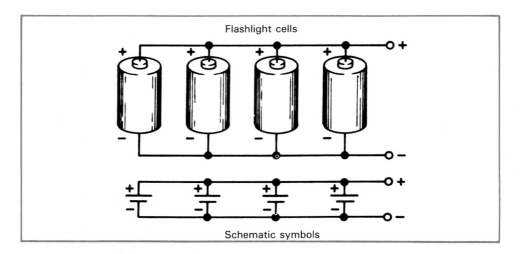

Figure 2-12. A battery of four parallel cells.

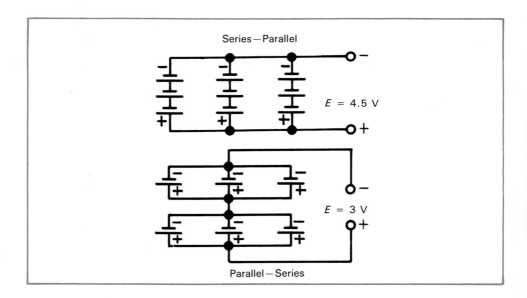

Figure 2-13. Combination groups of 1.5 V cells.

voltage drop across the *internal resistance* of the cell. (All cells have an **internal resistance**, which, in a cell, is the resistance met by current as it moves through the electrolyte.) This will reduce the output voltage available at the cell terminal. To compensate, it may be necessary to increase battery voltage, also, by going to a parallel *and* series combination.

Lead-acid batteries

The **lead-acid storage battery**, used in the automobile, is composed of a series of cells. These batteries come in sizes of 6 V and 12 V. The 6-V battery is made up of three secondary cells in series; the 12-V, of six cells in series. Figure 2-14 is a cutaway view of a typical 12-V battery.

Figure 2-15 offers a simple graphic explanation of how a lead-acid cell works. A basic cell is formed when two dissimilar plates and one separator are placed in an electrolyte. In all automotive-type batteries, the dissimilar plates are made of spongy lead (Pb) and lead peroxide (PbO_2). The electrolyte is a solution of sulfuric acid (H_2SO_4) and distilled water. Due to the chemical reaction that occurs between the electrolyte and the dissimilar plates, a voltage of about 2 V exists between the two plates.

When the plates are connected to a 2-V bulb, current runs from one plate, through the electrolyte and highly porous separator, to the other plate, then through the bulb to complete the circuit. Thus, when the bulb is lit, the cell is discharging and is converting chemical energy to electrical energy.

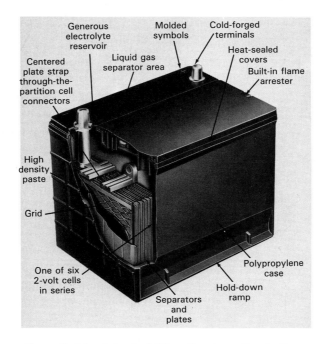

Figure 2-14. A typical 12-volt automotive battery. (Delco-Remy Div., General Motors Corp.)

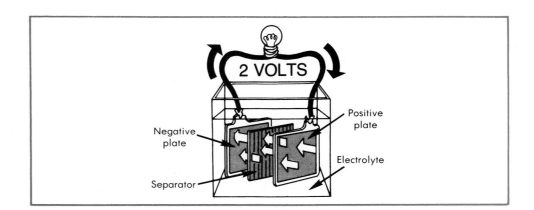

Figure 2-15. Basic operation of a lead-acid cell. (Delco-Remy Div., General Motors Corp.)

LESSON IN SAFETY: During the battery charging process, highly explosive hydrogen gas may be present. Do not light matches near charging batteries. Charge only in a well-ventilated room. Observe correct polarity of charging leads and battery terminals. Also, charger should be disconnected from power supply before connecting or disconnecting battery terminals. Otherwise, sparks might occur during connection, which could ignite the hydrogen gas and cause an explosion.

Plate construction. Both positive and negative plates found in the cells of the lead-acid storage battery are made by pressing a paste of very finely ground lead on a lead-alloy grid. The grid lends mechanical strength to the plates. These paste plates are dried to form a strong, thin plate of active material. They are then assembled in groups of alternate negative and positive plates. The polarities of the plates are *formed* in the manufacturing process and, later, *re*formed by recharging and reaction with the acid electrolyte. (**Forming** is a process that involves application of voltage to produce a desired change in electrical characteristics.) In the forming process, the plates are changed to lead peroxide (positive plate) or spongy lead (negative plate).

Each group of plates (cell) has one additional negative plate; for example, a 13-plate cell has 7 negative and 6 positive plates. Each plate is separated from neighboring plates by *separators* of porous plastic or rubber. These separators electrically insulate one plate from another. Yet, they are sufficiently porous to provide free passage of electrolyte.

After the cells are set in the compartments of the battery case, they are connected in series by interconnecting links of a lead alloy. The cover is sealed in place. Electrolyte is poured through the cell vents and the vents are capped.

Dry-charged batteries. Batteries are frequently shipped dry from the manufacturer. A dry-charged battery is convenient to ship and store. The battery is shipped with charged plates, but without electrolyte. It is made ready for use by filling with the correct grade of electrolyte. The dry-charged battery should be activated with a boost charge prior to installation. This is especially true in cold weather.

Cell chemistry. The chemical action within the lead-acid cell is graphically described in Figure 2-16. The chemical reaction is rather involved. Briefly, what is happening is:

- In the charged state, negative plates are spongy lead and positive plates are lead peroxide.
- During discharge, electrolyte increases in water content and decreases in sulfuric acid content. (This is why a battery would freeze quite easily when discharged. [The freezing point of a solution is lower than the freezing point of the pure solvent — water, in this case.])
- During discharge, spongy lead and lead peroxide plates are changed to lead sulfate.
- In recharging the cell, a reverse current is applied. This reverses the chemical reaction. The lead sulfate changes back to spongy lead and lead peroxide; the electrolyte increases in sulfuric acid content and decreases in water content.

Specific gravity. In discussing battery electrolyte, *specific gravity* becomes an important consideration. In general, **specific gravity** is the ratio of the density of a substance to the density of another substance, usually water, taken as a standard. For our purposes, it is the ratio of the weight of a liquid in respect to the weight of an equal volume of water. The specific gravity of water, naturally, is 1.000. Sulfuric acid is almost twice as heavy as water. It

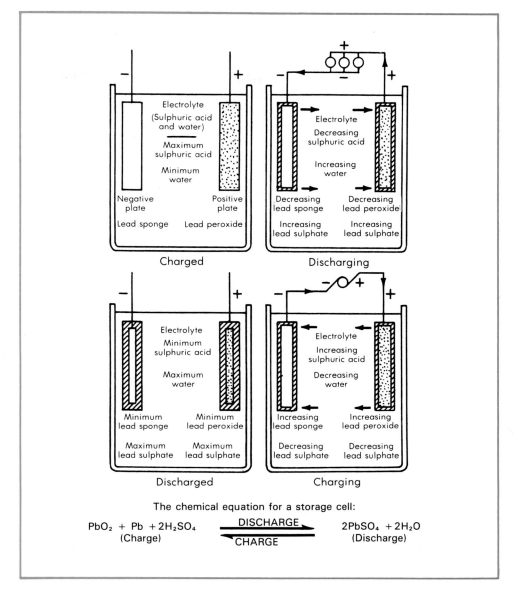

Figure 2-16. A diagram of the chemical action within a lead-acid cell. (ESB, Inc. Automotive Div.)

has a specific gravity of 1.835. The electrolyte used in the lead-acid battery is a solution of acid and water. The weight of this solution is about 1.3 times heavier than water.

During the discharge of a lead-acid cell, the electrolyte changes to almost pure water. When recharging, the electrolyte becomes richer in acid. The specific gravity, once again, approaches 1.300. This information becomes very useful since the *state of charge* of a lead-acid cell can be determined by measuring the specific gravity of the electrolyte. This is done by means of a **hydrometer**, Figure 2-17. The state of charge, broken down by specific gravity readings, is as follows:

State of Charge	Specific Gravity
100%	1.260 +
75%	1.230
50%	1.200
25%	1.170

A hydrometer like that shown in Figure 2-17 is used by squeezing the bulb to obtain a sample of the electrolyte. The float inside the outer tube is actually the hydrometer. The specific gravity is read from its scale. The hydrometer works on **Archimedes' principle**, which states:

A floating *object will sink into a liquid until the weight of the liquid it displaces is equal to the weight of the* floating *object.*

Figure 2-17. A hydrometer can be used to measure the state of charge of a lead-acid cell.

If the cell is in a fully charged state, the electrolyte liquid is heavier, so the float in the hydrometer will not sink as far. The distance that it does sink is calibrated in specific gravity on the scale. This can be read as the state of charge of the cell.

LESSON IN SAFETY: When checking the state of charge of a battery, you are handling acid. Be very careful that drops of acid clinging to the end of the hydrometer do not drop off onto your skin or clothes. It will burn your skin and eat holes in your clothes. It is wise to have some baking soda nearby to neutralize any acid accidently spilled. Also, you should wear safety goggles when working with acid to protect your eyes.

Maintenance-free batteries. The unique feature of a **maintenance-free battery,** Figure 2-18, is that water never needs to be added to the cells. These batteries can be easily identified by the absence of holes and filler caps in the sealed-on cover. By contrast, a battery that requires periodic maintenance has holes and filler caps in the cover so that water can be added at necessary intervals.

What makes a battery maintenance-free? Calcium is used in making the battery plates. The calcium reduces the production of battery gases. As a result, chemicals are not carried out of the battery with battery gas. Water is not lost and corrosive chemicals do not accumulate on the outside of the battery. Maintenance-free batteries have a longer service life.

Current rating. The *capacity* of a battery, or its ability to deliver power continuously over a given time is an indication of the quality of the battery. A common rating is *ampere-hours (A·h).* For example, a battery rated at 100 A·h could continuously supply 5 amperes for 20 hours (5 × 20 = 100 A·h). Roughly speaking, the same battery could deliver 10 amperes for 10 hours (10 × 10 = 100 A·h) or 1 ampere for 100 hours (1 × 100 = 100 A·h). A battery will not perform exactly in this manner since the *rate of discharge* also affects the capacity of the battery. If rapidly discharged, it will not live up to its amp-hour rating.

In buying a battery for a car, you should compare ampere-hour ratings between batteries. Also, the number of plates and the kind of separators used will indicate the quality of the battery. A well-known manufacturer with a good reputation is one of your best assurances of getting the most for your money.

ELECTRICITY FROM LIGHT ENERGY

The action of discrete quantities of light energy (photons) on certain atoms will result in the conversion of light energy into electrical energy. This leads to many interesting electrical phenomena. **Photoelectric cell**, or **photocell**, devices are the conversion media. Many think there is only a single classification of these; however, these people are mistaken. Actually, there are three general classes of photoelectric cells. These are:

- Photovoltaic cell.
- Photoconductive cell.
- Photoemissive cell.

Figure 2-18. A cutaway view of a typical maintenance-free lead-acid battery. (Delco-Remy Div., General Motors Corp.)

PHOTOVOLTAIC CELL

Since we are studying sources of electrical potential, the *photovoltaic cell* will be considered first. See Figure 2-19 and Figure 2-20. When a light shines on certain solid materials, electric charges will move across a barrier and establish a potential difference. A device that generates a potential difference upon absorption of this radiant energy is a **photovoltaic cell**, or **solar cell**.

In Figure 2-20, which shows the structure of a photovoltaic cell, light energy causes electrons from the selenium (Se) to move across the barrier layer and collect on the transparent front electrode. This action will make the front electrode negative and leave the selenium with a deficiency of electrons, or positive. If an external circuit is now connected to the terminals of the cell, there will be a small current. A typical solar cell is shown in Figure 2-21. You can probably name at least a half dozen devices that use solar cells.

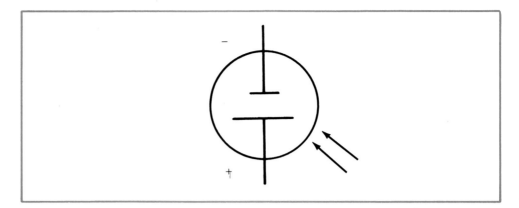

Figure 2-19. Symbol for a photovoltaic cell.

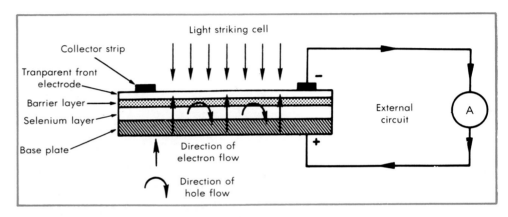

*Figure 2-20. Typical cell structure of a photovoltaic cell.
(International Rectifier Corp.)*

*Figure 2-21. A typical photovoltaic cell or solar cell.
(International Rectifier Corp.)*

Solar cells play an important role in technology by converting energy from the sun to useful electrical energy. A common use of the solar cell may be found in the light meter used by a photographer to measure illumination. Since the voltage developed depends upon the light, it is a simple matter to have the meter indicate the intensity of the light.

Figure 2-22 shows another use of photovoltaic cells. Each photovoltaic cell collects solar energy and converts it into electricity. The electricity that is produced is distributed for residential and commercial use. Right now, electricity produced this way merely supplements that produced by conventional means. Someday, these cells may produce a large part of the electricity we use; however, at the present time, they are too expensive.

Solar cell technology may yet provide a great source of power in future years, however. Right now, many of our orbiting satellites have banks of photovoltaic cells that supply energy to operate communication equipment and other instruments. Many common devices are now solar-powered. New products are developed every day. See Figure 2-23. The field is still relatively young.

Figure 2-22. This photovoltaic array is one of many in a field of collectors called a solar orchard. *The array uses* concentrating lenses *to concentrate the sun's energy and a* tracking system *to follow the sun across the sky. Peak output of one array is about 2400 W. (U.S. Department of Energy)*

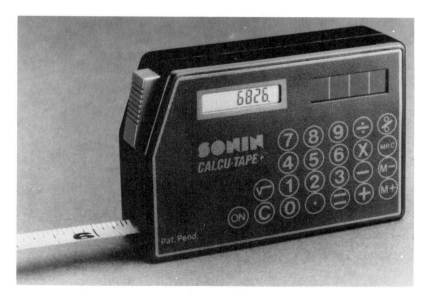

Figure 2-23. A measuring tool kit featuring a solar-powered calculator. (Sonin, Inc.)

PHOTOCONDUCTIVE CELL

Another class of photoelectric device is the *photoconductive cell*. A **photoconductive cell**, Figure 2-24, also known as a **photoresistive cell**, or **photoresistor**, is a semiconductor device whose resistance varies inversely as the intensity of light falling upon it. In other words, as light increases, resistance decreases and vice versa. As part of a circuit, changing the resistance, then, will change the current. The principle of operation is that light energy causes the formation of electron-hole pairs, which frees electrons for conduction purposes. This phenomenon will be studied in detail in Chapter 3.

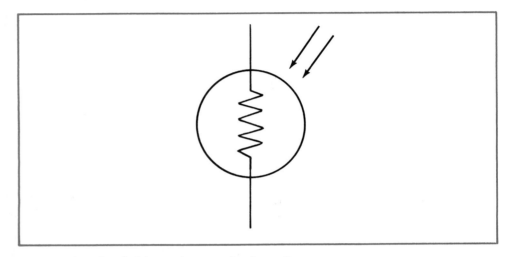

Figure 2-24. Symbol for a photoconductive cell.

Photoconductive cells are typically made of cadmium compounds like cadmium selenide (CdSe) and cadmium sulfide (CdS). Figure 2-25 shows the structure of a CdS cell. This device depends upon a small cadmium sulfide crystal that is doped with an impurity called an **activator**. Silver, antimony, and indium are used as impurities. The ratio of the values of resistance in the dark (**dark resistance**) to bright light is on the order of 600 to 1. The CdS cell is a very practical device. It can be used to directly operate switching relays without amplification. It is very sensitive to minute changes of light intensities. A simple circuit using a CdS cell and relay is shown in Figure 2-26. A more complex circuit is shown in Figure 2-27.

Another type of photoconductive cell is the **photodiode**. This cell differs from other photoconductive cells in that it only allows current in one direction. It will be necessary to delay the understanding of this cell until you study Chapter 3. At this time, it is only necessary to realize that light energy directed on the diode will increase its conduction.

PHOTOEMISSIVE CELL

The third photocell is called a **photoemissive cell**. When light energy is directed on such materials as sodium, cesium, potassium, and alkali earths like strontium and barium, electrons will be ejected from the surface of these materials. Laws governing photoemission state that emission is directly proportional to light intensity. The photoemissive cell is also frequency sensitive, which makes it useful in the measurement of electromagnetic radiation.

In vacuum tubes, a photoemissive cathode is enclosed in a tube with a plate that is held at positive potential. Emitted electrons are collected by the plate, which means that the tube is conducting. Incident light on the emitter surface will vary the conduction of the tube.

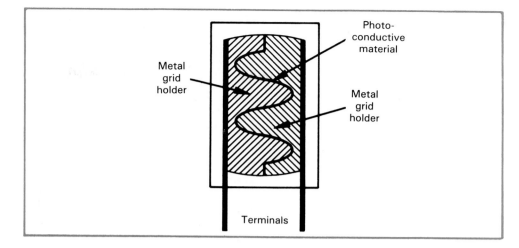

Figure 2-25. A cadmium sulfide photoconductive cell.

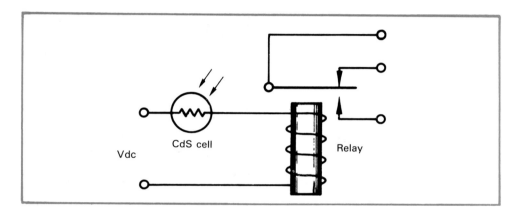

Figure 2-26. Light directed toward the CdS cell will decrease its resistance and permit current in circuit. The current will activate the relay.

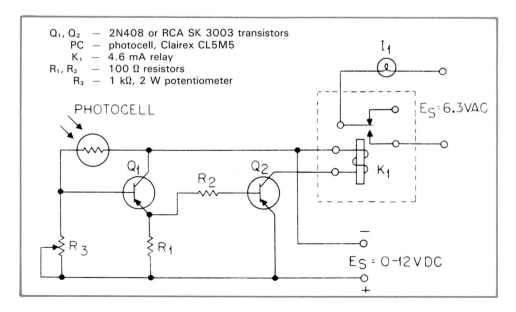

Figure 2-27. Photoconductive cell circuit. Light shining on photoresistor (photocell) drives transistor Q_1 into conduction. Charge is amplified by Q_2, which activates relay K_1. The light, I_1, or other electrical device, can be turned on or off. R_3 is a sensitivity control.

ELECTRICITY FROM HEAT ENERGY

In Figure 2-28, two dissimilar wires are twisted together. One is a copper wire, the other is iron. When heat is applied to the twisted connection, a small potential difference is developed between the two wires. This voltage is directly proportional to the heat intensity. Such a device is called a **thermocouple** and is widely used as a temperature sensing device and as a temperature control device. Many combinations of metals are used for thermocouples, including copper-constantan, iron-constantan, and platinum-platinum rhodium. Figure 2-29 shows a thermocouple. Figure 2-30 shows a thermocouple digital thermometer.

Several thermocouples may be connected in series to increase the output voltage. A device composed of several thermocouples is called a **thermopile**.

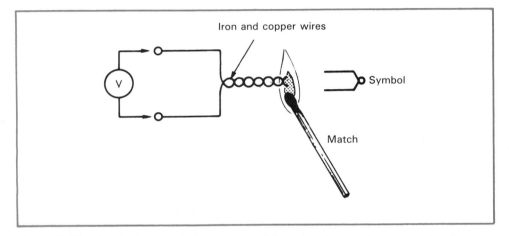

Figure 2-28. *Dissimilar metals in contact with each other produce a small voltage when heated.*

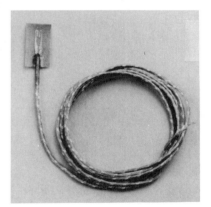

Figure 2-29. *A thermocouple.*

ELECTRICITY FROM MECHANICAL ENERGY

Electricity can be generated from mechanical energy. One way this is accomplished is through mechanical stress. Another way is via magnetism and a rotating machine called a generator. These will be discussed in the next few paragraphs.

Figure 2-30. A thermocouple digital thermometer. (Omega Engineering, Inc.)

MECHANICAL STRESS

When certain crystalline substances such as quartz, tourmaline, Rochelle salt, or barium titanate are subjected to mechanical stress or pressure, an electrical potential will develop across certain points of their crystal structure. Application of voltage between certain faces of a crystal produces mechanical distortion of the material—the reverse effect. This phenomenon is called the **piezoelectric effect**. In the first case, the voltage developed across the crystal surfaces is proportional to the applied distortion or stress on the crystal.

Piezoelectricity is very useful. Many of our modern conveniences depend upon it. A good example is the *pickup* located in the arm of a record player. A phonographic **pickup**, or **cartridge**, is used to convert mechanical movements into electrical impulses in the reproduction of sound. A **piezoelectric**, or **crystal, pickup** is seen in Figure 2-31. The phonograph needle is attached to one end of the thin crystal. As the needle rides in the grooves of a record, it vibrates back and forth. This vibration will flex and bend the crystal to produce a varying voltage, which is amplified and reproduced as the music or sound in the speaker. (A more popular cartridge is the **magnetic pickup**. It works by electromagnetism, which is discussed in later chapters.)

Another familiar application is the typical **crystal microphone**. This action is illustrated in Figure 2-32. The alternate *condensations* and *rarefactions* of the sound waves cause the diaphragm to move in and out. A mechanical linkage between the diaphragm and the crystal causes the crystal to also flex inward and outward. This develops a voltage in proportion to the amplitude and frequency of the sound wave that strikes the diaphragm. The voltage is then amplified, and its output may be heard from a speaker.

Frequently, crystals are joined together either in series, parallel, or series parallel to produce a desired output. One arrangement, Figure 2-33, is the *bimorph cell*. A **bimorph cell** consists of two crystals cemented together in such a way that an applied voltage causes one to expand and the other to contract. The cell then bends in proportion to the applied voltage. In the reverse effect, applied pressure generates double the voltage of a single cell. Bimorph cells are used in phonograph pickups and microphones.

Another very important feature of crystals is their ability to precisely control frequency. This characteristic will be discussed in Chapter 16 concerning oscillators.

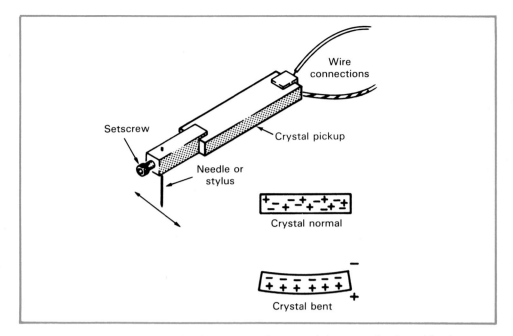

Figure 2-31. As needle rides in the groove of a record, it moves back and forth accor-ding to the cut groove. This movement causes crystal to flex and produce a voltage.

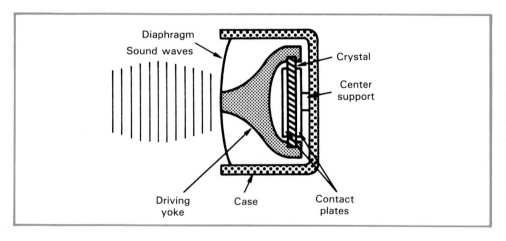

Figure 2-32. A sketch showing the construction of a crystal microphone.

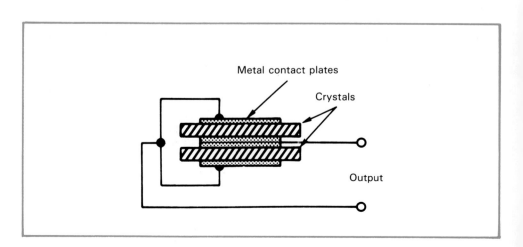

Figure 2-33. The construction of a crystal bimorph cell.

GENERATORS AND MAGNETISM

Moving a conductor through a magnetic field will induce a voltage in the conductor. This is the principle under which a generator operates. A generator is a rotating machine that converts mechanical energy into electrical energy. The electricity that is generated may be dc, or it may be *ac,* depending on machine design. Dc was defined earlier; **alternating current (ac)** is electricity that periodically changes direction of flow.

This source of electricity requires detailed study. Chapter 6 of this text is devoted to various types of generators and their operation.

SUMMARY

- Six basic sources of electrical voltage are friction, chemical action, light, heat, pressure, and magnetism.
- Potential difference, or voltage, is the force behind the electron, moving it along in a conductor. Usually potential difference is called voltage.
- The unit for voltage is the volt.
- Electric current is the movement of electrons in a conductor, and the unit for current is the ampere. Direct current (dc) is unidirectional; alternating current (ac) periodically changes direction.
- The cell is the basic unit for producing voltage by chemical action.
- Batteries are combinations of cells connected in either series or parallel.
- A primary cell cannot be recharged while a secondary cell can.
- The photovoltaic cell, or solar cell, can be used to convert light energy to electrical energy.
- The thermocouple is a device that converts heat energy to electrical energy.
- The piezoelectric effect makes it possible to directly convert mechanical energy through pressure to electrical energy.
- A generator produces electrical energy from mechanical energy via a magnetic field and motion.

KEY TERMS

Each of the following terms has been used in this chapter. Do you know their meanings?

activator	electrode	photoemissive cell
alkaline cell	electrolyte	photoresistive cell
alternating current (ac)	electrolytic solution	photoresistor
ampere (A)	electromotive force (emf)	photovoltaic cell
Archimedes' principle	electron flow	pickup
battery	forming	piezoelectric effect
bimorph cell	galvanism	piezoelectric pickup
capacity	hydrometer	polarization
carbon-zinc cell	internal resistance	potential
cartridge	joule (J)	potential difference
cell	lead-acid storage battery	primary cell
conductor	lithium cell	secondary cell
conventional current	magnetic pickup	silver oxide cell
crystal microphone	maintenance-free battery	solar cell
crystal pickup	mercury cell	specific gravity
current *(I)*	NICAD cell	storage cell
dark resistance	nickel-cadmium cell	thermocouple
depolarizer	photocell	thermopile
direct current (dc)	photoconductive cell	volt (V)
dissociate	photodiode	voltage *(E), (V)*
dry cell	photoelectric cell	work *(W)*

TEST YOUR KNOWLEDGE

Please do not write in this text. Place your answers on a separate sheet of paper.

1. What is the principle of the voltaic cell?
2. Polarization speeds up cell chemical action when hydrogen bubbles collect around the positive electrode. True or False?
3. Can your experimental zinc and carbon electrode voltaic cell be recharged?
4. Primary cells cannot be recharged. True or False?
5. A/an _____ cell can be recharged by applying reverse voltage.
6. Lead-acid batteries should be connected to the charger _____ the power is applied.
7. What is used as a depolarizer in a flashlight cell?
8. Give three advantages that a mercury cell has over a typical flashlight cell.
9. What is the major advantage of a nickel-cadmium cell?
10. _____ _____ is a classification of liquids according to their weight in respect to the weight of an equal volume of water.
11. Which combination of cells produces the highest terminal voltage?
    a. Series.
    b. Parallel.
12. Which combination of cells should be used for a heavy load?
    a. Series.
    b. Parallel.
13. Name five sources of electrical energy.
14. A voltage is generated by putting pressure on or flexing a crystal. True or False?
15. How could you use a CdS cell to turn on your house lights when darkness arrives?

Chapter 3

INSULATORS, SEMICONDUCTORS, CONDUCTORS

There are many important materials to discuss in electricity and electronics. Three of the most important are insulators, semiconductors, and conductors.

After studying this chapter, you will be able to:
- ☐ *Discuss the theory behind transferring electrical energy.*
- ☐ *Give characteristics of insulators, semiconductors, and conductors.*
- ☐ *List the factors affecting resistance.*
- ☐ *Determine the value of various color-coded resistors.*

INTRODUCTION

Methods of producing an electrical potential were discussed in the previous chapter. Potential difference, you will recall, is measured in voltage. Voltage has been established as a source of energy *at rest*, but ready to perform useful work if connected to a circuit containing wires and components. The amount of work that can be extracted depends to a large degree on the properties of the circuit materials with respect to their ability to conduct electricity. We classify these materials into one of three groups. These are as:

- **Insulators**—materials that have very high opposition to current.
- **Semiconductors**—materials having current conductive properties falling between insulators and conductors. Resistance changes with applied voltage.
- **Conductors**—materials through which current will pass with little opposition or resistance.

ENERGY DIAGRAMS

A simple **energy diagram**, Figure 3-1, shows how some materials conduct and others do not. In the diagram of the insulator, the electrons are held in the **valence band**, and a large amount of energy is needed to cause electrons to jump the **forbidden band**, or **energy gap**, and find freedom of movement in the **conduction band**. Note that electrons in this band are no longer tied to the atom; they are **free electrons**. Also, note that no current carriers can exist in the forbidden band region of the energy diagram. A discrete quantity of energy must be supplied to make the carriers "jump the gap." Electrons that have jumped the gap are in an excited state. Finally, an electron going from an excited state to its original energy level will give up energy in the form of light or heat.

Some materials are neither good insulators nor good conductors. They fall in the class of semiconductors. Note from the diagram that less energy is required to make them conduct—the forbidden band is narrow when compared to the insulator. (Semiconductor electrons require less energy to jump the gap than insulator electrons—about 39% less for silicon and 61% less for germanium.)

The diagram of a conductor shows much less energy is required to free electrons for conduction (about 78% less than insulators, in actuality). A material that is a good conductor has many free electrons ready to transfer energy by conduction.

A resistance chart is shown in Figure 3-2. It compares ability of conductors, semiconductors, and insulators to conduct electricity. Note that the lower the value of ohms per centimeter, the better the conductor.

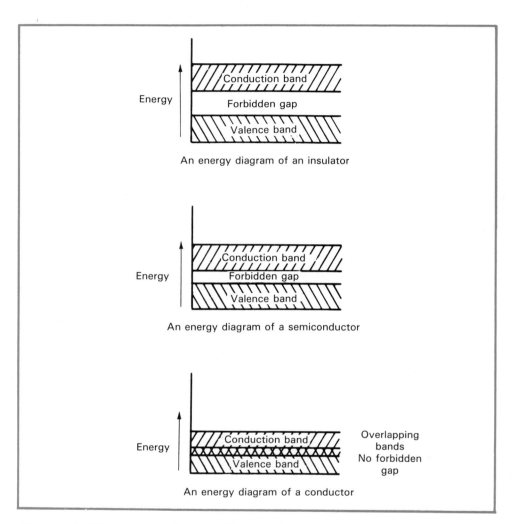

Figure 3-1. These energy diagrams illustrate the difference between an insulator, semiconductor, and conductor.

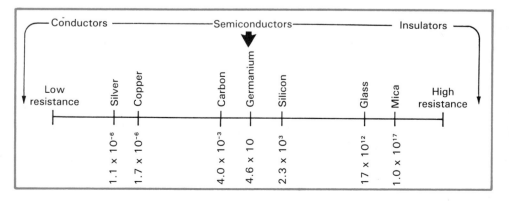

Figure 3-2. Resistance chart of certain conductors, semiconductors, and insulators (in ohms/cm).

INSULATORS

You learned in Chapter 1 reading about atomic theory that electrons in some materials are held securely, close to the nucleus, and require large amounts of energy to break loose for conduction. These materials are insulators. Insulating materials can be made to conduct by adding large amounts of energy. A very high potential could force the electrons to move; however, this could destroy the material.

You will remember that silicon, in its pure form, has all of its valence electrons securely locked in covalent bonds. None are free to circulate. In Figure 3-3, note that very little conduction takes place in a crystal of silicon. Silicon, you might think, should be a good insulator because of its covalent bonds; however, it is not. This is because relatively small amounts of energy are needed to break its bonds and set electrons free. You will learn more about silicon when you read about semiconductors.

Insulators serve many useful purposes. They have an extremely high resistance to the flow of electricity. Insulating materials are used to cover and support wires and components. They protect against electric shock and prevent short circuits between wires and parts. Examples of insulators include glass, rubber, Bakelite™, plastic, air, and many other special materials sold under manufacturers' trade names. Asbestos is another insulator; however, it is considered a hazardous material and so is no longer used in manufacturing or construction.

SEMICONDUCTORS

Semiconductors may be classified as *intrinsic* or *extrinsic*. For now, we will confine our discussion to intrinsic semiconductors. An **intrinsic semiconductor** is made of atoms of all the same kind. Pure germanium or silicon are examples of this type. Refer again to Figure 3-1. Note that energy is required to make semiconductors conduct. When heat, light, or electrical energy is added to the crystal, electrons jump from the valence band to the conduction band. This leaves a positive site, or **hole**, in the valence band.

Each time a hole is produced, a free electron is produced. (We speak of these as **electron-hole pairs**. They do not, however, stay together as *pairs* after they are produced, as the name implies.) Under the influence of heat or light energy, electron-hole pairs are always forming; there is always movement back and forth across the forbidden band. So the intrinsic crystal will conduct to some extent and is classified as a *semi*conductor.

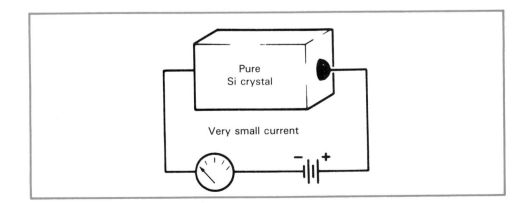

Figure 3-3. A pure silicon crystal is not a good conductor.

CONDUCTION BY FREE ELECTRONS

In Figure 3-4, a silicon crystal is connected to a source of potential. Remember, there are some free electrons that have moved up to the conduction level. Electrons leave the negative terminal of the power source and enter the crystal. For every electron entering the crystal, one must leave the positive end. The leaving electron is attracted to the positive source. The crystal *always* maintains an overall net charge of zero.

Further explanation is offered in Figure 3-5. In this sketch, a copper wire with many free electrons is used. An electron forced on one end of the wire causes an electron to leave the other end. This illustration oversimplifies the conduction phenomenon as the electrons appear as a string of balls in a tube. Actually, current is a drift pattern, Figure 3-6. This pattern is shown in an enlarged view as an exchange, or transfer, of energy between atoms.

The speed of energy transfer is interesting to note. It approximates the *speed of light*, which is 186,000 miles per second (3×10^8 meters per second). However, this is not the electron speed. Electrons move much slower than this.

CONDUCTION BY HOLES

As mentioned, when electrons become free in the intrinsic crystal, *holes* are created in the valence band. Conduction can also be accomplished by these holes. (Holes and conduction electrons are sometimes called *carriers*. **Carriers** are the elements capable of carrying an electric charge through a solid.) See Figure 3-7. It shows an electron, attracted from the crystal by a positive source,

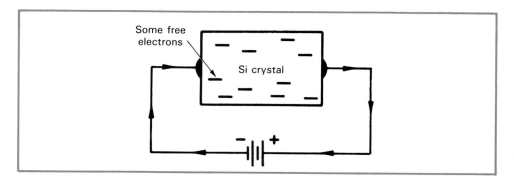

Figure 3-4. Conduction takes place by means of electrons.

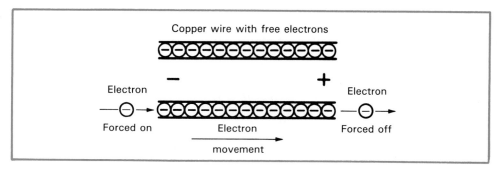

Figure 3-5. Energy exchanged between electrons constitutes a current.

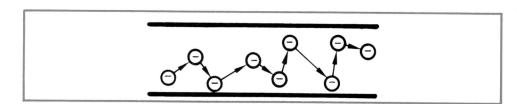

Figure 3-6. Transfer of energy is actually a drift pattern.

leaving behind a hole that needs to be filled. The electron from the next atom jumps over to fill the hole and leaves a hole behind. Holes are deserted and occupied in random fashion. This process of electrons breaking free and then finding new holes to fill is called **recombination**. The process proceeds until the hole arrives at the negative terminal of the crystal where it is finally filled by an electron from the source. Remember:

- The crystal always maintains a net charge of zero.
- Hole movement is in the opposite direction of electron movement.
- The current in the external circuit is always electron flow and moves from negative to positive.

A crystal containing both electrons and holes is connected in a circuit in Figure 3-8. Note the direction of movement in each case.

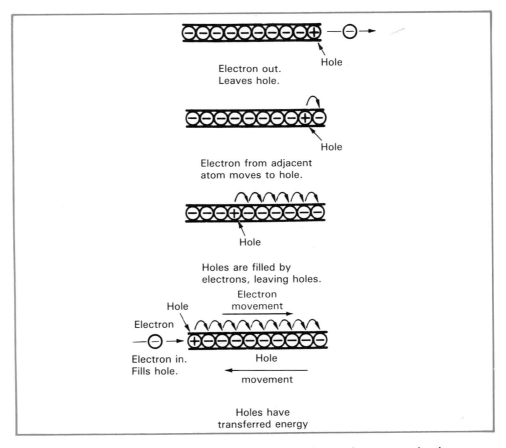

Figure 3-7. Conduction by hole movement is opposite to electron conduction.

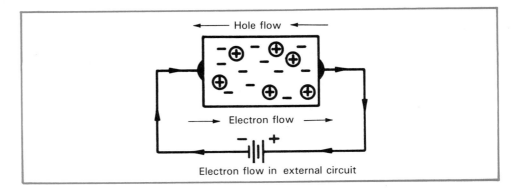

Figure 3-8. Conduction in a crystal containing both electrons and holes.

N-TYPE CRYSTAL FORMATION

In order to make the intrinsic crystal a better conductor, certain impurities are added to the crystal during crystallization. Remember that both silicon and germanium have *four* valence electrons and, in their crystal lattice structure, these electrons are held in covalent bonds. In the **n-type crystal**, impurities are selected that have *five* valence electrons. These include phosphorus (P), arsenic (As), bismuth (Bi), and antimony (Sb), which are pentavalents.

The process of adding impurities to crystal is called **doping**, and the impurities are called **dopants**. Only a small amount of dopant is used, such as one atom of impurity to 10 million atoms of silicon. Semiconductors that have doping are classified as **extrinsic semiconductors**. The conduction characteristic of extrinsic versus intrinsic semiconductors is quite different.

Figure 3-9 shows an n-type crystal formation. An arsenic atom with five valence electrons is added to the pure silicon. Four of the valence electrons of the arsenic join in covalent bonds with the silicon. Since the outer ring of the silicon is now satisfied, the extra electron from the arsenic is free to move about the crystal. This electron becomes a conduction electron since it is not attached to any atom. It is a current carrier.

Electrons are the **majority carriers** in an n-type crystal. A few holes do exist in the crystal due to thermally induced electron-hole pairs, and these are **minority carriers**. The minority carrier current is very small. Conduction in the n-type crystal is shown in Figure 3-10. Impurities that add free electrons to a crystal are called **donor impurities**. Since the donor impurity atom is shy of its full complement of electrons in the outer ring, it has a net positive charge, and this charge is immobile, or *bound*.

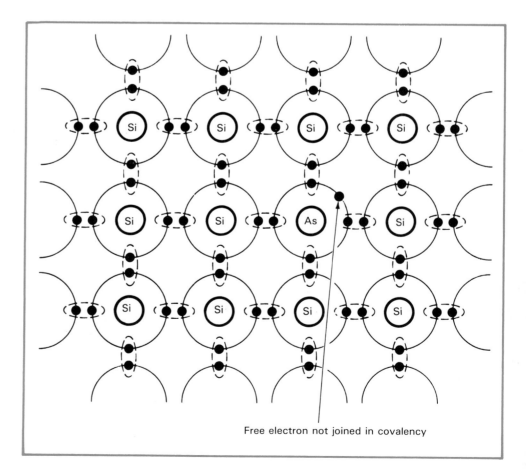

Free electron not joined in covalency

Figure 3-9. Doping the crystal with a pentavalent will produce free electrons.

P-TYPE CRYSTAL FORMATION

Beginning again with a pure crystal of silicon or germanium, a trace amount of trivalent impurity is added during crystallization to make a **p-type crystal**. Common trivalents, which have three valence electrons, are aluminum (Al), indium (In), gallium (Ga), and boron (B).

Figure 3-11 shows a p-type crystal formation. The three valence electrons of the indium join in covalent bonds with the silicon. The silicon atom is not satisfied, however. Another electron is needed to complete the covalent bonding structure. There remains a hole, which has a strong attraction for an electron. The trivalent dopant creates many of these holes, and conduction through this crystal is by holes.

Holes are the majority carriers in a p-type crystal. A few electrons freed by thermally induced electron-hole pairs will generate a very small current.

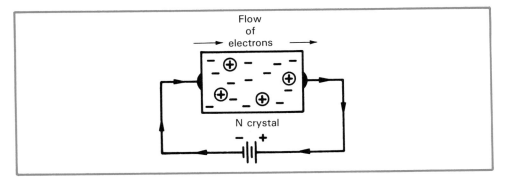

Figure 3-10. *Conduction by electrons in an n-type crystal.*

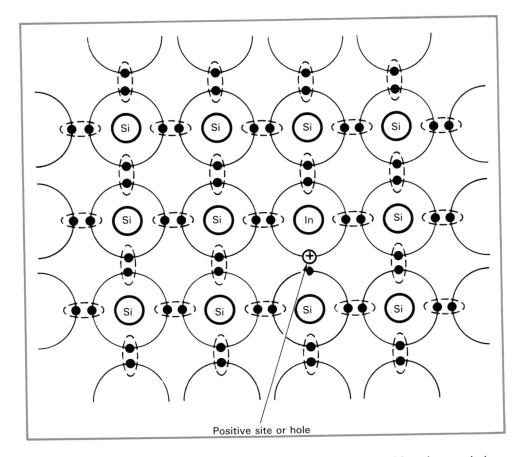

Figure 3-11. *Doping the crystal with a trivalent will produce positive sites, or holes.*

Electrons are the minority carriers. Conduction in the p-type crystal is shown in Figure 3-12. Since the dopant added to this type crystal creates holes, which will accept electrons, it is called an **acceptor impurity**. Since the acceptor impurity atom will take on an extra electron in its outer ring, it has a net negative charge. This charge, like the donor ion, is also immobile. Only the holes generated by the atom are mobile.

ENERGY LEVELS IN DOPED CRYSTALS

Why does a crystal with an impurity conduct better than a pure crystal? From the mechanical point of view, the answer includes the introduction of carriers such as electrons or holes. Take a closer look at the energy diagram in Figure 3-13. In the case of the *donor* impurity, which supplies free electrons, the free electron cannot fit into the valence band since that band is already full. You might assume that the electron should be found in the conduction band, but it does not have quite enough energy to make it to this level. It is customary to indicate a new discrete energy level for these electrons slightly below the conduction band.

In the case of the *acceptor* impurity, which creates holes, a new discrete energy level for holes will exist slightly above the valence band. This is due to a natural tendency to form covalent bonds with the valence electrons.

With donor impurities, the materials will conduct since only a small amount of energy is required to move the electrons into the conduction band. With the acceptor impurities, only a small amount of energy is required to move an electron from the valence band to a positive site (hole). This leaves a hole in the valence band. Conduction in a p-type crystal is by holes in the valence band. The conditions of added energy to the doped crystals are shown in Figure 3-14.

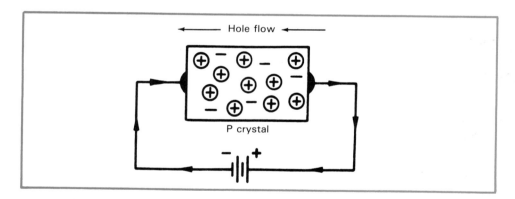

Figure 3-12. Conduction by holes in a p-type crystal.

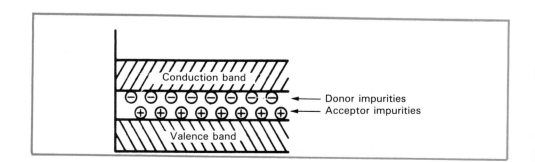

Figure 3-13. Electrons from donor impurities are located below the conduction band. Holes from acceptor impurities form a new energy level just above the valence band.

GALLIUM ARSENIDE

Gallium arsenide (GaAs) is assuming more importance as a major semiconductor material. Favorites have been germanium and silicon; however, GaAs is gaining popularity as the "souped-up" semiconductor material of the 90s. The primary advantage of GaAs is its speed and low operating power. GaAs is much faster than silicon in being able to switch computer circuits on and off.

CONDUCTORS

Electrons in the valence band of an atom are associated with the nucleus of the atom. In order to move these electrons into the conduction band to become current carriers, it is necessary to add sufficient energy to cause the electrons to jump the energy gap.

Look back at Figure 3-1 to the energy diagram for a conductor. It has no forbidden band. The valence band and conduction band overlap. Large numbers of electrons are immediately available for conduction. The quality of a conductor is determined by the number of free carriers available. To express it another way, the *resistance* to current in copper is very low. In Figure 3-15, the *specific resistance* of several metals (and carbon) is tabulated. **Specific resistance**, or **electrical resistivity**, is the resistance of a conductor expressed in one of two ways. These are: in circular-mil ohms per foot (cmil·Ω/ft.); in ohm-centimeters (Ω·cm), where this value is based on a standard cross-sectional area of 1 cm². (A *circular mil* is a measure of area pertaining to a wire conductor. It is discussed in greater detail later in this chapter.) The symbol for specific resistance is the Greek letter ϱ (rho).

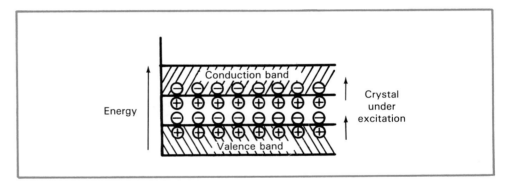

Figure 3-14. Excitation causes electrons to move to the conduction band and other electrons to move from the valence band to holes.

Material	Specific resistance, ϱ (cmil·Ω/ft.)
Aluminum	17.
Carbon	2500-7500 x copper
Constantan	295.
Copper	10.4
Gold	14.
Iron	58.
Nichrome	676.
Silver	9.8
Steel	100.
Tungsten	33.8

Figure 3-15. Table showing specific resistance of several materials at 68°F (20°C).

SUPERCONDUCTORS

Practically since the discovery of electricity, people have dreamed of eliminating resistance in conductors so that current could move unopposed. One method that researchers have used to achieve this has been to "freeze" circuits to near *absolute zero*.

Absolute zero is considered the temperature at which molecular motion vanishes and a body is without heat energy. It is equal to $-460°F$, or $-273°C$). At this temperature, the property of **super***conductivity*, which is the absence of electrical resistivity, can exist. (*Conductivity* [σ{sigma}] is the inverse of resistivity; something with zero resistivity would have great conductivity.)

The problem with the research has been that it is difficult and costly to achieve such very low temperatures. It normally requires use of **cryogenic fluids**. These are liquids that boil at very low temperatures and, as such, are very cold. One cryogenic fluid that is used is liquid helium. It has a temperature of near $-452°F$.

Scientific research has recently made it possible to produce materials that exhibit superconductivity, called **superconductors,** that can operate at much higher temperatures. The breakthrough has been heralded as great as the invention of the transistor. It could lead to great improvements both in the ease of use and in the cost of superconductor materials. In the future, it may be possible to have superconductivity occurring at *near* room temperature.

In an ordinary circuit using a conductor such as copper, a voltage source like a battery is used to overcome electrical resistance. One characteristic of current in superconductors is that electrons flow in pairs making it easier to overcome resistance. A current, once established, can continue apparently indefinitely without the use of a power source, and these freely moving currents can be large, thus producing big magnetic fields.

New superconductors are made from materials such as copper oxide and metals such as lanthanum, strontium, and barium. The applications for superconductors appear to be many. Some of these may be faster operating computers and more efficient large-scale transmission of electric power. Also, we may see better magnets, which could be used for a number of items including magnetically *levitated* trains, Figure 3-16.

RESISTANCE

Resistance (R) is opposition exhibited to electrical current. The unit of resistance is the **ohm**, which is symbolized by the Greek letter Ω (omega). The value of 1 ohm is the resistance when there is 1 ampere of current with 1 volt applied. The name "ohm" comes from the German physicist, Georg Simon Ohm, who developed the basic mathematical relationship between potential difference, electrical current, and resistance.

Materials with greater resistance require greater energy to raise electrons to the conduction band. Consequently, energy is consumed by resistance and appears as heat. In fact, one of the primary uses of resistance is to convert electrical energy into heat energy. Such an application is found in toasters and electric stoves, ovens, and heaters.

Resistance finds another primary use in producing light. The resistance filament in a light bulb becomes white hot and converts electrical energy to light energy. Resistance units find countless uses in electrical and electronic circuits. Resistance is represented in circuit drawings by the symbol shown in Figure 3-17.

Resistance R in a metal conductor is directly proportional to its length l and inversely proportional to the cross-sectional area A. That is:

$$R = \varrho \frac{l}{A}$$

Figure 3-16. Superconducting magnets are used to suspend magnetically levitated trains just above a guideway and to power the train. (Japan Railways Group)

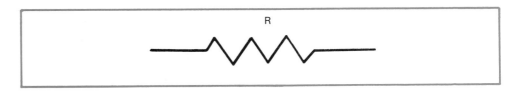

Figure 3-17. The schematic symbol for resistance.

Material	Temperature coefficient, α $°C^{-1}$
Silver	0.0061
Copper	0.0068
Aluminum	0.00429
Tungsten	0.0045
Iron	0.00651
Nichrome	0.0004
Carbon	−0.0005
Germanium	−0.05
Silicon	−0.07

Figure 3-18. Table showing temperature coefficient of resistance of several materials at 68°F (20°C).

where ϱ, the constant of proportionality, is the specific resistance of the metal.

Specific resistance depends somewhat on temperature. For metals, it increases nearly linearly with temperature. However, in semiconductors, specific resistance can decrease with increasing temperature. Whether electrical resistivity increases or decreases with temperature is indicated by a factor called the **temperature coefficient of resistance** (α [alpha]). Some values of α are given in Figure 3-18. Note that semiconductors have a negative temperature coefficient, which indicates that the material will decrease in resistance with increasing temperature.

For the most part, then, resistance in a metal conductor is influenced by four factors. These are:
- Type of material.
- Cross-sectional area (size).
- Length.
- Temperature.

Resistance and conductor material

As we have established, the ability of a conductor to conduct depends upon the material from which it is made. By referring back to Figure 3-15, it is apparent that copper has about one-tenth the resistance of steel. Aluminum has only slightly more resistance than copper. Silver offers the least resistance of all materials, but it is expensive. Conducting wires generally are made of copper, except in cases where mechanical strength is needed. Particular note should be made of carbon. It is a highly resistive material. Compounds consisting of carbon and other materials are used in small resistors found in electrical and electronic circuits.

Resistance and conductor size

It is just common sense to expect that a larger wire will conduct electricity more freely and in greater quantity than a smaller wire. There are many more free carriers to do the work. You have seen that the resistance of a wire is inversely proportional to its size. The larger the wire, the less the resistance; the smaller the wire, the greater the resistance.

The size of a wire is specified by a number. Figure 3-19 shows an American standard wire gauge. The larger the number on the gauge, the smaller the diameter of the wire. Numbers 20 and 22 are commonly used for hookup wire in electronic equipment. Numbers 12 and 14 will be found in household lighting circuits. Number 6 wire would be used to connect an electric range. The size of a wire is determined by the amount of current the wire is required to carry. If a wire is too small, it will heat up excessively due to the loss of energy as it overcomes the resistance.

A common way of expressing cross-sectional area of a round wire is in *circular mils*. The number of **circular mils (cmil)** in a cross-sectional area of wire (conductor) is equal to the square of the diameter (d^2), where d is in mils. (One **mil** is 0.001 inch.) Therefore, a wire with a diameter of 1 mil has an area of 1 cmil ($1^2 = 1$). A wire with a diameter of 2 mils has an area of 4 cmil ($2^2 = 4$). A diameter of 15 mils gives an area of 225 cmil ($15^2 = 225$). In Figure 3-20, wire gauge number and circular-mil area are given.

The circular mil is a more convenient way of expressing the diameter of a conductor than the fractional inch. An example will help illustrate this point.

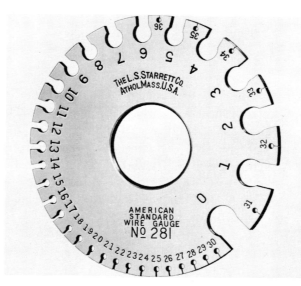

Figure 3-19. A gauge used to determine wire size.   (L.S. Starrett Co.)

According to our formula, a wire with a diameter of 50 mils (0.05 in.) has an area of 2500 cmil. The area of a circle is:

$$A = \pi r^2 = \pi \left(\frac{d}{2}\right)^2 = \pi \frac{d^2}{4} = 3.1416 \frac{d^2}{4} = 0.7854 d^2$$

Computing the area of this wire in square inches:

$$
\begin{aligned}
A &= 0.7854 d^2 \\
&= 0.7854(0.05 \text{ in.})^2 \\
&= 0.0019635 \text{ in.}^2
\end{aligned}
$$

No doubt, you will find it easier to work with 2500 cmil than 0.0019635 in.²!

Area of a wire given in **square mils** is equal to $0.7854 d^2$, where d is again in mils. There is 0.7854 *square* mil in 1 *circular* mil. *Multiply* circular mils by 0.7854 to get square mils. *Divide* square mils by 0.7854 to get circular mils.

There is another term common to wire. It is the *circular mil-foot*, the unit conductor. A **circular mil-foot** is a wire having a cross-sectional area of 1 circular mil and a length of 1 foot.

Problem:

What is the resistance of 100 feet of No. 22 copper wire?

Solution:

Using the equation for resistance in a conductor and values from Figure 3-15 and Figure 3-20:

$$
\begin{aligned}
R &= \varrho \frac{l}{A} \\
&= \frac{10.4 \text{ cmil·}\Omega}{\text{ft.}} \times \frac{100 \text{ ft.}}{642.4 \text{ cmil}} \\
&= 1.62 \ \Omega
\end{aligned}
$$

All units cancel except for ohms.

Conductors are made in a wide variety of sizes, both bare and insulated, to fit the various needs. Some wires are stranded with either 7, 19, or 37

GAUGE NO.	DIAM. MILS	CIRCULAR MIL AREA	RESISTANCE OHMS PER 1,000 FT. OF COPPER WIRE AT 25°C	GAUGE NO.	DIAM. MILS	CIRCULAR MIL AREA	RESISTANCE OHMS PER 1,000 FT. OF COPPER WIRE AT 25°C
1	289.3	83,690	0.1264	21	28.46	810.1	13.05
2	257.6	66,370	0.1593	22	25.35	642.4	16.46
3	229.4	52,640	0.2009	23	25.57	509.5	20.76
4	204.3	41,740	0.2533	24	20.10	404.0	26.17
5	181.9	33,100	0.3195	25	17.90	320.4	33.00
6	162.0	26,250	0.4028	26	15.94	254.1	41.62
7	144.3	20,820	0.5080	27	14.20	201.5	52.48
8	128.5	16,510	0.6405	28	12.64	159.8	66.17
9	114.4	13,090	0.8077	29	11.26	126.7	83.44
10	101.9	10,380	1.018	30	10.03	100.5	105.2
11	90.74	8,234	1.284	31	8.928	79.70	132.7
12	80.81	6,530	1.619	32	7.950	63.21	167.3
13	71 96	5,178	2.042	33	7.080	50.13	211.0
14	64.08	4,107	2.575	34	6.305	39.75	266.0
15	57.07	3,257	3.247	35	5.615	31.52	335.0
16	50.82	2,583	4.094	36	5.000	25.00	423.0
17	45.26	2,048	5.163	37	4.453	19.83	533.4
18	40.30	1,624	6.510	38	3.965	15.72	672.6
19	35.89	1,288	8.210	39	3.531	12.47	848.1
20	31.96	1,022	10.35	40	3.145	9.88	1,069.

Figure 3-20. Copper wire table.

separate wires twisted together. This provides the necessary cross-sectional area to carry the current. It also provides flexibility and ease of handling.

Resistance and conductor length

A wire increases in total resistance as its length is increased. Resistance varies directly with length. If a given wire has a resistance of 1 ohm per foot, the resistance of 10 feet of the wire would be 10 ohms. The length of wire was a factor in computing the previous problem involving circular-mil area and resistivity.

Resistance and temperature

Most conductors increase in resistance at increased temperatures. This is not in conflict with the energy theory. At higher temperatures, the atoms are moving more rapidly. They are arranged in a less orderly fashion so they might be expected to interfere with the flow of electrons. Also, with increased temperatures, there are more free electrons for conduction. The drift path becomes congested, and there are many more collisions. Thus, the wire does not conduct as well.

This is not the case with semiconductors. Increased temperature will bring many electrons into the conduction band, and the resistance of the crystal decreases to where it takes on the properties of a conductor.

Types of resistors

Resistors are probably the most common components in electrical and electronic equipment. The purpose of using a resistor in a circuit is either to reduce

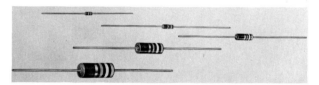

Figure 3-21. Various sizes of carbon-composition resistors. (Ohmite Mfg. Co.)

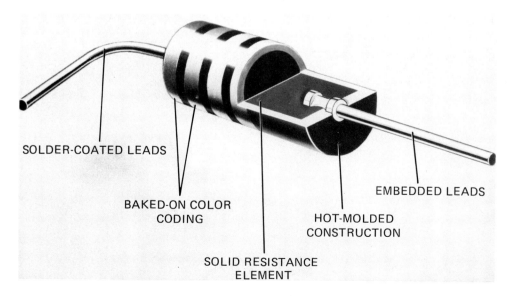

SOLDER-COATED LEADS

BAKED-ON COLOR CODING

EMBEDDED LEADS

HOT-MOLDED CONSTRUCTION

SOLID RESISTANCE ELEMENT

Figure 3-22. A cutaway view of a carbon-composition resistor. (Allen-Bradley Co.)

current to a specific value or to provide a desired voltage drop. Resistors come in various composition, resistance values, and physical size. Some different types of resistors include:

- *Carbon-composition resistors.*
- *Power resistors.*
- *Variable resistors.*
- *Film resistors.*

Carbon-composition resistors. These are the most common of all resistors used in electronic circuits. They are manufactured in a wide range of values, from a fraction of an ohm to millions of ohms. Also, resistors of a specified value are made in various sizes such as 1/4 watt, 1/2 watt, 1 watt—up to 2 watts. The larger the wattage, the larger the physical size of the resistor. (Wattage will be discussed in greater detail in Chapter 4.) The larger sizes give a greater surface area for dissipation of heat. They can carry higher currents without being damaged. An assortment of resistors is shown in Figure 3-21. Figure 3-22 shows the construction of a carbon-composition resistor in a cutaway view.

Power resistors. These are usually *wire-wound resistors* on an insulating ceramic core. See Figure 3-23. Their large size is indicative of their use; that is, in high power circuits (5 W or more) with considerable current. Power resistors are manufactured with either two fixed taps or with two fixed taps and an adjustable slider. This special type of power resistor is shown in Figure 3-24. *Adjustable power resistors* are useful in voltage divider circuits, which will be discussed later on in this text.

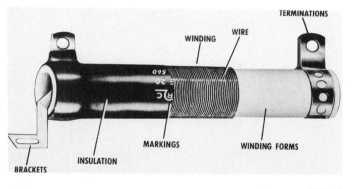

Figure 3-23. A cutaway view of a power resistor.   (IRC, Inc.)

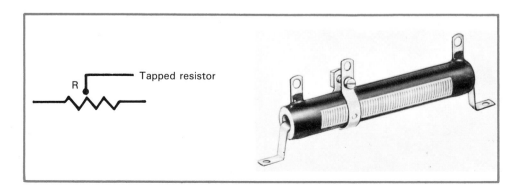

Figure 3-24. An adjustable power resistor and its schematic symbol. (Ohmite Mfg. Co.)

There is another type of resistor worthy of mention here. This is the *metal-oxide resistor*. See Figure 3-25. Metal-oxide resistors are also used for high voltage and wattage applications.

Variable resistors. These resistors have wide application in the field. One type was just discussed—the adjustable power resistor. Another type, the **potentiometer**, is shown in Figure 3-26. The potentiometer, or **pot** for short, is a popular variable resistor. The resistance unit of this device may be a molded-carbon ring for low power or a wire-wound circular form for high power applications.

In the potentiometer, a contact arm, controlled by a shaft and knob, rides in contact with a resistance unit. A desired amount of resistance may be selected by turning the knob. There are three connections to this component; one at each end of the total resistance and a third attached to the moving arm. The voltage source is connected across the total resistance. The moving arm is used to vary the voltage division between the center terminal and the ends.

Potentiometers are made in a wide range of ohmic values and sizes to meet circuit needs. See Figure 3-27. A precision type of potentiometer with an end screwdriver adjustment is shown in Figure 3-28. The dime shows the relative size. These potentiometers, also made in a wide range of ohmic values, come with terminals and mountings to fit most any application.

Closely related to potentiometers are *rheostats*. A **rheostat** is a variable resistor with *two* terminals. These terminals are connected in series with a load. The purpose of the rheostat is to vary the amount of current; whereas

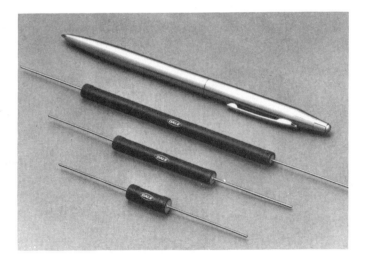

Figure 3-25. Metal oxide resistors. The pen shows their relative size.
(Dale Electronics)

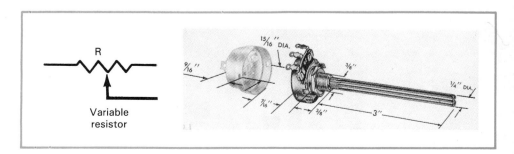

Figure 3-26. A typical potentiometer and its schematic symbol.   (Centralab)

the pot is used to vary the voltage. Potentiometers may be used as rheostats; however, rheostats may not be used as pots.

Film resistors. These are sometimes called *thin-film resistors*. The mechanics are simple. See Figure 3-29. A ceramic rod is coated with a thin film of various resistive alloys. End caps are press-fitted. The film is cut in a helix around the cylindrical rod to increase resistance and leads are welded to the end caps. The body is coated for environmental protection and color banded for identification.

Figure 3-27. A wire-wound potentiometer. (Ohmite Mfg. Co.)

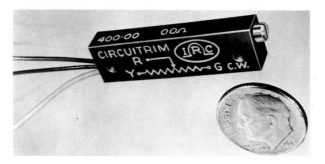

Figure 3-28. A precision screwdriver-adjust trimmer potentiometer. (IRC, Inc.)

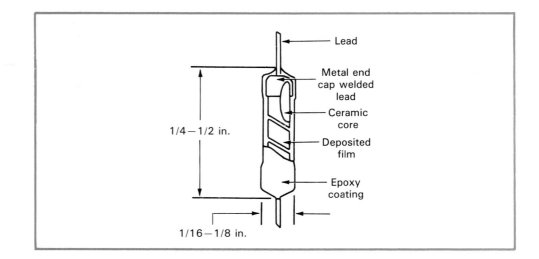

Figure 3-29. A standard precision film resistor.

Film resistors are very accurate. The film makes possible improved precision. Film resistors tend to have lower levels of electrical noise. They also tend to be more efficient heat dissipators. This means that more power can be handled by smaller-sized resistors.

Resistor color coding

Most low wattage (2 W or less) resistors are identified by color codes. Refer to Figure 3-30. Note the bands appearing around the body of the resistor. The values of these bands of color are standard; they have been adopted by the Electronics Industries Association (EIA) and the military. You should memorize the color code.

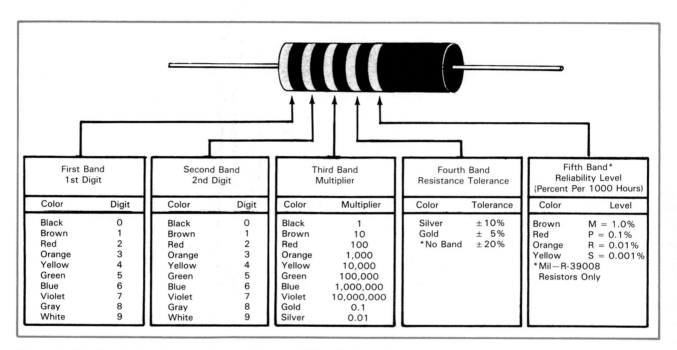

First Band 1st Digit		Second Band 2nd Digit		Third Band Multiplier		Fourth Band Resistance Tolerance		Fifth Band* Reliability Level (Percent Per 1000 Hours)	
Color	Digit	Color	Digit	Color	Multiplier	Color	Tolerance	Color	Level
Black	0	Black	0	Black	1	Silver	±10%	Brown	M = 1.0%
Brown	1	Brown	1	Brown	10	Gold	± 5%	Red	P = 0.1%
Red	2	Red	2	Red	100	*No Band	±20%	Orange	R = 0.01%
Orange	3	Orange	3	Orange	1,000			Yellow	S = 0.001%
Yellow	4	Yellow	4	Yellow	10,000			*Mil—R-39008	
Green	5	Green	5	Green	100,000			Resistors Only	
Blue	6	Blue	6	Blue	1,000,000				
Violet	7	Violet	7	Violet	10,000,000				
Gray	8	Gray	8	Gold	0.1				
White	9	White	9	Silver	0.01				

Figure 3-30. Standard color code for resistors.   (Allen-Bradley Co.)

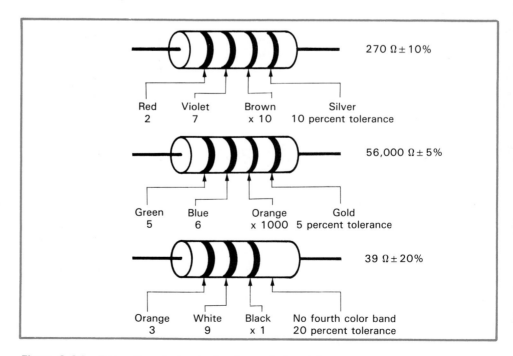

Figure 3-31. Examples of standard color-coded resistors.

In reading the value, the resistor is held so that the color bands are on the left. Colors in the two leftmost bands stand for certain digits. For example, gray means an 8; white means a 9. The third band is the multiplier of the first two bands. Here, different colors represent different multipliers. The fourth band gives the tolerance. Normally, resistors have only four color bands. However, a fifth color band may appear on some resistors to indicate reliability level. Examples of reading resistor values using the color code are shown in Figure 3-31.

SUMMARY

- Electrical materials may be classified according to their conductive properties as either insulators, semiconductors, or conductors.
- Insulators require a great deal of energy to move electrons from valence band to conduction band; semiconductors require less energy; conductors require minimal energy to move electrons to the conduction band since valence and conduction bands overlap.
- Insulators have a high resistance to the flow of electricity.
- Intrinsic semiconductors are poor conductors of electricity.
- Doping semiconductors improves their electrical conductivity.
- Conduction in semiconductors is by free electrons and by holes.
- Pentavalent dopants are used in n-type semiconductors. Electrons are the majority carriers in this type.
- Trivalent dopants are used in p-type semiconductors. Holes are the majority carriers in this type.
- Resistance to current in conductors is very low.
- Once established, current will continue in a superconductor indefinitely without aid of a power source.
- Four factors affecting resistance in a conductor are material type, size, length, and temperature.
- The purpose of using a resistor in a circuit is either to reduce current to a specific value or to provide a desired voltage drop.
- Color-coding provides a standardized way of marking and determining the ohmic value of a resistor.

KEY TERMS

Each of the following terms has been used in this chapter. Do you know their meanings?

absolute zero	energy diagram	power resistor
acceptor impurity	energy gap	p-type crystal
carbon-composition resistor	extrinsic semiconductor	recombination
carrier	film resistor	resistance (R)
circular mil (cmil)	forbidden band	rheostat
circular mil-foot	free electron	semiconductor
conduction band	hole	specific resistance (ϱ)
conductivity (sigma)	insulator	square mil
conductor	intrinsic semiconductor	superconductivity
cryogenic fluid	majority carrier	superconductor
donor impurity	mil	temperature coefficient of
dopant	minority carrier	resistance (α)
doping	n-type crystal	valence band
electrical resistivity	ohm (Ω)	variable resistor
electron-hole pair	potentiometer (pot)	

TEST YOUR KNOWLEDGE

Please do not write in this text. Place your answers on a separate sheet of paper.

1. A/an _____ semiconductor is one made of atoms all of the same kind.
2. A/an _____ semiconductor is pure crystal with certain impurities added during crystallization.
3. Impurities that add free electrons to a crystal are called _____ impurities, while those that create holes that will accept electrons are called _____ impurities.
4. What causes the formation of electron-hole pairs?
5. Direction of current in the external circuit depends on the type of semiconductor used. True or False?
6. Hole movement is in the opposite direction to electron movement. True or False?
7. Name four pentavalents.
8. What are the majority and minority carriers in an n-type crystal?
9. What are the majority and minority carriers in a p-type crystal?
10. Which of the following are trivalent, or acceptor, impurities?
    a. Phosphorus.   d. Gallium.
    b. Arsenic.   e. Bismuth.
    c. Indium.   f. Boron.
11. Superconductors are maintained in very warm environments. True or False?
12. The kind of a material used in a conductor affects its resistance. Name three other factors that affect resistance.
13. What is the resistance of 50 feet of No. 24 copper wire?
14. Why is a 100-Ω fixed carbon resistor made in several sizes?
15. What are the color codes for the following resistors?
    a. 82 Ω, 10 percent tolerance.
    b. 1.6 MΩ, 20 percent tolerance.
    c. 36,000 Ω, 5 percent tolerance.
    d. 470 Ω, 10 percent tolerance
    e. 7500 Ω, 5 percent tolerance.
    f. 0.62 MΩ, 20 percent tolerance.

Chapter 4   RESISTIVE CIRCUITS

Resistance is that property in a circuit that opposes electric current. In some circuits, resistance is desired. In other circuits, it is undesired.

After studying this chapter, you will be able to:
- ☐ *State the relationship between voltage, current, and resistance.*
- ☐ *Explain the difference between resistance and conductance.*
- ☐ *Explain how electrical energy is converted to power in resistive circuits.*
- ☐ *List features of a series circuit.*
- ☐ *List features of a parallel circuit.*
- ☐ *Find equivalent resistance of a combination circuit.*
- ☐ *Discuss different types of switches and how they work.*

INTRODUCTION

The fundamental relationship between voltage, current, and resistance is basic to all studies of electricity and electronics. There is perhaps no lesson in this text that is more important to understand and apply. This relationship, known as *Ohm's law,* is the foundation of circuit design and service.

OHM'S LAW

When a voltage is applied to a metal conductor, the current that moves through the conductor is directly proportional to the applied voltage. Exactly how much current there is depends not only on the voltage but also on the resistance the wire offers to the flow of electrons. This relationship was established by Georg Simon Ohm, the German scientist, during the 19th century. It is stated mathematically by the formula:

$$I = \frac{E}{R}$$

where I is intensity of the current in amperes, E is electromotive force in volts, and R is resistance in ohms. This formula is known as **Ohm's law**. It states that:

- *The current in any circuit is directly proportional to the applied voltage.*

  In Figure 4-1, for example, the circuit (left) has a voltage of 10 V and a current of 1 A. If the voltage is increased to 15 V, the current increases to 1.5 A. If the voltage is reduced to 5 V, the current decreases to 0.5 A. In each case, the ratio of voltage to current is 10:1. (Note that the symbol Ⓐ is used for **ammeter**, which is an instrument used to measure current. All of the current through the circuit also goes through the meter.)

- *The current in a circuit is inversely proportional to its resistance.*

  In Figure 4-2, for example, the circuit (left) has 10 Ω of resistance and a current of 1 A. If the resistance is increased to 20 Ω, the current decreases to 0.5 A. If the resistance is reduced to 5 Ω, the current increases to 2 A.

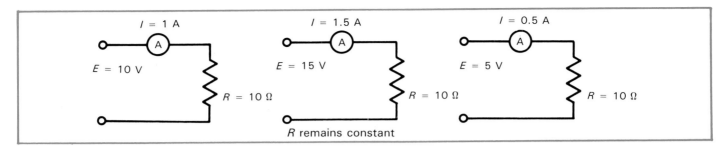

Figure 4-1. *Current in a circuit is directly proportional to applied voltage.*

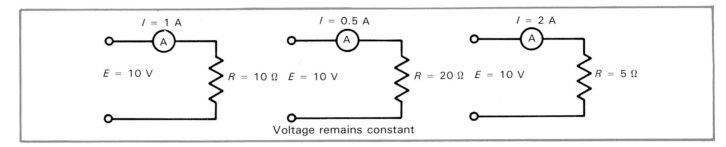

Figure 4-2. *Current in a circuit is inversely proportional to resistance.*

As is true of inverse proportions, the product of the two inverse quantities, current and resistance, is constant. Also, as resistance increases, current decreases and vice versa.

● The formula for Ohm's law may be arranged in three different ways. Knowing any two of the three values, the third may be found.

$$I = \frac{E}{R} \qquad\qquad E = I \times R \qquad\qquad R = \frac{E}{I}$$

Note that Ohm's law is valid for *metal conductors*. Their resistances are constant and do not depend upon the applied voltage. Ohm's law depicts a linear relationship between voltage and current. It does not apply generally for other substances such as semiconductors and vacuum tubes. Circuits and devices that follow Ohm's law are said to be *ohmic*. Those that do not are said to be *nonohmic*, or *nonlinear*. Ohm's law can be applied to nonohmic cases; however, *R*, in these cases, would not be constant. It would depend on the value of *E*.

Problem:

A current of 0.1 A flows in a circuit that has a resistance of 1000 Ω. What is the applied voltage?

Solution:

$$E = I \times R = 0.1 \text{ A} \times 1000 \text{ } \Omega = 100 \text{ V}$$

Problem:

A circuit has a current of 0.1 A when the applied voltage is 100 V. What is the circuit resistance?

Solution:

$$R = \frac{E}{I} = \frac{100 \text{ V}}{0.1 \text{ A}} = 1000 \text{ } \Omega$$

Important: When using Ohm's law, all quantities must be in the same basic units: E in volts, I in amperes, R in ohms. For example, if a quantity such as ohms is given in kilohms, it must be changed to ohms. If current were given in milliamperes, it must be changed to amperes. See Appendix A for metric prefixes and their conversions.

Problem:

A circuit with a resistance of 1.5 kΩ has a current of 50 mA. What is the applied voltage?

Solution:

Step 1. Convert kilohms to ohms.

$$1.5 \text{ k}\Omega = 1.5 \times 10^3 \ \Omega = 1500 \ \Omega$$

Step 2. Convert milliamps to amps.

$$50 \text{ mA} = 5 \times 10^{-2} \text{ A} = 0.05 \text{ A}$$

Step 3. Substitute in values and solve.

$$E = I \times R = 0.05 \text{ A} \times 1500 \ \Omega = 75 \text{ V}$$

Note: The use of scientific notation, or powers of 10, is explained in Appendix 1.

CONDUCTANCE

The ability to conduct is opposite to the ability to resist. Either may be used in the computation of circuit values. **Conductance (G)** of a circuit is the reciprocal of its resistance. The unit of conductance is the **siemen (S)**, formerly, the *mho* (*ohm* spelled backwards). The formula for computing conductance is:

$$G = \frac{1}{R}$$

Problem:

What is the conductance of a 10-Ω resistor?

Solution:

$$G = \frac{1}{R} = \frac{1}{10 \ \Omega} = 0.1 \text{ S}$$

Conductance can be used in Ohm's law in place of resistance. Remember that since it the reciprocal, or inverse, of resistance, when you substitute conductance for resistance, you must invert the formula. For instance, from Ohm's law, $R = E/I$. Conductance, then, would be $G = I/E$.

Conductance should be remembered. It is used in later studies of transistor circuit parameters. These are the controlling elements of a circuit such as voltage, current, resistance, inductance, and capacitance.

POWER

Energy is the ability to do work, and **power** is the rate of doing work. However, what is work? It was mentioned briefly in Chapter 2. Work W is the product of a force F and distance d:

$$W = F \times d$$

where work, in U.S. Conventional units, is measured in **foot-pounds (ft·lb)**.

If you lifted a 1 pound weight 1 foot, you would accomplish 1 foot-pound of work. No mention has been made of the time required to lift the weight.

If you lifted the weight 1 foot once each second, your rate of doing work, or power, would be 1 foot-pound per second. Power *P* is work divided by time *t*:

$$P = \frac{W}{t}$$

In mechanics, the power of an engine may be rated in **horsepower (HP)**, where:

$$1 \text{ HP} = 33,000 \text{ ft·lb/min.} = 550 \text{ ft·lb/sec.}$$

In electricity, we have learned that voltage is a force, and current is the movement of charge per unit time. We have learned that the amount of work done in a circuit is the product of voltage and the quantity of charge moved. The power in an electrical circuit would then be voltage times charge per unit time, or:

$$P = I \times E$$

where power is in **watts (W)**, named after the Scottish mechanical engineer and inventor, James Watt. Watt is credited with the invention of the condensing steam engine. He also invented the centrifugal governor for controlling the speed of an engine. Mechanical power equates to electrical power by the following conversion:

$$1 \text{ HP} = 746 \text{ W}$$

The basic power formula, referred to at times as the **power law**, may be arranged in three different ways. Knowing any two of the three values, the third may be found.

$$P = I \times E \qquad E = \frac{P}{I} \qquad I = \frac{P}{E}$$

Problem:

A device operates on 110 V and draws 5 A of current. What is its power?

Solution:

$$P = I \times E = 5 \text{ A} \times 110 \text{ V} = 550 \text{ W}$$

Problem:

How much current is drawn by a 100 W light bulb with voltage at 110 V?

Solution:

$$I = \frac{P}{E} = \frac{100 \text{ W}}{110 \text{ V}} = 0.91 \text{ A}$$

By simple algebraic manipulation, it is possible to combine Ohm's law and the power law so that an unknown may be found directly if any two of the other values are known. The original laws and methods used in combining laws are given in Figure 4-3. The formulas are also arranged in a wheel-shaped chart for ready reference.

FUNDAMENTALS OF A CIRCUIT

Electrical circuits are complete pathways for electric current. Three factors are needed for a circuit to operate. These factors are *voltage source*, *load*, and *conductive pathway*. Many circuits have a fourth factor. This would be some type of *control* of current.

A

FORMULA	SOURCE
1. $E = IR$	Ohm's law
2. $E = \dfrac{P}{I}$	Power law
3. $E = \sqrt{PR}$	By transposing Formula 12 and taking the square root
4. $I = \dfrac{E}{R}$	Ohm's law
5. $I = \dfrac{P}{E}$	Power law
6. $I = \sqrt{\dfrac{P}{R}}$	By transposing Formula 9 and taking the square root
7. $R = \dfrac{E}{I}$	Ohm's law
8. $R = \dfrac{E^2}{P}$	By transposing Formula 12
9. $R = \dfrac{P}{I^2}$	By transposing Formula 11
10. $P = IE$	Power law
11. $P = I^2 R$	By substituting IR from Formula 1 for E
12. $P = \dfrac{E^2}{R}$	By substituting $\dfrac{E}{R}$ from Formula 4 for I

B

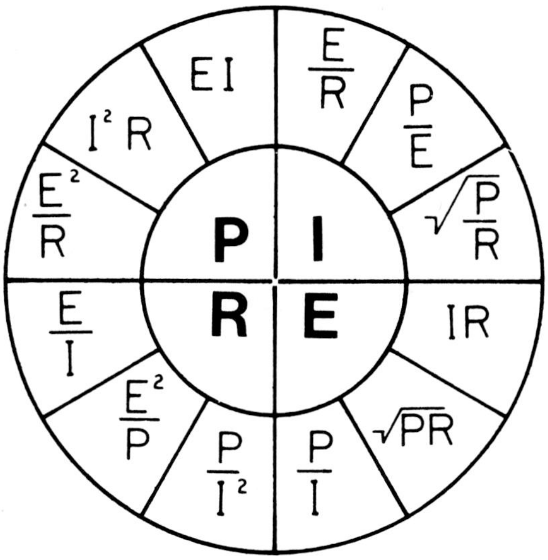

Figure 4-3. Basic formulas used to solve problems. A—Formulas summarized in table form. B—The same formulas summarized in wheel-shaped chart. To use, find the unknown represented in one of four quadrants in center of the wheel. In outer part of the same quadrant, find the letters of your two known circuit values. The wheel tells you what operation to perform to solve for the unknown.

SERIES CIRCUITS

When components are connected end to end so that all the circuit current is through each component, the circuit is a **series circuit**. In Figure 4-4, a series circuit of three resistors is drawn schematically. The number subscripts by each resistor are used to identify a particular resistor.

There are several rules that apply to series circuits. These include:

- When resistors are connected in series, the total resistance of the circuit is equal to the sum of the individual values of all resistors. Refer again to Figure 4-4.

$$R_T = R_1 + R_2 + R_3 + \ldots$$

- *The current is the same at all points in a series circuit.* This is a simplified form of **Kirchhoff's current law** for series circuits. The current in the circuit is found using Ohm's law. See Figure 4-5 and Figure 4-6.

$$I_T = I_{R_1} = I_{R_2} = I_{R_3} = \ldots$$

- As current runs through each resistor in a series circuit, an amount of electrical energy is consumed. (It is converted to heat energy and radiated to the surroundings.) This is called **voltage drop**. Voltage drop for each resistor is found by Ohm's law. *The sum of the voltage drops around a series circuit is equal to the applied voltage:*

$$E_S = E_{R_1} + E_{R_2} + E_{R_3} + \ldots$$

This is a simplified form of **Kirchhoff's voltage law** for series circuits. See Figure 4-7 and Figure 4-8.

- Resistors consume power at a certain rate, in watts. Consider the series circuit in Figure 4-9:

$$R_T = R_1 + R_2 = 500 \ \Omega + 1000 \ \Omega = 1500 \ \Omega$$

$$I_T = \frac{E_S}{R_T} = \frac{150 \ V}{1500 \ \Omega} = 0.1 \ A$$

$$P_T = I_T \times E_S = 0.1 \ A \times 150 \ V = 15 \ W$$

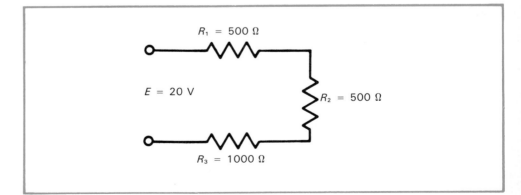

Figure 4-4. *The total resistance R_T of a series circuit is equal to the sum of the individual resistances.*

$$R_T = R_1 + R_2 + R_3 + \ldots$$
$$R_T = 500 \ \Omega + 500 \ \Omega + 1000 \ \Omega = 2000 \ \Omega$$

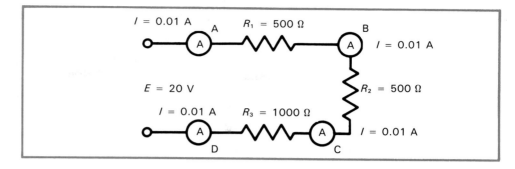

Figure 4-5. *In a series circuit, the current is the same value regardless of the point at which it is measured. Four meters have been connected in the circuit. Meters A, B, C, and D all read exactly the same value of current.*

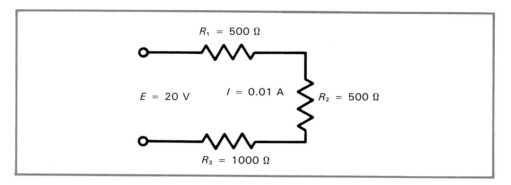

Figure 4-6. *The current in a series circuit is equal to the applied voltage E divided by the total resistance R_T. If 20 V is applied to the circuit, the current can be computed:*

$$I = \frac{20\ V}{2000\ \Omega} = 0.01\ A$$

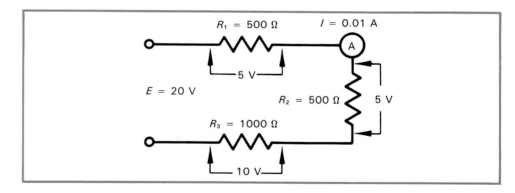

Figure 4-7. *Voltage drop may be found for each resistor by Ohm's law using the resistance value of each resistor for each drop. In this circuit, voltage drops are computed.*

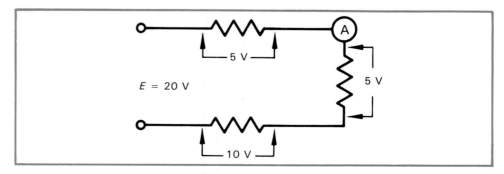

Figure 4-8. *The sum of the voltage drops around a series circuit is equal to the applied voltage E_S.*

$$E_S = E_{R_1} + E_{R_2} + E_{R_3} + \ldots$$
$$E_S = 5\ V + 5\ V + 10\ V = 20\ V$$

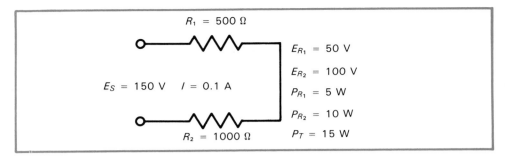

Figure 4-9. Power is computed by finding the product of the voltage and the current.

The power expended by the individual resistors can be computed by first finding the voltages across these resistors:

$$E_{R_1} = I_T \times R_1 = 0.1 \text{ A} \times 500 \text{ }\Omega = 50 \text{ V}$$

$$E_{R_2} = I_T \times R_2 = 0.1 \text{ A} \times 1000 \text{ }\Omega = 100 \text{ V}$$

Then:

$$P_{R_1} = I_T \times E_{R_1} = 0.1 \text{ A} \times 50 \text{ V} = 5 \text{ W}$$

$$P_{R_2} = I_T \times E_{R_2} = 0.1 \text{ A} \times 100 \text{ V} = 10 \text{ W}$$

and

$$P_T = P_{R_1} + P_{R_2} = 5 \text{ W} + 10 \text{ W} = 15 \text{ W}$$

RULE OF TEN TO ONE—SERIES CIRCUITS

Often, practical adjustments are made to scientific laws if the error created is insignificant. For example, look at the top circuit in Figure 4-10. Total resistance is 1100 Ω; circuit current is 0.01 A. Below this, the circuit is redrawn leaving out the 100-Ω resistor. Since the ratio between the two is ten to one, the smaller one is dropped. Total resistance is considered as 1000 Ω only. The current only changed 1 mA to 0.011 A. The error created is hardly worth the time and effort to compute the values exactly. After all, the resistors may not be precisely made. A 10% tolerance resistor for 1000 Ω could be anywhere between 900 and 1100 Ω. Simplify your mathematics unless you are working with extremely accurate circuits—use this *rule of ten to one.*

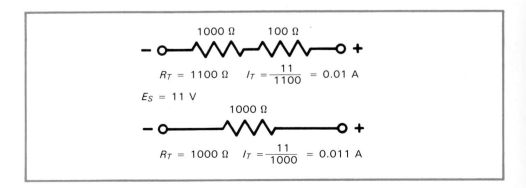

Figure 4-10. When the ratio of one resistor to a second one in series is ten to one or more, the smaller resistor may be ignored for practical purposes.

VOLTAGE DROP POLARITIES

The study of voltage *polarities*, whether negative or positive, is of extreme importance in transistor and other semiconductor circuits. The situation is severely aggravated by disagreements, in theory and practice, about the direction of current. Engineering and technical texts remain divided between teaching electron flow and conventional current.

As mentioned, in this text, an electric current will always be considered the movement of electrons, which is from negative to positive. Since most commercial test instruments follow this theory, it will spare a lot of grief and equipment damage.

In Figure 4-11, a comparison is made between potential, or voltage, levels and a staircase. In this illustration, the reference level has been taken as zero. On the staircase:

- Level A is at zero reference level.
- Level B is 1 foot higher or more positive than zero level A.
- Level C is 2 feet higher or more positive than zero level A.
- Level D is 3 feet higher or more positive than zero level A.
- Level E is 4 feet higher or more positive than zero level A.

In Figure 4-12, the same staircase is used but the zero reference level is taken at level C. As a result:

- Level A is 2 feet lower or more negative than zero level C.
- Level B is 1 foot lower or more negative than zero level C.
- Level C is at zero reference level.
- Level D is 1 foot more positive than zero level C.
- Level E is 2 feet more positive than zero level C.

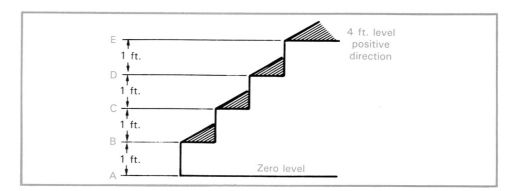

Figure 4-11. Using a staircase to show the meaning of voltage level.

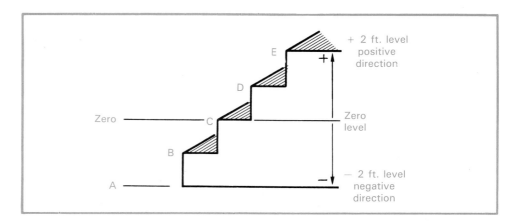

Figure 4-12. A change in reference level gives both positive and negative values.

In both of these illustrations, the total difference in height has been 4 feet. Compare this with the circuit of four resistors in Figure 4-13. Each resistor equals one step, or level. Polarities are indicated. In the circuit:

- Level B is 1 V more positive than zero level A.
- Level C is 2 V more positive than zero level A.
- Level D is 3 V more positive than zero level A.
- Level E is 4 V more positive than zero level A.

The circuit is redrawn in Figure 4-14, using the exact same voltages. However, all measurements will be made from level C, the zero reference level. As a result:

- Level A is 2 V more negative than zero level C.
- Level B is 1 V more negative than zero level C.
- Level D is 1 V more positive than zero level C.
- Level E is 2 V more positive than zero level C.

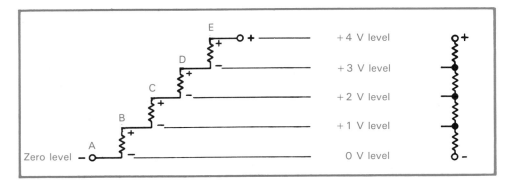

Figure 4-13. With four equal resistors in series, voltage levels compare with equal steps.

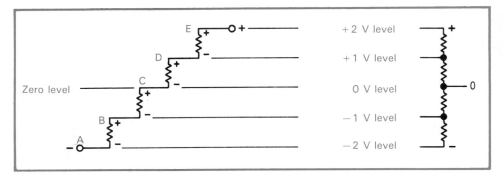

Figure 4-14. Zero reference is changed to level C to provide a negative voltage.

The circuit of Figure 4-7 is redrawn in Figure 4-15 to show the polarity of voltage drops. The heavy arrows indicate the negative connection, or black test lead, of a meter used to measure the voltage. Orienting the meter leads this way will give a proper and positive reading on the meter. Polarity is easy to determine. Between any two points in a circuit, the point nearer the negative terminal is more negative; the point nearer the positive terminal is more positive.

Note that the polarity of voltage across each resistor opposes the source voltage. (Voltage drop and source voltage values are both indicated positive.

Polarities *appear* to be the same. This was accounted for by the proper placement of meter leads across the resistors.) Rearranging Kirchhoff's voltage law from before, we see by the opposite algebraic signs that voltage drops and source voltage have opposite polarity:

$$E_S - E_{R_1} - E_{R_2} - E_{R_3} = 0 \text{ V}$$

Let us use values from our circuit of Figure 4-15 to see if this rule is valid:

$$20 \text{ V} - 5 \text{ V} - 5 \text{ V} - 10 \text{ V} = 0 \text{ V}$$

Our answer supports Kirchhoff's voltage law. This is no different than previously explained. It tells us that *the algebraic sum of all voltages around a circuit, including the source voltage, is zero.*

Note also in Figure 4-15 that all the current to Point A is equal to the current leaving Point A. This is another way of stating Kirchhoff's current law. That is, *the sum of the currents into a point or junction in a circuit is equal to the sum of the currents leaving that junction.*

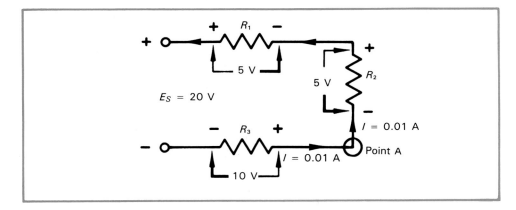

Figure 4-15. This circuit illustrates the polarity of voltage drops. Heavy arrows indicate negative, or black, test lead of the voltmeter. Arrows on circuit diagram show direction of electron flow.

Ground potential

A **ground** is an electrical connection between a circuit and the earth or a metallic object that takes the place of the earth. It is the reference potential. It is a path for current and is often the lowest potential in the circuit. **Ground potential** is zero, relative to the earth, or ground. If the ground is achieved by a physical connection to the earth, it is called an **earth ground**. See Figure 4-16. This type is used in residential wiring, usually by connection to a buried metallic water pipe or a rod driven into the ground about 8-10 feet.

Not all grounds are at earth potential. At times, the ground may be the potential of, say, the metal frame of a car, the metal chassis of a radio receiver, or a conducting foil on a printed circuit board of an electronic circuit. Grounds not connected to earth are called **chassis grounds**, or **equipment grounds**. In each case, voltages of grounded systems are measured relative to ground.

Remember that a ground is a path for current. Some grounded systems use a one wire electrical system. This means that, rather than a wire, a ground serves as a return path for current to complete the circuit. Automobiles use this type of system, where the metal chassis provides the return path. Figure 4-17 shows a schematic of a one-wire system.

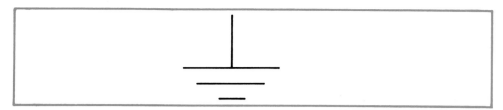

Figure 4-16. The symbol for a ground.

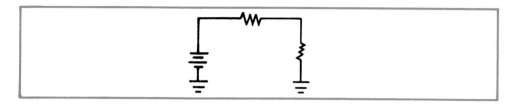

Figure 4-17. A one-wire series circuit. Current in this circuit is from the source, through the ground, then back to the source through the wire conductor.

Grounding is the most important element in wiring safety. It protects by limiting the possibility of damage to electrical equipment and conductors and by preventing shock to people contacting electrical equipment. Without proper grounding, incidental high currents are carried through the electrical system, which can be harmful to people and equipment. Equipment may have a **ground fault**, that is a live wire in contact with the housing or frame. With proper grounding, errant or excess current is rapidly directed through a path of low resistance to the lowest potential—the ground—sparing people and equipment.

LESSON IN SAFETY: Do not work on wet floors with electrical equipment. Your contact resistance is substantially reduced and an otherwise harmless shock may become serious. Work on a rubber mat or insulated platform if you are testing high voltage circuits.

Never remove the grounding prong of a three-wire appliance plug. By its removal, you have eliminated the grounding feature of the equipment or instrument. This could create a serious shock hazard. By removing the low resistance path to ground, current will seek the next lowest resistance path. This could be you! Some types of instruments will not work correctly if the ground is removed.

PROPORTIONAL VOLTAGE IN SERIES CIRCUITS

Voltage varies directly with resistance. Under constant current conditions, voltage drops are proportional to resistance values of resistors in series. Each resistor provides a voltage drop equal to its proportional part of the applied voltage. A useful *voltage divider formula* may be set up to determine the voltage drop across any resistor in the series string:

$$E_R = \frac{R}{R_T} \times E_S$$

In Figure 4-18, the total resistance of the circuit is 100 kΩ, and E_S is 50 V. To find voltage drop across the resistors:

$$E_{R_1} = \frac{R_1}{R_T} \times E_S = \frac{10 \text{ k}\Omega}{100 \text{ k}\Omega} \times 50 \text{ V} = 5 \text{ V}$$

$$E_{R_2} = \frac{R_2}{R_T} \times E_S = \frac{40 \text{ k}\Omega}{100 \text{ k}\Omega} \times 50 \text{ V} = 20 \text{ V}$$

$$E_{R_3} = \frac{R_3}{R_T} \times E_S = \frac{50 \text{ k}\Omega}{100 \text{ k}\Omega} \times 50 \text{ V} = 25 \text{ V}$$

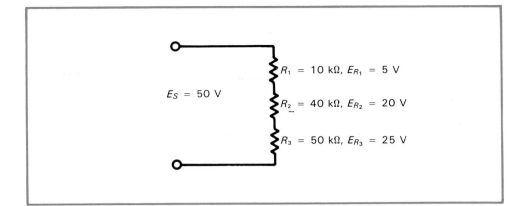

Figure 4-18. Voltage drops are proportional to resistance.

Note that kilohms were not converted to basic units prior to performing the math operation. Since units are the same in numerator and denominator, they cancel. Therefore, we did not need to worry about converting them first.

PARALLEL CIRCUITS

In a **parallel circuit**, components are connected side by side, rather than end to end, providing multiple pathways for current. See Figure 4-19. Resistance R_1 and R_2 are connected in parallel to a common voltage source. The voltage across R_1 is equal to E_S. The voltage across R_2 is also equal to E_S. This must be true since both resistors are connected to a common source.

There are several rules that apply to parallel circuits. These include:
* *The voltages across all branches of a parallel circuit are equal.* This is a simplified form of Kirchhoff's voltage law for parallel circuits.

$$E_S = E_{R_1} = E_{R_2} = E_{R_3} = \ldots$$

* *Total current in a parallel circuit is equal to the sum of the branch currents.* This is a simplified form of Kirchhoff's current law for parallel circuits.

$$I_T = I_{R_1} + I_{R_2} + I_{R_3} + \ldots$$

Referring to Figure 4-19:

$$I_{R_1} = \frac{E_S}{R_1} \text{ and } I_{R_2} = \frac{E_S}{R_2}$$

The total current is the sum of these currents:

$$I_T = I_{R_1} + I_{R_2} = \frac{E_S}{R_1} + \frac{E_S}{R_2}$$

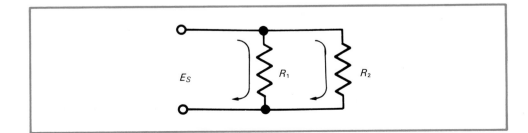

Figure 4-19. Parallel circuits provide multiple paths for current.

• Total power consumed in parallel circuits is determined the same as it is for series circuits, by totaling power consumed by each resistor:

$$P_T = P_{R_1} + P_{R_2} + P_{R_3} + \ldots$$

or by the power formulas:

$$P_T = I_T \times E_S$$

RESISTANCE IN PARALLEL CIRCUITS

It may seem surprising, but each resistor added in parallel to a circuit *decreases* the total resistance of the circuit. Actually, each resistor provides an additional current path, increasing the total cross-sectional area for current — decreasing total resistance. With a constant voltage, a decrease in total resistance will mean an increase in total current.

Equal resistors in parallel

In Figure 4-20, two *equal* resistors are connected in parallel. When two or more equal resistances are connected in parallel, the total circuit resistance R_T may be found by this formula:

$$R_T = \frac{R}{n}$$

where R is the value of any single resistor, and n is the number of resistors in parallel.

In the circuit of Figure 4-20:

$$R_T = \frac{R}{n} = \frac{100\ \Omega}{2} = 50\ \Omega$$

Current in the circuit may be found by:

$$I_T = \frac{E_S}{R_T} = \frac{100\ \text{V}}{50\ \Omega} = 2\ \text{A}$$

To prove that the total current is equal to the sum of the branch currents, we will compute the individual branch currents:

$$I_{R_1} = \frac{E_S}{R_1} = \frac{100\ \text{V}}{100\ \Omega} = 1\ \text{A}$$

$$I_{R_2} = \frac{E_S}{R_2} = \frac{100\ \text{V}}{100\ \Omega} = 1\ \text{A}$$

Total current equals:

$$I_T = I_{R_1} + I_{R_2} = 1\ \text{A} + 1\ \text{A} = 2\ \text{A}$$

This agrees with our previous answer.

Total power in the circuit is:

$$P_T = I_T \times E_S = 2\ \text{A} \times 100\ \text{V} = 200\ \text{W}$$

Unequal resistors in parallel

In Figure 4-21, *unequal* resistors are connected in parallel. When two or more unequal resistances are connected in parallel, the total circuit resistance may be found by this formula:

$$\frac{1}{R_T} = \frac{1}{R_1} + \frac{1}{R_2} + \frac{1}{R_3} + \ldots$$

This formula says that for any number of resistors in parallel, the *reciprocal* of the total resistance equals the *sum of the reciprocals* of their individual resistances. Note that this could be stated another way—the total *conductance* of a circuit equals the sum of the individual *conductances*.

In the circuit of Figure 4-21:

$$\frac{1}{R_T} = \frac{1}{R_1} + \frac{1}{R_2} + \frac{1}{R_3} = \frac{1}{5\ \Omega} + \frac{1}{10\ \Omega} + \frac{1}{30\ \Omega}$$

The least common denominator of the fractions is 30:

$$\frac{1}{R_T} = \frac{6}{30\ \Omega} + \frac{3}{30\ \Omega} + \frac{1}{30\ \Omega} = \frac{10}{30\ \Omega}$$

To get R_T, you must take the reciprocal:

$$R_T = \frac{1}{\dfrac{10}{30\ \Omega}} = \frac{30\ \Omega}{10} = 3\ \Omega$$

To prove this is total resistance, we will compute the individual branch currents:

$$I_{R_1} = \frac{E_S}{R_1} = \frac{30\ V}{5\ \Omega} = 6\ A$$

$$I_{R_2} = \frac{E_S}{R_2} = \frac{30\ V}{10\ \Omega} = 3\ A$$

$$I_{R_3} = \frac{E_S}{R_3} = \frac{30\ V}{30\ \Omega} = 1\ A$$

Total current equals:

$$I_T = I_{R_1} + I_{R_2} + I_{R_3} = 6\ A + 3\ A + 1\ A = 10\ A$$

Calculating total resistance using Ohm's law:

$$R_T = \frac{E_S}{I_T} = \frac{30\ V}{10\ A} = 3\ \Omega$$

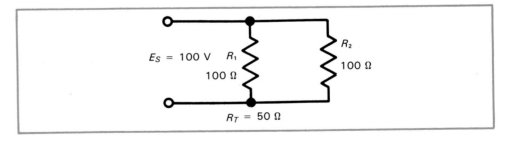

Figure 4-20. *Equal resistors in parallel.*

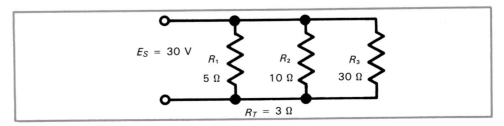

Figure 4-21. *Unequal resistors in parallel.*

This agrees with our previous calculation.

Power in each resistor is:

$$P_{R_1} = I_{R_1} \times E_S = 6 \text{ A} \times 30 \text{ V} = 180 \text{ W}$$

$$P_{R_2} = I_{R_2} \times E_S = 3 \text{ A} \times 30 \text{ V} = 90 \text{ W}$$

$$P_{R_3} = I_{R_3} \times E_S = 1 \text{ A} \times 30 \text{ V} = 30 \text{ W}$$

Total power in the circuit is:

$$P_T = P_{R_1} + P_{R_2} + P_{R_3} = 180 \text{ W} + 90 \text{ W} + 30 \text{ W} = 300 \text{ W}$$

To prove this is total power:

$$P_T = I_T \times E_S = 10 \text{ A} \times 30 \text{ V} = 300 \text{ W}$$

or:

$$P_T = I_T{}^2 \times R_T = (10 \text{ A})^2 \times 3 \text{ }\Omega = 300 \text{ W}$$

These agree with our previous answer.

Two unequal resistors in parallel

In Figure 4-22, *two* unequal resistors are connected in parallel. A simplified formula may be used to find the total resistance of a parallel circuit having two unequal resistors:

$$R_T = \frac{R_1 R_2}{R_1 + R_2}$$

In the circuit of Figure 4-22:

$$R_T = \frac{R_1 R_2}{R_1 + R_2} = \frac{(20 \text{ }\Omega)(30 \text{ }\Omega)}{20 \text{ }\Omega + 30 \text{ }\Omega} = \frac{600 \text{ }\Omega^2}{50 \text{ }\Omega} = 12 \text{ }\Omega$$

Important: The total resistance of any parallel circuit must be *less* than the value of any branch resistance in the parallel circuit. When working problems, remember this statement as you check your math.

To prove this is total resistance, we will compute the individual branch currents:

$$I_{R_1} = \frac{E_S}{R_1} = \frac{120 \text{ V}}{20 \text{ }\Omega} = 6 \text{ A}$$

$$I_{R_2} = \frac{E_S}{R_2} = \frac{120 \text{ V}}{30 \text{ }\Omega} = 4 \text{ A}$$

Total current equals:

$$I_T = I_{R_1} + I_{R_2} = 6 \text{ A} + 4 \text{ A} = 10 \text{ A}$$

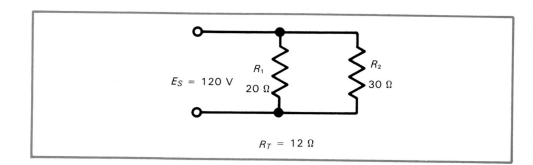

Figure 4-22. Two unequal resistors in parallel.

Calculating total resistance using Ohm's law:

$$R_T = \frac{E_S}{I_T} = \frac{120\text{ V}}{10\text{ A}} = 12\ \Omega$$

This agrees with our previous computation.

Power in each resistor, using P = I^2R, *is:*

$$P_{R_1} = I_{R_1}{}^2 \times R_1 = (6\text{ A})^2 \times 20\ \Omega = 720\text{ W}$$

$$P_{R_2} = I_{R_2}{}^2 \times R_2 = (3\text{ A})^2 \times 30\ \Omega = 480\text{ W}$$

Total power in the circuit is:

$$P_T = P_{R_1} + P_{R_2} = 720\text{ W} + 480\text{ W} = 1200\text{ W}$$

To prove this is total power:

$$P_T = I_T \times E_S = 10\text{ A} \times 120\text{ V} = 1200\text{ W}$$

This agrees with our previous answer.

RULE OF TEN TO ONE — PARALLEL CIRCUITS

The rule of ten to one does not apply just to series circuits. It applies to parallel circuits as well. Again, it is used in order to simplify math and when approximate answers are sufficient.

Consider the circuit in Figure 4-23. The total resistance of this circuit is:

$$R_T = \frac{R_1 R_2}{R_1 + R_2} = \frac{(100\ \Omega)\,(1000\ \Omega)}{100\ \Omega + 1000\ \Omega} = \frac{10^5\ \Omega^2}{1100\ \Omega} = 90.9\ \Omega$$

This is only 9.1 Ω less than a circuit using R_1 only. This error is acceptable in the practical design of a circuit and is actually less than might be caused by using resistors with a 10% tolerance. It is interesting to note that if the ratio is greater than ten to one, the approximate answer becomes more accurate.

PROPORTIONAL CURRENT IN PARALLEL CIRCUITS

Current varies inversely with resistance. Under constant voltage conditions, branch currents are inversely proportional to resistance values of resistors in parallel. In other words, the total current divides among parallel resistors in a manner inversely proportional to the resistance values. A useful formula to determine the current through any branch of a parallel circuit is:

$$I_{R_x} = \frac{R_T}{R_x} \times I_T$$

where *x* is the branch resistor number. This formula applies to a parallel circuit with any number of branches.

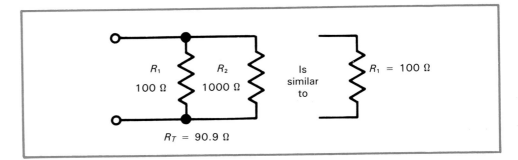

Figure 4-23. If the ratio of one resistor to a second one in parallel is ten to one or more, the larger resistor may be ignored for practical purposes.

Two parallel resistors are often found in practical circuits. For this reason, we modify the previous formula. For the special case of two parallel branches:

$$I_{R_1} = \frac{R_2}{R_1 + R_2} \times I_T$$

and

$$I_{R_2} = \frac{R_1}{R_1 + R_2} \times I_T$$

In Figure 4-24, the total resistance of the circuit is 12 Ω, and, when applied voltage equals 12 V, the total current is 1 A. To find the branch currents:

$$I_{R_1} = \frac{R_2}{R_1 + R_2} \times I_T = \frac{30\ \Omega}{50\ \Omega} \times 1\ A = 0.6\ A$$

$$I_{R_2} = \frac{R_1}{R_1 + R_2} \times I_T = \frac{20\ \Omega}{50\ \Omega} \times 1\ A = 0.4\ A$$

COMBINATION CIRCUITS

The technician and engineer are often required to compute the total resistance of series-parallel circuits, or **combination circuits**. This total resistance is referred to as the **equivalent resistance** of the circuit. It can be determined by using both series and parallel resistance formulas. Equivalent resistance R_{eq} is represented by a single resistor, which represents the total load.

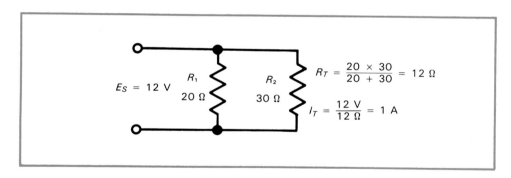

Figure 4-24. Branch currents of this circuit can be found using the proportional current formulas. Refer to text.

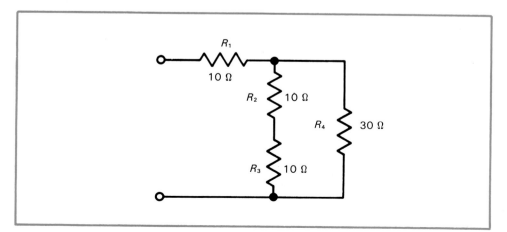

Figure 4-25. This complex circuit is to be reduced to R_{eq}.

In Figure 4-25, a combination circuit is schematically drawn. The first step in solving this circuit is to combine series resistors R_2 and R_3:

$$R_{2,3} = R_2 + R_3 = 10 \; \Omega + 10 \; \Omega = 20 \; \Omega$$

Redraw the circuit with R_2 and R_3 combined. See Figure 4-26.

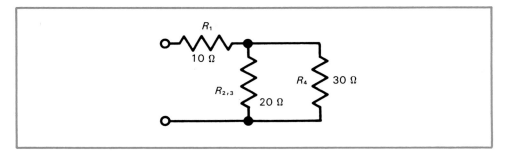

Figure 4-26. Equivalent resistance of R_2 and R_3 shown by $R_{2,3}$.

The next step in solving this circuit is to combine parallel resistances $R_{2,3}$ and R_4:

$$R_{2,3,4} = \frac{R_{2,3}R_4}{R_{2,3} + R_4} = \frac{(20 \; \Omega) \, (30 \; \Omega)}{20 \; \Omega + 30 \; \Omega} = \frac{600 \; \Omega^2}{50 \; \Omega} = 12 \; \Omega$$

Redraw the circuit with $R_{2,3}$ and R_4 combined. See Figure 4-27.

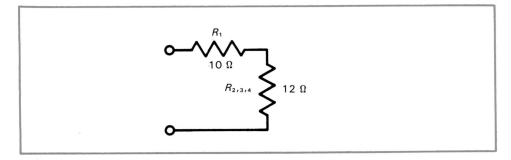

Figure 4-27. Equivalent resistance of $R_{2,3}$ and R_4 shown by $R_{2,3,4}$.

The final step is to combine series resistances $R_{2,3,4}$ and R_1. This is the value of R_{eq}.

$$R_{eq} = R_{1,2,3,4} = R_{2,3,4} + R_1 = 12 \; \Omega + 10 \; \Omega = 22 \; \Omega$$

Redraw the equivalent circuit. See Figure 4-28. Notice that the equivalent circuit is a circuit with only one resistor. This 22-Ω resistance is the load that the power source sees.

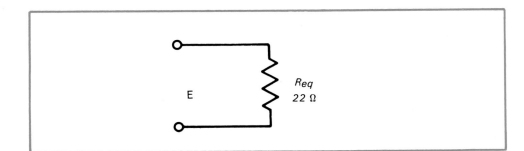

Figure 4-28. Remaining series resistances are added to find R_{eq}.

Look at another example, Figure 4-29. The first step in solving this circuit is to combine parallel resistors R_2 and R_3:

$$R_{2,3} = \frac{R_2 R_3}{R_2 + R_3} = \frac{(2\ \Omega)\ (3\ \Omega)}{2\ \Omega + 3\ \Omega} = \frac{6\ \Omega^2}{5\ \Omega} = 1.2\ \Omega$$

Redraw the circuit with R_2 and R_3 combined. See Figure 4-30.

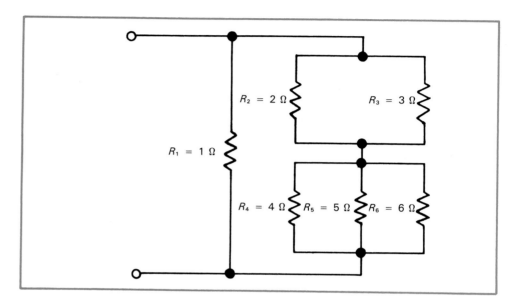

Figure 4-29. Reduce this combination circuit to R_{eq}.

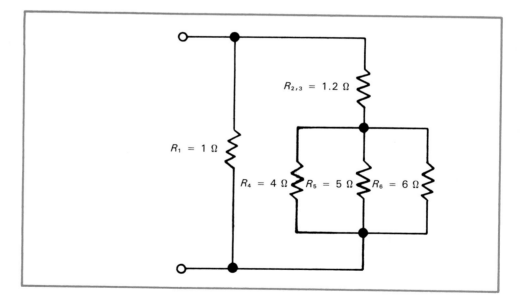

Figure 4-30. Equivalent resistance of R_2 and R_3 shown by $R_{2,3}$.

The next step in solving this circuit is to combine parallel resistors R_4, R_5, and R_6:

$$\frac{1}{R_{4,5,6}} = \frac{1}{R_4} + \frac{1}{R_5} + \frac{1}{R_6} = \frac{1}{4\ \Omega} + \frac{1}{5\ \Omega} + \frac{1}{6\ \Omega} = 0.6167\ \Omega$$

$$R_{4,5,6} = 1/0.6167\ \Omega = 1.62\ \Omega$$

Redraw the circuit with R_4, R_5, and R_6 combined. See Figure 4-31.

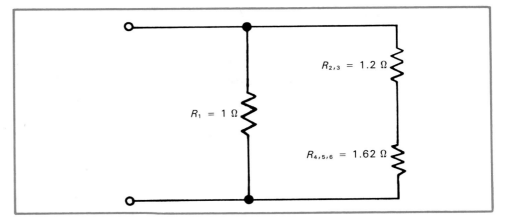

Figure 4-31. Equivalent circuit showing R_1, $R_{2,3}$ and $R_{4,5,6}$.

The next step in solving this circuit is to combine series resistances $R_{2,3}$ and $R_{4,5,6}$:

$$R_{2,3,4,5,6} = R_{2,3} + R_{4,5,6} = 1.2\ \Omega + 1.62\ \Omega = 2.82\ \Omega$$

Redraw the circuit with $R_{2,3}$ and $R_{4,5,6}$ combined. See Figure 4-32.

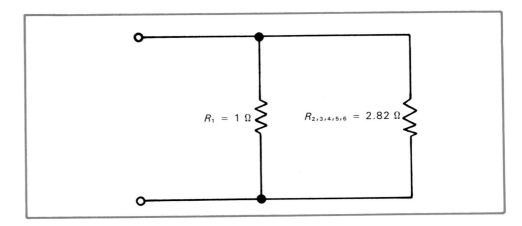

Figure 4-32. Equivalent circuit showing R_1 and $R_{2,3,4,5,6}$.

The final step is to combine parallel resistances $R_{2,3,4,5,6}$ and R_1. This is the value of total resistance, or R_{eq}.

$$R_{eq} = \frac{R_1 R_{2,3,4,5,6}}{R_1 + R_{2,3,4,5,6}} = \frac{(1\ \Omega)\ (2.82\ \Omega)}{1\ \Omega + 2.82\ \Omega} = \frac{2.82\ \Omega^2}{3.82\ \Omega} = 0.738\ \Omega$$

Redraw the equivalent circuit. This is the total resistance of the circuit. See Figure 4-33.

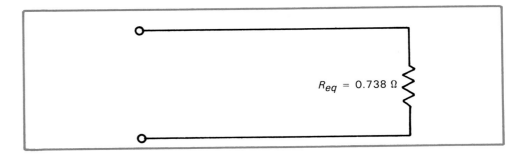

Figure 4-33. Original circuit reduced to an equivalent resistance. This is total circuit resistance.

WHEATSTONE BRIDGE CIRCUITS

The *bridge circuit* is widely used in electrical measurement devices and in other electronic circuits. There are many types of bridge circuits. One type is the **Wheatstone bridge**. It is used to measure unknown resistance. The basic Wheatstone bridge circuit is shown in Figure 4-34. The two input terminals of the bridge are connected to a voltage source. The meter connected between A and B is a sensitive **galvanometer**, which measures magnitude and direction of current. R_1 and R_2 represent the *ratio arm* of the bridge. The value of these resistors is set for some known ratio. R_S is a variable standard resistor, and R_X is unknown.

The bridge is said to be *in balance* when the voltage drops across R_S and R_2 are equal, and there is proportional division of voltage across the bridge; that is, voltage division is the same ratio in the R_X-R_S arm as the ratio arm. In this "null" condition, the voltage at A is equal to voltage at B. The meter reads *zero* since there is no difference in potential. A slight change in resistance R_X would cause an unbalanced condition, and the proportional voltage division would be upset. The differences in voltages between A and B would cause the meter to deflect. A small adjustment of R_S can bring the bridge back into balance.

The proportion of the voltages set up by the bridge can be written as follows:

$$\frac{E_{R_X}}{E_{R_S}} = \frac{E_{R_1}}{E_{R_2}}$$

This can be written:

$$\frac{I_A R_X}{I_A R_S} = \frac{I_B R_1}{I_B R_2}$$

The currents cancel to give:

$$\frac{R_X}{R_S} = \frac{R_1}{R_2}$$

If the bridge is used to find the value of an unknown resistor R_X, solving for R_X:

$$R_X = R_S \times \frac{R_1}{R_2}$$

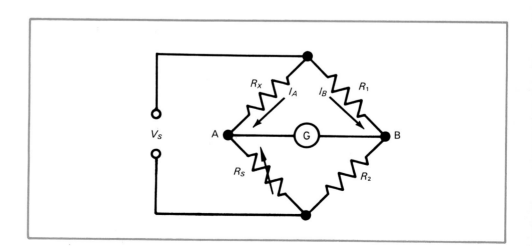

Figure 4-34. The Wheatstone bridge for measuring unknown resistance.

If R_X is substituted with some resistive component that changes its resistance slightly because of temperature, light, pressure, humidity, or other physical effects, the slight change will be indicated on the meter. The meter can be calibrated and the dial marked to read directly in the physical quantity being measured.

SWITCHES

A **switch** is a device used to break or complete a current path. In other words, it is a device used to open or close a circuit. Switches are also used for diverting current to a different path. Switches may be manually operated or automatic. You will find one most anywhere it is necessary to disconnect power from a circuit or piece of equipment. Switches are manufactured in hundreds of different sizes and types, including *slide, push-button, rotary,* and *toggle switches.* Further, these switches may be classified as one of the following:
* *Single-pole, single-throw.*
* *Single-pole, double-throw.*
* *Double-pole, single-throw.*
* *Double-pole, double-throw.*

Some of the more common switches are shown in Figures 4-35 through 4-42. Schematic symbols are also pictured.

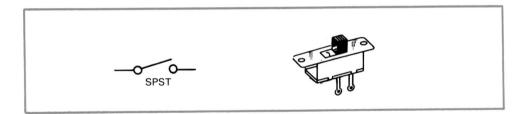

Figure 4-35. **Single-pole, single-throw (SPST) switch.** *This switch disconnects one side of a line or a single wire circuit.*

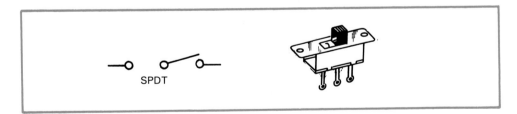

Figure 4-36. **Single-pole, double-throw (SPDT) switch.** *This switch actually has two ON positions. In the center, it is OFF. This switch finds wide usage in the switching from one circuit to another. It operates in a single-wire circuit.*

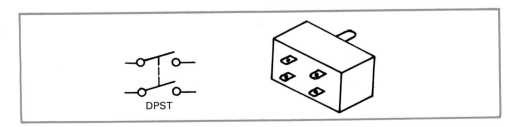

Figure 4-37. **Double-pole, single-throw (DPST) switch.** *This switch is like the SPST except both sides of a two-wire line may be switched at once. It may also be used as two single-pole switches acting together.*

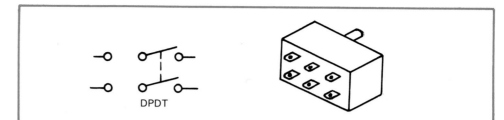
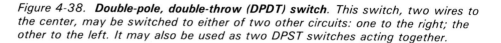

Figure 4-38. **Double-pole, double-throw (DPDT) switch.** *This switch, two wires to the center, may be switched to either of two other circuits: one to the right; the other to the left. It may also be used as two DPST switches acting together.*

Figure 4-39. Push-button switch, NO (normally open). Used to sound bells or alarms or momentarily close a circuit.

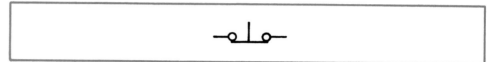

Figure 4-40. Push-button switch, NC (normally closed). Used to momentarily open a circuit.

Figure 4-41. Push-button switch assemblies.   (Centralab)

Figure 4-42. Typical rotary switch.   (Centralab)

ROTARY SWITCHES

Today, many electronic devices use rotary switches. A rotary switch you might be familiar with is the channel selector switch on older model television sets. Figure 4-42 shows a front view of a typical rotary switch. There are many different types of rotary switch circuits. A single-pole rotary switch symbol and standard section circuit diagram are shown in Figure 4-43.

BIMETAL THERMOSTATS

There are many types of automatic switches. Some are activated by pressure, some by liquid level, others by fluid flow. One of the most common is activated by heat. It is the **bimetal thermostat**. This type of switch takes advantage of the different expansion rates of different metals.

The construction of a bimetal thermostat features two dissimilar metals fastened together. See Figure 4-44. A rise in temperature will cause metal A to expand more than metal B. As a result, the bimetal starts to curl up, which opens the switch contacts. A decrease in temperature will again close the switch and reactivate the circuit.

Some thermostats use a **mercury switch** to serve as the contact points. See Figure 4-45. Basically, the movement of the bimetal causes the mercury switch

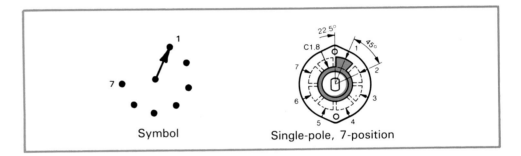

Figure 4-43. Rotary switch symbol and standard section circuit diagram. (Centralab)

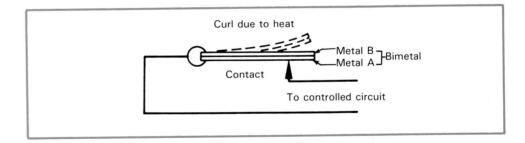

Figure 4-44. The bimetal curls up (dotted lines) when there is an increase in temperature.

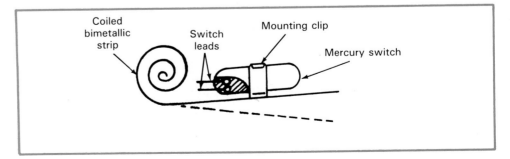

Figure 4-45. A mercury switch combined with a bimetallic strip to form a thermostat.

to tilt. This causes the pool of mercury inside the glass capsule to flow. Mercury conducts electricity. When the switch tilts one way, the contacts are made. When it tilts the other way, they are broken. Bimetal thermostats are widely used in temperature control. You may have one in your home to control your furnace.

FUSES

Fuses are used in many circuits to provide circuit protection. A **fuse** is a link of metal with a low melting point. Under normal conditions, the current through the fuse will not produce excessive heat. If the circuit is overloaded, the excessive current will produce enough heat to melt the fusible link. It must be replaced before the circuit is again operative. You must investigate the cause of the high current before replacing a blown fuse. Figure 4-46 shows a simple circuit that includes the symbol for a fuse.

Fuses are manufactured in a wide variety of shapes and current ratings. A special type is the **time-delay (slow-blow) fuse**, which will withstand surges of high current without blowing. It is a delayed fuse. Figure 4-47 shows some typical cartridge fuses.

RESISTOR SUBSTITUTION BOX PROJECT

A fairly common device used by designers and technicians for circuit design and testing is the **resistance box**, or **resistor substitution box**. See Figure 4-48. The box provides a convenient way to try different resistance values in a circuit. Rather than having to wire and unwire a circuit, a resistance box enables the user to select different resistances with the turn of a knob.

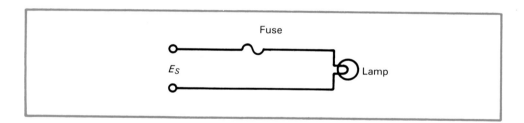

Figure 4-46. A simple circuit with fuse protection.

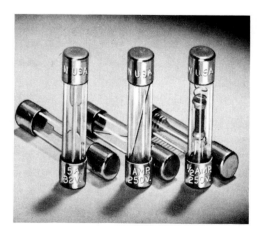

Figure 4-47. Cartridge fuses are used in many electronic devices.   (Littelfuse)

The resistor substitution box shown in Figure 4-48 uses material covered in this chapter on resistive circuits and switches. A schematic and parts list is given in Figure 4-49. Note that the specific resistors may vary, depending on what is available and what is needed. Also, a rotary switch may be used that has more than 12 contacts if available.

Figure 4-48. Resistor substitution box.

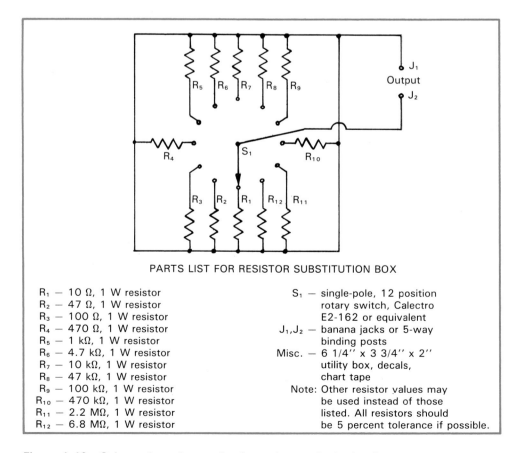

PARTS LIST FOR RESISTOR SUBSTITUTION BOX

R₁ — 10 Ω, 1 W resistor
R₂ — 47 Ω, 1 W resistor
R₃ — 100 Ω, 1 W resistor
R₄ — 470 Ω, 1 W resistor
R₅ — 1 kΩ, 1 W resistor
R₆ — 4.7 kΩ, 1 W resistor
R₇ — 10 kΩ, 1 W resistor
R₈ — 47 kΩ, 1 W resistor
R₉ — 100 kΩ, 1 W resistor
R₁₀ — 470 kΩ, 1 W resistor
R₁₁ — 2.2 MΩ, 1 W resistor
R₁₂ — 6.8 MΩ, 1 W resistor

S₁ — single-pole, 12 position rotary switch, Calectro E2-162 or equivalent
J₁,J₂ — banana jacks or 5-way binding posts
Misc. — 6 1/4" x 3 3/4" x 2" utility box, decals, chart tape
Note: Other resistor values may be used instead of those listed. All resistors should be 5 percent tolerance if possible.

Figure 4-49. Schematic and parts list for resistor substitution box.

The resistor substitution box project can be "dressed-up" by using rub-on decals to identify the resistor values and other information. Chart tape or printed circuit tape can be used to separate certain portions of the front panel from other areas.

LESSON IN SAFETY: A soldering gun or iron can be a frequent cause of injury. Never leave a hot iron on the bench where an unsuspecting party might touch the hot end. Put a hot iron someplace where you will not accidentally come in contact with it.

LESSON IN SAFETY: Remember that the metal edges and sharp corners on chassis and panels can cut and scratch. File them smooth.

SUMMARY

- Ohm's law is a mathematical formula stating the relationship between voltage, current, and resistance in a circuit. The formula for Ohm's law is $E = I \times R$.
- Power is the time rate of doing work. Electrical power is the rate of doing electrical work.
- The power formula states the relationship of power, voltage, and current in a circuit. It is $P = I \times E$.
- There are definite relationships between Ohm's law and the power formula.
- Conductance is the reciprocal of resistance.
- There are three basic types of circuits. These are the series, the parallel, and the combination circuit.
- The condition of a circuit or component to have a negative and positive value is called polarity.
- A series circuit is one where there is only one pathway for current.
- In a series circuit, the following facts are known:
  Current is equal throughout.

$$(I_T = I_{R_1} = I_{R_2} = I_{R_3} = \ldots)$$

Resistances add up to the total resistance.

$$(R_T = R_1 + R_2 + R_3 + \ldots)$$

The total voltage is equal to the sum of the voltage drops across the resistors (Kirchhoff's voltage law).

$$(E_T = E_{R_1} + E_{R_2} + E_{R_3} + \ldots)$$

- Parallel circuits have more than one pathway for current.
- In a parallel circuit, the following facts are known:
  Voltage across parallel branches is the same.

$$(E_T = E_{R_1} = E_{R_2} = E_{R_3} = \ldots)$$

Total current equals the sum of the branch currents. Also, the amount of current entering a junction must be the same as the current leaving that junction (Kirchhoff's current law).

$$(I_T = I_{R_1} + I_{R_2} + I_{R_3} + \ldots)$$

Total resistance equals the reciprocal of the sum of the reciprocals of the resistance values.

$$(1/R_T = 1/R_1 + 1/R_2 + 1/R_3 + \ldots)$$

- Combination circuits are made of certain components connected in series and others connected in parallel.
- Switches are used to make or break a circuit.
- Fuses are used to protect circuits from overcurrent.

KEY TERMS

Each of the following terms has been used in this chapter. Do you know their meanings?

ammeter
bimetal thermostat
chassis ground
combination circuit
conductance *(G)*
DPDT switch
DPST switch
earth ground
electrical circuit
energy
equipment ground
equivalent resistance
foot-pound (ft·lb)

fuse
galvanometer
ground
ground fault
ground potential
horsepower (HP)
Kirchhoff's current law
Kirchhoff's voltage law
mercury switch
Ohm's law
parallel circuit
power *(P)*

power law
resistance box
resistor substitution box
series circuit
siemens (S)
SPDT switch
SPST switch
switch
time-delay (slow-blow) fuse
voltage drop
watts (W)
Wheatstone bridge

TEST YOUR KNOWLEDGE

Please do not write in this text. Place your answers on a separate sheet of paper.

1. Four lamps of the same kind are connected in series across a 6 Vdc power source. (See illustration.) Current draw of the lamps is 0.6 A. Are they all equal in brightness?

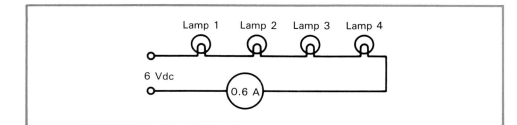

2. Are they as bright as a single lamp connected across the 6 V? Explain.
3. What is the resistance of one lamp?
4. What is the voltage drop across each lamp?
5. What power is being used by all four lamps?
6. How much power is used by lamp 1?
7. Connect a jumper wire across lamps 3 and 4. Do lamps 1 and 2 burn brighter or dimmer?
8. Four lamps of the same kind are connected in parallel across a 6 Vdc power source. (See illustration.) Do all lamps burn at equal brightness?

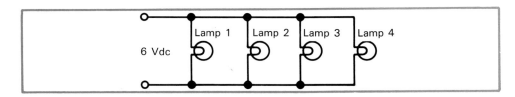

9. Do all lamps burn brighter, dimmer, or the same compared to a single lamp connected to the source?
10. What is the voltage across each lamp?
11. If each lamp has a resistance of 4 Ω, what is the total circuit resistance?
12. What is the current through one lamp?
13. What is the total current?
14. What power is used by one lamp? By all lamps?

Compute the following problems. Make conversions when necessary.

15. $E = 100$ V $I = 2$ A $R =$ _____
16. $I = 100$ mA $R = 100$ Ω $P =$ _____
17. $I = 10$ mA $E = 50$ V $P =$ _____
18. $P = 500$ W $E = 250$ V $I =$ _____
19. $P = 50$ W $R = 2$ Ω $I =$ _____
20. $E = 50$ V $R = 10$ kΩ $I =$ _____
21. $I = 0.01$ A $R = 100$ Ω $E =$ _____
22. $P = 100$ W $I = 2$ A $R =$ _____
23. $I = 20$ mA $E = 100$ V $R =$ _____
24. $P = 10$ W $I = 1$ A $R =$ _____
25. What is the total circuit resistance with 1000 ohms, 1.2 kilohms, 5.6 kilohms, and 10 kilohms in series?
26. What is the total circuit resistance with 1200 ohms, 5 kilohms, and 10 kilohms in parallel?

Applications of Electronics Technology:
This compact VHS camcorder features electronic image stabilization (EIS).
EIS helps to minimize the effects of camcorder movement by digitally stabiliz-
ing the picture. (RCA)

Section I

ELECTRICAL ENERGY AND POWER

SUMMARY

Important Points

☐ Movement of electrons in a conductor is electric current, or electricity.

☐ Matter is anything that occupies space and has mass.

☐ Elements are basic or pure forms of matter.

☐ Compounds are made of two or more chemically combined elements.

☐ Atoms are the smallest particles of an element still having the properties of that element.

☐ Molecules are the smallest particles of a molecular compound still having the properties of that compound.

☐ Electrons have a negative ($-$) charge.

☐ Protons have a positive ($+$) charge.

☐ Neutrons have a neutral charge.

☐ Static electricity is electricity that is at rest or not moving.

☐ Like charges repel each other, while unlike charges attract each other.

☐ A coulomb is 6.24×10^{18} electrons.

☐ Voltage is the force that moves electrons along in a conductor. It is measured in volts (V).

☐ Current is the movement of electrons in a conductor. It is measured in amperes (A).

☐ The measure of 1 ampere is equal to 1 coulomb of charge (6.24×10^{18} electrons) moving past a given point in a conductor in 1 second.

☐ The electron theory of current states that current in the external circuit always flows from negative to positive.

☐ Primary cells cannot be recharged; secondary cells can.

☐ Connecting cells in series increases their voltage rating. Connecting cells in parallel increases their current rating.

☐ Electrical energy can be created from light by a solar cell, or photovoltaic cell.

☐ A thermocouple converts heat to electrical energy.

☐ Converting pressure to electrical energy is called piezoelectricity.

☐ Insulators do not have free electrons and have a high resistance to current.

☐ Semiconductors lie between insulators and conductors in their resistance to current.

☐ Conductors have free electrons and low resistance. Current flows easily through a conductor.

☐ Superconductors are materials that offer little or no resistance to current.

☐ Pentavalent elements have five electrons in their outermost shell.

☐ Trivalent elements have three electrons in their outermost shell.

☐ Resistance is the opposition to current. The unit for resistance is the ohm (Ω). Conductance is opposite and the reciprocal of resistance. The unit for conductance is the siemen (S).

☐ Ohm's law states the mathematical relationship between voltage, current, and resistance.

☐ Electrical power is the time rate of doing electrical work. It is measured in watts.

☐ The power law states that electrical power equals the product of voltage and current in a circuit.

☐ A circuit is a complete pathway in which current flows. It consists of a voltage source, a pathway, a load, and, usually, a control.

☐ A series circuit has only one pathway in which current flows.

☐ A parallel circuit has more than one pathway in which current flows.

☐ A combination circuit has components connected both in series and in parallel.

1. Solve the following:
   a. $E = 100$ V, $I = 2$ A, $R = $ _____.
   b. $E = 50$ V, $R = 1000$ Ω, $I = $ _____.
   c. $I = 0.5$ A, $R = 50$ Ω, $E = $ _____.
   d. $E = 10$ V, $I = 0.001$ A, $R = $ _____.
   e. $I = 0.05$ A, $R = 1000$ Ω, $E = $ _____.
   f. $P = 10$ W, $I = 2$ A, $E = $ _____.
   g. $E = 100$ V, $I = 0.5$ A, $P = $ _____.
   h. $P = 500$ W, $E = 250$ V, $I = $ _____.
   i. $I = 0.01$ A, $R = 100$ Ω, $E = $ _____.
   j. $P = 100$ W, $I = 2$ A, $R = $ _____.
   k. $E = 10$ V, $P = 10$ W, $R = $ _____.
   l. $E = 500$ V, $I = 2$ A, $R = $ _____.
   m. $E = 100$ V, $R = 1000$ Ω, $P = $ _____.
   n. $I = 0.5$ A, $R = 50$ Ω, $P = $ _____.
   o. $I = 4$ A, $R = 10$ Ω, $P = $ _____.
   p. $I = 10$ mA, $E = 50$ V, $P = $ _____.
   q. $I = 20$ mA, $E = 100$ V, $R = $ _____.
   r. $P = 10$ W, $I = 1$ A, $R = $ _____.
   s. $E = 1000$ V, $R = 100$ Ω, $I = $ _____.
   t. $I = 100$ mA, $R = 100$ Ω, $E = $ _____.

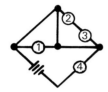

2. Which light bulb in the circuit (left) burns the brightest? (All of the bulbs are the same type.)

3. The value of a resistor that has green, blue, brown, and gold color bands is _____.

4. Convert the following units.
   a. 420 mA = _____ A
   b. 1.2 MΩ = _____ Ω
   c. 50 kV = _____ V
   d. 0.05 W = _____ mW

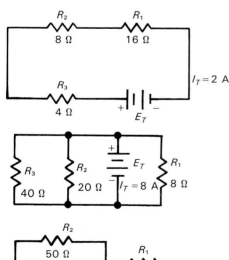

5. Find the unknown quantities in the series circuit shown (left).
   a. R_T d. E_{R2}
   b. E_T e. E_{R3}
   c. E_{R1} f. P_T

6. Find the unknown quantities in the parallel circuit shown (left).
   a. R_T e. I_{R2}
   b. E_T f. I_{R3}
   c. E_{R1} g. P_T
   d. I_{R1}

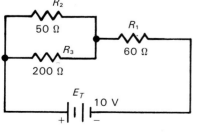

7. Find the unknown quantities in the combination circuit shown (left).
   a. R_T f. I_{R1}
   b. I_T g. I_{R2}
   c. E_{R1} h. I_{R3}
   d. E_{R2} i. P_T
   e. E_{R3}

103

Section II

APPLICATIONS OF MAGNETISM

Magnetism is the most common source of energy used for generating electricity. Over 95% of all our electricity that we consume is produced by generators, which operate on the principle of magnetism. Magnetism is also one of our oldest sources of electricity, dating back to the early use of magnets, or lodestones, for direction-seeking devices, or compasses. Magnetism is also the basis for operation of transformers, relays, motors, solenoids, and computer memories.

In Chapter 5, the fundamentals of the important source of electrical energy — magnetism — are discussed. Electromagnetism, or producing a magnetic field from an electric current, is presented. Some applications of magnetic energy are given.

Chapter 6 explains the fundamentals of the generator, both ac and dc. An introduction to ac is also given.

The operating principles of the ac and dc motor are discussed in Chapter 7. Many common types of motors are explained in detail. An exciting project is given at the end of this chapter for the reader.

Chapter 8 explains the basic test instruments used in electricity and electronics. Included in the discussion are the multimeter, the digital meter, and the oscilloscope.

Applications of Electronics Technology:
An integrated answering machine/cordless telephone is shown here. Cordless telephones are popular because of their convenience, and answering machines have become common household items. (PhoneMate)

Chapter 5

MAGNETISM

Magnetism is the best known and most common method of generating electricity. Yet, for the average person, it is probably the least understood source of electricity.

After studying this chapter, you will be able to:

☐ *Discuss the nature of and state the properties of permanent and electromagnetic fields.*

☐ *Define terms associated with magnetic circuits.*

☐ *Name common units of measurement associated with magnetism.*

☐ *Describe the conversion of electrical energy to mechanical energy by electromagnetic fields.*

☐ *Cite practical applications of magnetism and electromagnetism.*

INTRODUCTION

The mysteries of magnetism are legendary folklore passed from generation to generation. Shepherds before the Christian era were mystified by small pieces of stone found in their pastures. These odd stones possessed an invisible, attractive and repulsive force. Early navigators found that the same end of such a stone, when attached to a floating block of wood in a vessel of water, would always turn in a northerly direction. They used it as a compass.

Since the magnetic stones were used in navigation, they were called *leading stones*, or **lodestones**. These natural magnets are now a curiosity. They have been replaced by manufactured magnets that are stronger and even longer-lasting. We have come a long way from the dark ages of mystery and superstition when the magnet was believed to contain the devil.

What are the sources of magnetism? If we are to make use of magnetism, where do we find magnets? One place mentioned is in nature. The traditional lodestone came from a naturally occurring substance called *magnetite*. A very large magnet found in nature is the earth's core. Manufactured magnets are another source. Such magnets are called *permanent magnets*. A third source of magnetism comes from electricity. This is called *electromagnetism*. All of these sources will be discussed in this chapter.

THE EARTH AS A MAGNET

Why does a compass point in a northerly direction? This question led William Gilbert in 1600 to conclude that the earth is like a giant magnet with its poles near the geographic poles. Gilbert was an English scientist. He is noted for his studies in electricity and magnetism and is credited with coining the term *electricity*.

Gilbert was correct—the earth acts like a magnet. Like any magnet, it is surrounded by a **magnetic field**. This magnetic field emerges from each end of the earth and is made up of *invisible* **magnetic lines of force**. These lines seem to connect north and south in one great continuous magnetic circuit. Figure 5-1 shows the earth and its magnetic field.

A compass anywhere on the earth's surface will line up parallel to the earth's magnetic lines of force. Also, by definition, it will point toward **magnetic north**. Magnetic north is not *true* north, the north *geographic* pole, but they are in the same proximity. The angular difference between them is called the **variation**, or **declination**.

From the scientific point of view, magnetic north actually has *south* magnetic polarity. This is why the north pole of a compass will point to magnetic north instead of being repelled. Magnetic lines of force flow from north to south magnetic polarity. Flow direction is indicated in Figure 5-1.

PERMANENT MAGNETS

A **permanent magnet** is a magnet that retains its magnetism after removal of the magnetizing force. Magnets are manufactured from certain ferrous materials by placing a bar of this material in a strong magnetic field. To observe a magnetic field, or *magnetic flux* as it is also called, place a magnet under a sheet of cardboard. While tapping the cardboard gently, sprinkle the surface with iron filings. The iron particles will align themselves according to the individual lines of force around the magnet. See Figure 5-2.

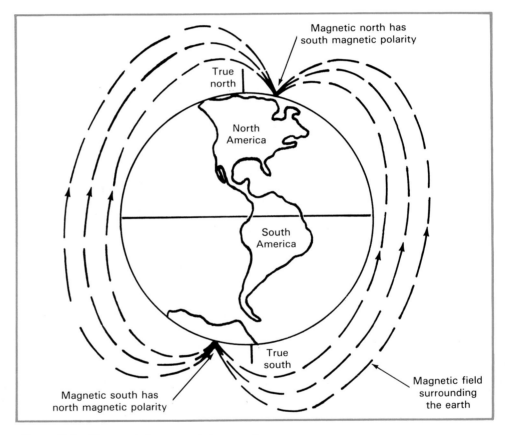

Figure 5-1. The earth is one great magnet.

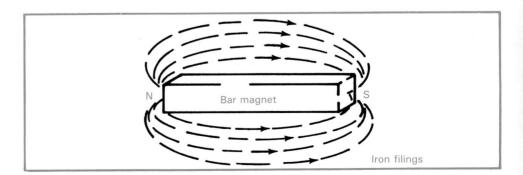

Figure 5-2. A magnet is surrounded by a field of many magnetic lines of force. By definition, the lines are always directed from the north pole to the south pole of a magnet.

In Figure 5-2, notice the concentration of force lines at the ends, or poles. Look closely and see that each line is continuous and that no two lines cross. The lines at each pole are very dense; however, in the space between the poles, they tend to expand and separate. Each line repels adjacent lines and tends to keep them apart. The number of lines in the field will depend upon the strength of each magnet.

Law of magnetism

The *law of magnetism* applies to any magnet — be it natural, permanent, or an electromagnet. The **law of magnetism** states:

- *Unlike poles attract each other.*
- *Like poles repel each other.*

These actions are demonstrated by working with two magnets suspended from wire stands as shown in Figure 5-3. When hung with *like* poles together adjacent to each other, either north or south, the magnets are repelled further apart. When hung with *unlike* poles adjacent, they are attracted together.

Iron filings and cardboard can be used to further prove this law of repulsion and attraction. Place the cardboard over two magnets with unlike poles in close proximity but fixed in place so they cannot move. Sprinkle the surface with iron filings. They should form the pattern shown in Figure 5-4.

In the figure, note the dense concentration of force lines between N and S at the center. Also, note the force lines in the space between extreme ends of the pair of magnets. The pair appears as one magnet. In fact, if not fixed in position, they would snap together and form a single magnet.

The fields of magnets in a repulsive relationship may be studied in the same manner. Place two poles close together in a fixed position beneath the cardboard. See Figure 5-5. Note that the gap between the like poles is void of force lines. The lines from each magnet repel each other and do not get close unless forced to do so. If the magnets were released from their fixed positions, they would jump apart.

The force between magnets varies according to the distance between the magnets. Specifically, the attractive *or* repulsive force varies inversely as the distance squared. Hold two magnets in an attractive position but separated by several inches so they will not come together. Move one magnet slowly toward the other. A point will be reached when the magnets will move together. Beyond this point, you might notice that the closer the magnets get, the greater the attractive or repulsive forces seem to be.

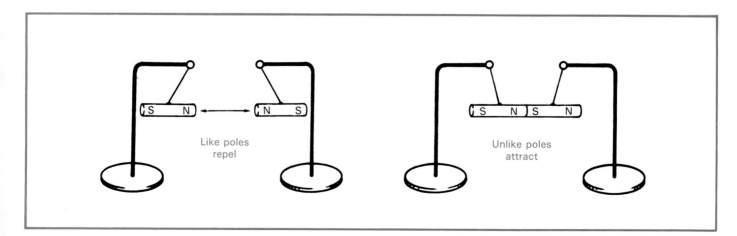

Figure 5-3. Demonstration of the law of magnetism.

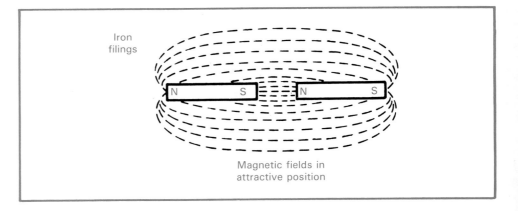

Figure 5-4. *Magnetic fields between bar magnets are shown in an attractive relationship.*

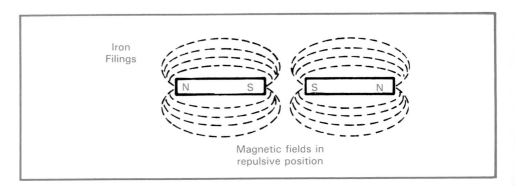

Figure 5-5. *Magnetic fields of magnets in a repulsive relationship.*

Theory of magnetism

The exact nature of magnetism in a magnetic field has been investigated by scientists for years. A modern theory of magnetism is the *domain theory*. This theory relates to the atomic structure of elements discussed in Chapter 1. You will recall that an atom consists of a nucleus and a number of orbiting electrons. The **domain theory** suggests that each electron in an atom is also spinning on its own axis, as well as orbiting around its nucleus. If an equal number of electrons are spinning in opposite directions, the atom is in an unmagnetized state. However, if more electrons are spinning in one direction than the other, the atom is in a magnetized state and is surrounded by a magnetic field.

When a number of these magnetized atoms exist in a material, they interact with adjacent atoms and form domains of atoms having the same magnetic polarity. These domains exist in random patterns throughout the material. Under the influence of a strong external magnetic field, the domains become aligned, and the entire material is magnetized. The strength of the retained magnetic field depends upon the number of domains that have been lined up. See Figure 5-6. Magnetic materials contain many domains. About 10 million tiny domains may be contained in a cubic centimeter of magnetic material.

Magnetic materials

There are several types of magnetic materials. In electricity and electronics, **ferromagnetic materials** are the most important. These materials are relatively easy to magnetize and greatly assist the passage of magnetic lines of force. Iron, nickel, cobalt, and their alloys are ferromagnetic materials. One such alloy has the trade name *Permalloy*. It is a nickel/iron alloy. Another such alloy is *alnico*. It is an alloy of aluminum, nickel, and cobalt. Alnico is most

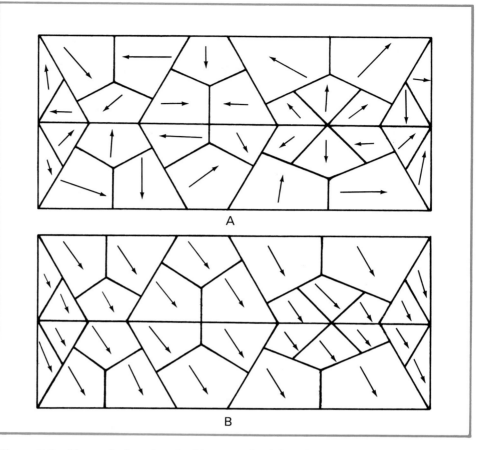

Figure 5-6. Magnetic domains. A—Unmagnetized domains arranged in random fashion. B—Aligned domains in a magnetized object.

satisfactory for permanent magnets and holds its magnetism for long periods of time. *Ferrite* is another type of ferromagnetic material. It is made of a certain type of ceramic.

Paramagnetic materials assist the passage of magnetic lines of force. These materials will become only slightly magnetized, even in a strong magnetic field. Paramagnetic materials include aluminum, chromium, and platinum. Any magnetism of a paramagnetic material will be in the same direction as the magnetizing force.

A **diamagnetic material** can also be slightly magnetized, but it will assume a polarity opposite to the polarity of its magnetizing force. Materials with this property are repelled by a magnet and tend to position themselves at right angles to magnetic lines of force. Relative to vacuum, these materials would slightly reduce the magnetic flux per unit area if used as a core material. Diamagnetic materials include copper, silver, gold, and mercury.

Induced magnetism

Why does a magnet pick up a nail? See Figure 5-7. When the nail is brought into the magnetic field, the lines of force will pass through the nail. The nail becomes a part of the total magnetic circuit. It becomes temporarily magnetized with north and south poles. In the figure, the south pole is identified at the end of the nail nearest the north pole of the magnet. The nail, therefore, is attracted to the magnet. Had the south pole of the magnet contacted the nail, the head of the nail would have had north polarity. This explains why either end of the magnet will pick up magnetic material. Magnetization of an object while it is in the magnetic field of the magnet is called **induced magnetism**.

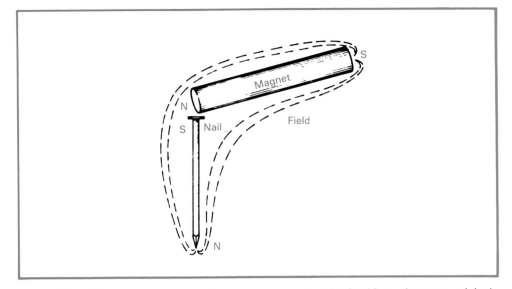

Figure 5-7. *A bar magnet temporarily induces magnetic poles in objects that were originally unmagnetized. The nail, in this example, will itself be made into a temporary magnet by the flux.*

ELECTROMAGNETISM   The relationship between electricity and magnetism was first discovered in 1819 by Danish scientist, Hans Christian Oersted. The discovery was made while doing some experiments with electrical circuits in the laboratory. A compass lying on the bench near the circuits showed very peculiar behavior when the circuit was turned on or off. This led Oersted to the conclusion that an electric current in a conductor produced a magnetic field. This type of magnetism, produced by an electric current rather than a magnet, is called **electromagnetism.**

In 1831, English physicist Michael Faraday discovered the principle of **electromagnetic induction.** The principle states that a changing magnetic flux produces an electric field. Then, in 1864, Scottish mathematician and physicist James Clerk Maxwell unified in one great theory the theories of electricity and magnetism. Maxwell showed that all electric and magnetic phenomena could be described using only four equations. These equations summarizing electromagnetic phenomena are known as *Maxwell's equations.*

The relationship between electricity and magnetism is demonstrated in Figure 5-8. A conductor is passed vertically through a hole in a horizontal piece of cardboard. A compass is placed on the cardboard and moved about the conductor, which has current through it. Upward direction of current (electron flow) through the wire shows compass needle pointing in a clockwise direction. Since the direction of magnetic lines of force is defined as the direction in which a north pole points, the circular magnetic field produced is clockwise. With a downward current, a counterclockwise field is produced.

The **left-hand rule for conductors** can be applied to determine the direction of a magnetic field. Grasp the conductor in your left hand with your thumb extended and pointing in the direction of the current. Your fingers, encircling the conductor, point in the direction of the magnetic field.

Figure 5-9 shows cross sections of current-carrying conductors. At left, the black dot in the center represents the point of an arrow, showing that the current is toward you. The direction of the field is shown. The cross on the right represents the feathers at the butt end of an arrow, indicating that current is away from you. Use the left-hand rule to prove that the magnetic fields are correctly drawn.

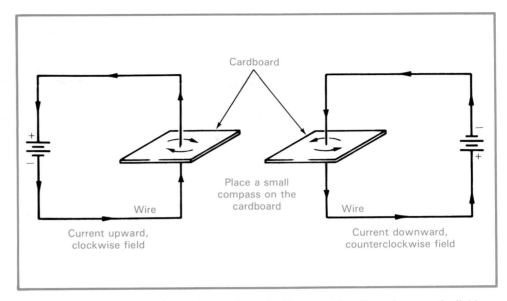

Figure 5-8. A compass placed on cardboard will show direction of magnetic field.

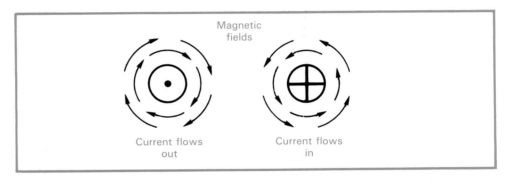

Figure 5-9. These conventions are used to illustrate current direction and associated magnetic fields.

In Figure 5-10, two conductors are illustrated. One drawing shows the magnetic fields when currents are in opposite directions. In the second drawing, currents in the conductors are in the same direction. In opposite directions, the fields repel one another; in the same direction, they attract. Pay special attention to the magnetic field patterns.

Solenoids and electromagnets

When current travels through a wire wound into a coil, Figure 5-11, the magnetic fields around the wires join and reinforce each other. The coil assumes a polarity, just like a magnet. One end becomes north, the other end, south. The **left-hand rule for coils** can be applied to determine the north pole of the coil. Grasp the coil with your left hand, pointing your fingers in the direction of the current. Your extended thumb will point toward the north pole of the coil. Note that some coils are wound in the shape of doughnuts. For these **toroidal coils**, as they are called, north and south poles cannot be identified.

A coil wound without a core is called a **solenoid**. A coil wound on a core of magnetic material is called an **electromagnet**. With the core, the total magnetic flux of the coil is greatly increased. The left-hand rule is valid for both solenoids and electromagnets. An electromagnet is shown in Figure 5-12. Some electromagnets have *movable* cores. These, though they have cores, are called *solenoids*.

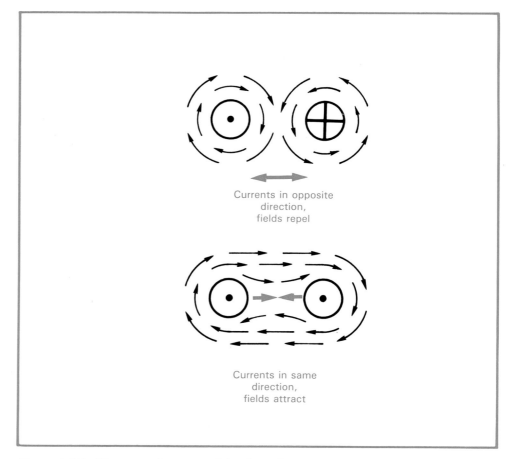

Figure 5-10. *The attractive or repulsive force between magnetic fields depends upon current direction in the conductors.*

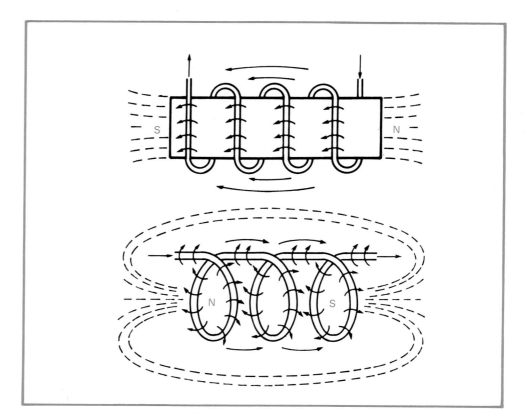

Figure 5-11. *The magnetic fields around the wires of a coil reinforce each other and give the coil magnetic polarity.*

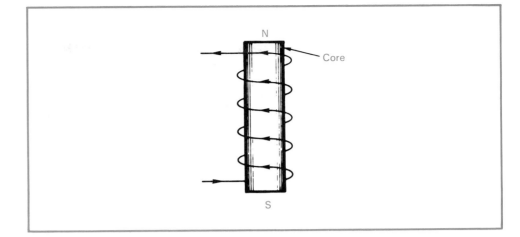

Figure 5-12. A coil wound on a core is an electromagnet.

Systems of units

A most confusing system for units of measurement in magnetic circuits is now in use. Some terms are not universally accepted. To make it more difficult, at least four systems of measurement are found in texts and engineering reports. The four systems are the U.S. Conventional, the SI Metric, the centimeter-gram-second (cgs), and the meter-kilogram-second (mks). The cgs system defines small units, whereas the mks system defines larger, more practical units. For all practical purposes, the mks and the SI Metric systems are the same. Most texts now use the SI Metric system of units.

Magnetic flux and magnetic flux density

The total lines of force in a magnetic field are referred to as **magnetic flux**. It is represented by the symbol Φ (phi). The mks units of magnetic flux are *newton-meters per ampere* (N·m/A). In honor of German physicist Wilhelm Weber, 1 N·m/A is called 1 *weber*. The **weber (W)** is defined as the magnetic flux which, linking a circuit of one turn, induces an emf of 1 volt when the flux is reduced to zero at a constant rate in 1 second. In the cgs system, the unit of magnetic flux is the **maxwell (Mx)**, and 1 maxwell equals 10^{-8} weber.

The flux per unit cross-sectional area A of a magnetic field is called the **magnetic flux density *(B)***. It is also called the **magnetic induction**. Magnetic flux density is given by the formula:

$$B = \frac{\Phi}{A}$$

The mks unit of magnetic flux density is the *tesla*. One **tesla (T)** is equal to the flux density of 1 weber per square meter. The cgs unit of flux density is the *gauss*. One **gauss (G)** is equal to 1 maxwell per square centimeter. See Figure 5-13. One gauss also equals 1×10^{-4} tesla.

Magnetomotive force and magnetic field intensity

We have referred to the flowing, closed loops of magnetic lines of force as a *magnetic circuit*. The magnetic circuit is similar to the electrical circuit. We can call the force that makes the lines flow the **magnetomotive force (mmf)**, which is like the emf of an electrical circuit. We can compare the total magnetic lines of force (Φ) to current in an electrical circuit. Finally, a magnetic circuit will offer a resistance to the flow of magnetic lines. This resistance is similar to the resistance in electrical circuits and is called **reluctance ($\mathcal{R}$)**. The value depends on the material of the magnetic circuit. Air has a high reluctance; iron has a low reluctance.

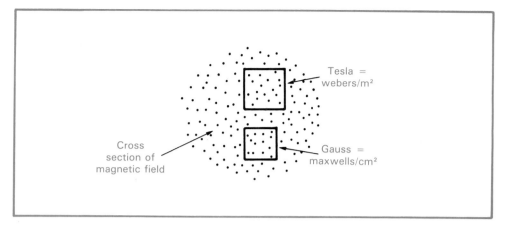

Figure 5-13. A comparison (not to scale) of mks and cgs units of measuring magnetic flux density.

The three factors — magnetic flux, magnetomotive force, and reluctance — are related by the following formula:

$$\Phi \ = \ \frac{\text{mmf}}{\mathcal{R}}$$

This formula is known as **Ohm's law for magnetic circuits**. It corresponds to $I = E/R$. The mks units for mmf are *ampere-turns* (A·t). The cgs unit is the **gilbert (Gb)**. One ampere-turn is equal to 1.256 gilberts. The mks units for reluctance are *ampere-turns per weber*. The cgs units are *gilberts per maxwell*.

In the case of a solenoid or any other current-carrying coil, magnetomotive force is related to the current and the number of turns of wire. Specifically, it is given by the formula:

$$\text{mmf} = N \times I$$

where N is the number of turns in the wire, and I is the current in amperes. From this, it follows that any increase in mmf by increasing the current through a coil will increase the magnetic flux.

The *intensity* of the magnetic field depends upon how long the coil is. The intensity is properly called the **magnetic field intensity (H)**. It is also called the **magnetic field strength**, or **magnetizing force**. The magnetic fields around each loop of a coil are additive. By increasing the number of turns, the number of flux lines increases, and the intensity of the total magnetic field is increased. Also, since the magnetic fields around a wire are a function of current, a larger current produces stronger individual fields. Thus, the total field intensity is increased.

The magnetizing force is basically magnetomotive force per unit length. It is given by the formula:

$$H = \frac{NI}{l}$$

where l is the length in meters. The mks units of magnetic field intensity are *ampere-turns per meter*. The cgs units are *gilberts per centimeter*, or **oersteds (Oe)**. One ampere-turn per meter equals 0.0126 oersted. The field intensity of a coil will remain uniform throughout its cross section if the length of the coil is ten or more times greater than the diameter. If not, the calculation is much more complex.

Problem:

An emf is applied to a 10-centimeter coil having 400 turns of wire. The current through the coil is 2 A. What is the magnetomotive force produced? What is the magnetic field intensity? What is the total flux with a reluctance of 400 × 10⁶ ampere-turns per weber?

Solution:

Step 1. Determine magnetomotive force.

$$\text{mmf} = N \times I = (400 \text{ turns})(2 \text{ A}) = 800 \text{ A·t}$$

Step 2. Determine magnetic field intensity.

$$H = \frac{\text{mmf}}{l} = \frac{800 \text{ A·t}}{0.1 \text{ m}} = \frac{8000 \text{ A·t}}{\text{m}}$$

Step 3. Determine the total flux.

$$\Phi = \frac{\text{mmf}}{\mathcal{R}} = \frac{800 \text{ A·t}}{400 \times 10^6 \text{ A·t/W}} = 2 \ \mu\text{W}$$

The coil of the sample problem has an mmf of 800 ampere-turns with 2 amperes of current through it. Note that the same force might be produced by 4 amperes of current in a 200-turn coil. Likewise, 8 amperes in a 100-turn coil would produce the same mmf. Any number of current and coil combinations would give this mmf. However, most coils that you will work with have a fixed number of turns. Therefore, you will vary the magnetomotive force by varying the current through the coil. This is not theory but a practical matter. You will be required to understand this before you can work on generators, motors, relays, and many other devices.

Permeability and relative permeability

The measure of the ease with which magnetic flux may be established is known as **permeability**. It is represented by the symbol μ (mu). Permeability reflects a material's ability to concentrate flux. It is the magnetic analog of electrical *conductivity* and is inversely proportional to reluctance. It is the ratio of magnetic flux density to magnetic field intensity and is expressed by the formula:

$$\mu = \frac{B}{H}$$

In the mks system, units are *webers per ampere-meter*, or, simply, *henrys per meter*. (The units are equivalent. We will not justify the conversion, but know that the henry is the unit of inductance. It is discussed in Chapter 9 in detail.) The permeability of air or vacuum is called the **permeability of free space**. This is designated μ_0. The value of μ_0 is $4\pi \times 10^{-7}$ W/A·m.

Materials such as iron or steel will greatly increase magnetic flux when inserted into a current-carrying coil. These materials have a much greater permeability than air. The permeability of these materials is conveniently given as a ratio of the permeability of the material to the permeability of air. This ratio is called the **relative permeability**. It is designated μ_r. Formulas for relative permeability are:

$$\mu_r = \frac{\mu}{\mu_0} = \frac{B}{\mu_0 H}$$

Relative permeability is a unitless number. In the case of air, $\mu_r = 1$. The relative permeability of iron or steel may range from 200 to 2500 or more, depending on the particular type. In general, permeability is not a constant quantity; it very much depends upon magnetic field intensity.

The formula for permeability, $\mu = B/H$, may be rearranged:

$$B = \mu H$$

Substituting for H, the magnetic field intensity:

$$B = \frac{\mu NI}{l}$$

Examination of this formula will explain how a magnetic material added to a coil will increase the magnetic flux density of the coil. Note, however, that adding a core material does nothing to magnetomotive or magnetizing forces.

Problem:

Calculate the magnetic field intensity and flux density of a solenoid carrying a current of 0.1 A. The 3000-turn coil measures 15 centimeters in length.

Solution:

Step 1. Calculate the magnetic field intensity.

$$H = \frac{NI}{l} = \frac{(3000 \text{ turns}) (0.1 \text{ A})}{0.15 \text{ m}} = \frac{2000 \text{ A·t}}{\text{m}}$$

Step 2. Calculate the flux density.

$$B = \mu_r \mu_0 H = (1)(4\pi \times 10^{-7} \text{ W/A·m})(2000 \text{ A·t/m}) = 2.5 \times 10^{-3} \text{ T}$$

Problem:

What is the magnetic field intensity and flux density of the coil in the previous problem when an iron core is inserted into the coil? Assume relative permeability of the core is 200.

Solution:

Step 1. Calculate the magnetic field intensity.

$$H = \frac{NI}{l} = \frac{(3000 \text{ turns}) (0.1 \text{ A})}{0.15 \text{ m}} = \frac{2000 \text{ A·t}}{\text{m}}$$

Step 2. Calculate the flux density. Rearranging equation for relative permeability:

$$B = \mu_r \mu_0 H = (200)(4\pi \times 10^{-7} \text{ W/A·m})(2000 \text{ A·t/m}) = 0.5 \text{ T}$$

Now compare magnetic field intensity of this problem to that of the previous problem. The answer is the same. Comparing the flux densities we see that the answers are not the same. With 0.1 A of current through the coil, the iron core increases the magnetic flux density by a factor of 200.

Hysteresis

The characteristics of a magnetic material within a coil may be plotted on a curve. Figure 5-14 is a plot of what is called a **hysteresis loop**. The x-axis represents the field intensity measured in ampere-turns per meter. The y-axis represents flux density in webers per meter² (teslas). As H increases, the flux density increases to a point of **saturation**. This means that the magnetic core material is so full of flux lines that no more can be produced by further increasing field intensity.

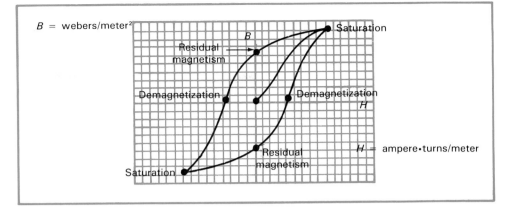

Figure 5-14. The hysteresis loop of a magnetic material.

From saturation, the field intensity *H* is returned to *zero*. Note, however, that the curve did not return to zero. This point marks the level of **residual magnetism**. This is the magnetism retained in a magnetic material after the magnetizing force is removed. The degree of residual magnetism left behind depends upon the **retentivity** of the material. This property is a measure of a ferromagnetic material's ability to retain magnetism. For permanent magnets, a material of high retentivity is required. On the other hand, a core in a *relay*, which is an electromagnetic switch, should retain little magnetism when not activated. It should have a low retentivity.

A certain amount of force *H* in the opposite polarity is required to demagnetize the material. Continuation of the curve is shown as *H* is applied in the opposite polarity to saturation. The flux density is returned to saturation, this time by continuing to apply *H* in the opposite polarity. Continuing around the loop, we see that the changing levels of flux density lag behind the changing levels of magnetic field intensity. This lagging effect is called **hysteresis**.

The hysteresis loop graphically shows the characteristics of a magnetic material with a magnetizing force alternating in polarity. You will need to understand this theory in order to intelligently work with chokes, transformers, and many other devices and circuits.

LESSON IN SAFETY: Electrical energy dissipated in any kind of resistance produces heat. Resistors, vacuum tubes, transistors, and many other components can become very hot. Burns can be painfully serious. Be cautious!

Magnetic shielding

Magnetism will pass through any material, including glass, water, and insulation. To prevent this, the high permeability of some materials can be put to good use. That is, such materials can be used as a means of conducting a magnetic field *around* a circuit or delicate instrument, for example. This action is called **shielding**.

In Figure 5-15, for example, a circular shield of magnetic material surrounds a meter. Since the disturbing magnetic field will take the easiest path from N to S, the shield redirects the magnetic lines around the meter. This is something like a short circuit for magnetic lines of force. This type of shielding is used to protect against a *steady* magnetic field.

Not all shielding is accomplished with magnetic materials. Sometimes, *conductive* materials like copper and aluminum are used. This type of shielding does its job in a different manner. It is used to shield against induction caused

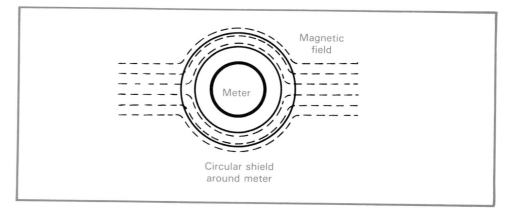

Figure 5-15. A shield of high permeability conducts magnetic flux around meter.

by a *varying* magnetic field. The shield has induced currents that oppose the inducing field. The effect of the inducing field is then canceled. Thus, any material surrounded by the shield is unaffected by the varying magnetic field. Coaxial cable is an example of this type of shielding.

APPLICATIONS OF SOLENOIDS AND ELECTROMAGNETS

Solenoids and electromagnets serve a multitude of purposes. For control applications they are invaluable. A few of these applications are presented in this section. However, these are just a few. There are many more that are not discussed here.

Solenoids

Figure 5-16 shows an electromagnet with a movable core (solenoid). The core is placed in the hollow center of the solenoid. Figure 5-16A shows the position of the core in an unenergized coil. When the coil is energized, a magnetic field appears, as illustrated. The magnetic lines seek the shortest possible path between poles. The lines seem to be elastic. They exert a *sucking force* on the movable core and pull it into the center of the coil. See Figure 5-16B. When the coil is de-energized, a spring returns the core to its original position.

The solenoid just described is a common means of converting electrical energy to mechanical movement via electromagnetism. Devices that convert energy this way are called **electromechanical devices**. A lever or rod attached to the core of a solenoid may be linked to levers, gears, and switches to perform all kinds of operations. Such a device is used in many automobile starters. It moves the starter's pinion gear, which meshes with the flywheel gear to start the car.

Solenoids are used in dishwashers and washing machines to change the cycles of operation and to turn hot and cold water on and off. They are used in air conditioning units to activate valves that start and stop refrigerant flow. See Figure 15-17. In Figure 5-18, a solenoid is used in an electric door chime. The push-button switch is depressed, completing the circuit and energizing the coil. The core then strikes and rings the chime.

Relays

A **relay** is an electromagnetic switch, Figure 5-19. By design, a movable element called an **armature** is mounted above the core of an electromagnet. The armature is attached to a spring. The relay has one or more sets of contact points, or **contacts**. When the coil is energized, the armature is attracted and the contacts open or close, depending on the arrangement. **Normally closed (NC) contacts** open when a relay is energized; **normally open (NO) contacts** close. When the energizing potential is removed, spring action returns the arma-

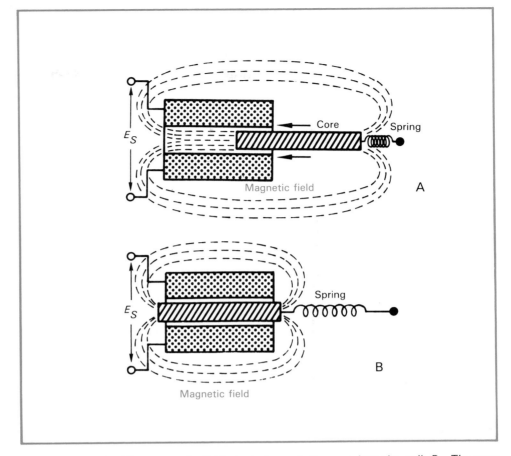

Figure 5-16. A—The magnetic field starts to suck the core into the coil. B—The core is at rest in the center of the coil.

*Figure 5-17. A **solenoid valve** uses an electromagnet with a movable core (valve stem) to control the flow of gases and liquids. Shown here is a 3-way solenoid valve. (ASCO)*

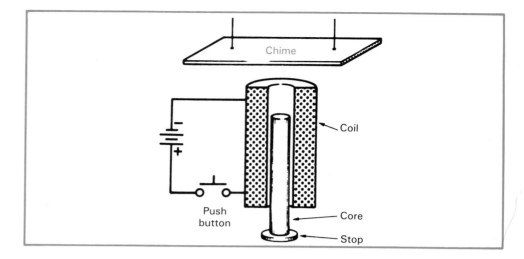

Figure 5-18. Sketch shows the construction of a door chime.

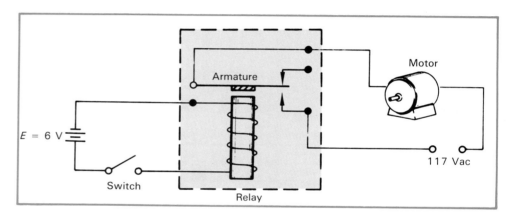

Figure 5-19. A diagram showing the application of a relay.

ture to its original state. Switch points may be set up for single-pole switching, or there may be several contacts for double-pole and more complicated switching operations.

Relays eliminate the need for manual switching. They also enable rapid and positive switching control of machinery and devices from remote locations. This provides safety for operating personnel. This is because relay operating voltages and currents can be relatively small when compared to levels required for running an apparatus. Relays also permit the use of small wires and low currents for switching a machine. The contacts only have to be heavy enough to carry the line current to run the machine. Relays eliminate the need for heavy wires. Some relays are made for heavy-duty service, however. They are called **contactors**.

When a relay is selected to do a specific job, consideration should be given to coil specifications. The *dc resistance* of the coil is the resistance of the wire from which the coil is wound. The *operating current* or *closing voltages* are indicative of the sensitivity of the relay and must conform to the particular need. The contact points must be sufficiently large and heavy to carry the required currents. Further, either an ac or dc relay must be specified, depending on the application. Some terms you may run across include:

• **Coil voltage**—the minimum voltage needed to operate the relay. It is also called the **pick-up voltage**.

- **Coil current**—the amount of current needed to operate the relay.
- **Holding current**—the minimum current required to keep a relay operating.
- **Drop-out voltage**—the maximum voltage at which a relay no longer operates.
- **Contact voltage rating**—maximum voltage that relay contacts are capable of switching safely.
- **Contact current rating**—maximum current that relay contacts are capable of switching safely.
- **Surge current**—maximum current that relay contacts can withstand for short periods of time without damage.
- **Contact voltage drop**—voltage drop across contacts of relay when contacts are *made*, or connected.

Voltage relays. Relays may be connected either in parallel or in series. See Figure 5-20. In the *shunt*, or parallel, connection, the coil is placed across the voltage source. Since current through the shunt coil connection is limited only by the resistance of the coil, it usually is made with many turns of relatively fine wire to limit the operating current. It is called a **voltage relay**.

Current relays. In the series connection, all the current to a specific load travels through the coil. The coil is considered to be current operated. The windings must be heavy enough to carry the load current. These are called **current relays**. In projects you will construct, relays that employ both of these methods of connection will be used.

Lock-up relays. A start/stop switching device employing a *lock-up relay*, Figure 5-21, is commonly used on machinery around a machine shop. Red and black push buttons in a small enclosure are located near the operating position. This allows the machine to be turned off rapidly. **Lock-up relays** are relays that are locked in the energized position magnetically or electrically, rather than mechanically, until the circuit is interrupted.

Look at the circuit diagrams for the lock-up relay in Figure 5-22. The circuit basically consists of a normally open ON button, a normally closed OFF button, a relay, and a load. It is connected to a 120 Vac source. Note that *H* on the diagram designates the *hot* side and *N*, the neutral side. Both pushbutton switches are *momentary*. **A momentary switch** is designed to return to its normal position when the actuating force is removed. In contrast, a push button may be *maintained*. **A maintained switch** is designed to remain at the operated condition after the actuating force is removed. In order to return to normal position, these must be actuated a second time.

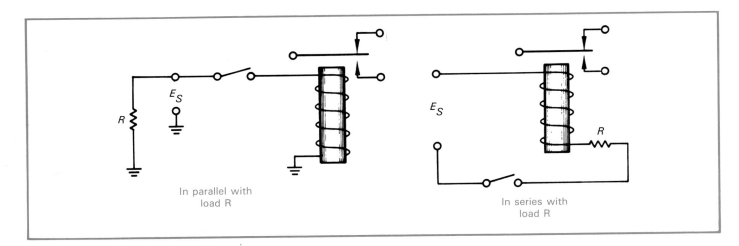

In parallel with
load R

In series with
load R

Figure 5-20. Relays are made to be voltage or current sensitive. They may be connected in parallel or series.

Figure 5-21. Push-button magnetic safety switch using a lock-up relay circuit.

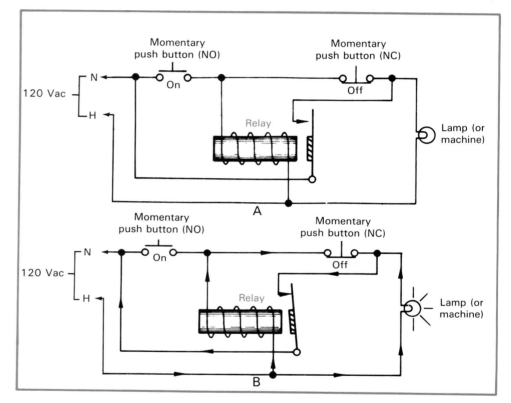

Figure 5-22. A—Circuit for a lock-up relay shown in the OFF position. B—Lock-up relay after pushing the momentary ON button.

Figure 5-22A shows the relay in the OFF position. The lamp (or other device) is off. Upon pressing the momentary ON button, the relay becomes energized, Figure 5-22B. The energized coil pulls in the relay armature. This provides a complete path for current through the load; it turns on. The current also continues through the relay, the OFF switch, then back through its armature to the neutral. Follow the arrows on the schematic. The relay is *locked-up* and will remain so until the OFF button is depressed. This will interrupt the current, the relay armature will drop out, and the load will shut off.

Overcurrent relays. These are relays designed to protect equipment (and life) from hazardous overcurrent conditions. One type of overcurrent relay is the magnetic **circuit breaker.** This is a special type of current relay. In Figure 5-23, a schematic of a simple circuit breaker is shown. The coil is connected in series, and all the current to the load flows through the coil and points.

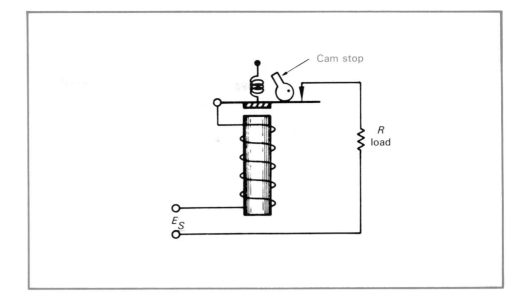

Figure 5-23. The circuit of a simple circuit breaker.

Normal load currents will not activate the relay. A larger-than-normal current, due to a short circuit or overload, will activate the relay. This will open the circuit to the load. A stop of some kind falls in place, which prevents the contacts from reclosing. The breaker has a time-delay feature to accommodate temporary overloads. This keeps the breaker from tripping from something such as a motor starting current.

Some breakers use a bimetal element to break the circuit. This type must have a *stop*, which will fall in place when a rise in temperature opens the points. Once tripped, the stop must be removed, or reset, in order to operate the circuit again.

Breakers prevent destruction of motors and devices by excessive currents. They also eliminate fire hazards from conductors and devices overheated by overload currents. Further, a tripped breaker can be used again, merely by resetting. Of course, the reason for the excessive load current should first be corrected.

Another type of overcurrent relay is the **thermal overload relay**. This type is designed primarily to protect motors. They are found in **magnetic motor starters**—contactors with thermal overload protection. See Figure 5-24. Thermal overload relays are desirable because they require time to operate and open a control circuit. This time delay makes these relays worthwhile for motor overload protection. It allows heavy inrush starting currents to pass without causing the relay to open the control circuit before a motor has reached running speed.

A thermal overload relay, Figure 5-25, consists of a metal element called a **heater** and an NC contact. The heater coil is connected in series with the motor. This way, current through the motor also flows through the coil. The coil is wound around a tube assembly that is filled with a **eutectic alloy**. This is an alloy having a fixed temperature at which it will go from a solid to a liquid state. A loaded spring tries to open the relay contacts; however, they are held closed by a locking device that is connected to the tube assembly. If excess current is drawn by the motor, the additional heat will melt the alloy. With the alloy melted, the shaft within the tube assembly is free to turn. The locking device will no longer hold the loaded spring. The spring, then, will force the NC contacts open.

A   B

*Figure 5-24. A—This combination magnetic starter provides motor protection and com-
plete motor control. Thermal overload relays are found in the magnetic motor starter seen
in the lower left corner of this enclosure. B—Motor control centers (MCCs) such as this
are often used to locate motor controls, like starters, at one central location.   (Furnas
Electric Co.)*

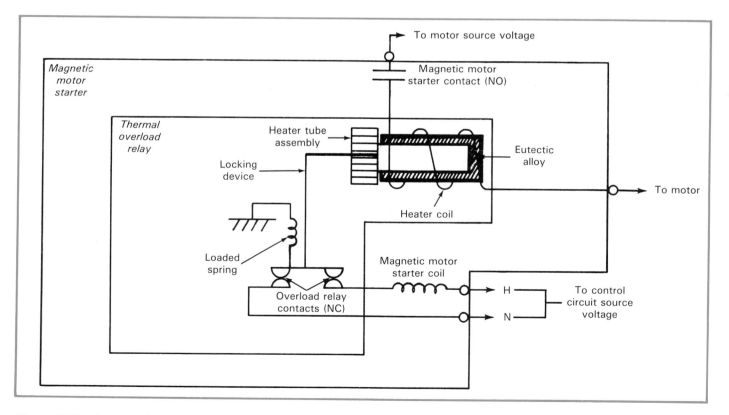

*Figure 5-25. A magnetic motor starter showing thermal overload relay. For simplicity,
the starter is used here for a single-phase motor; however, 3-phase is standard.*

The normally closed contacts are connected in series with the coil of the motor starter used to control the motor. When the overload contacts open, the motor starter coil de-energizes and disconnects the motor from the source voltage. Of course, this shuts the motor off. See the ladder diagram in Figure 5-26.

Reed relays. The reed relay is an interesting type of electromagnetic switch. These relays are made up of two reeds (SPST) or three reeds (SPDT), which serve as switch contacts. The reeds are sealed in a glass capsule filled with an inert gas. Figure 5-27 shows a cross-sectional view of a single-pole, single-throw reed relay. The glass capsule is inserted in a bobbin on which a dc coil is wound. When a dc current is fed through the coil, the switch contacts open or close, depending on the arrangement (NO or NC). Reed relays switch very fast. Therefore, they are not suited for ac current due to its fluctuating nature.

Advantages of the reed relay over other general-purpose relays are:
- Absence of moving parts—except for movable reeds.
- Sealed capsule—eliminates oxidation and contamination.
- Low operating power—usually in milliwatt range.
- Long contact life—usually 20 million minimum.
- Speed—100 operations per second is common.
- Plug-in version available for circuit board applications—see Figure 5-28.

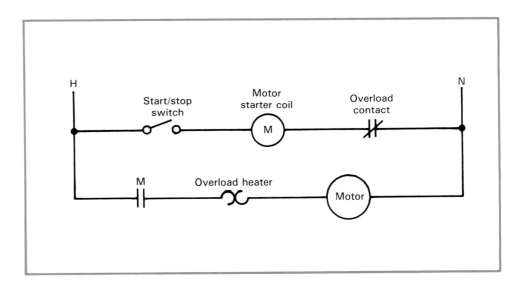

Figure 5-26. A ladder diagram showing wiring of a motor and its control circuit.

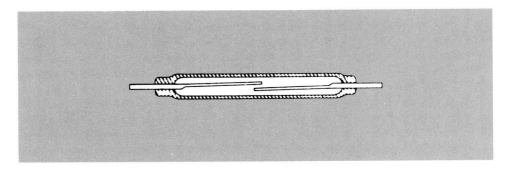

Figure 5-27. The glass capsule of an SPST reed relay. This capsule would be inserted in a bobbin wound with a dc coil. (Magnecraft)

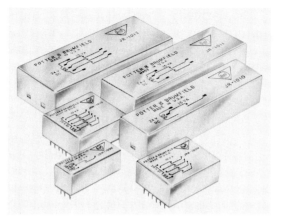

Figure 5-28. *Reed relays are available for circuit board applications.*
(Potter & Brumfield)

Buzzers and choppers

The circuit of an electrical **buzzer** is drawn in Figure 5-29. When the buzzer is energized, the magnetic field of the coil opens the armature contacts. Since these contacts are in series with the coil, the coil circuit also opens. This causes the magnetic field to collapse and close the contacts. This, in turn, energizes the coil. The operation repeats itself, and the device produces a buzzing noise, which is caused by the vibrating contacts. If desired, an extension with a striker can be attached to the vibrating armature to ring a bell. The buzzer then becomes a doorbell.

Since a buzzer starts and stops the current repeatedly at a specified frequency, it may be used as a **chopper**. This is a switch that produces a modified square-waveform output. Choppers may be electromechanical or solid state. They are widely used where it is necessary to sample a voltage or signal at prescribed intervals.

A type of chopper called a *vibrator* was used in the power supply of the old-fashioned automobile radio. The **vibrator power supply** consisted of a vibrator, step-up transformer, rectifier, and filters. The vibrator changed dc to the modified square-wave ac. The transformer, which works only on ac, then would change the low voltage ac to a high voltage ac. The rectifier and filter would change the ac back to a smooth dc.

The vibrator power supply for high voltage is no longer used in cars. It has been replaced by the *chopper amplifier*. The **chopper amplifier** is a solid state device. In it, two transistors operate alternately. Acting as switches, they *chop* a dc input to get an ac signal. This permits amplification of the signal by an ac amplifier. The chopper amplifier is usually followed by a filter to remove the ac component and restore dc. Solid state choppers are desirable; they are not subject to mechanical vibration and wear.

OTHER APPLICATIONS OF MAGNETISM

Until now, discussion of magnetism has focused on permanent magnets, solenoids, and electromagnets. Magnetism has other applications as well. They are discussed in detail in later chapters, but are presented here for your information.

Rotating equipment

Magnetism is made use of by *rotating equipment* such as *generators* and *alternators*. These are devices that convert mechanical energy into electrical energy. Simply put, they work by rotating a conductor through a magnetic field. *Motors* also make use of magnetism. This type of rotating equipment converts electrical energy into mechanical. A motor works by the interaction between the magnetic field of a current-carrying conductor and a fixed magnetic field. This interaction provides the rotational force that drives the motor.

Magnetic storage of information

Magnetism is also employed by computers in the magnetic storage of information. Among *magnetic storage devices* are magnetic tapes and disks, Figure 5-30. Also, the magnetic strip found on the back of a credit card is used to store information.

LESSON IN SAFETY: Many people are killed by supposedly "unloaded guns." Many technicians are injured by supposedly "dead circuits." Be sure your equipment is turned off and disconnected from the main power source before you make circuit changes and replace components. See Figure 5-31.

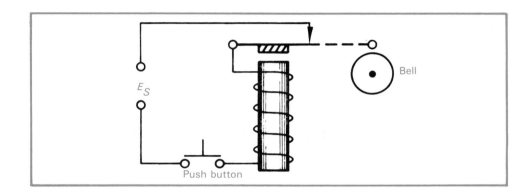

Figure 5-29. The circuit of a buzzer or doorbell.

Figure 5-30. The 3.5'' diskettes pictured in the foreground use magnetism to store data and programs like the scheduling program being used here.   (Waterloo Manufacturing Software)

Figure 5-31. Blockouts for circuit breakers lock breaker in OFF position and prevent circuit from being energized. When used with the tagout tags shown, they demonstrate compliance with OSHA Tagout Regulations. (IDESCO Corp.)

SUMMARY

- The earth acts like a magnet and is surrounded by a magnetic field.
- A permanent magnet retains its magnetism after removal of the magnetizing force.
- The law of magnetism states that like poles repel and unlike poles attract.
- Spinning electrons are surrounded by a magnetic field. An atom is magnetized when more electrons spin in one direction than another. Under the influence of an external magnetic field, the atoms of an element align in one direction, and the entire material becomes permanently magnetized — the domain theory.
- Ferromagnetic materials are the most important magnetic material used in electricity and electronics.
- Induced magnetism is the temporary magnetization of an object in a magnetic field.
- Magnetism produced by an electric field is called electromagnetism.
- A coil wound without a core is a solenoid. A coil wound on a core is an electromagnet. However, a coil wound around a *movable* core is a solenoid.
- The total lines of force in a magnetic flux is the magnetic flux. The flux per unit cross-sectional area is the magnetic flux density.
- Magnetomotive force in a magnetic circuit is the analog of electromotive force in an electric circuit. Magnetic flux is the analog of current. Reluctance is the analog of resistance.
- Magnetic field intensity is magnetomotive force per unit length.
- The measure of the ease with which magnetic flux may be established is permeability. The ratio of the permeability of a material to the permeability of air is the relative permeability.
- Changing levels of flux density lag behind changing levels of magnetic field intensity. This lagging effect is called hysteresis.
- Conducting a magnetic field around a circuit or device is called shielding.
- An electromagnet with a movable core, called a solenoid, is used to convert electrical energy into mechanical movement.
- A relay is an electromagnetic switch with a coil and one or more sets of contacts.
- There are numerous types of relays. Among them are voltage relays, current relays, lock-up relays, overcurrent relays, and reed relays.
- Buzzers and choppers are electromechanical switches that repeatedly energize and de-energize.
- Magnetism is employed by rotating equipment and by computer storage devices.

KEY TERMS _____

Each of the following terms has been used in this chapter. Do you know their meanings?

armature
buzzer
chopper
choper amplifier
circuit breaker
coil current
coil voltage
contact current rating
contact voltage drop
contact voltage rating
contactor
contacts
current relays
declination
diamagnetic material
domain theory
drop-out voltage
electromagnet
electromagnetic induction
electromagnetism
electromechanical device
eutectic alloy
ferromagnetic material
gauss (G)
gilbert (Gb)
heater

holding current
hysteresis
hysteresis loop
induced magnetism
law of magnetism
left-hand rule for coils
left-hand rule for conductors
lock-up relays
lodestone
magnetic field
magnetic field intensity *(H)*
magnetic field strength
magnetic flux density *(B)*
magnetic flux (Φ)
magnetic induction
magnetic lines of force
magnetic motor starter
magnetic north
magnetizing force
magnetomotive force (mmf)
maintained switch
maxwell (Mx)
momentary switch
normally closed (NC) contacts
normally open (NO) contacts
Ohm's law for magnetic circuits

oersteds (Oe)
overcurrent relay
paramagnetic material
permanent magnet
permeability (μ)
permeability of free space (μ_0)
pick-up voltage
reed relay
relative permeability (μ_r)
relay
reluctance $(\mathcal{R})$
residual magnetism
retentivity
saturation
shielding
solenoid
solenoid valve
surge current
tesla (T)
thermal overload relay
toroidal coil
true north
variation
vibrator power supply
voltage relays
weber (W)

TEST YOUR KNOWLEDGE _____

Please do not write in this text. Place your answers on a separate sheet of paper.

1. Invisible magnetic lines of force are called a magnetic field, or _____.
2. The angular difference between magnetic north and true north is called the _____.
3. State the law of magnetism.
4. _____ is the measure of the ease with which magnetic flux may be established.
5. If a crane used a large electromagnet to pick up junk, should the core of the electromagnet have a high retentivity?
6. A meter can be protected from a magnetic field by installing _____.
7. Direction of a magnetic field around a current carrying conductor is determined by the _____-_____ _____.
8. Indicate the north and south poles of these electromagnets.

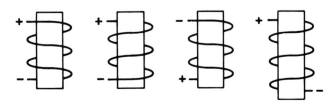

9. _____ is the ability of a material to hold its magnetism after the material has been removed from a magnetizing force.
10. Which of the following metals are paramagnetic materials?
    a. Aluminum.
    b. Silver.
    c. Chromium.
    d. Mercury.
    e. Copper.
    f. Platinum.
11. Which of the following metals are diamagnetic materials?
    a. Aluminum.
    b. Silver.
    c. Chromium.
    d. Mercury.
    e. Copper.
    f. Platinum.
12. Which theory of the exact nature of magnetism suggests that each electron in an atom is spinning on its own axis, as well as orbiting around its nucleus?
    a. Domain theory.
    b. Weber's molecular theory.
13. Magnetization of a nail while it is in the flux field of a magnet is called _____ _____.
14. A coil of wire without a core and connected to an energy source is called a solenoid. True or False?
15. A coil of wire wound on a core of magnetic materials is a/an _____.
16. _____ is a term used to describe the number of flux lines per unit cross-sectional area.
17. A coil of 600 turns has a current of 0.5A. What is the magnetomotive force in gilberts?
18. Draw a circuit to remotely operate an alarm system.
19. State several advantages of relays over manual switching.
20. A reed relay is an electromagnetic _____.
21. Give the magnetizing force and magnetic flux of an electromagnet at a current of 0.5A. The 1000-turn coil measures 16 centimeters in length and has a cross-sectional area of 2 cm². The permeability of the core is 2.51×10^{-4} W/A·m.

Applications of Electronics Technology:
Cellular telephones are becoming increasing common. This portable cellular telephone weighs only 10 oz. and provides approximately 70 minutes of talk time. (AT&T)

Chapter 6

GENERATING ELECTRICITY

In Chapter 5, you learned how magnetism is used to convert mechanical energy to electrical energy. You saw how a conductor moving through a magnetic field produced a potential source of energy. This knowledge will now serve to open the door to many useful applications.

After studying this chapter, you will be able to:
☐ *Cite the nature and properties of alternating current.*
☐ *Describe how a generator works to induce a voltage.*
☐ *Explain the methods used to control a generator's voltage and current output.*
☐ *Discuss how a generator may produce a dc output.*

INTRODUCTION

During the early part of the 19th century, English scientist Sir Humphrey Davy was asked, "What do you consider your greatest discovery?" His simple and forthright reply was, "Michael Faraday!"

Faraday made significant contributions to the field of electricity. You may recall that he is credited with the discovery of electromagnetic induction. He also invented the electric **generator**, or **dynamo**, the machine that produces electricity from mechanical energy. Indeed, he deserves at least some credit for the modern-day *wind turbines* of Figure 6-1. In his honor, Faraday has two units of measurement named after him—the faraday, a unit of electrical charge, and the farad, a unit of capacitance. Faraday deserves the credit for making it possible to have the electric power we use today at home and in industry. Most certainly, all nations depend upon this power to turn the wheels of industry and commerce, Figure 6-2.

A   B   C

Figure 6-1. Wind turbines, designed to generate electricity. The wind drives a propeller or turbine, which drives a generator, which produces electric current. A—Horizontal-axis wind turbines. B—Vertical-axis wind turbines. C—Close-up of a vertical-axis wind turbine. (Parker Hannifin Corp., U.S. Department of Energy)

Figure 6-2. *Byron Station nuclear generating facility. This nuclear power plant, located in northern Illinois, generates electricity for residential and commercial use. Total capacity is 2240 megawatts. Nuclear reactors, which produce steam to drive turbine generators, are seen in background. Large structures seen in foreground (right) are cooling towers. (Commonwealth Edison)*

ELECTROMAGNETIC INDUCTION

Faraday discovered electromagnetic induction. He found that if a conductor was moved through a magnetic field, a voltage was induced in the conductor. More specifically: *A voltage is induced in a conductor in a magnetic field if there is relative motion between the field and the conductor.* The conductor may be moved, or the magnetic field may be moved. Also, if the conductor forms a complete circuit, a *current* is induced as well.

DIRECTION OF INDUCED CURRENT

In Figure 6-3, a coil is connected to a **galvanometer**, an instrument for measuring very small values of electric current. The meter shown will indicate a current in either direction. A permanent magnet pushed into the hollow coil will cause the meter to deflect in one direction. When the magnet is pulled out, the meter will deflect in the opposite direction. When there is no motion, there is no current. Without motion, the meter will not deflect. If you hold the magnet in a fixed position and move the coil up and down, the same effect is observed; a current is produced first in one direction, then in the other. Relative motion between the coil and the magnetic field must always be present.

We have learned that the direction of current in a conductor moving in a magnetic field depends upon the direction of movement. Figure 6-4A illustrates a single conductor moving *down* through a magnetic field at a right angle. The dot in the center of the conductor (front view) means that electron flow in the conductor is toward you. In Figure 6-4B, the conductor is moving up through the field. The induced current is away from you. This is indicated by the cross on the end of the conductor.

The **left-hand rule for generators** may be used to determine the direction of electron flow. Extend your thumb and first two fingers of your left hand at right angles to one another. See Figure 6-5. Point your thumb in the direction that the conductor is moving. Point your first finger in the direction of the magnetic field (toward the south pole). Your second, or middle, finger will point in the direction of electron flow.

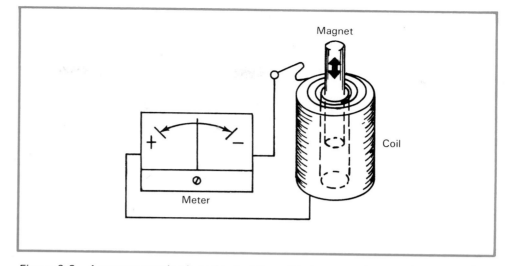

Figure 6-3. A magnet moving in and out of the coil induces a voltage and current, which will cause the meter to deflect.

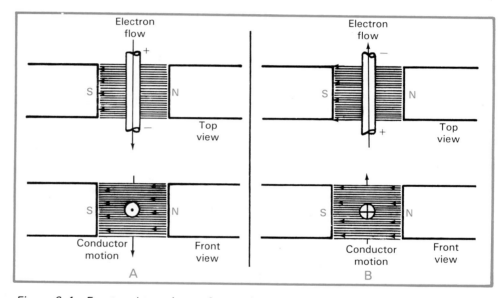

Figure 6-4. Front and top views of a conductor moving through a magnetic field. A—Conductor moving downward through field causes current to flow outward. B—Conductor moving upward through field causes current to flow inward.

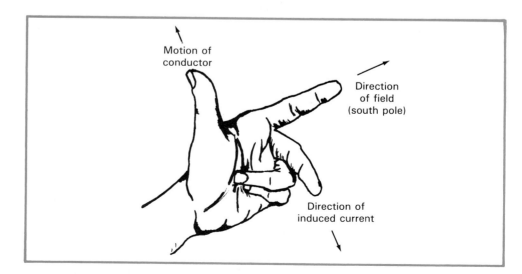

Figure 6-5. Left-hand rule for generators.

POLARITY OF INDUCED VOLTAGE

The previous discussion on direction of current assumed a closed circuit. This is because in order to have current, there must be a complete circuit. However, you should understand that the relative motion of a wire and magnetic field *always* induces a voltage. When the wire is alternately moved in one direction and then in the other, the polarity of the induced emf alternates with each movement.

Looking back to Figure 6-4, voltage polarity is indicated in the top views with the + and − signs. Since we have induced a voltage, we can call what we have a voltage *source*. Electron flow *inside* a voltage source is positive to negative. We can use our left-hand rule to determine polarity. Our middle finger will point to the negative terminal.

AMOUNT OF INDUCED VOLTAGE

How much of a voltage is induced when a conductor is moved through a magnetic field? This will depend upon four factors:

- The *speed* at which the conductor cuts the magnetic field. If the conductor cuts the field at a greater rate of speed, the induced force on the electrons within the conductor is greater. Consequently, a greater voltage is developed.
- The *amount of magnetic flux.* A denser field, or a greater flux density, means more magnetic lines for the conductor to cut. If the conductor is a rotating loop, the flux will increase, too, as the area in the plane of the loop is increased. Generated output voltage is in direct proportion to magnetic flux. If a conductor cuts through a field of 1 weber in 1 second, a voltage of 1 volt is induced. If the magnetic flux were to increase, so would the voltage.
- The *number of turns* of the conductor cutting the field. If a certain voltage is developed in one conductor as it cuts across a field, then more conductors would add substantially to the total generated voltage. Ten conductors would produce ten times more, and so on.
- The *angle* at which the conductor cuts across the field. No voltage is induced in a conductor as it moves parallel to the magnetic field. Maximum voltage is induced when the conductor cuts directly across at right angles to the flux lines. At cutting angles between 0° and 90°, a voltage between zero and maximum will be induced. See Figure 6-6.

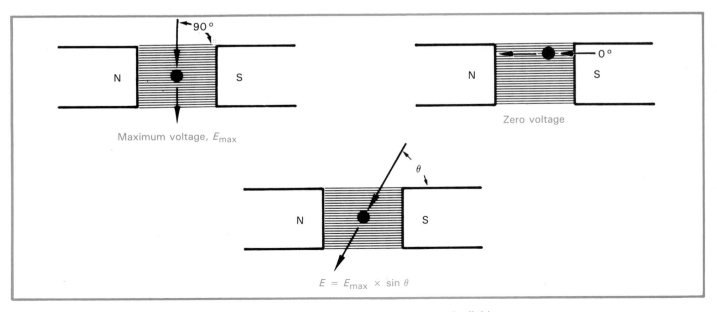

$$E = E_{max} \times \sin \theta$$

Figure 6-6. The voltage induced in a conductor as it moves across a magnetic field depends upon the cutting angle.

Moving conductors

In a conductor moving through a uniform magnetic field, induced voltage *e*, in volts, is determined by the formula:

$$e = Blv \sin \theta$$

where:

B = magnetic flux density in teslas
l = length of conductor in meters
v = velocity of conductor in meters per second
θ (theta) = angle between velocity vector and magnetic field

Problem:

A 15-centimeter conductor moves through a magnetic field. It moves 80 centimeters per second at an angle of 60° relative to the field. The flux density of the field is 0.1 tesla. What emf is induced in the conductor?

Solution:

Using the formula for emf induced in a moving conductor:

$$
\begin{aligned}
e = Blv \sin \theta &= (0.1 \text{ T})(0.15 \text{ m})(0.8 \text{ m/s}) \sin 60° \\
&= (0.1 \text{ Wb/m}^2)(0.15 \text{ m})(0.8 \text{ m/s}) \sin 60° \\
&= (0.012 \text{ Wb/s})(0.866) \\
&= 1.04 \times 10^{-2} \text{ Wb/s} \\
&= 1.04 \times 10^{-2} \text{ V}
\end{aligned}
$$

To justify units:

$$\frac{\text{weber}}{\text{second}} = \frac{\text{newton-meter/ampere}}{\text{second}} = \frac{\text{newton-meter/(coulomb/second)}}{\text{second}}$$

$$= \frac{\text{newton-meter}}{\text{coulomb}} = \frac{\text{joule}}{\text{coulomb}} = \text{volts}$$

Rotating loops

In a rectangular loop rotating through a uniform magnetic field, induced voltage *e*, in volts, is determined by the formula:

$$e = E_0 \sin \theta$$

where *e* is instantaneous voltage at a time *t*, E_0 is *amplitude*, or maximum voltage E_{max}, and θ is the angle of the rotating loop relative to the field at time *t*. (At $\theta = 0°$, the *plane* of the loop will be perpendicular to the field.)

The angle θ in *radian measure* is equal to the product of the loop's *angular velocity*, ω (omega), and time *t*. In other words:

$$\theta = \omega t$$

Angular velocity (ω), or **radian frequency** as it is sometimes called, is the rate at which an angle changes. Expressed in radians per second, it is equal to the *frequency* of rotation, or *revolutions per second*, multiplied by 2π. In other words:

$$\omega = 2\pi f$$

where *f* is the frequency of rotation. Now we can write our formula for induced voltage in a rotating loop as follows:

$$e = E_0 \sin \theta = E_0 \sin \omega t = E_0 \sin (2\pi f)t$$

You might be wondering where E_0 comes from and what it is equal to. Go back to our formula for induced voltage in a moving conductor, $e = Blv \sin \theta$. By substitution of quantities, it is determined that:

$$E_0 = \omega NBlw$$

where l is the length, and w is the width of the rectangular loop. The factor N has been included to apply to a loop having N turns rather than a single turn of wire. (The derivation of this formula is left as an exercise for the student.)

Problem:

A rectangular coil 10 centimeters long and 8 centimeters wide is wound with 200 turns of wire. It rotates 30 times per second in a uniform magnetic field of 0.1 tesla. What does the instantaneous voltage equal at 5/16 of a second?

Solution:

Step 1. Calculate ω for the rotating coil.

$$
\begin{aligned}
\omega &= 2\pi f \\
&= (2\pi \text{ rad/rev})(30 \text{ rev/s}) \\
&= 188.5 \text{ rad/s} \\
&= 188.5 \text{ s}^{-1}
\end{aligned}
$$

Step 2. Calculate voltage amplitude.

$$
\begin{aligned}
E_0 &= \omega NBlw \\
&= (188.5 \text{ s}^{-1})(200)(0.1 \text{ T})(0.1 \text{ m})(0.08 \text{ m}) \\
&= 30.16 \text{ V}
\end{aligned}
$$

Step 3. Calculate induced voltage at $t = 5/16$ second.

$$
\begin{aligned}
e &= E_0 \sin \omega t \\
&= (30.16 \text{ V}) [\sin (188.5 \text{ s}^{-1})(0.3125 \text{ s})] \\
&= (30.16 \text{ V})(0.706) \\
&= 21.29 \text{ V}
\end{aligned}
$$

Note that the angle (ωt) is given in radian measure. Therefore, a calculator must be in the *radian* mode to find the correct sine of the angle.

Dc generators

Some generators produce a dc output. These will be discussed in this chapter. There is a useful formula for dc generators having multiple windings. It is derived from the formula of instantaneous voltage in rotating loops. This useful formula gives total dc voltage appearing at the generator output under no-load conditions. It is:

$$E_O = \frac{Zn\Phi}{60} = K_G \Phi n$$

where:

E_O = no-load dc voltage at output
Z = total number of rotating conductors
n = speed of rotation in revolutions per minute
Φ = flux per pole in webers
K_G = winding constant for a given generator

Problem:

The armature of a six-pole dc generator has 45 slots. At no load, the generator rotates at 800 rpm. Each coil has 8 turns. The flux per pole is 0.03 W. What is the value of the induced voltage at no load?

Solution:

Step 1. Determine the total number of conductors. Each turn has two active conductors, and 45 coils are required to fill 45 slots. Therefore, the total number of conductors is:

$$Z = (45 \text{ coils})\left(\frac{8 \text{ turns}}{\text{coil}}\right)\left(\frac{2 \text{ conductors}}{\text{turn}}\right) = 720 \text{ conductors}$$

Step 2. Calculate induced voltage at no load.

$$E_O = \frac{Zn\Phi}{60} = \frac{(720)(800)(0.03)}{60} = 288 \text{ V}$$

Note that the induced voltage is the generator terminal **no-load voltage**. It is specified this way because with no load (open circuit), load, or circuit, current is zero. With no load current, voltage drop caused by the internal resistance, r_i, of the generator is zero. In this case, full generated voltage is available across the generator output terminals.

LENZ'S LAW

The beginning student sometimes feels that a generator *creates* energy in the form of electricity. This is not true. The **law of conservation of energy** states that energy can be transformed from one form to another, but the total amount remains constant. From this it follows that a generator does not create electrical energy, but rather *converts* it from mechanical energy.

In 1834, Russian physicist Heinrich Lenz formulated a law based on the energy conservation principle. **Lenz's law**, as it is called, states:

An induced emf always generates a current whose magnetic field opposes the change in the existing magnetic field.

In other words, the induced field around a moving conductor is opposed by the existing field.

Consider what would happen if this were not the case. What if induced current would produce a flux in the *same* direction as the existing field? A *greater* change in flux would result. This would produce an even larger current followed by an even greater change in flux, and so on. The current would continue to grow indefinitely. Even after the initial impulse was removed, power would continue to be produced. This would amount to perpetual motion, which would violate the conservation of energy. Lenz's law as stated, however, does not.

In generators, a *counter torque* is developed because of the opposing fields. Turning the *armature*, or rotating loop, of a generator takes little effort when the generator is not connected to an external circuit. This is because there is no current. However, if the generator is connected to a current-drawing device, a current will flow in the armature coils. Since this coil is in a magnetic field, there will be a torque exerted on it that opposes its motion. This is the **counter torque**. The greater the load—that is, the more current drawn—the greater the counter torque. Hence, the *external* applied torque will have to be greater to keep the generator turning. This is the reason for the huge turbine generators we see at our hydroelectric, fossil fuel, and nuclear power plants.

THE SIMPLE GENERATOR

Now that you have acquired some background information, we are ready to construct a simple ac generator (**alternator**). In Figure 6-7, the single moving conductor is replaced by a single, rotating loop (armature). The rotation is due to an external driving force (not shown). Note that as one side of the coil, or loop, moves upward through the field, the other side moves downward. External connections to the armature are made through **slip rings** and **brushes**.

In Figure 6-7, when the plane of the loop is parallel to the magnetic field, the conductors of the loop are cutting the field at right angles. Maximum voltage is induced at this point. Current flows out from side A of the rotating loop and across the load in the direction shown. In one quarter of a revolution, the conductors will be moving parallel to the field. At this point, no voltage will be induced. In another quarter turn, the loop will be cutting the field at 90° again. Now, however, side A will be moving upward and side B, downward through the field. As a result, the induced voltage will be opposite in polarity with respect to sides A and B, and the current will be in the opposite direction. It will now flow out from side B, and thus, across the load resistor in the opposite direction. In Figure 6-8, the armature is illustrated in four positions. Output waves show magnitude and polarity of the induced voltage.

DEVELOPMENT OF A SINE WAVE

Generally, the rise and fall and polarity of an induced voltage is represented by means of a *rotating vector*, or *phasor*. **Vectors** and **phasors** are quantities that have both magnitude *and* direction, requiring an angle to give the value completely. However, a vector quantity has direction in space, while a phasor quantity varies in time. A mechanical force is an example of a vector quantity. An ac current or voltage is an example of a phasor quantity. Vector and phasor quantities are represented by directed line segments. The magnitude is given by line length. The angle of the arrow with respect to a horizontal reference gives the direction.

In Figure 6-9, the phasor represents a quantity of 10 V. Using a scale of 1/4″ equals 1 V, it should be 2 1/2″ long. By convention, the phasor is rotated counterclockwise. To the right of the diagram is a graph. Instantaneous voltage is plotted against time for one revolution. From before, the voltage at any point is a function of the sine of the angle of rotation ($e = E_{max} \sin \theta$). For example, at 30°, the voltage in Figure 6-9 would be 5 V (10 V × sin 30° = 5 V).

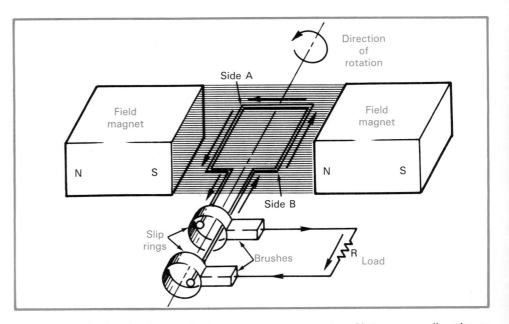

Figure 6-7. A sketch of a simple alternator, or ac generator. Note current direction as indicated by arrows.

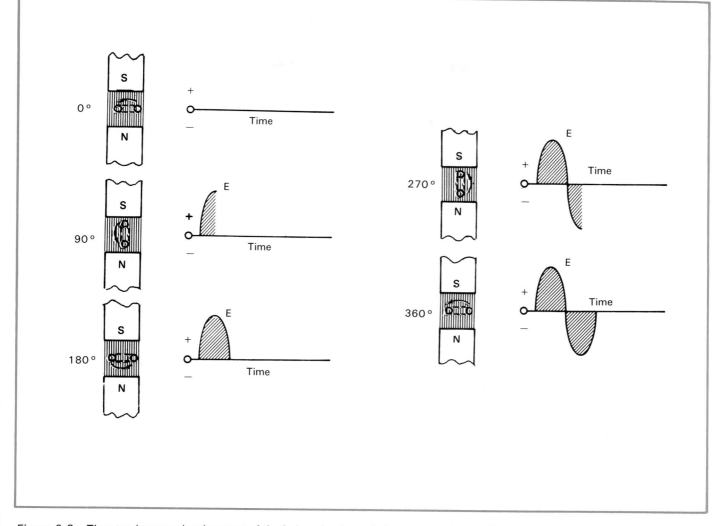

Figure 6-8. The step-by-step development of the induced voltage during one revolution of an armature in a magnetic field. The alternating current waveform produced is a sine wave.

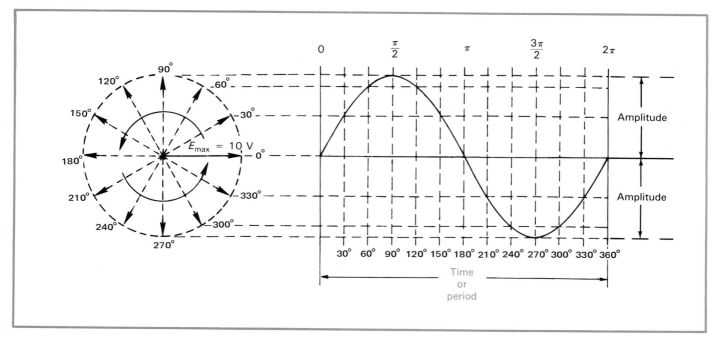

Figure 6-9. Development of a sine wave.

Waveform characteristics

We have now constructed a graph of an *alternating current* **sine wave**. Ac differs from dc in that ac is continually changing amplitude and polarity. Some appropriate definitions will help to describe the ac wave.

Amplitude. The *magnitude* of variation in a sine wave measured from the zero line to the maximum (positive or negative) instantaneous value. Refer back to Figure 6-9.

Cycle. Starting at zero, the wave rises to maximum at one polarity, drops to zero, rises to maximum in the opposite polarity, then returns to zero. One cycle has occurred. These cycles may continue to repeat and would represent the output of a simple ac generator. One cycle represents on complete rotation of an armature—that is, 2π radians, or 360°.

Period. The time required for one complete cycle of a sine wave is said to be the *period (T)* of the ac wave. Each period is the same as all other periods for a given sine wave.

Frequency. The number of cycles per unit of time is termed the *frequency (f)* of the ac wave. The standard unit of time is the second. Originally, frequency was measured in cycles per second (cps). In 1967, a new and shorter unit of measurement called the **hertz (Hz)** was adopted. The term honors Heinrich Hertz, the German scientist who demonstrated the existence of electromagnetic waves. His work on the electromagnetic nature of light is perhaps the greatest achievement of 19th century science. Radio waves are called Hertzian waves in his honor. Conversion of hertz to cycles per second is as follows:

$$1 \text{ hertz (Hz)} = 1 \text{ cycle per second (cps)} = 1 \text{ s}^{-1}$$
$$1 \text{ kilohertz (kHz)} = 1000 \text{ cps} = 1000 \text{ s}^{-1}$$
$$1 \text{ megahertz (MHz)} = 1{,}000{,}000 \text{ cps} = 1{,}000{,}000 \text{ s}^{-1}$$

Frequency and period are related to each other. Mathematically stated, the frequency of a wave is the reciprocal of its period. In other words:

$$f = \frac{1}{T} \quad \text{or} \quad T = \frac{1}{f}$$

Consider the following examples:

- The period of a certain sine wave is 20 milliseconds (20 ms). The frequency is 50 Hz.

$$f = \frac{1}{T} = \frac{1}{20 \text{ ms}} = \frac{1}{20 \times 10^{-3} \text{ s}} = 50 \text{ Hz}$$

- A certain sine wave has a frequency of 100 Hz. It has a period of 0.01 seconds.

$$T = \frac{1}{f} = \frac{1}{100 \text{ Hz}} = 0.01 \text{ s}$$

- A certain sine wave has a frequency of 1 kHz. It has a period of 1 ms.

$$T = \frac{1}{f} = \frac{1}{1 \times 10^3 \text{ Hz}} = 1 \times 10^{-3} \text{ s} = 1 \text{ ms}$$

Angular velocity. Angular velocity (ω) is often used in electronic computations. As discussed, it is another way of expressing the speed of rotation. It is the time rate of change of angular displacement expressed in radians per second. **A radian** is an angle of about 57.3°. In one complete revolution of a phasor, the angular displacement will be 2π radians, or 360°. (The factor π has a constant value of 3.1416.) Assuming a frequency of 100 Hz, an ac phasor would rotate 100 revolutions per second. If there is an angular displacement of 2π radians for each revolution, then the phasor velocity would be:

$$\omega = 2\pi f = (2\pi \text{ rad/rev})(100 \text{ rev/s}) = 628.3 \text{ rad/s}$$

Wavelength. The frequency of a wave may also be designated by its length in meters. The length of the wave is measured from the peak of one cycle to the corresponding peak of the next cycle. See Figure 6-10.

The Greek letter λ (lambda) has been assigned to wavelength. Wavelength is determined by the velocity of transmission v and the frequency of the wave:

$$\lambda = \frac{v}{f}$$

For electromagnetic radio waves in air, the velocity is equal to the speed of light: 3×10^8 m/s, or 186,000 mi./sec. For sound waves (such as from a loudspeaker) in air under average conditions, the velocity is equal to 344.4 m/s, or 1130 ft./sec.

For radio waves, λ may be equated as:

$$\lambda \text{ (in meters)} = \frac{3 \times 10^8}{f \text{ (in Hz)}}$$

$$= \frac{3 \times 10^5}{f \text{ (in kHz)}}$$

$$= \frac{3 \times 10^2}{f \text{ (in MHz)}}$$

$$\lambda \text{ (in feet)} = \frac{984}{f \text{ (in MHz)}}$$

If the wavelength is known, the frequency of a radio wave may be found by transposing the equations to read:

$$f \text{ (in Hz)} = \frac{3 \times 10^8}{\lambda \text{ (in meters)}}$$

$$f \text{ (in kHz)} = \frac{3 \times 10^5}{\lambda \text{ (in meters)}}$$

$$f \text{ (in MHz)} = \frac{3 \times 10^2}{\lambda \text{ (in meters)}}$$

$$f \text{ (in MHz)} = \frac{984}{\lambda \text{ (in feet)}}$$

Problem:

What is the wavelength of a 3-MHz radio wave?

Solution:

$$\lambda = \frac{3 \times 10^2}{3} = 100 \text{ m}$$

Problem:

What is the frequency of a 15-meter radio wave?

Solution:

$$f = \frac{3 \times 10^2}{15} = 20 \text{ MHz}$$

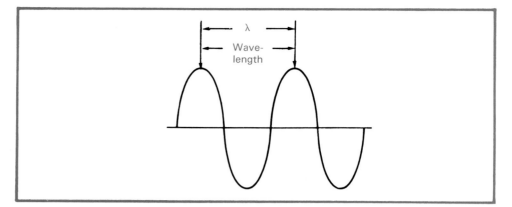

Figure 6-10. Wavelength is measured from one peak to the corresponding peak of the next cycle.

Sine wave voltage and current values

We saw that current in a circuit having an alternating voltage source changes its direction as the polarity of the voltage changes. Current, too, produces a sine wave. Both voltage and current values depend on the amplitude of its wave. These values can be described in terms of:

• Instantaneous value.
• Peak value.
• Peak-to-peak value.
• Average value.
• Effective, or rms, value.

Instantaneous value. These values vary with time. They change from one instant to the next. Instantaneous values are expressed as lowercase letters — e or v for voltage, i for current. We discussed how to calculate e earlier in the chapter.

Peak value. The peak value of the ac wave is measured from the zero line to the positive and negative maximum points. The positive peak value occurs at the positive maximum; the negative peak value, at the negative. In this book, peak voltage is denoted E_{peak}, E_{max}, or E_0. Peak current is denoted I_{peak}, or I_{max}. The peak value is an instantaneous value, like any other value on the sine wave. However, in general, we use capital letters to denote a peak value. This is because capital letters are used for constant values, and, for a given sine wave, the peak value is a constant.

Peak-to-peak value. This is the value of the wave from the maximum positive peak, to the maximum negative peak. It is always twice the peak value. In this book, peak-to-peak voltage is denoted E_{p-p}. Peak-to-peak current is denoted I_{p-p}.

Average value. One ac cycle is made up of two half-cycles. If a sine wave was averaged for one complete cycle, the average value would be zero. This is because the positive half-cycle would cancel the negative. Therefore, for the average value to be useful, it is defined over a half-cycle. The average value is found by adding up all instantaneous values in a half-cycle and dividing by the number of values. The exact value is determined using calculus. It turns out to be $2/\pi$ times, or 63.7 percent of, the peak value.

A meter with a moving coil cannot keep up with changes in instantaneous voltage. It will average out the fluctuations. A meter, therefore, will measure average value. However, the ac signal must first be *rectified* so that the negative half-cycle is changed to positive polarity, Figure 6-11. If not, the average value would always be zero. In this book, average voltage is denoted E_{avg}. Average current is denoted I_{avg}.

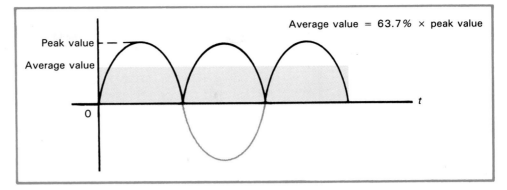

Figure 6-11. Graph of a full-wave rectified signal and its average value.

The following formulas may be used to find unknown values:

$$E_{avg} = 0.637 \times E_{peak} \qquad I_{avg} = 0.637 \times I_{peak}$$

$$E_{peak} = 1.57 \times E_{avg} \qquad I_{peak} = 1.57 \times I_{avg}$$

Root-mean-square (rms), or effective, value. The rms value of a sine wave is a direct comparison of ac to dc. It corresponds to the same amount of direct current or voltage in heating power. For instance, an alternating voltage of 120 V_{rms} has the same heating effect upon the filament of a light bulb as 120 V from a steady dc source. Most meters you will use, although average responding, are calibrated for rms. Further, when working with ac, you can assume a voltage or current is rms unless it is stated otherwise. In this book, effective voltage is denoted E_{rms}, or E_{eff}. Effective current is denoted I_{rms}, or I_{eff}.

You may be wondering about the origin of the term *root-mean-square*. It is a mathematical process that involves *squaring* each instantaneous value in one half- (or full) cycle of a sine wave. The *mean*, or average, of these values is then found. Finally, the square *root* is taken of the average squared value. Through calculus, the root-mean-square value is found to be 0.707 times the peak value of a sine wave. Voltage and current rms values are useful in relationships for power in ac circuits. This will be discussed shortly. For now, let us look at some of the relationships between effective and other voltage and current values:

$$E_{rms} = 0.707 \times E_{peak} \qquad I_{rms} = 0.707 \times I_{peak}$$

$$E_{peak} = 1.414 \times E_{rms} \qquad I_{peak} = 1.414 \times I_{rms}$$

Note that $1.414 = \sqrt{2}$ and $0.707 = 1/1.414 = 1/\sqrt{2}$. By substitution, these formulas may be rewritten as:

$$E_{rms} = \frac{E_{peak}}{\sqrt{2}} \qquad I_{rms} = \frac{I_{peak}}{\sqrt{2}}$$

$$E_{peak} = \sqrt{2} \times E_{rms} \qquad I_{peak} = \sqrt{2} \times I_{rms}$$

We can go a step further and change the formulas, using a peak-to-peak value for the sine wave. We have:

$$E_{peak} = \frac{E_{p\text{-}p}}{2} \qquad I_{peak} = \frac{I_{p\text{-}p}}{2}$$

Therefore:

$$E_{rms} = \frac{E_{p\text{-}p}}{2\sqrt{2}} \qquad I_{rms} = \frac{I_{p\text{-}p}}{2\sqrt{2}}$$

Sine wave power values

Instantaneous power p in an ac circuit is:

$$p = i \times e$$

where i is instantaneous current, and e is instantaneous voltage. Based on values of i and e at every instant, we can plot the graph of power. See Figure 6-12. Notice that the power wave is always positive. This is because the product of i and e is positive when they are both positive or both negative. This shows that energy, while not a constant rate, is always supplied to the resistor. It does not matter in what direction the current is.

Instantaneous power from a practical viewpoint is not very useful. What is of more use is *average* power of the sine wave. As is easily seen, the power curve is symmetrical about a value equal to one-half its maximum ordinate, which is I_{peak} times E_{peak}. When current and voltage are *in phase*, we can write average power P as:

$$P = \frac{1}{2} I_{peak} E_{peak}$$

Substituting quantities for I_{peak} and E_{peak}:

$$P = \frac{1}{2} (I_{rms}\sqrt{2})(E_{rms}\sqrt{2})$$

This allows us to write average power in terms of rms values:

$$P = I_{rms} \times E_{rms}$$
$$P = I_{rms}^2 \times R$$
$$P = E_{rms}^2/R$$

(Average power when current and voltage are out of phase is discussed in Chapter 9. Refer to the discussion on *true power*.)

Problem:

A 10-ohm resistance is connected to an ac source. Peak voltage is 10 V. What power is consumed in the resistance? With what dc voltage would the resistor consume the same amount of power?

Solution:

Step 1. Find the power consumed in the resistance.

$$P = \frac{1}{2} I_{peak} E_{peak} = \frac{1}{2} \times \frac{E_{peak}^2}{R} = \frac{(10 \text{ V})^2}{2(10 \text{ }\Omega)} = 5 \text{ W}$$

Step 2. Find the rms voltage. A dc voltage of the same value would consume the same amount of power.

$$E_{rms} = 0.707 \times E_{peak} = 0.707 \times 10 \text{ V} = 7.07 \text{ V}$$

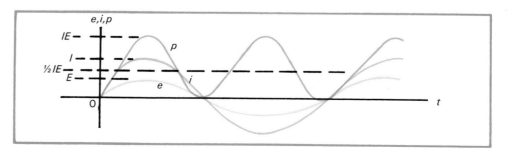

Figure 6-12. Instantaneous power through a resistor ($p = i \times e$). Average power is $1/2(I \times E)$.

SINGLE-PHASE AND POLYPHASE GENERATORS

From earlier discussion, output of a generator can be increased. One way is by increasing the magnetic field of the generator. In earlier examples, only permanent magnets of limited strength were used. These field magnets can be made of electromagnets. They can either be excited, then, by a part of the generator output (**self-excited generator**) or by a separate power source (**separately excited generator**).

Output may also be increased by increasing the number of conductors moving through the field. This is easily done by winding the armature with many turns of relatively small wire and connecting the coils to appropriate slip rings.

In addition, a generator can be made to have an output of more than one *phase*. In other words, it would produce more than one sine wave (of like frequency), and these would be **out of phase**. This means the waves would cross the zero line at different times. The amount of variation would be given by the **phase angle**. This would be the angular difference between them. (Sine waves having a phase angle of 0° are said to be **in phase**.)

SINGLE-PHASE GENERATORS

Although not commonly used, the **single-phase generator** produces a voltage output of a single sine wave. The output is just like that of the generator described in Figure 6-7. However, a more complex and efficient generator is used.

TWO-PHASE GENERATORS

By placing two separate windings on the armature of the generator and orienting the coils at an angle of 90°, two separate outputs can be obtained. In other words, a **two-phase generator** is made. See Figure 6-13. Coil A produces a sine wave output, and coil B produces a sine wave output, but the two waves are 90° out of phase. Study Figure 6-14.

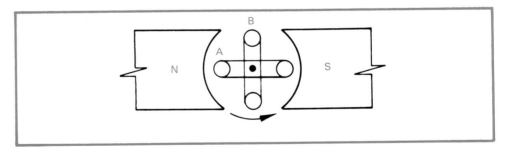

Figure 6-13. Separate out-of-phase voltages are produced by each of the coils 90° apart.

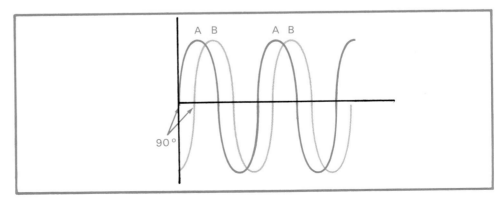

Figure 6-14. Two-phase output sine waves.

A single-phase current may be obtained from either of the two windings. Otherwise, the two windings may be connected in series to produce a peak output voltage of 1.414 times the peak voltage of a single coil.

THREE-PHASE GENERATORS

The most practical and widely used generator is the **three-phase generator**. Three coils are placed on the armature. The angle between each coil is 120°, Figure 6-15. Each coil produces its own sine wave voltage. Three separate out-of-phase voltages are generated. These can be used singularly or together. If the three phases are used together, the windings may be *delta* connected or *wye* connected. A delta connection offers only one voltage output. A wye connection, however, offers two. You can get the coil voltage or 1.73 times the coil voltage depending on the method of connection. Delta and wye connections will be discussed in more detail shortly.

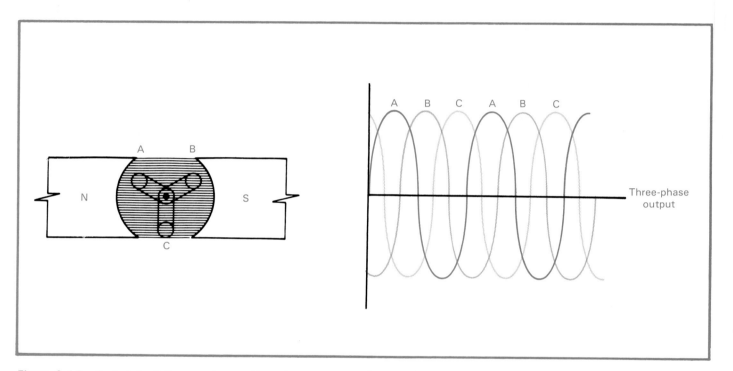

Figure 6-15. A sketch of the armature coils and output wave forms of a three-phase generator.

ALTERNATORS

Alternator is another word for ac generator. There are two types. One is the **rotating-armature alternator.** This type has an armature that rotates inside a stationary field. These have a distinct disadvantage. High voltages and currents are produced in the rotating armature. It must be connected to the external circuit by some form of sliding contacts. Slip rings and brushes are used. Sparking and burning can result. More mechanical problems can be expected.

We have talked about armatures throughout this chapter; however, we have never defined the term. The **armature** of a generator is the main current-carrying winding in which an emf is induced. Until now, armatures we have discussed have been rotating. However, this is not true of all armatures. In fact, in the ac generator, it is better to have a fixed, rather than a rotating, armature. The **rotating-field alternator** serves this purpose.

The fixed armature of the rotating-field alternator is called the **stator**. The ac voltage is induced in fixed coils. These are called **stator windings**. The magnetic field is rotated inside the stator. The rotating device is called a **rotor**. The magnetic field is created when dc is supplied to the **rotor windings**. Slip rings and brushes are used to conduct dc to the rotor windings. Figures 6-16 and 6-17 show the stator and rotor of a typical alternator. Figure 6-18 is a typical alternator used in automotive electrical systems.

As mentioned, the rotor coils are excited by an external dc source to produce a magnetic field. See Figure 6-19. As the rotor passes each of the three stator poles, a voltage is induced. It is apparent that the three voltages of stator coils A, B, and C are 120° out of phase with each other.

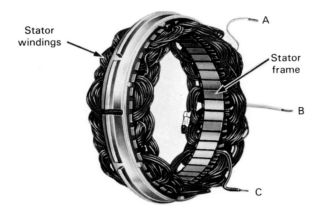

Figure 6-16. The stator coils of a typical automotive alternator. (Delco-Remy Div., General Motors Corp.)

Figure 6-17. The rotor of a typical alternator. (Delco-Remy Div., General Motors Corp.)

WYE VERSUS DELTA CONNECTIONS

The six output leads of Figure 6-19 may be connected so that they will not interfere with each other. In Figure 6-20, two methods are illustrated. In the **wye connection**, sometimes called the **star connection**, all coils are connected to a common point. This serves as the ground return for each phase. An advantage of this hookup is that the voltage between any two of the phase outputs is equal to 1.73 times the voltage from any one phase to neutral.

In the **delta connection**, a single-phase voltage may be taken from any two points of the circuit. With this setup, the line current increases to 1.73 times the current produced by a single phase. In summary, the wye connection gives increased voltage; the delta connection gives increased current.

Figure 6-18. *The complete alternator found on many cars. (Delco-Remy Div., General Motors Corp.)*

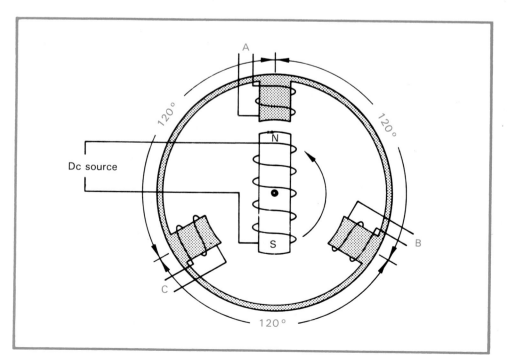

Figure 6-19. *A sketch showing the inside of a three-phase alternator.*

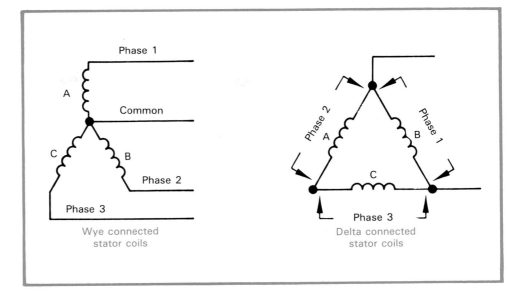

Figure 6-20. Wye and delta connections for stator coils.

ALTERNATOR REGULATION

Current output of an alternator is controlled automatically by the principles of induction. However, voltage developed by an alternator increases with the speed of the rotor. At high speeds, the voltage could increase enough to overcharge the car's battery and damage the accessories. For this reason, alternator **regulators** are used to hold the voltage output at a constant level. Basically, the regulator senses a change of voltage at the alternator output. It then produces a corresponding and corrective change in the current through the rotor. This, in turn, varies the magnetic field strength, which regulates the voltage.

There are different types of regulators. Regulators of the past were electromechanical. Presently, we have transistorized, Figure 6-21, and integral regulators. Integral regulators are essentially the same as transistorized except they are mounted inside the alternator. Now, in certain applications, voltage regulation has become a part of the engine computer control system.

We need to make one other point before we go on. That is, automotive electrical equipment is made to operate on dc. The battery, which is charged by the alternator, also requires dc. Further, dc output is needed for the alternator field coils. The ac developed by the alternator must therefore be *rectified* to dc. To do this, *diodes* are mounted in the alternator. You will study diodes in Chapter 12.

Figure 6-22 is a schematic of a transistorized voltage regulator combined with an electromechanical field relay. Look at the circuit to see how it operates. When ignition switch SW is closed, relay K_1 is closed. The alternator rectified voltage output is applied to the collector and base of transistor Q_1. The transistor is *forward biased* and conducts. It is in series with the field coils, and collector current is the field current.

Consider what happens if alternator voltage output should rise above a preset value. The voltage coil, then, in relay K_2 is sufficiently energized to open the points of K_2. This disconnects voltage to the base of the transistor. In turn, this cuts off the transistor and the collector current, reducing alternator field current. The output of the alternator is also reduced. The points of K_2 will reclose. These points vibrate many times per second and maintain the alternator voltage output at a fixed value. The diodes D_1 and D_2 prevent damage to the circuit that might be caused by transient spikes of voltage.

That is the basic operation. At this point in your studies of electronics, you may not entirely follow what is happening. Study of later chapters in this text on transistor action should help. You may want to return to this circuit at that time for further understanding.

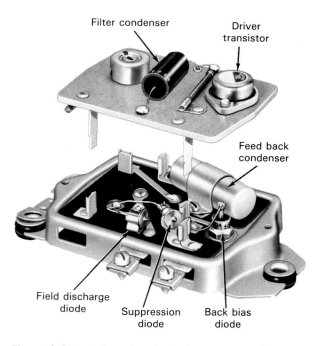

Figure 6-21. A transistorized alternator regulator. (Delco-Remy Div., General Motors Corp.)

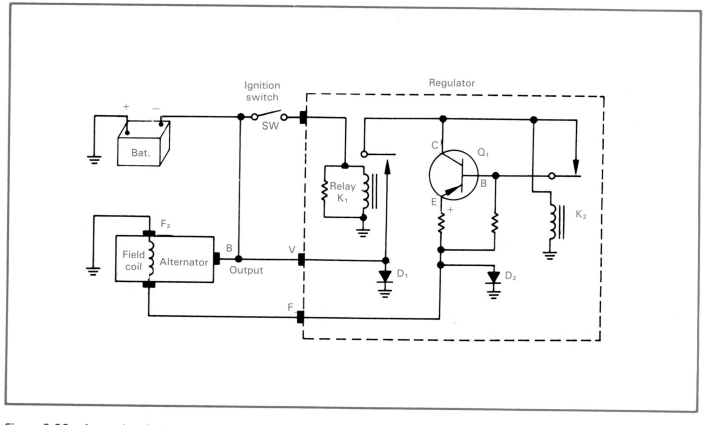

Figure 6-22. A transistorized regulator circuit for an automotive alternator.

DC GENERATORS

Basically, the principle of a dc generator is the same as the ac generator. Each develops alternating voltage in the armature. The major difference is a mechanical switching device on the dc generator called a **commutator**. It consists of two semi-cylindrical pieces that rotate with the armature. Brushes connect the commutator to the external circuit. The resulting action produces a pulsating direct current in the output, rather than ac. (It is a *direct* current because it does not change polarity.) This is *mechanical* rectification. Diodes in an alternator perform the same function electronically.

Now, follow dc generator action in Figure 6-23. In the top illustration, brush A is in contact with commutator section A. Brush B is in contact with commutator section B. Current in the armature is indicated by arrows. Current flows *in* from section A. It flows out through section B. As the armature turns 180°, or one-half revolution, Figure 6-23, bottom, the current reverses in the armature. It now flows in from section B and out through section A. This reverse current does not appear in the external circuit. That is because now section A contacts brush B, and section B contacts brush A. By the switching action of the commutator, the output current appears as shown in Figure 6-24.

Further improvements may be made on the dc generator by adding more coils to the armature. Each coil has its own commutator sections. The output becomes less pulsating and approaches a constant value dc. In Figure 6-25,

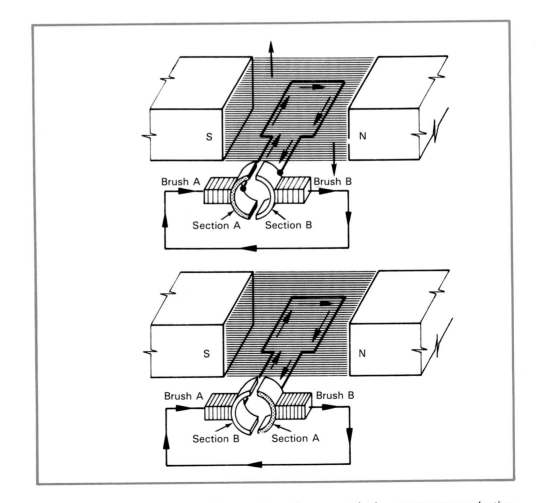

Figure 6-23. The commutator changes alternating current in the armature to a pulsating direct current in the external circuit.

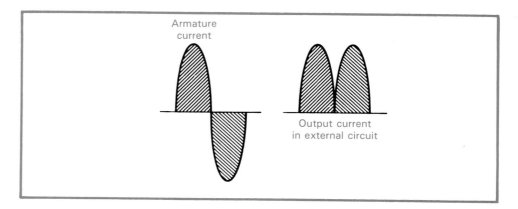

Figure 6-24. Output of the dc generator is pulsating dc current.

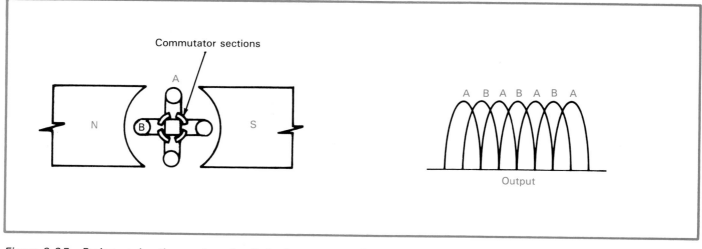

Figure 6-25. By increasing the number of coils in the armature, the output approaches steady direct current.

a generator with two coils and its output waveform is shown. In Figure 6-26, a dc generator is shown. This type of generator was used in cars before alternators were. Automotive dc generators use many coils. See field coil and armature in Figure 6-26.

The magnitude of generator output primarily depends upon the strength of the magnetic field. In order to increase the field strength and control it, the permanent magnets are replaced with electromagnets. These are called **field coils,** or **field windings.** The field coils are mounted on magnetic iron cores called **pole shoes.**

Figure 6-27 illustrates a *shunt generator.* A **shunt generator** has the field coils *shunt* across, or in parallel with, the armature. It is self-excited. This means that the shunt generator supplies its own *exciting current.* A small amount of residual magnetism in the pole shoes creates a weak magnetic field. As the armature is rotated, a small voltage is induced. This causes a small current to flow to the output and through the field coils (also connected to the output). This current increases the field strength. The action is cumulative and the generator quickly builds up to maximum output.

Two more types of generators are the *series* and the *compound.* See Figure 6-28. The **series generator** has the field coils connected in series with the armature windings. The **compound generator** has both the series and the shunt field wound on the same pole pieces. The particular performances of these types will not be discussed in this text.

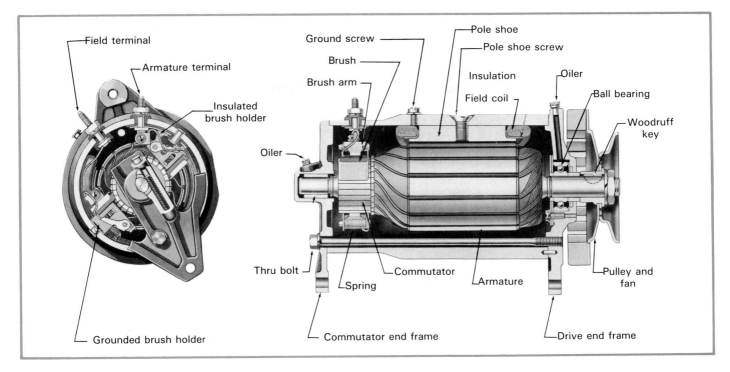

Figure 6-26. Cutaway view of an automotive dc generator.
(Delco-Remy Div., General Motors Corp.)

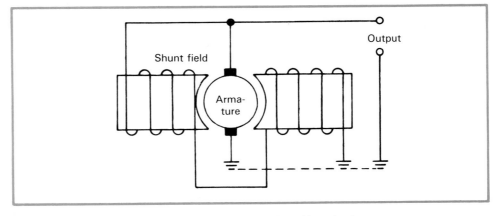

Figure 6-27. Drawing of a shunt generator. It is self-excited.

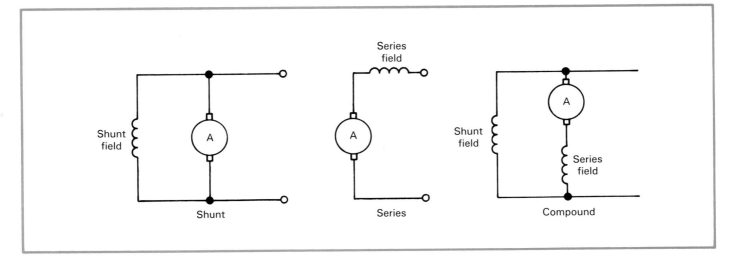

Figure 6-28. Schematics for shunt, series, and compound generators.

DC GENERATOR REGULATION

The **voltage regulation** of a power source, whether a generator (dc *or* ac) or a power supply, may be defined as the percentage of voltage drop between no load and full load. Mathematically, it may be expressed as:

$$\% \text{ Regulation} = \frac{E_{\text{no load}} - E_{\text{full load}}}{E_{\text{full load}}} \times 100\%$$

To explain this formula, assume that the voltage of a generator with no load applied is 100 volts. Under full load, the voltage drops to 95 volts. Therefore:

$$\% \text{ Regulation} = \frac{100 \text{ V} - 95 \text{ V}}{95 \text{ V}} \times 100\% = \frac{5 \text{ V}}{95 \text{ V}} \times 100\% = 5.3\%$$

In most cases, the output generator voltage should be maintained at a fixed value under varying load conditions. The output voltage of the generator depends upon the amount of flux. The amount of flux depends upon the field current. There is a practical method of maintaining both a constant output voltage *and* current for the dc generator. The method is made possible by controlling the current through field coils and, thus, controlling magnetic flux.

When we wish to control current, we immediately think of resistance. Any automatic device that will vary the resistance of the field circuit when needed will act as a means of regulation. If either voltage or current output exceeds a preset value, the regulator will switch a resistance in the field circuit and lower the generator output. With no resistance in the field, the generator will produce its maximum output.

Today, most regulators are electronically controlled. Figure 6-29, however, shows an electromechanical regulator circuit. It is one like that found in the old automotive generator. Follow the action. The generator output terminal *G* is joined to field coils, car battery, and winding of a magnetic relay. The voltage produced by the generator causes a current in the relay coil. If the voltage exceeds a preset value, the increased current will provide enough magnetism to open the relay contacts.

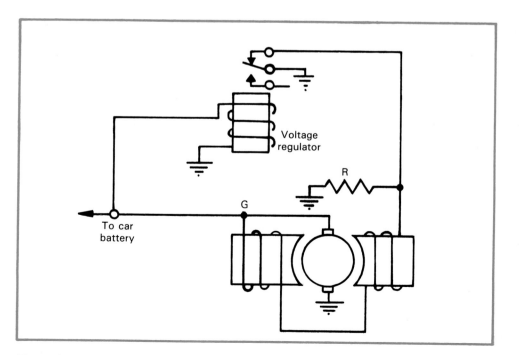

Figure 6-29. A typical dc generator regulator circuit.

Notice that the generator field is grounded through the relay contacts. When they open, the field current must pass through resistance *R* to ground. This reduces the current, which reduces the field strength and reduces the terminal voltage. When the voltage is reduced, the relay contact closes, permitting maximum field current. Terminal voltage rises. In operation, the contact points vibrate. They alternately cut resistance in and out of the field circuit. In this way, they maintain a constant voltage output of the generator.

GENERATOR LOSSES

All of the mechanical power used to turn the generator is not converted into useful electrical power. There are some losses in the generator. These losses occur in several ways. They are largely attributed to:

* Copper loss.
* Eddy current loss.
* Hysteresis loss.

COPPER LOSS

When current is through an armature or field coil, power is consumed due to the resistance of the wire. This power has served no useful purpose. It is considered as lost to the generator output. It causes undesirable heat in the generator coils and may eventually cause destruction of insulation. Losses associated with the resistance of the wire in the coils are called **copper losses.**

Copper loss is also called I^2R **loss.** This is because power loss *P* in any resistance is equal to I^2R. The fact that power loss increases as the square of the current is of special concern. For example, if the current in the generator doubles, the power loss is 2^2, or four times more. As a result, the limiting factor in generator output is usually wire size.

EDDY CURRENT LOSS

Currents are not only induced in the windings of the rotating armature of the generator. They are also induced in the iron core on which the armature coils are wound. These currents, called **eddy currents**, contribute nothing to the generator output. They only cause the core to heat.

To reduce loss due to eddy currents, the core is made up of thin sections, or layers, of iron that are insulated from each other by varnish, or lacquer. These layers are called **laminations**. The laminations form a high resistance path for any eddy currents. Thereby, they reduce eddy current loss to a minimal value.

HYSTERESIS LOSS

A third loss is termed **hysteresis loss.** This is loss due to molecular friction. The armature in a magnetic field becomes magnetized. The molecules tend to be lined up in a magnetized state. As the armature rotates, there is considerable friction between molecules as they attempt to follow the direction of the magnetizing force. This friction produces heat and power loss. Special core irons of silicon steel with special treatment are used to reduce this loss.

LESSON IN SAFETY: Do not work with electricity when alone. It is good practice to have someone around to shut off the power in case trouble is experienced.

SUMMARY

- To produce an induced current by magnetism, there must be a magnetic field, a conductor in a closed circuit, and relative movement between the field and the conductor.
- A generator is a device that converts mechanical energy into electrical energy.
- Amount of induced voltage depends upon number of turns and speed of the conductor, amount of magnetic flux, and the angle at which the conductor cuts across the field.
- Lenz's law states that the polarity of an induced emf is such that it generates a current, the magnetic field of which always opposes the change in the existing field.
- Alternating current (ac) is current that periodically flows one way in a conductor and, then, reverses direction.
- A sine wave is described by its *amplitude, cycle, period,* and *frequency,* as well as its *angular velocity* and *wavelength.*
- Sine wave voltage and current values may be given in terms of *instantaneous, peak, peak-to-peak, average,* and *effective values.*
- Ac power is usually related in terms of *average* power.
- Two sine waves are *in phase* when they are of the same frequency, and they go through the zero points at the same time. They are *out of phase* when they have the same frequency but go through the zero points at different times.
- Alternators have slip rings, while dc generators have commutators.
- Generators may be single-, two-, or three-phase.
- Windings in three-phase generators may be delta or wye connected. Current output of the delta and voltage output of the wye are 1.73 times greater than produced by a single winding.
- Regulators are used to hold voltage output at a constant level.
- Dc generators include shunt, series, and compound.
- Regulation of a power source, whether it is a generator or a power supply, may be defined as the percentage of voltage drop between no load and full load.
- Generators have I^2R, eddy current, and hysteresis losses that reduce their efficiency.

KEY TERMS

Each of the following terms has been used in this chapter. Do you know their meanings?

alternator
amplitude
angular velocity (ω)
armature
average value
brushes
commutator
compound generator
copper loss
counter torque
cycle
delta connection
dynamo
eddy current loss
effective value
field coils (windings)
frequency (*f*)
galvanometer
generator
hertz (Hz)

hysteresis loss
I²R loss
in phase
instantaneous value
laminations
law of conservation of energy
left-hand rule for generators
Lenz's law
no-load voltage·
out of phase
peak value
peak-to-peak value
period (*T*)
phase angle
phasor
pole shoes
radian
radian frequency
regulator
root-mean-square (rms) value

rotating-armature alternator
rotating-field alternator
rotor
rotor windings
self-excited generator
separately excited generator
series generator
shunt generator
sine wave
single-phase generator
slip rings
stator
stator windings
three-phase generator
two-phase generator
vector
voltage regulation
wavelength (λ)
wye (star) connection

TEST YOUR KNOWLEDGE

Please do not write in this text. Place your answers on a separate sheet of paper.

1. What was Faraday's major discovery?
   a. Electromagnetic induction.
   b. Radio waves.
   c. Electron flow.
2. Lenz's law states that the induced field around a moving conductor is opposed by the existing _____.
3. A conductor moves through a magnetic field as shown. Should electron flow through the conductor be indicated with a cross (+) or a dot (·)?

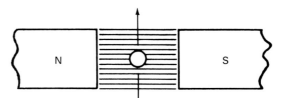

4. In a generator, what four factors determine the magnitude of its output?
5. In the study of electricity, the rise and fall and polarity of an induced voltage is represented by means of a _____ _____.
6. Alternating current differs from direct current in that ac is continually changing _____ and _____.
7. Frequency of an ac sine wave is the number of cycles occurring per _____.
8. What is the average value of an 18-V peak-to-peak wave?
9. What is the rms value of an 18-V peak-to-peak wave?

10. What is the wavelength in meters of a 7-megahertz radio wave?
    a. 23 meters.
    b. 33 meters.
    c. 43 meters.
11. The period of a 60-Hz wave is _____ milliseconds.
12. Using graph paper, construct a sine wave for a 100-volt peak ac wave. Plot the wave at intervals of 30°.
13. What is the angular velocity of a 1-kHz wave?
14. What three types of losses are associated with generators?
15. In an automotive alternator, what advantages are realized by rotating the magnetic field and taking the output from the stator?
16. Why is alternating current from an automotive alternator rectified?
17. A rectangular coil 10 centimeters long and 5 centimeters wide is wound with 500 turns of wire. It rotates in a uniform magnetic field of 0.1 tesla. With a frequency of 60 hertz, what is peak voltage?
18. Average current in a certain ac circuit is 637 mA. Average voltage is 63.7 V. What power is delivered to the circuit?

Chapter 7   ELECTRIC MOTORS

Electric motors—devices that convert electrical energy into mechanical energy. Of all the applications of electricity, this is one of the most useful.

After studying this chapter, you will be able to:

☐ *Explain how reaction between magnetic fields will produce mechanical motion.*

☐ *Discuss the principles of operation of the dc motor.*

☐ *Explain how an alternating current can be used to operate a motor.*

☐ *Cite methods used to control the speed and turning power of a motor.*

INTRODUCTION

How many electric motors do you have in your home? They are used to keep food cool in your refrigerator. They are used to circulate warm air from your heating system and to operate the washing machine and clothes dryer. Motors are used to mix our foods, brush our teeth, and ventilate the kitchen. Just look at industry! The wheels of production would remain at rest, if it were not for the motor. You will learn more about the electric motor as you read this chapter.

MOTOR ACTION

Basic to the understanding of all electric motors is the reaction between a fixed magnetic field and a current-carrying conductor in that field. Before examining Figure 7-1, you may wish to study Figures 5-9 and 5-10 in Chapter 5. You may be assured that there is a circular magnetic field surrounding a current-carrying conductor. The direction of this magnetic field of force will depend upon the direction of current. Use the left-hand rule for conductors.

In Figure 7-1, the current in the conductor is assumed to be toward you. This causes the circular magnetic field around the conductor to oppose the direction of the fixed field below the conductor. It also reinforces the field above the conductor, causing the conductor to move *downward* toward the weakened field.

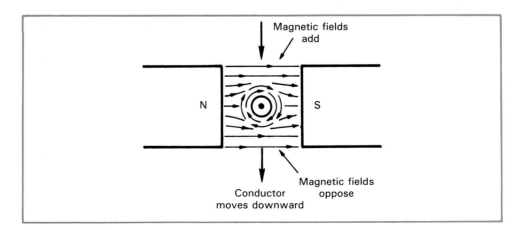

Figure 7-1. The field of the conductor opposes the fixed field below the conductor. The conductor moves toward the weakened field.

In Figure 7-2, the current in the conductor is assumed to be away from you. The circular field is in the opposite direction. In this case, the weakened field appears above the conductor, and the conductor will move *upward*.

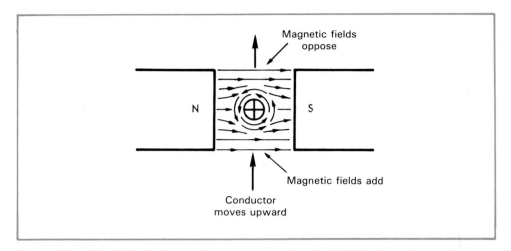

Figure 7-2. The field of the conductor opposes the fixed field above the conductor. The conductor moves upward.

Dc motor action project

A demonstration of this action, called **motor action**, is simulated in the project shown in Figure 7-3. This setup is only a suggestion. It can be constructed in many other ways. The switch is a DPDT, 3-position (center off) in order to reverse the current in the single conductor. A No. 8 copper wire is bent in a large U-shape. Two small brackets have holes in them, which permit the copper U to swing back and forth. Any two magnets may be used, but they must be fixed in place.

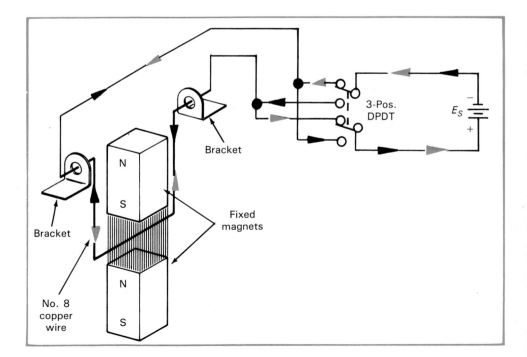

Figure 7-3. The construction of a ''motor action'' demonstrator.

To demonstrate the motor action, toggle the switch from one ON position to the other. The copper U pivots back and forth as the direction of current changes. The arrows in the drawing indicate current in each position of the switch. Do *not* let the circuit remain on for more than a few seconds. It is a direct short circuit across your battery or supply. Use the OFF position of your switch.

To determine the direction of movement of a conductor in a field, use this simple **right-hand rule for motors**. Extend your thumb and first two fingers of your right hand at right angles to one another. See Figure 7-4. Point your first finger in the direction of the magnetic field (toward the south pole). Point your second, or middle, finger in the direction of electron flow. Your thumb will point in the direction that the conductor will move. Try this rule with the diagrams of Figures 7-1 through 7-3, and prove to yourself that the right-hand rule for motors is valid.

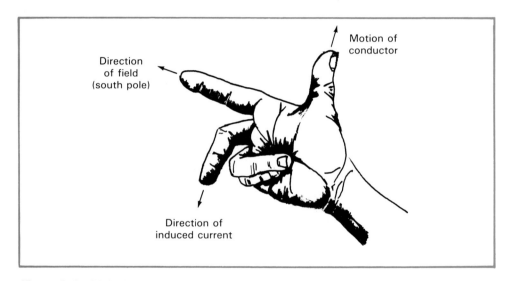

Figure 7-4. Right-hand rule for motors.

THE DC MOTOR

Construction of a simple dc motor is shown in Figure 7-5. We will use this setup to describe motor action. Note that the single-loop coil is arranged to rotate in the magnetic field. The rotating loop or element of the dc motor is the *armature*. Like the dc generator, the ends of the coil are connected to a commutator. This is a metal ring separated into two half sections, or semicylinders, by an insulating material. The commutator rotates with the armature.

Brushes of carbon provide a sliding contact to the commutator sections. This brings electric current from the dc source into the armature coil. Note the current as indicated by the arrows in Figure 7-5. In particular, note that current flows in from section A and out through section B.

The plane of the loop, as shown, is parallel to the magnetic field. As a result of *motor action*, the interacting fields will cause side C of the armature to move upward and side D to move downward. The coil rotates to where the plane is perpendicular to the field ($\theta = 0°$). At this point, the brushes short circuit across the commutator sections, and the value of current in the loop is zero. *Inertia* (objects in motion tend to stay in motion) carries the loop past the vertical position.

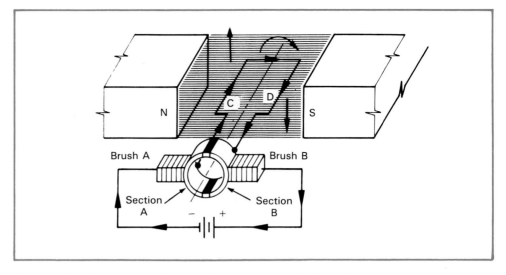

Figure 7-5. The construction and theory of a simple dc motor.

Now, commutator section A comes in contact with brush B, and section B comes in contact with brush A. This time, current flows in from section B and out through section A. The current in the armature has reversed its direction. Coil side C moves downward, and coil side D moves upward. As the armature continues to rotate, the direction of armature current is changed each half rotation. See Figure 7-6.

A practical motor armature will have many coils placed in slots around the armature core. Each set of coils will have its own pair of commutator sections. This arrangement produces continuous torque and power, and inertia of the armature, although present, is no longer required to maintain rotation. Figure 7-7 is a simplified drawing showing construction of a dc motor.

CHARACTERISTICS OF MOTORS

Motors, whether dc or ac, have certain characteristics associated with them. These include *torque, counter,* or *back emf, armature current, motor speed,* and *speed regulation.* (Many of these are typically plotted as curves called *load characteristic curves.* These will give the conditions of a given motor's operation as the applied voltage is held constant.)

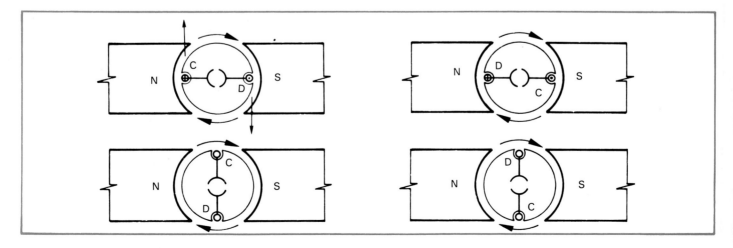

Figure 7-6. The simple dc motor continues to rotate. The direction of current in the armature is changed each half rotation.

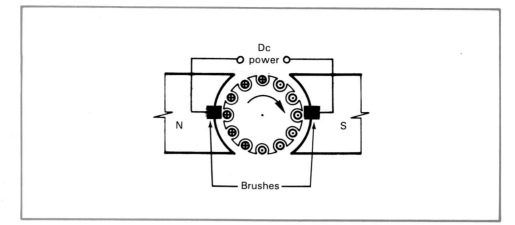

Figure 7-7. Many armature coils are placed in slots around the armature core.

Motor torque

In the case of the simple motor of Figure 7-5, only the force between one current-carrying conductor in the fixed field was used to produce rotation. If this coil was replaced by another with many turns, the turning force, or **torque**, would greatly increase. In the case of Figure 7-7, many coils are contributing to the torque. The total torque of the motor would be the sum of all the contributing torques of individual conductors and coils.

As stated, movement may be produced by the interaction of a fixed field and the field around a current-carrying conductor. It is only reasonable to assume, then, that this force can be increased by increasing the magnetic flux of the fixed magnetic field. In order to do this, the permanent magnets previously used will be replaced with electromagnets. Mathematically speaking, the torque of a motor is in direct proportion to its magnetic flux and its armature current. In a dc motor, it is stated as:

$$T = K\Phi I_a$$

where:

T = torque in newton-meters
K = a constant for a particular motor based on the number of windings in the coils and the number of coils.
Φ = magnetic flux per field pole of the motor in webers
I_a = armature current in amps

In Figure 7-8, the field windings of a parallel-, or shunt-wound, field motor are illustrated. The number of field coils is also increased to four, or two pair, and two sets of brushes are used. These improvements all contribute to increasing the torque and power of the motor.

Counter emf

A comparison of a motor and a generator reveals many similarities. In fact, a generator can act as a motor, and a motor act as a generator if turned by mechanical power. However, certain design and construction features prohibit interchange for practical purposes. To understand motor theory, you must realize that as a motor is rotating, the armature and conductors are cutting across the magnetic field. In the process, they induce a voltage. This is called a **back**, or **counter, emf (cemf)**. The cemf is proportional to the *speed* of the motor and its *magnetic flux*.

Consider why we are concerned with cemf. Since it is in opposition to the source voltage used to drive the motor, the net voltage applied to the armature

E_a is the algebraic sum of these two voltages. In other words:

$$E_a = E_S - E_{cemf}$$

where E_S is source voltage, and E_{cemf} is the counter emf generated by the motor.

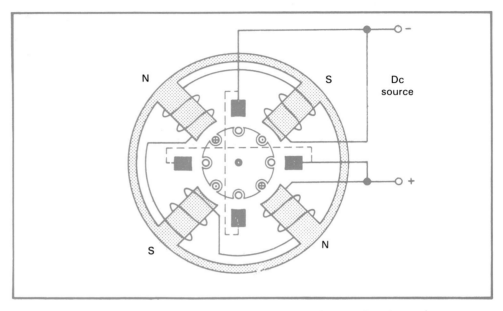

Figure 7-8. The permanent field magnets are now replaced with electromagnets. Field coils are in parallel with armature.

Refer to Figure 7-9. The current in any circuit is found by Ohm's law. The dc resistance of the armature coils R_a depends on the length and size of the wire used for winding. The current through the armature I_a is directly proportional to the net voltage applied to the armature:

$$I_a = \frac{E_S - E_{cemf}}{R_a} = \frac{E_a}{R_a}$$

The voltage drop across the motor armature is:

$$E_a = I_a \times R_a = E_S - E_{cemf}$$

By rearranging the formula, we find that:

$$E_S = E_a + E_{cemf}$$

Finally, induced voltage, or counter emf, depends upon speed of rotation, the number of conductors, and the amount of flux. The value and polarity of the counter emf are just like those obtained when a machine operates as a dc generator. Referring back to Chapter 6, the formula, then, for counter emf in a dc motor is:

$$E_{cemf} = \frac{Zn\Phi}{60}$$

This theoretical discussion is extremely valuable from a practical point of view. Read the following questions, and try to "think through" the correct answer. Then, read the explanation to check your knowledge.

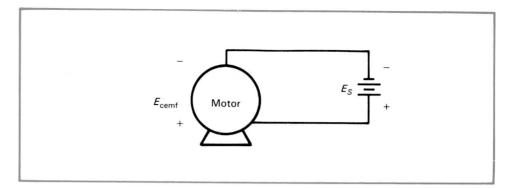

Figure 7-9. The net voltage applied to the motor is the source voltage minus the cemf.

1. When does a motor draw the greatest amount of current from the line? Why?

   Answer: A motor when stopped is developing no cemf, and the armature current is at maximum, limited only by the resistance of the armature windings. For this reason, current draw is greatest when starting a motor.

2. When does a motor draw the least current from the line?

   Answer: When the motor is at full speed (under no load). Then, the cemf is almost equal to the line voltage and the current is minimum.

3. As motor load increases, the rotor slows down. Why, then, does a motor under load use more power?

   Answer: When a motor slows down under load, the cemf drops, and the armature current increases. Power varies as the square of the current.

4. Why does an overloaded motor become excessively hot?

   Answer: The increased current due to lower cemf produces heat in flowing through the resistance of the armature windings. This heat can eventually destroy the motor by burning off insulation.

5. Why is it necessary to start a motor with a series resistance between the motor and the source voltage?

   Answer: The series resistance limits the high starting currents. (The resistance is switched out of the circuit when the motor reaches its speed, and the cemf is sufficient to limit the current.)

Motor speed

We have seen that cemf in a dc motor is directly proportional to speed and the strength of the magnetic field. We can rearrange the equation for counter emf to find motor speed:

$$n = \frac{E_{cemf} \times 60}{Z \times \Phi}$$

We find that speed is inversely proportional to magnetic flux. This will present some interesting theory as we continue our studies of methods of connecting motors.

Speed regulation

The term **speed regulation** is used to describe the speed change of a motor between full-load and no-load conditions. It is expressed in percent of full-load speed. It is desirable, of course, for a motor to maintain a constant speed. Speed regulation may be obtained from:

$$\% \text{ Regulation} = \frac{\text{no-load speed} - \text{full-load speed}}{\text{full-load speed}} \times 100\%$$

Assume a motor runs at 1700 rpm and drops to 1600 under load. Its speed regulation is:

$$\% \text{ Regulation} = \frac{(1700 - 1600) \text{ rpm}}{1600 \text{ rpm}} \times 100\% = 6.25\%$$

A low speed regulation means that the motor operates at somewhat constant speed, regardless of the load applied. Also, the term does not refer to speed *control*, which is the varying of speed by external circuitry and devices.

TYPES OF DC MOTORS

There are three different types of dc motors — *shunt, series,* and *compound.* The construction of each type is about the same, but the fields in each type are connected differently in relation to the armature. This results in different operating characteristics for each.

The shunt motor

The **shunt motor,** as you might expect, has its field coils and armature in parallel. See Figure 7-10. Assuming that a constant voltage is applied to this motor, the magnetic flux tends to remain constant. If a load is applied and the speed of rotation is reduced, the counter emf decreases, and the armature current increases. This, then, increases the torque of the motor. When the torque matches the load, the motor remains at constant speed. The shunt motor maintains a fairly constant speed, even under varying-load conditions; hence, it is commonly called a **constant-speed motor.** These motors are used in driving machine tools and other machines needing relatively constant speed.

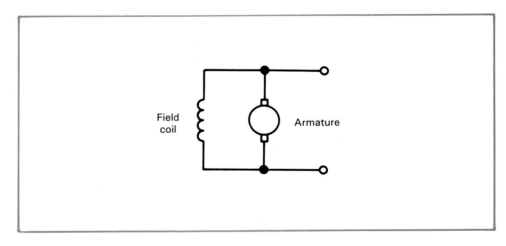

Figure 7-10. The circuit of a shunt motor.

Although the shunt motor is constant speed, its speed may be changed by inserting resistance in the field circuit. This will change the field current and flux density. See Figure 7-11. An increase in resistance will decrease the field current and increase the motor speed.

Another method of varying the speed of the shunt motor is with a variable series resistance in the armature circuit. Look at the formulas:

$$n = \frac{E_{\text{cemf}} \times 60}{Z \times \Phi}$$

and

$$E_{\text{cemf}} = E_S - I_a R_a = E_S - E_a$$

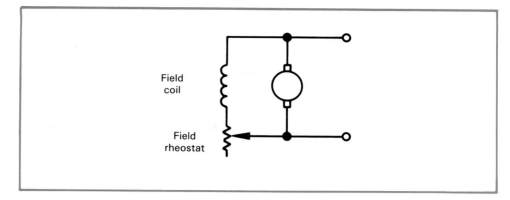

Figure 7-11. Resistance in the field of a shunt motor will change its speed.

You can see that a change in speed will result from a change in armature voltage. This is easily accomplished by reducing the armature current by a series resistance as in Figure 7-12.

Although this is a widely used method of speed reduction, the high currents resulting from reduced cemf produce appreciable power losses in the series resistance. To limit currents, *silicon controlled rectifiers* may be used. In this text, you will construct a light dimmer/motor speed control using these devices.

The series motor

In the **series motor,** the field coils are in series with the armature circuit, and the magnetic flux is proportional to the armature current. Look at Figure 7-13.

The series motor must always have a load connected to it. Why? As the speed of the motor increases, the cemf also increases, and the armature current decreases. A decrease in armature current also decreases the field current and the magnetic flux. A decreasing magnetic field causes an increase in speed. The effect is cumulative. The rpm will rise to a point of self-destruction. It may literally fly apart.

The series motor is most useful for turning heavy loads from a dead stop. At these speeds, tremendous torque is developed due to high current through the armature and field. These motors are used on heavy equipment. They are always geared to the load. No belts are used. A broken belt would allow excessive motor speed — to the point of self-destruction. The torque of this motor actually varies as the square of its armature current.

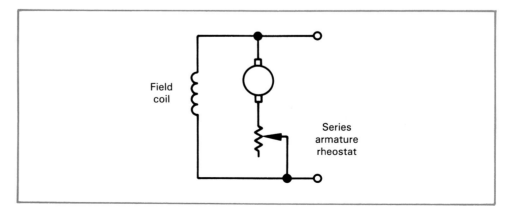

Figure 7-12. Resistance in the armature circuit will change the speed of the shunt motor.

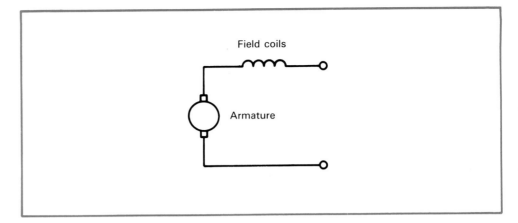

Figure 7-13. The diagram of a series motor.

The compound motor

Motors may be wound and connected to have both series and shunt field coils. These are called **compound motors**. If the series winding is connected so that its field reinforces, or aids, the shunt winding, the motor is called a **cumulative compound motor**. If the series and shunt windings have opposing fields, the motor is called a **differential compound motor**. The circuits of both types are shown in Figure 7-14.

The reason for manufacturing compound motor circuits is to combine the desirable features of both series and shunt motors. The cumulative compound motor develops high starting torque. It is used where heavy loads are applied and some variance in speeds can be tolerated. The load may be safely removed from this motor. The differential compound motor behaves much like the shunt motor. The starting torque is low. It has good speed regulation if loads do not vary greatly. Detailed characteristics of compound motors can be found in more advanced texts on dc machinery.

MOTOR STARTERS

There are many different types of motor starters. The purpose of each dc motor starter is to insert a resistance in the motor circuit. This limits the armature current until the motor reaches its operating speed and develops sufficient counter emf to hold the line current at a low value. Some starters are manually operated; others are automatic. In Chapter 5, we discussed the magnetic motor starter. This automatic starter, however, is for ac motors, which may be started with full line voltage without damage.

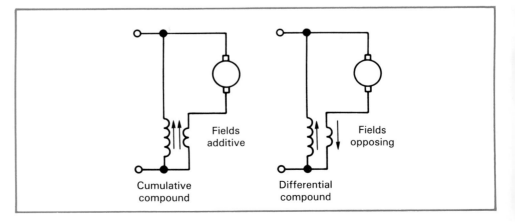

Figure 7-14. The circuit diagrams of cumulative and differential compound motors.

A simple step-resistance, manual motor starting circuit for a series motor appears in Figure 7-15. With the lever in position 1, there is maximum resistance in the motor armature circuit. As the speed of the motor progressively builds, the operator moves the lever to position 2, then 3 and 4.

The lever is held at position 4 by the magnetic coil, which also is in series with the line. If the line voltage should fail, the lever would be released and returned to the OFF position by spring action. This is a safety feature. To start the motor again, it would be necessary to proceed through each position of resistance.

THE UNIVERSAL MOTOR

Can a dc motor be operated with ac current? Yes, a series motor can with some minor design changes. You will recall the action of a conductor in a magnetic field. If the current in the conductor is reversed, the force and movement of the conductor is also reversed. However, if the field polarity is reversed at the same time the armature current is reversed, then original conditions will exist. Torque, therefore, will act in the same direction, allowing a motor to rotate. This is what happens in the series motor, where both field and armature currents change approximately together.

The dc series motor will have some serious defects when operated on alternating current. A chief disadvantage is the high current induced in the armature and field windings by a rapidly changing flux. This causes excessive sparking at the brush contacts and commutator. This disadvantage is minimized by making field windings of only a few turns of heavy wire to weaken the flux. (Reactance in such windings will be negligible; so, field current would be the same whether ac or dc was applied.) To compensate for loss of torque force, more turns are added to the armature windings. Further, to limit high induced currents and prevent sparking, resistors are frequently placed in the brush leads.

Eddy current and hysteresis losses are greater with ac operation. To reduce these losses and to prevent field core overheating, the field core is laminated. Core losses will also be reduced by the weaker flux of the modified field winding.

The modified dc series motor just described is called a **universal motor**. Sometimes it is referred to as an **ac series motor**. These motors find wide usage in small power tools such as drills, grinders, and ventilating fans.

The shunt dc motor does not operate well on ac. The reactance of the shunt field causes out-of-phase conditions that seriously impair the motor operation.

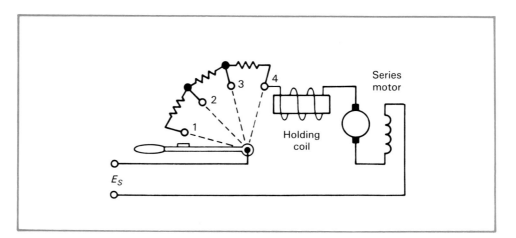

Figure 7-15. A step-resistance motor starting circuit for a series motor.

THE AC MOTOR

Ac motors are, of course, different than dc motors. Shunt, series, and compound dc motors are all **doubly excited motors**. This means that both the armature and the field have *excitation currents*. The **ac induction motor** is a **singly excited motor**. That is, it has a single source of power that is brought to the fixed *stator* coil. Excitation of the *rotor* is achieved by induction, or transformer action. A rotating-magnetic field sustains the motion of the rotor.

Since alternating current is readily available from the local power company, most motors used at home and in industry are ac motors. In order to understand the principles of induction and rotating magnetic fields, more background information must be provided.

MOTOR ACTION

To illustrate what happens in the ac induction motor, look at Figure 7-16. It shows a cross section of a magnetic field and a conductor loop. In this case, the conductor is *fixed*, and the field is passed across it in order to induce a current in the conductor. If the field moves from right to left, this is the same as the conductor moving from left to right. Apply the left-hand rule for generators to determine the direction of the induced current. Point your thumb in the direction the conductor is moving. Point your first finger toward the south pole. Your second, or middle, finger will point in the direction of electron flow as indicated.

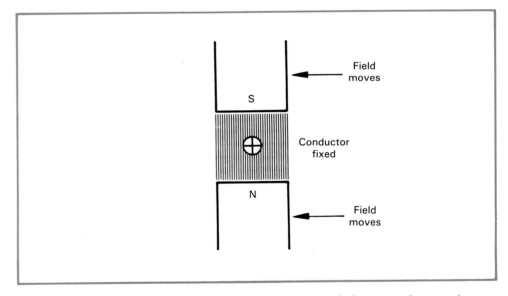

Figure 7-16. A moving field cutting across a conductor induces a voltage and current in the conductor as indicated.

Now, suppose we free the conductor so that it can move. Look at Figure 7-17 to discover the motor action caused by the reaction between a moving field and its induced current. This time use the right-hand rule for motors. Point your first finger toward the south pole. Point your second, or middle, finger in the direction of electron flow. Your thumb will point in the direction that the conductor will move. Observe that the conductor follows the moving field.

Compare this to the action of a dc motor. With the dc motor, current through the conductor was provided by an external voltage source. We had motor action caused by the reaction between a current-carrying conductor

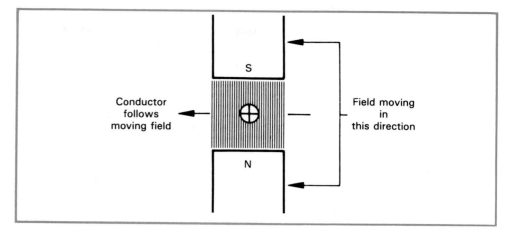

Figure 7-17. The conductor tends to follow the moving field.

and a *fixed* magnetic field. With the ac motor, current through the conductor is induced by a moving magnetic field. Now we have a reaction between a current-carrying wire and a *moving* magnetic field. We find that the conductor follows the moving field.

Animated ring project

An interesting project is developed around induced ac fields. Look at Figure 7-18. An *animated ring*, which demonstrates motor action, is shown. The ring consists of a coil, a NO push-button switch, and washers. The coil is wound with 325 turns of No. 18 enameled copper wire. The core of the coil is an 8-inch-long, low carbon steel rod. Make copper, aluminum, and iron washers to fit over the core. Connect the ring to a 12-Vac source. Select one of the washers and slide it down the core rod. Press the push button and observe the action. Repeat for the other two washers. How do you explain any difference among actions of the copper, aluminum, and iron washers?

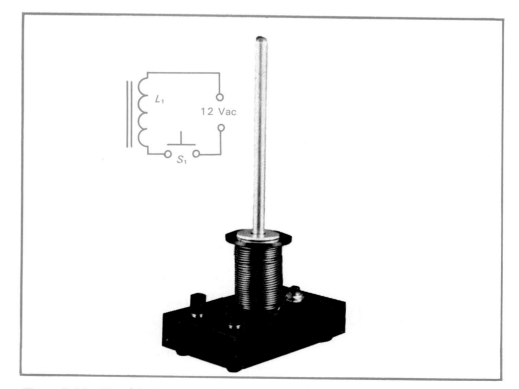

Figure 7-18. The animated ring demonstrates motor action.

ROTATING MAGNETIC FIELDS

It would be easy to make a conductor or armature rotate, if it were possible to make a magnetic field rotate. This can be done with three-phase ac. See Figure 7-19. Phase 1 current is connected to pole A. Phase 2 current is connected to pole B. Phase 3 current is connected to pole C. When phase 1 rises to maximum, the rotor is attracted to pole A. As phase 1 starts to decrease, phase 2 rises to maximum. It attracts the rotor to pole B. As phase 2 starts to decrease, phase 3 rises to maximum. It attracts the rotor to pole C. Starting again with pole A and phase 1, the field rotates around the stator and attracts the rotor with it.

THREE-PHASE AC MOTORS

There are two main types of three-phase motors—*induction*, and *synchronous*. Both are induction-type motors. However, while the rotor speed of the induction motor runs slower than the speed of the rotating field, the rotor of the **synchronous motor** is synchronized with the rotating field. This is accomplished by applying a dc current to the rotor once it reaches running speed. Beyond this, the synchronous motor will not be discussed in this text.

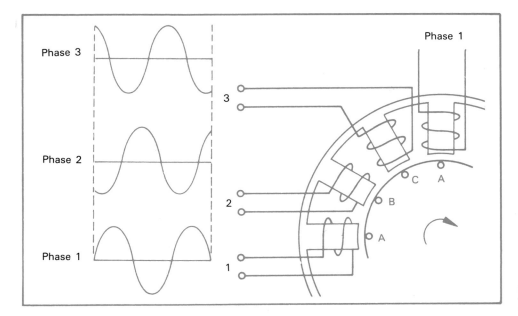

Figure 7-19. A three-phase current will produce an ac rotating magnetic field.

The induction motor

To build the three-phase, or *polyphase*, induction motor, it is necessary to place a rotating armature in the rotating field. These armatures, or *rotors*, may be of two types. Either type will be built around a laminated iron core. Induction motors are further classified, then, according to the type of rotor they have. This includes:

Squirrel-cage induction motor. This type of motor has a rotor with heavy, single copper wires embedded in an iron core. The construction resembles a squirrel cage. See Figure 7-20.

Wound-rotor induction motor. This type of motor has coils of wire fixed in slots around the rotor core, similar to the dc motor armature.

Voltages and currents induced in the rotor coils of these motors will cause them to move with the rotating field. But consider this! In order to induce a voltage, there must be *relative motion* between field and conductor. If the conductor moves with the field, there is no relative motion. As a result, the

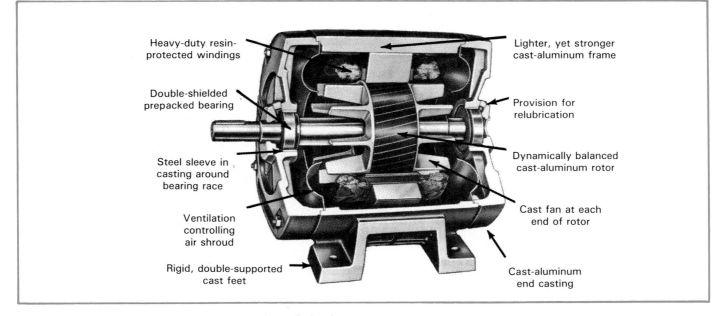

Heavy-duty resin-protected windings

Double-shielded prepacked bearing

Steel sleeve in casting around bearing race

Ventilation controlling air shroud

Rigid, double-supported cast feet

Lighter, yet stronger cast-aluminum frame

Provision for relubrication

Dynamically balanced cast-aluminum rotor

Cast fan at each end of rotor

Cast-aluminum end casting

Figure 7-20. Photo of a squirrel-cage rotor in an induction motor. (W.W. Grainger, Inc.)

conductors and rotor must rotate slightly slower than the field. This lag in speed is called **slip**.

When no load is on the rotor, the rotor speed can approach the speed of the rotating field **(synchronous speed)**. As the motor is loaded, it slows down slightly. The slip increases, and larger currents are induced in the rotor windings. This, in turn, increases the torque of the motor.

Motor speed. The speed of the induction motor is dependent upon the frequency of the applied voltage and the number of poles around the field, or stator, windings. Synchronous speed S_S in rpm is given as:

$$S_S = \frac{120\,f}{P}$$

where f is the frequency of the applied stator voltage in hertz, and P is the number of stator poles.

If the percent of slip at full load is known, we can determine the speed of the rotor S_R at full load. The formula is:

$$S_R = \frac{120\,f}{P}\,(1 - \text{slip})$$

Problem:

A two-pole, three-phase induction motor is connected to a 60-Hz source. It has a full-load slip of 22%. What is the speed of the rotating magnetic field and the full-load speed of the rotor?

Solution:

Step 1. Determine the speed of the rotating field.

$$S_S = \frac{120\,f}{P} = \frac{120 \times 60}{2} = 3600 \text{ rpm}$$

Step 2. Determine the speed of the rotor.

$$S_R = \frac{120\,f}{P}(1 - \text{slip}) = S_S\,(1 - \text{slip}) = 3600\,(1 - 0.22) = 2808\ \text{rpm}$$

Motor torque. The formula for torque in an induction motor is very similar to the formula for torque in a dc motor. Now, however, the subscript R is used to denote rotor quantities. Also, since the flux and current are not exactly in phase, a factor must be applied to the formula. This factor is the cosine of what is called the *rotor power factor angle*. Note that current and flux in the induction motor are both sinusoidal, and that rotor current lags the flux by some angle. This is the **rotor power factor angle**. The magnitude, then, of the torque in an induction motor is:

$$T = K\Phi I_R \cos \theta_R$$

where:

T = torque in newton·meters
K = a constant for a particular motor based on the number of windings in the coils and the number of coils.
Φ = magnetic flux per field pole of the motor in webers
I_R = rotor current in amps
θ_R = rotor power factor angle

SINGLE-PHASE AC MOTORS

The most popular type of *fractional* horsepower motor is the **single-phase motor**. The theory of operation is similar to the three-phase motor; however, a single-phase current cannot, in itself, produce a rotating field. If we could get the single-phase motor up to synchronous speed, it would continue to run just like the three-phase motor. The problem is getting it started.

The starting problem is solved by placing two sets of windings in the stator. These windings have unequal *reactance* (ac resistance) due to the number of turns in each coil and the size of wire used. Now, unequal reactance will result in lagging current. In other words, two out-of-phase currents will be produced. These out-of-phase currents produce the rotating field that is so necessary to get the motor started.

There are five general types of single-phase motors. These include the *split-phase, capacitor-start, shaded-pole, repulsion,* and *series motors*. The series motor was presented in the discussion about the universal motor. The others we will discuss here.

The split-phase motor

A **split-phase motor** is shown in Figure 7-21. A schematic of its internal circuitry is shown in Figure 7-22. When the motor reaches speed, the centrifugal switch disconnects the start winding. The motor operates, then, only on its main, or running, winding.

A *phasor diagram* of the currents in the winding of this split-phase motor is drawn in Figure 7-23. A **phasor diagram** shows phase relationships using phasors. The length of the arrow represents effective current or voltage magnitude. The angle of the arrow with respect to a horizontal reference phasor gives the phase angle.

As you know, magnetic flux is the result of current. In Figure 7-23, we can see that currents are out of phase and, therefore, reach peak value at different times. From this, we know that the magnetic fields of each stator coil will also reach maximum at different times. This causes the rotating field.

*Figure 7-21. Fractional horsepower split-phase motor.
(Delco Products Div., General Motors Corp.)*

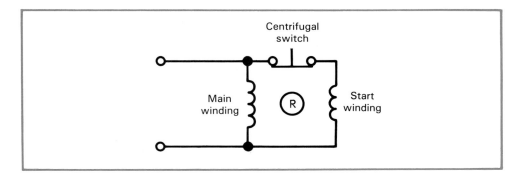

Figure 7-22. Diagram of a split-phase motor showing starting and running windings.

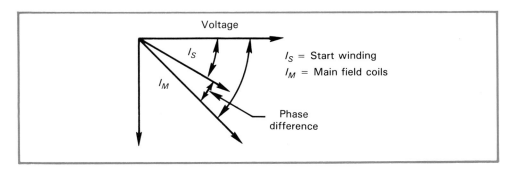

Figure 7-23. Phasor diagram of phase displacement between currents in start and main windings of split-phase motor.

The capacitor-start motor   If the phase difference between start and running windings can be increased, the starting torque of the motor can be greatly improved. Basic theory tells us that *capacitance* (stored charge) in a circuit causes the current to *lead* the source voltage. This is where the current phasor is *above* voltage — the horizontal reference phasor. *Inductance* (opposition to current change) causes the current to *lag*. This is where the current phasor is *below* the voltage reference phasor. Both of these effects, which are discussed at length in later chapters, are incorporated in the *capacitor-start* motor. A **capacitor-start motor** is a type of split-phase motor with a capacitor in series with the starting resistance. See Figures 7-24 and 7-25. Capacitors used for this application range between 80 and 400 microfarads for motors under 1 horsepower.

The shaded-pole motor

Another method of getting an ac motor started is by using a **salient-pole stator**. This is a stator, each pole of which has a separate projection as shown in Figure 7-26. Around part of each salient pole is placed a short-circuiting ring called a **shading ring**, or **coil**. This low resistance ring is made up of heavy copper wire—usually of single turn. A motor constructed as just described

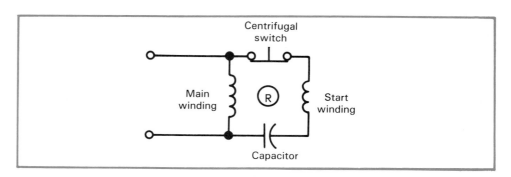

Figure 7-24. Diagram of a capacitor-start motor.

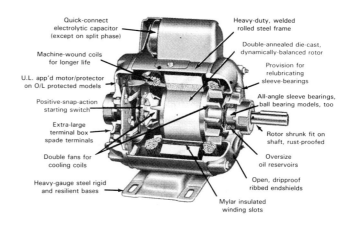

Figure 7-25. A commercial capacitor-start motor.   (W.W. Grainger, Inc.)

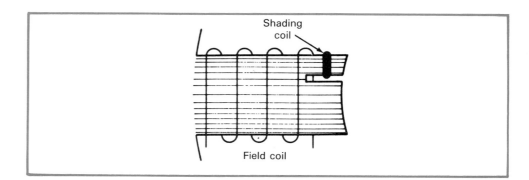

Figure 7-26. Partial view of a field coil showing salient pole and shading coil.

is called a **shaded-pole motor.** Like the capacitor-start motor, the shaded-pole is a type of split-phase motor.

Refer to Figure 7-27. A rising current in the coil in position A causes an increasing magnetic field. A current is induced in the shading coil that produces an opposing magnetic field (Lenz's law). Therefore, the field on this side is weaker.

Position B is at the peak of the incoming ac wave. At this point, the change in flux has dropped to zero. There is no current induced in the shading coil and, therefore, no field induced by the coil. Its effect has disappeared. The magnetic field is *equal* across the pole face.

Position C shows the effect of a decreasing current and magnetic field. The induced field in the shading coil opposes the reducing field. Therefore, the field becomes stronger than the left side of the pole where there is no opposition. The entire action caused by the shading coil sets up a nonuniform rotating flux effect. It is enough to start the rotor turning.

After starting, the motor runs as an induction motor, approaching synchronous speed similar to the split-phase motor. The shaded-pole motor develops very little starting torque. It is used in electric clocks and small fans. A diagram of a typical two-pole shaded-pole motor used in a clock is shown in Figure 7-28.

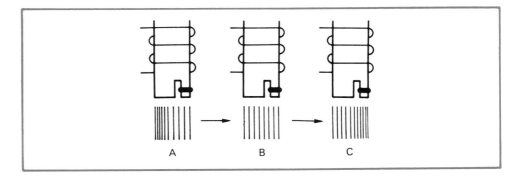

Figure 7-27. The field on the salient-pole stator moves from left to right because of the shading coil.

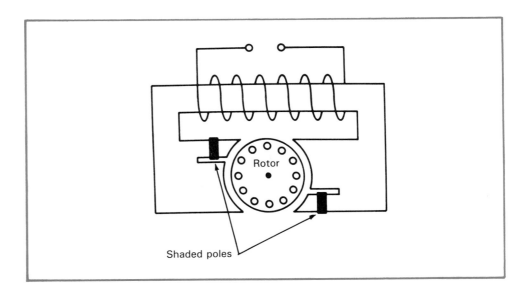

Figure 7-28. A diagram of a two-pole shaded-pole motor.

The repulsion motor

The **repulsion motor** resembles a dc motor in construction because it has a commutator and brushes. However, in the repulsion motor, the brushes are connected to each other, rather than to a source of power. There are three types of repulsion motors—*repulsion, repulsion-induction*, and *repulsion-start induction-run*. These motors have become obsolete for the most part. However, they may still be found in some older installations.

Of the three types of repulsion motors, the latter has been the most common. It is better known as the **repulsion-start induction motor**, Figure 7-29. In operation, this motor starts with the brushes on the commutator as any repulsion motor. The two opposite brushes are shorted together during starting. Since the brushes are shorted, an armature current will flow due to induction. A field is developed, and poles are created. These poles oppose the stator poles, and rotation is started by repulsion. When the motor reaches about 75% of full speed, a centrifugal switch short circuits all the commutator sections. At this point, the motor runs as an induction motor. In general, this motor has been replaced by the capacitor-start motor.

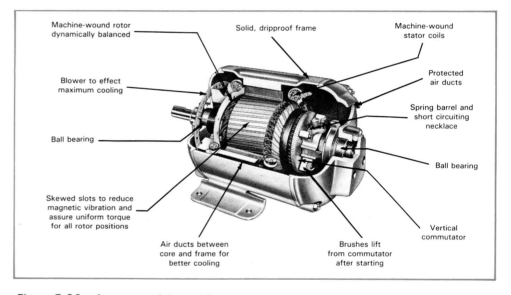

Figure 7-29. A commercial repulsion-start induction motor.   (W.W. Grainger, Inc.)

Motors come in all shapes, sizes, and types, and the technology is ever improving, Figure 7-30. The study of motors in this chapter is intended as a brief introduction of these popular machines. There is much to be learned about them, and keeping abreast of the changes is an ongoing process. If your interest is high, it is suggested that you study texts devoted entirely to the subject of electric motors.

SCR MOTOR SPEED-CONTROL PROJECT

A useful project that applies many of the lessons in this text is the Silicon Controlled Rectifier (SCR) Motor Speed Control. This unit, when completed, can be used to vary the speed of a universal motor. This is the type of motor found in most drills, table saws, and hand grinders. You should *not* use this speed control unit for other types of motors. To understand the operation of an SCR, you will need to refer to Chapter 12. Figure 7-31 gives the schematic

*Figure 7-30. Shown here is a sample of the many types of electric motors that are available. Also shown are a number of **electronic motor drives**, which are used to control the motor speed and direction.   (Baldor)*

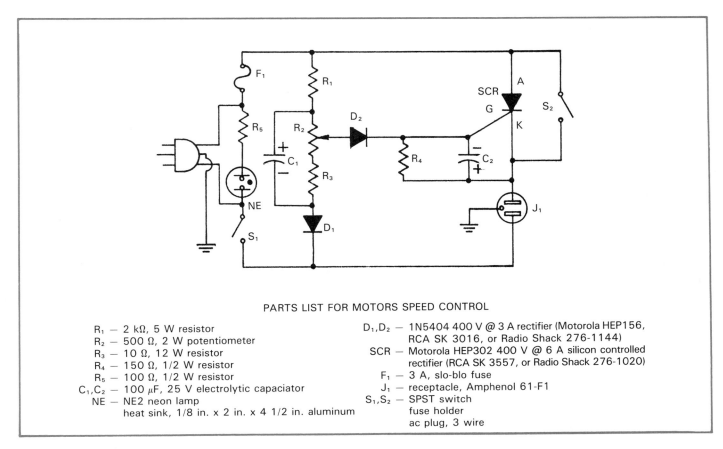

PARTS LIST FOR MOTORS SPEED CONTROL

R₁ — 2 kΩ, 5 W resistor
R₂ — 500 Ω, 2 W potentiometer
R₃ — 10 Ω, 12 W resistor
R₄ — 150 Ω, 1/2 W resistor
R₅ — 100 Ω, 1/2 W resistor
C₁,C₂ — 100 μF, 25 V electrolytic capaciator
NE — NE2 neon lamp
    heat sink, 1/8 in. x 2 in. x 4 1/2 in. aluminum

D₁,D₂ — 1N5404 400 V @ 3 A rectifier (Motorola HEP156, RCA SK 3016, or Radio Shack 276-1144)
SCR — Motorola HEP302 400 V @ 6 A silicon controlled rectifier (RCA SK 3557, or Radio Shack 276-1020)
F₁ — 3 A, slo-blo fuse
J₁ — receptacle, Amphenol 61-F1
S₁,S₂ — SPST switch
    fuse holder
    ac plug, 3 wire

Figure 7-31. Schematic and parts list for SCR Motor Speed Control.   (Graymark Electronics)

diagram and parts list for the SCR Motor Speed Control. The completed project is shown in Figure 7-32.

The construction is not critical, and you should design your own circuit layout. Switch S_2 can be set for straight-through operation or controlled operation. The current through the SCR is controlled by the voltage applied to the gate, which depends upon the time constant of the circuit and the setting of R_2. This circuit will control up to 3 amps maximum.

LESSON IN SAFETY: Wait for solder joints to cool. The practice of cooling soldered joints with a wet finger is ill-advised.

Figure 7-32. SCR Motor Speed Control or Light Dimmer.

SUMMARY

- A motor is a device for changing electrical energy into mechanical energy.
- Counter emf, or cemf, is the induced voltage that opposes the applied voltage.
- The rotation of a motor produces turning, or twisting, power called torque.
- The term *speed regulation* is used to describe the speed change of a motor between full-load and no-load conditions.
- A series motor has field windings connected in series.
- Shunt motors have field windings connected in parallel.
- Compound motors have field windings connected in series and parallel.
- The purpose of a dc motor starter is to insert a resistance in the motor circuit. This limits the armature current until the motor reaches its operating speed and develops sufficient counter emf to hold the line current at a low

value. In general, ac motors may be started with full line voltage without damage.

- In an induction motor, field poles and windings remain stationary.
- Induction motors need starting circuits. They have start and running windings.

KEY TERMS

Each of the following terms has been used in this chapter. Do you know their meanings?

ac induction motor
ac series motor
back emf
capacitor-start motor
compound motor
constant-speed motor
counter emf (cemf)
cumulative compound motor
differential compound motor
doubly excited motor
electric motor
electronic motor drive

motor action
phasor diagram
repulsion motor
repulsion-start induction motor
right-hand rule for motors
rotor power factor angle
salient-pole stator
series motor
shaded-pole motor
shading ring (coil)
shunt motor

single-phase motor
singly excited motor
slip
speed regulation
split-phase motor
squirrel-cage induction motor
synchronous motor
synchronous speed
torque
universal motor
wound-rotor induction motor

TEST YOUR KNOWLEDGE

Please do not write in this text. Place your answers on a separate sheet of paper.

1. A motor converts _____ energy to _____ energy, and a generator converts _____ energy to _____ energy.
2. Current flows through a conductor in a magnetic field as shown. Will the conductor move upward or downward?

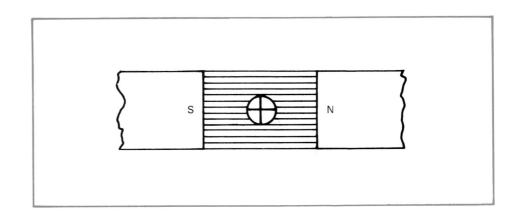

3. Motor action is based on _____ toward the weakened field.
4. Draw diagrams of shunt and series dc motors.
5. Cemf is counter voltage induced by a rotating armature. True or False?
6. Why does a motor heat up if excessively loaded?
7. What is the desirable characteristic of a shunt motor?
8. Why does a series motor always have to be connected to its load?
9. A distinct advantage of the series motor is that it will attain high _____ from a dead stop.

10. What is the purpose of a dc motor starter?
11. How does a split-phase motor start?
12. What is the speed of a rotating field in a four-pole induction motor connected to a 60-Hz power source?
    a. 1200 rpm.
    b. 1800 rpm.
    c. 2400 rpm.
13. What advantage is there to the capacitor-start motor, and how is this accomplished?
14. Compound motors may be wound and connected to have both _____ and _____ field coils.
15. The net voltage applied to a motor is the source voltage minus the _____.
16. The term "speed regulation" is used to describe the performance of a motor under _____ and _____ conditions.

Chapter 8

INSTRUMENTS FOR ELECTRICAL MEASUREMENT

The techniques of understanding and using instruments are among the most important lessons in your study of electricity and electronics. Instruments are the windows through which the operation and performance of electronic circuitry may be observed.

After studying this chapter, you will be able to:

☐ *Discuss the basic principles of ac and dc meter operation.*

☐ *Explain how one meter movement can be used to measure voltage, current, and resistance.*

☐ *Demonstrate the use of the oscilloscope.*

INTRODUCTION

The technical and scientific progress of any nation can be based upon its ability to measure. The success of a technician or engineer is judged by the ability to measure precisely and to interpret the results in circuit performance. Years ago, a measurement to the hundredth of a unit was considered very accurate. Today, measurements are made with sophisticated instruments to the millionth of a unit.

In electricity and electronics, we frequently need to obtain values of voltage and current by measuring. These values may be ac or dc. Further, we need to know values of resistance. There are many different methods and instruments for measuring these quantities. Voltage measurements are made with such varied devices as electromechanical voltmeters, digital voltmeters, and oscilloscopes. Current measuring methods use ammeters. Some of these operate by actually sensing current. Others determine current indirectly from an associated variable like voltage. Resistance meters actually measure current but are calibrated to display resistance directly.

Meters that measure voltage, current, and/or resistance can be split into two classes — *analog* meters and *digital* meters. Those that use electromechanical movements and pointers to display the quantity being measured along a continuous scale are **analog meters**. Those that indicate the quantity being measured by a numerical display are **digital meters**. Further, these meters can be split into those that measure dc values, ac values, or both. For those that measure ac voltage or current, additional circuitry — basically, a rectifier — is needed.

THE BASIC METER MOVEMENT

Most analog meters in use today employ what is called the **moving-coil**, or **D'Arsonval**, **movement**. This setup depends upon the interaction of a moving magnetic coil in a fixed magnetic field. To understand this action, look at the diagrams in Figure 8-1 and review Figure 5-9 in the chapter on magnetism.

As a review, current in a conductor produces a magnetic field around the conductor. When this current-carrying conductor is placed in a fixed magnetic

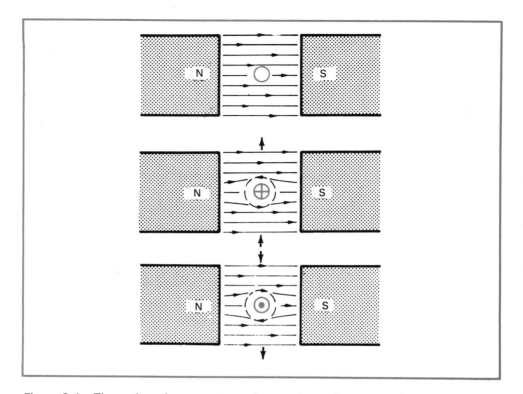

Figure 8-1. The action of a current-carrying conductor in a magnetic field.

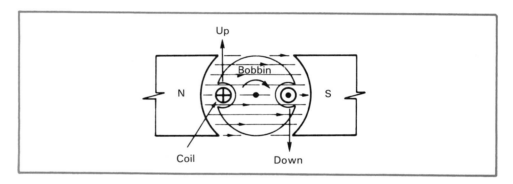

Figure 8-2. Current in the coil in the indicated direction will cause the coil to rotate in a clockwise direction.

field, the two fields add together on one side. On the opposite side of the conductor, the fields oppose each other. As a result, the conductor will move by motor action toward the weakened field. If a conductor is wound on a rotating core, or bobbin, as shown in Figure 8-2, the coil will rotate in a clockwise direction.

Now, examine the phantom view of a real meter movement in Figure 8-3. (Note the use of the shortened word, *movement*.) A rotating coil is mounted between the poles of a permanent magnet. The coil is prevented from complete rotation by springs, which also provide the electrical connections to the coil. Meter leads bring current to the coil. For a dc quantity, **proper meter polarity** must be observed. In other words, the negative lead of the meter must be connected to the more negative side of the circuit; the positive lead to the more positive side. Usually, the negative lead is black and the positive lead is red.

Current through the coil will cause it to react against the field of the magnet. With correct polarity, the coil will move clockwise against the tension of the

Figure 8-3. The D'Arsonval movement.   (Weston Instruments, Inc.)

springs. The stronger the current, the greater the rotating force against the springs and, therefore, deflection of the coil. When current ceases, the springs will return the moving coil to its original position.

Now, note that an indicating needle is attached to the moving coil. It will deflect from left to right as the coil turns. A scale placed under the indicating needle will measure the amount of rotation. Since current through the coil is a function of applied voltage, the scale can be made to read in volts.

There are two important values associated with meter movements. One is the **full-scale deflection current (I_M)**. This is the amount of current needed to deflect the pointer to the last mark on the scale. The other is the **internal resistance of the moving coil (r_M)**. This is the dc resistance of the meter movement. You will soon learn the importance of these two values.

THE DC VOLTMETER

As its name implies, a **voltmeter** is used to measure voltage. A dc voltmeter consists basically of a moving coil in series with a high resistance. The high resistance is called a **multiplier**. The multiplier is used to limit the current through the moving coil to I_M for some specified maximum voltage.

Since a voltmeter has high resistance, it must be connected in *parallel* to measure potential difference across two points in a circuit. Otherwise, if connected in series, the circuit current would be very much reduced by the high-resistance multiplier. Connected in parallel, the high resistance does not have too much effect on circuit current.

MULTIPLIERS

A meter movement responds only to *current* in the moving coil. Knowing that voltage is proportional to current, we can calibrate the meter scale to measure voltage. Suppose we want to measure voltages between 0 V – 10 V with a moving coil. The I_M of our movement is given as 1 mA. From Ohm's law, we determine that we need 10,000 Ω of total resistance to limit I_M to 1 mA. With this resistance and a voltage of 10 V, there would be 1 mA of current through the movement. This would cause a full-scale deflection. At this point, we would mark 10 V on a scale. If the voltmeter is connected across a 5-V potential difference, the current in the movement is 0.5 mA. The deflection is one-half of full scale. We would mark 5 V at this point on our scale, etc.

As stated, we consider a voltmeter to be made up of a moving coil and a multiplier. To find what value is needed for the multiplier, we first figure total voltmeter resistance, R_V. The resistance value of the multiplier, R_{mult}, then, is R_V minus r_M, the internal resistance of the moving coil. In other words:

$$R_{mult} = R_V - r_M$$

Since $R_V = R_T = E_T/I_T$, and $I_T = I_M$, we can expand this to:

$$R_{mult} = \frac{E_T}{I_M} - r_M$$

$$= \frac{\text{max. voltage of range}}{\text{full-scale current}} - r_M$$

Problem:

Assume that a meter movement requires 1 mA for full-scale deflection. Its internal resistance is found from the manufacturer's specifications to be 50 Ω. If we wish to use this meter movement as a voltmeter to measure 0 V − 1 V, what value of multiplier do we need?

Solution:

$$R_{mult} = \frac{E_T}{I_M} - r_M = \frac{1 \text{ V}}{0.001 \text{ A}} - 50 \text{ Ω} = 950 \text{ Ω}$$

We could think of the previous problem in terms of voltage drop. In other words, what voltage applied across the moving coil would cause a full-scale deflection current in the coil? We could then determine the necessary drop across the multiplier in order not to exceed the coil voltage, E_M. Working the problem this way, Figure 8-4, we would have:

$$E_M = I_M \times r_M = 0.001 \text{ A} \times 50 \text{ Ω} = 0.05 \text{ V}$$

At no time can more than 0.05 V be applied across the meter movement without damage to it. Since we wish to measure up to 1 V, we must take a

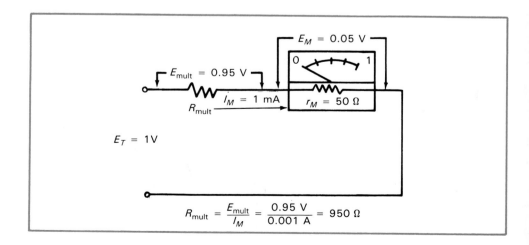

$$R_{mult} = \frac{E_{mult}}{I_M} = \frac{0.95 \text{ V}}{0.001 \text{ A}} = 950 \text{ Ω}$$

Figure 8-4. Computation of a multiplier resistor to measure potential differences of 0 to 1 V.

drop across the multiplier of 0.95 V. The multiplier resistance that would give us this voltage drop would be:

$$R_{\text{mult}} = \frac{0.95 \text{ V}}{0.001 \text{ A}} = 950 \ \Omega$$

Problem:

Using the same meter movement as used before, compute multiplier resistor needed to measure 0 V – 50 V. (Use voltage drop method.) Refer to Figure 8-5.

Solution:

$$R_{\text{mult}} = \frac{(50 - 0.05) \text{ V}}{0.001 \text{ A}} = 49,950 \ \Omega$$

Note that either procedure may be used to compute the multiplier for any range. Also, if the multiplier resistance is greater than 100 times the meter resistance, it generally is not necessary to subtract the meter resistance from the total resistance of the circuit. In the problem just presented, the multiplier may be considered as 50,000 Ω. Figure 8-6 shows a lab setup for determining multiplier values.

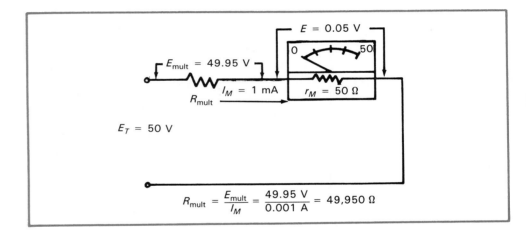

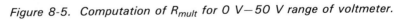

$$R_{\text{mult}} = \frac{E_{\text{mult}}}{I_M} = \frac{49.95 \text{ V}}{0.001 \text{ A}} = 49,950 \ \Omega$$

Figure 8-5. Computation of R_{mult} for 0 V–50 V range of voltmeter.

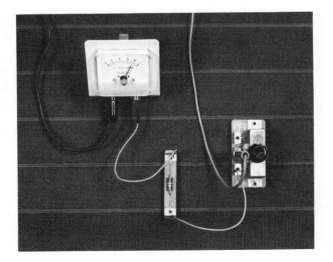

Figure 8-6. A breadboard setup for determining the value of multiplier resistors. (Lab-Volt)

Voltmeter ranges

Figure 8-7 is a schematic showing internal circuitry of a voltmeter. The voltmeter shown has three multipliers and, hence, three ranges. A switch is used to change the meter range. When the range is changed, the appropriate scale on the face of the meter must be used. For example, for 0 V – 1 V and 0 V – 10 V ranges, read the 0 – 10 scale on the meter face. For 0 V – 50 V, use the 0 – 50 scale. A lower voltage may be read using a higher range; however, it is best that the pointer be in the middle third of the scale when reading the meter.

SENSITIVITY AND CIRCUIT LOADING

A voltmeter will draw current from the circuit it is measuring. The **sensitivity** of a voltmeter, expressed in **ohms per volt**, is used to indicate how much a meter will load a circuit. It is equal to the reciprocal of the current required for full-scale deflection:

$$\text{sensitivity} = \Omega/V = \frac{1}{I_M}$$

As an example, the moving coil of the previous problem was a 1-mA movement. The sensitivity of a meter having this full-scale deflection current, then, is:

$$\text{sensitivity} = \frac{1}{0.001} = 1000 \ \Omega/V$$

A higher sensitivity means a higher voltmeter resistance. This is desired over a lower sensitivity. If a low-sensitivity meter is used to measure voltage across a high resistance, the meter will act like a shunt. This will reduce the equivalent resistance of the branch, and an unreliable reading will result.

If the sensitivity of a meter and the meter resistance are known, the multiplier value for any range may be found. Multiply the maximum voltage for any range by the Ω/V. This gives total resistance. Subtract the known resistance of the meter movement. The number remaining is the multiplier value. For example, looking at our 50-Ω moving coil, for a 0 V – 10 V range:

$$R_V = \left(\frac{1000 \ \Omega}{V}\right)(10 \ V) = 10,000 \ \Omega$$

and

$$R_{\text{mult}} = 10,000 \ \Omega - 50 \ \Omega = 9950 \ \Omega$$

Circuit loading by a meter causes incorrect measurements and should be avoided. This is part of knowing how to use a meter. The loading effect can

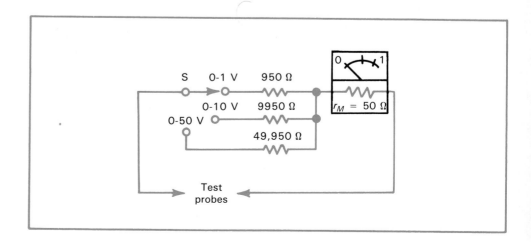

Figure 8-7. This combined circuit permits the voltmeter to measure in three ranges.

be demonstrated by the simple circuit of Figure 8-8. Given source voltage and resistance values, it is a simple matter to compute voltages across R_1 and R_2. Voltage drops are proportional to resistance values of resistors in series. Knowing this, we easily determine that voltage drops across R_1 and R_2 are both 5 V.

What if we were to measure these voltage drops? Let us use our previously constructed meter with the 1000 Ω/V sensitivity. To measure these drops, we will use the 10-V scale. In this range, the resistance of the meter equals 10,000 Ω (10 V × 1000 Ω/V = 10,000 Ω).

In Figure 8-9, the meter is connected to measure the voltage across R_2. It is in parallel with R_2. Therefore, the equivalent resistance of R_2 and R_V in parallel equals:

$$R_{eq} = \frac{R_2 \times R_V}{R_2 + R_V} = \frac{5 \text{ k}\Omega \times 10 \text{ k}\Omega}{5 \text{ k}\Omega + 10 \text{ k}\Omega} = \frac{50 \text{ k}\Omega^2}{15 \text{ k}\Omega} = 3.33 \text{ k}\Omega$$

Using the formula for proportional voltage in a series circuit:

$$E_R = E_S \left(\frac{R}{R_T}\right)$$

The voltage across R_2 would become:

$$E_{R_2} = E_S \left(\frac{3.33 \text{ k}\Omega}{3.33 \text{ k}\Omega + 5 \text{ k}\Omega}\right) = 10 \text{ V} \left(\frac{3.33 \text{ k}\Omega}{8.33 \text{ k}\Omega}\right) = 4 \text{ V}$$

Note that the loading of the circuit by the voltmeter caused a 1-V error in measurement. Suppose we repeat this procedure using a high-quality meter with a sensitivity of 20,000 Ω/V. The meter resistance in the 10-V range would be 10 V × 20 kΩ/V = 200 kΩ. Placing a 200-kΩ meter in parallel with a 5-kΩ

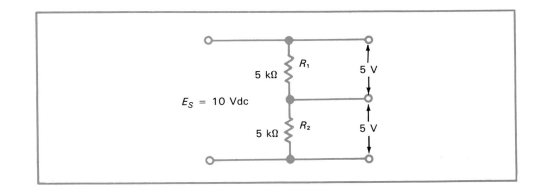

Figure 8-8. Voltage divides equally across equal resistors.

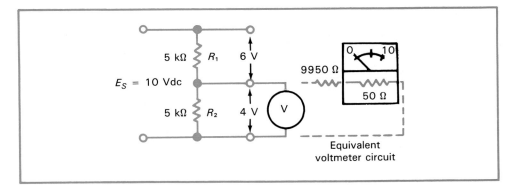

Figure 8-9. Voltmeter puts 10 kilohms in parallel with R_2.

resistor would make an insignificant change in equivalent resistance. The measured voltage would be very close to the expected 5 V. It pays to use good instruments!

LESSON IN SAFETY: Work with one hand only when measuring voltages. A current between two hands can be more dangerous than a current from one hand to your foot. The wise technician will learn to work on live equipment with one hand in a pocket.

THE DC AMMETER

In order to measure current, an **ammeter** is used. A dc ammeter is used to measure direct current. The circuit must be opened and the meter inserted in *series* with the other components in the circuit or branch so that all current passes through the meter. Therefore, it makes good sense to keep the resistance of an ammeter at a low value so that it will not upset the circuit performance. Do *not* put the meter leads across, or in parallel with, a circuit component. You will shunt the current through the low-resistance path of the meter. The resulting short-circuit path could lead to circuit and/or meter damage.

SHUNTS

Working with the same meter movement used for our voltmeter of previous examples, let us review what we know about it.
• Current required for full-scale deflection is 1 mA.
• Voltage required for full-scale deflection is 0.05 V.
• Resistance of meter movement is 50 Ω.

Our meter movement can be used as is to measure currents in the 0 mA−1 mA range. The circuit is shown in Figure 8-10. It is important to note that the

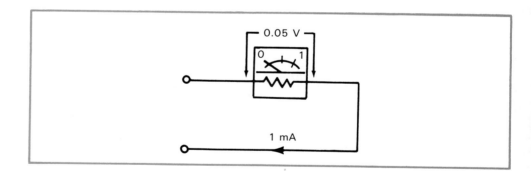

Figure 8-10. *This meter movement will measure from 0 mA−1 mA without extra circuitry.*

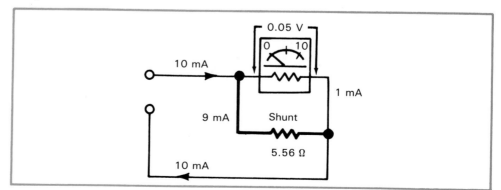

Figure 8-11. *The shunt included here extends current measuring capability to 10 mA.*

voltage drop across our meter is 0.05 V. At no time can 1 mA (or 0.05 V) be exceeded, or the meter will be damaged.

In order to extend the range of the meter to 0 mA – 10 mA, it is necessary to provide an alternate path, or shunt for 9 mA of current. A shunt is added to the circuit shown in Figure 8-11. Since we know shunt current, I_{sh}, and shunt voltage drop, E_{sh}, we can easily calculate shunt resistance, R_{sh}:

$$R_{sh} = \frac{E_{sh}}{I_{sh}} = \frac{0.05 \text{ V}}{0.009 \text{ A}} = 5.56 \text{ }\Omega$$

In order to measure currents up to 50 mA, a different shunt must be used. See Figure 8-12. The shunt must carry 49 mA of current. The same voltage across the meter movement must be used. The shunt resistance is:

$$R_{sh} = \frac{0.05 \text{ V}}{0.049 \text{ A}} = 1.02 \text{ }\Omega$$

In the 0 mA – 100 mA range, the shunt must carry 99 mA. The shunt resistance is:

$$R_{sh} = \frac{0.05 \text{ V}}{0.099 \text{ A}} = 0.51 \text{ }\Omega$$

THE OHMMETER

Frequently, we may want to know the actual resistance value of a circuit component or branch. At other times, we may want to check a circuit for **continuity** to see if we have a broken connection somewhere in a circuit. In this case, we are looking for an open circuit (∞ Ω) or a continuous path (low or 0 Ω). For either application, an **ohmmeter** may be used. This meter is designed to measure the resistance of a circuit or component. (Another instrument used for very accurate resistance measurement is the Wheatstone bridge. The Wheatstone bridge circuit was discussed in Chapter 4.)

THE SERIES OHMMETER

Two types of ohmmeters are *shunt* and *series*. Their function is much the same but they differ in internal circuitry. We will limit discussion here to the series ohmmeter. A basic one-range series ohmmeter circuit is shown in Figure 8-13. The circuit consists of a battery, meter movement, rheostat, and scale resistor. Battery voltage will vary among meters. A 3-V battery is used in this

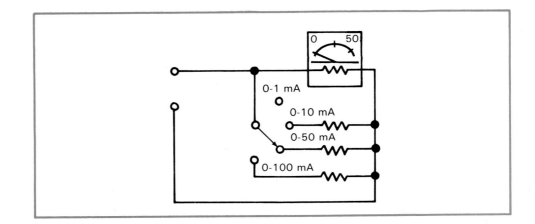

Figure 8-12. Each current range of a meter will require switching to its shunt.

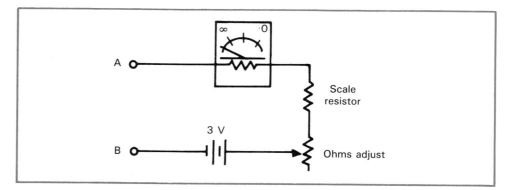

Figure 8-13. The basic circuit of a series ohmmeter.

circuit. The moving-coil assembly of this meter is the same type that has been used in all of the meters discussed so far. The rheostat is used as an *ohms adjust*. Its purpose will be discussed shortly.

The ohmmeter scale resistor is used to limit current through the meter. Since an ohmmeter will usually have several ranges, it will have several of these resistors of different value. By changing resistors with a range selector switch, we can adjust current through the meter to compensate for the effect of external resistance on current. Varying the resistance measuring range of the ohmmeter this way makes it possible to have mid-scale readings for both large and small resistances.

RESISTANCE READINGS

Before going further, remember that:
• Resistance of an open circuit is ∞ (infinite) ohms.
• Resistance of a closed (short) circuit is zero ohms.
Consequently, the meter in Figure 8-13 is carrying no current when test terminals A and B are open. The indicating needle would be far to the left at a point marked on the scale as ∞ Ω.

When test points A and B are shorted together, the needle should deflect to the far right, where 0 Ω is marked on the scale. If it does not, we must change our **ohms adjust** rheostat until it does. If this adjustment is not made, readings will be inaccurate. This procedure is called **zeroing the meter**. It should always be done prior to taking a resistance measurement.

Under what circuit conditions will the meter read 0 Ω? For our meter movement, only when there is 1 mA of current in the circuit. The current is limited to 1 mA by the series resistances. The total resistance R_T of the circuit for 1 mA of current must be:

$$R_T = \frac{3 \text{ V}}{0.001 \text{ A}} = 3000 \text{ } \Omega$$

The total resistance is combined resistance of the meter movement, scale resistor, and ohms adjust. Our meter movement has a resistance of 50 Ω. Therefore, the remaining resistance is 2950 Ω. With this resistance value, we have our required 0.001 A for full-scale needle deflection. Part of this resistance is made variable, so that any decrease in the battery voltage due to aging may be compensated. This is the reason for the ohms adjust.

Needle deflection of the ohmmeter is a function of current, and current is a function of circuit resistance. As a result, we can calibrate our scale in ohms and read the measured resistance directly. As discussed, the meter face is marked 0 Ω for full-scale deflection and ∞ Ω for no deflection. In-between values of resistance result in less than 1 mA through the meter movement. The corresponding deflection on the scale indicates how much resistance is across the ohmmeter.

OHMMETER USE

To measure resistance in a circuit, first turn off circuit power. The battery of an ohmmeter serves as the power source. Do not connect the ohmmeter to a live circuit. Doing so might destroy the meter movement. Place the meter leads across the unknown resistance. Circuit polarity is set by the battery; therefore, the ohmmeter reads upscale regardless of the polarity of the leads. Be cautious of what resistance you are actually measuring. Disconnect one terminal of the device or branch whose resistance you want to measure from the rest of the circuit. Otherwise, if you have a resistance somewhere in parallel, it will throw off your reading. When finished with the meter, turn it off to keep the battery from draining.

Measuring resistance of meter movement

At times, the r_M value of a meter *movement* is unknown, but we need to know its value. We cannot simply put an ohmmeter to it because we are likely to exceed its maximum ratings for current and voltage. To find the unknown value, the simple circuit of Figure 8-14 may be used. Connect a fresh flashlight cell in series with 5-kΩ rheostat R_1 and the meter movement. R_1 should initially be set for maximum resistance to avoid damage to the movement. Adjust R_1 until a half-scale deflection is achieved. Remove R_1 from the circuit and measure its resistance. Record the value as R_{hs}. Replace R_1 in the circuit and adjust it for full-scale deflection. Again, remove it from the circuit and measure its resistance. Record the value as R_{fs}.

We can calculate resistance of our meter movement using the following formula:

$$r_M = R_{hs} - 2R_{fs}$$

We derive our formula knowing two facts. One is that total resistance in our circuit is equal to the resistance of the movement plus the resistance of the rheostat. The other is that total resistance at half-scale deflection is double that at full-scale. In other words:

$$R_T = r_M + R_{fs}$$

and

$$2R_T = r_M + R_{hs}$$

Rearranging this second formula gives:

$$R_T = \frac{r_M + R_{hs}}{2} = \frac{1}{2} r_M + \frac{1}{2} R_{hs}$$

Substituting R_T from our first formula:

$$r_M + R_{fs} = \frac{1}{2} r_M + \frac{1}{2} R_{hs}$$

Transposing unknowns, we get:

$$\frac{1}{2} r_M = \frac{1}{2} R_{hs} - R_{fs}$$

Now, multiplying both sides by 2 to eliminate the fraction leaves our formula for meter movement resistance:

$$r_M = R_{hs} - 2R_{fs}$$

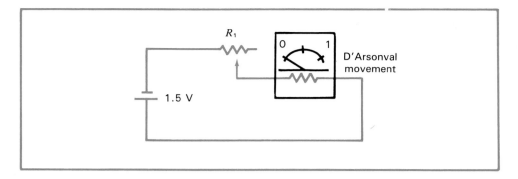

Figure 8-14. Circuit used to measure resistance of a D'Arsonval movement.

Assume that a meter movement is known to have a resistance of 100 Ω. When R_1 is set for full-scale deflection and then measured, it is found to have a resistance of 900 Ω. The total circuit resistance is $r_M + R_{fs}$, or 1000 Ω. In order for the scale to read one-half deflection, the total resistance must be double, or 2000 Ω. Since r_M is 100 Ω of this total resistance, then the value of R_{hs} must be 1900 Ω. Applying our formula:

$$r_M = 1900 \ \Omega - 2(900 \ \Omega) = 100 \ \Omega$$

The formula checks out.

THE MULTIMETER

The **multimeter**, or **volt-ohm-milliammeter (VOM)**, is a basic piece of test equipment found on every technician's bench. It is a very versatile instrument. This meter is three meters combined in one case, Figure 8-15. As its name implies, the three meters that make up the VOM are the voltmeter, the ohmmeter, and the ammeter. Most multimeters can measure ac as well as dc voltage.

AC VOLTAGE AND CURRENT METERS

In the voltmeter and ammeter circuits previously discussed, we have only considered direct current. It is possible to modify these meters for the measurement of alternating current and voltage. In order to do this, the alternating current must be *rectified*, or changed to a pulsating direct current. Measurements are taken just as they are in dc circuits, except that we are not concerned with polarity. You should be aware that the ac meters referred to here need a sinusoidal input. Nonsinusoidal waves will give incorrect readings.

The theory of rectification will be studied in some detail in Chapter 12. For purposes of this instruction, we only need to know that a *diode* is a unidirectional conductor. In other words, current will only travel in one direction through this device. The rectifier circuit is arranged to permit current in one direction and prevent it in the opposite direction. Therefore, an ac wave is converted to a pulsating dc wave. Meter circuits may employ a full-wave or half-wave rectifier circuit. The output of each is shown in Figure 8-16.

After rectification, the rest of the meter circuitry is much like that of the dc meter. The meter movement responds to the average value of the pulsating dc through the moving coil. The scales on the meter face, in most cases, will be calibrated to read the rms value of the ac signal. However, some meters will also have scales that read in average, peak, and peak-to-peak values. Study the meter you are about to use and determine the correct scales for your purpose.

Figure 8-15. Two different VOM models. (AEMC Corp., Triplett Corp.)

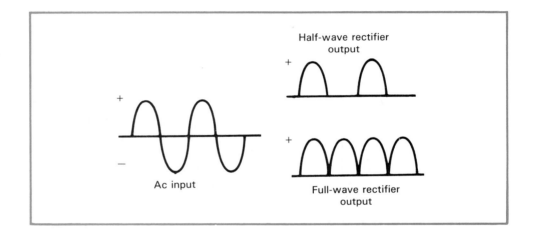

Figure 8-16. A comparison of full-wave and half-wave rectifier outputs. Both are pulsating dc.

The circuits of *half-wave rectifier* meters are drawn in Figure 8-17. Figure 8-17A is a simplified functional circuit where a dc meter is combined with a half-wave rectifier. On the positive half-cycle, point A is positive; point B is negative. There is current through the meter. On the negative half-cycle, point B is positive; point A is negative. There is a very small reverse current through the nonconducting diode. It is so small, however, that we say there is no conduction in this half-cycle.

Figure 8-17B shows another half-wave circuit combined with a dc meter. This circuit uses two diodes. When the ac input is positive, point A is positive; point B is negative. Diode D_1 conducts, and diode D_2 is cut off. On the negative half-cycle, point B is positive; point A is negative. Diode D_2 conducts, and diode D_1 is cut off. The reason for using this arrangement is that on the negative half-cycle all current is shunted through D_2. This way, the small reverse current mentioned in the previous circuit is not found here. The meter is virtually unaffected.

A *full-wave bridge rectifier* for a meter is shown in the circuit of Figure 8-18. This circuit is found in the more expensive meters. It has a greater dc output for any ac input.

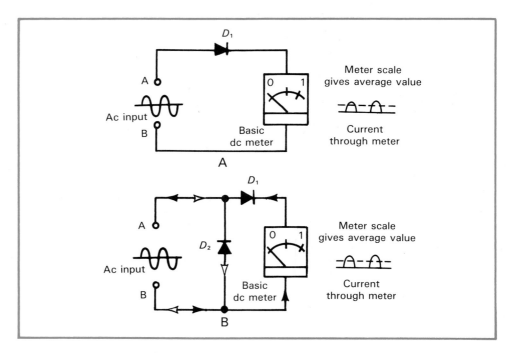

Figure 8-17. Circuits of rectifier-type meters. A—Simplified half-wave circuit. B—Practical half-wave circuit using double-diode arrangement.

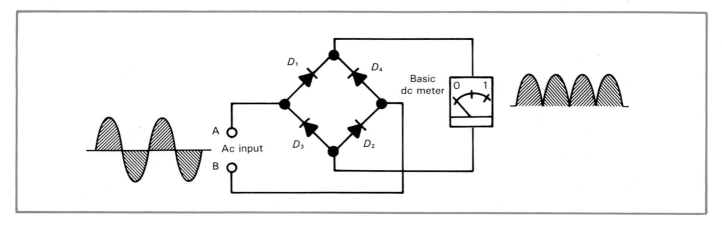

Figure 8-18. A full-wave bridge rectifier for an ac metering circuit.

When the ac input is positive, point A is positive; point B is negative. Diodes D_1 and D_2 will conduct. When the ac input is negative, point B is positive; point A is negative. Diodes D_3 and D_4 will conduct. As a result, there is a current pulse as indicated for each half-cycle of ac input. After rectification, the shunts and multipliers are used to adapt a particular meter movement to a selected range.

THE DIGITAL MULTIMETER

The **digital multimeter (DMM)**, Figure 8-19, is a multimeter with a numerical readout. The readout is either a seven-segment *light-emitting diode (LED)* or *liquid crystal display (LCD)*—similar to what a digital watch has. DMMs are very popular in the electronics field. They precisely measure ac and dc current, ac and dc voltage, and resistance. Advantages of DMMs include: greater accuracy; easy readability; automatic polarity and zeroing; measurement of both ac and dc current.

Figure 8-19. A digital multimeter.   (Beckman Instruments, Inc.)

Figure 8-20.   A typical oscilloscope used in electronics.   (Hickok Electrical Instruments, Inc.)

Measurements with the various functions of the digital meter are taken as they are with any meter. However, again we do not need to be concerned about circuit polarity. The internal principles of DMMs are much different from those of analog meters. These meters do not employ the D'Arsonval movement. The fundamental process involved is an analog-to-digital conversion of the incoming signal. More than this is beyond the scope of this text.

THE OSCILLOSCOPE

The basic instrument of measurement of ac signals is the **oscilloscope**, Figure 8-20. It is called a **scope** for short. The scope produces a picture of the input signal on a screen. Oscilloscopes are essential tools of electronic technicians and engineers and many research scientists. The oscilloscope will be used in many of your experiments and service procedures. Your familiarity with this instrument will increase to a great extent your understanding and use of electronic circuitry and equipment.

The great advantage of the oscilloscope is its ability to stop a rapidly oscillating wave of a given frequency. This allows the operator not only to measure the amplitude of the waveform but, also, to examine its shape. A significant advantage of measuring with a scope is its *high input impedance*, usually in the range of 10 MΩ or more. This means that the instrument will not seriously affect the measurements and performance of the circuit. It draws very little power from the circuit under test.

There are many special features of oscilloscopes that vary among makes, models, and years. We will discuss some of the common features in this section. Obviously, we cannot present every feature of every scope. For correct adjustment and operation of the oscilloscope you are using, you should consult the scope operating manual or your instructor. It is always a good idea to read the instruction manual supplied by the manufacturer before using a meter or any other test instrument.

THE CATHODE-RAY TUBE

The heart of the oscilloscope is the **cathode-ray tube (CRT)**. Patterns formed on the CRT screen result when a narrow beam of electrons is shot from an electron gun within the tube onto the screen. The screen is coated with **phosphor**, which glows where the beam strikes it. The waveform pattern appears as the beam is deflected horizontally and vertically by the instantaneous applied voltages. Even though the beam produces only a single spot of light on the screen, you see a waveform pattern. This is due to the *persistence* of the phosphor. **Persistence** is a quality whereby the phosphor continues to glow where it was struck by the beam. Further, the pattern is seen due to **persistence of vision**. This is where the eye retains an image for a short time after the image is gone.

A cathode-ray tube is depicted in Figure 8-21. The following offers a brief explanation. A voltage is applied to pins 1 and 2 to heat the cathode attached to pin 3. When heated, the *cathode* emits a cloud of electrons, which emerges from the *control grid* (pin 4) as a slightly divergent beam. *Intensity*, or *brightness*, of the screen pattern is controlled by a knob on this grid. The knob, called the **intensity control**, essentially controls the quantity of electrons in the beam—in other words, the *beam current*.

The diverging beam of electrons emerging from the control grid is converged and brought to a focus by the *focus* (pin 5) and *accelerating* (pin 6) *anodes*. A control element provides adjustment of the voltage on the focusing anode. This is the **focus control**. A beam in sharp focus at the center of the screen, however, may be out of focus near the edge. This happens when the lengths of the electron paths change as the beam is deflected. To adjust, some scopes have an **astigmatism control**. It sets the voltage on the accelerating anode of the CRT. This control gives a sharp focus over the entire screen.

Two sets of *deflection plates* follow. The first set consists of the **vertical deflection plates**. The signal voltage, sensed at its source of origin by an **oscilloscope probe**, is applied to pins 7 and 8 of these plates. For the more common *single-ended input*, the signal is applied to one plate; the other plate is grounded. The plates change the direction of the electron beam in the vertical direction. This happens as the negatively charged electron is attracted by a positively charged and repelled by a negatively charged plate. The beam is deflected in proportion to the amplitude of the input signal.

In a similar fashion, voltage applied to the next set of plates—the **horizontal deflection plates**—will deflect the electron beam from left to right. Voltage is applied to pins 9 and 10 using a **sweep oscillator** that produces a sawtooth waveform. See Figure 8-22. The sawtooth voltage increases in a linear manner. This way, it will move the electron beam uniformly to the right by some amount depending on its amplitude. This is called **sweep**. When the voltage falls rather abruptly to zero, the beam is returned almost instantly to the left edge of the screen. This is called **flyback**. The frequency of the sawtooth signal must be synchronized with that of the vertical signal. This way, the successive images observed on the screen will superimpose in exact step. A stable stationary pattern will result.

Control elements provide adjustment of voltages to the deflection plates. Adjusting them allows us to shift the center of a displayed waveform to any point on the screen. The **vertical position control** permits the operator to move

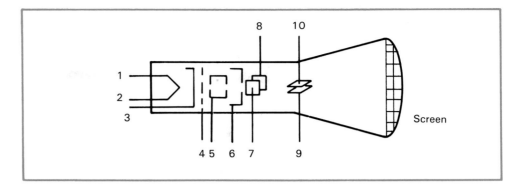

Figure 8-21. The construction of a cathode-ray tube.

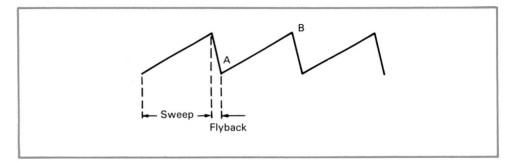

Figure 8-22. Sawtooth waveform from a sweep oscillator.

the trace up or down on the screen. The **horizontal position control** permits the operator to move the trace right or left on the screen.

TYPES OF OSCILLOSCOPES

Oscilloscopes may be roughly divided into two basic types — *recurrent sweep* and the more advanced *triggered sweep*. Of the two, the second is more practical for modern electronics work. As such, it is more widely used.

Recurrent-sweep oscilloscopes

In the **recurrent-**, or **free-running-**, **sweep oscilloscope**, the sweep signal voltage is generated by a sawtooth oscillator like that previously discussed. In order to be in sync, the frequency of the sawtooth must be equal to or be a submultiple of the vertical signal frequency. For example, a 60-Hz vertical input signal must be accompanied by a horizontal frequency of either 60 Hz, 30 Hz, 20 Hz, etc.

The oscillator is usually synchronized to the input signal, in part, by feeding it a sample of the input signal from the vertical circuits. This sample is obtained by internal wiring connections and a switch set for *internal sync*. A desired submultiple is chosen by adjusting a **sweep range control** that will allow a range of frequencies to be generated. Afterward, a **sweep vernier control** is adjusted for fine tuning.

A **vertical sensitivity**, or **volts/div, control** on the scope is used to set the height of the display. The vertical distance tells the voltage amplitude of the displayed waveform. This is done by multiplying the vertical spacing in divisions by the value of the volts/div control setting.

Recurrent-sweep oscilloscopes are used for the display of fairly simple waveforms. These scopes require manual adjustment to synchronize signals. More importantly, they are limited to displaying recurrent signals of constant frequency and amplitude. The advantage to these scopes has been their availability at low cost.

Triggered-sweep oscilloscopes

More advanced scopes employ the triggered-sweep, instead of recurrent-sweep, design. Note the *trigger* section on the front panel of the advanced digital oscilloscope shown in Figure 8-23. The sweep generator of the recurrent-sweep scope is present in the **triggered-sweep oscilloscope**; in addition, a *pulse generator* is added. A **pulse generator** is a device used for emitting controlled series of electrical pulses. The sweep generator is inactive until a trigger signal starts it operating. The signal comes from one of three sources — *internal, external,* or *line*. Internal is derived from the vertical input signal. External comes from an external source. Line is 60-Hz line voltage. These sources, or modes, are selected with the **trigger source switch**.

In the internal mode, when the trigger signal reaches a preselected slope (+ or −) and amplitude, the pulse generator delivers a pulse to the sweep generator. One cycle of sweep is then produced. The sweep generator then rests until the input is again at the preselected conditions. The conditions are set by the **trigger slope control** and **trigger level control** of the scope.

If the input signal is a continuous sine wave, a continuous sawtooth is generated, as in a recurrent-sweep scope. It is in sync with the input signal so that the display stands still. If there is no input signal, no sweep occurs. For the most part, triggered-sweep scopes automatically adjust to the frequency at which they are driven. As a result, the display remains stable in spite of variations in the frequency and amplitude. Further, they can be set so that they do not produce a trace in the absence of a vertical signal. If the input consists of random pulses, the sweeps occur only when there are pulses. There must be a pulse for the trace to show on the screen.

Like recurrent-sweep scopes, triggered-sweep scopes have a volts/div control to adjust vertical beam deflection. In addition, the triggered-sweep scope has a control to adjust horizontal beam deflection. This is the **sweep time**, or **time/div, control**. It sets the time it takes to move the electron beam horizontally across one division of the screen. Adjusting this control will make a waveform wider or narrower. Switching to a lower time setting will display fewer cycles. Switching to a higher time setting will display more cycles.

The calibration markings on the time/div control permit the elapsed time between any two points on the display to be determined. This is done by

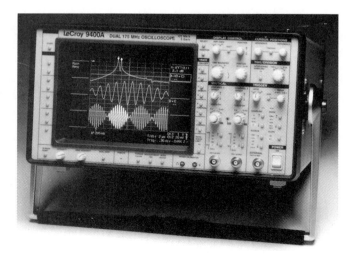

*Figure 8-23.   A **digital scope** converts an analog input waveform into a digital signal that is stored in memory. It then converts it back into analog form for display on a conventional CRT.   (LeCroy Corp.)*

multiplying the horizontal spacing in divisions by the value of the time/div control setting. This control can be used to figure the period and frequency of a waveform.

VOLTAGE AND CONTINUITY TESTER PROJECT

You can build a handy voltage and continuity tester using an LED. See Figure 8-24. This tester will show polarity. It will check for dc at any terminal or connector. It will test the continuity of a bulb, fuse, conductor, etc.

Examine the tester circuit in Figures 8-25 and 8-26. When the toggle switch is in the *voltage* position, the LED is in series with diode D_1 and limiting resistor R_1. In this position, you can check for voltage polarity and presence of voltage. For the LED to light, there must be a voltage present, and polarity with the circuit under test must be correct. The positive (red) test lead must be connected to the more positive, and the negative (black) lead, to the more negative terminal of the test circuit. If the leads are reversed, the LED will not light. The circuit will react to any voltage from 3 V to 15 V. Do not use the tester to check out circuits carrying more than 15 V or you will ruin the LED.

When the toggle switch is in the *continuity* position, a 9-V battery is added to the circuit. The LED will light when the test leads are connected to a bulb, fuse, conductor, etc., provided it is electrically sound. If the device is part of a circuit, check for voltage before turning the switch to the continuity position.

Important: When soldering LED devices, use standard transistor and IC techniques. That is, place a heat sink on the lead being soldered, between the device and the point at which the lead is being soldered.

LESSON IN SAFETY: When soldering or desoldering, do not shake off or brush off hot solder in such a way that it might hit you or your working partner. If you have ever had a drop of hot solder land on you, you would know that this rule of safety cannot be overemphasized.

Figure 8-24. LED voltage and continuity tester.

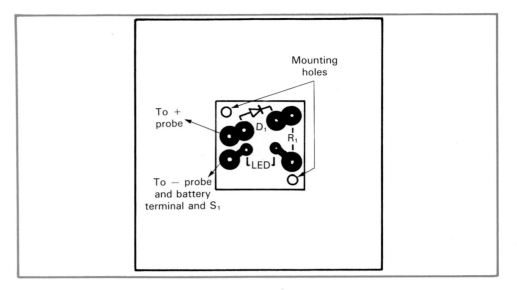

Figure 8-25. Printed circuit layout for LED tester.

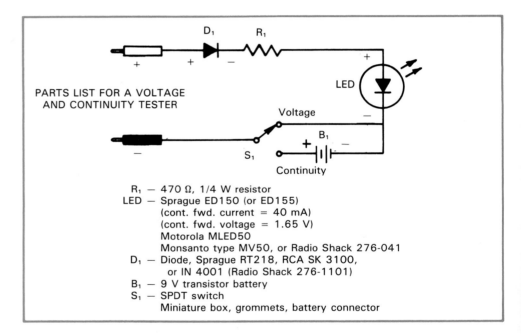

Figure 8-26. Schematic of LED voltage and continuity tester. *(Sprague)*

SUMMARY

- The basic meter movement used for many instruments is the D'Arsonval movement.
- Analog meters use a needle and scale with continuous variable values. Digital meters give values in exact amounts using digits 0 through 9.
- It is vital that you observe correct polarity in using analog meters for measuring dc current and voltage.
- Ammeters measure current and are connected in series in the circuit. Shunt resistors, connected in parallel with meter movements, increase available ranges for the ammeter.
- Voltmeters measure voltage and are connected in parallel in the circuit. Multiplier resistors, connected in series with meter movements, increase available ranges for the voltmeter.

- The sensitivity of a meter is an indication of its quality. Sensitivity is measured in ohms per volt.
- A multimeter is one instrument that will measure current, voltage, and resistance.
- Ac meters use rectifiers to convert ac to dc. Otherwise, the internal circuitry is like that of a dc meter.
- Oscilloscopes display waveforms, which show what is happening in a circuit.

KEY TERMS

Each of the following terms has been used in this chapter. Do you know their meanings?

ammeter
analog meter
astigmatism control
cathode-ray tube (CRT)
continuity
D'Arsonval movement
digital meter
digital multimeter (DMM)
digital scope
flyback
focus control
free-running-sweep oscilloscope
full-scale deflection current (I_M)
horizontal deflection plate
horizontal position control
intensity control
internal resistance of moving coil (r_M)

moving-coil movement
multimeter
multiplier
ohmmeter
ohms adjust
ohms per volt (Ω/V)
oscilloscope
oscilloscope probe
persistence
persistence of vision
phosphor
proper meter polarity
pulse generator
recurrent-sweep oscilloscope
scope
sensitivity
sweep

sweep oscillator
sweep range control
sweep time control
sweep vernier control
time/div control
trigger level control
trigger slope control
trigger source switch
triggered-sweep oscilloscope
vertical deflection plate
vertical position control
vertical sensitivity control
voltmeter
volt-ohm-milliammeter (VOM)
volts/div control
zeroing a meter

TEST YOUR KNOWLEDGE

Please do not write in this text. Place your answers on a separate sheet of paper.

1. Most meters in use today employ the _____ movement which depends upon the interaction of a moving magnetic coil in a fixed magnetic field.
2. The _____ of a voltmeter can be used to determine how much the meter will load the circuit when it is used to measure voltage.
3. In order to measure dc current in a circuit, the circuit must be opened and an ammeter inserted in series with other components. True or False?
4. The ohmmeter is designed to measure the _____ of a circuit or component.
5. The _____ is the basic instrument of measurement of ac voltage signals and waveforms.
6. Refer back to Figure 8-14 in this chapter. The measured resistance of R_1 when the meter movement is at full-scale deflection is 500 Ω. At half scale, the resistance is measured at 1200 Ω. What is the resistance of the movement?
7. A certain meter requires 1 mA for full-scale deflection. It has an internal resistance of 100 Ω. What is its sensitivity?
8. What value multiplier is required for the 50-V range of the previous meter?
9. What value multiplier is required to measure up to 300 V?
10. Using our previous movement, compute the shunt to measure current in the 100-mA range.

11. A voltage divider in the accompanying illustration has the resistance values indicated. Compute the voltage expected across resistor R_2.
12. Referring to the illustration, compute the voltage expected across R_2 if measured with a 5000 Ω/V meter. Use the 0 V – 10 V meter range.
13. Compute voltage expected across R_2 if measured with a 20,000 Ω/V meter and the 0 V – 10 V meter range.
14. What is the purpose of the focus control on an oscilloscope?
15. What is the purpose of the vertical position control on an oscilloscope?
16. What is the purpose of the astigmatism control on an oscilloscope?

Applications of Electronics Technology:
The engineer shown here is custom-designing a servo motor using a CAD system. (Custom Servo Motors, Inc.)

Section II

APPLICATIONS OF MAGNETISM

SUMMARY

Important Points

- [ ] The earth is a large magnet.
- [ ] Magnetism is the number one source of electricity.
- [ ] Magnets have two different poles—*north* and *south*.
- [ ] Like poles repel, and unlike poles attract each other.
- [ ] Lenz's law states that the induced voltage in any circuit is always in a direction that opposes the effect that produces it. This is sometimes called electrical inertia.
- [ ] Flux lines exist around magnetized materials. These lines represent the magnet's field of force.
- [ ] The stronger the magnet, the stronger the flux density.
- [ ] The ability of a material to conduct magnetic lines of force is permeability. The opposition to magnetic lines of force is called reluctance.
- [ ] Electromagnets are created by a current flowing through a coil. Their field strength can be increased and decreased by varying the current. Electromagnets can be turned on and off.
- [ ] A generator converts mechanical energy into electrical energy.
- [ ] A motor converts electrical energy into mechanical energy.
- [ ] To induce a current in a conductor, these factors must be present:
  - A magnetic field.
  - A closed conductive circuit.
  - Relative motion between conductor and field.
- [ ] Generators are generally very efficient. However, they have some inefficiency in the form of copper, eddy current, and hysteresis losses.
- [ ] Generators can output ac or dc by the use of either slip rings or commutators on the armature.
- [ ] The practical value for an ac sine wave is the effective, or root-mean-square (rms), value.
- [ ] Sine waves can be either in phase or out of phase, depending on the types and values of components in the circuit.
- [ ] A simple motor has an armature, a magnetic field, and some means for getting current to the armature such as a commutator or slip rings.
- [ ] Some basic types of dc motors are series, shunt, and compound motors.
- [ ] Ac induction motors are very popular motors today.
- [ ] Instruments are used to give exact measurement of specific values in a circuit.
- [ ] Ammeters are used in series and measure current. Voltmeters are connected in parallel and measure voltage.
- [ ] Ohmmeters have their own power source and measure resistance.
- [ ] Analog and digital meters are used as basic measurement devices in circuits.
- [ ] An oscilloscope is a test instrument. It uses a cathode-ray tube, which permits observation of the frequency and amplitude of waveforms.

1. 210 volts rms = _____ volts peak
2. 1520 volts peak = _____ volts rms
3. The basic unit of frequency is the _____.
4. What is the wavelength of a 92 megahertz radio wave, in feet? _____
5. One cycle has _____ peaks.
6. A relay is an electromagnetic _____.
7. Give the value for the shunt, R_{sh}, in the circuit below:
   R_{sh} = _____

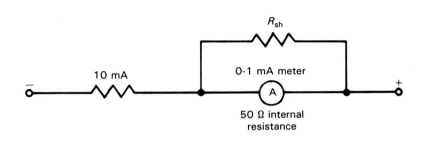

8. Compute the value of the multiplier resistor, R_{mult}, in the circuit below:
   R_{mult} = _____

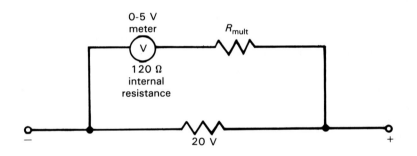

Section III

APPLICATIONS OF ALTERNATING CURRENT

Changing, or alternating, electron flow is the basis for many electrical devices today. The capacitor, inductor, and the transformer rely on this change for their operation. The total opposition offered to a change in current or voltage in an alternating current circuit is impedance.

Chapter 9 presents the theory and operation of inductors and transformers in electrical circuits. Also, this chapter discusses inductors coupled with resistors (RL circuits) to give reactance.

Capacitance is an interesting concept in electricity. It is covered in Chapter 10. Presented are the factors which affect capacitance, the types of capacitors, reactance, and RC circuits.

Chapter 11 discusses ac circuits which have resistors, capacitors, and inductors placed in them. Resonance and its effect in an ac circuit is presented. Also, the basis for oscillators, the tank circuit, is explained.

A thorough understanding of alternating-current theory is important for a solid understanding of semiconductor technology.

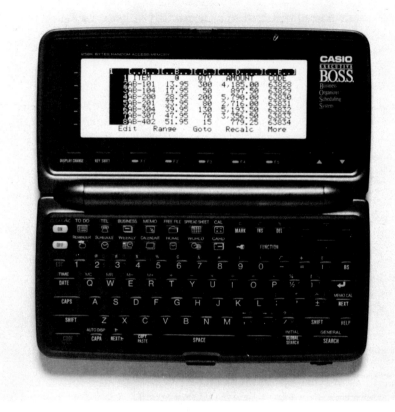

Applications of Electronics Technology:
This business organizer acts as an electronic assistant. It includes a built-in
Lotus compatible spreadsheet, a telephone director, a memo function, and
calculator, to name a few. (Casio)

Chapter 9   INDUCTANCE

Inductance is a most interesting and important topic in electricity and electronics. You will want to know a great deal about inductors, and that will be to your advantage. A thorough understanding of inductance will lead to competence as a technician or an engineer.

After studying this chapter, you will be able to:

☐ *Discuss self-inductance and mutual inductance.*
☐ *Identify various kinds of inductors and their uses in electronic circuits.*
☐ *Analyze the performance of inductors in dc and ac circuits.*
☐ *Explain how transformers work and name several applications.*

INTRODUCTION

Inductance *(L)* is the measure of the ability of a coil to establish an induced voltage as a result of a changing current. It is that property of a circuit that opposes a change in current. It may also be described as that property of a circuit in which energy is stored in a magnetic field. The basic unit of inductance is the **henry (H)**. It is named in honor of Joseph Henry, an American scientist. Henry observed induced currents even before Faraday. However, Faraday was first to publish his own independent findings. As a result, Faraday received credit.

For a review and further emphasis, a current through a coil of wire produces a magnetic field around the coil. Its strength and density will depend upon the number of turns of, and the current through, the coil. Its polarity is determined by the left-hand rule for coils.

A basic component of electrical and electronic circuits is the *inductor*. Basically, an **inductor** is a *coil* of wire, with or without a core; however, a single loop of wire will exhibit properties of inductance under certain conditions. Inductor symbols are shown in Figure 9-1. Other names for an inductor are *coil. choke.* and *reactor.* Examples are shown in Figure 9-2.

Chapter 6, you will recall, briefly highlighted Michael Faraday's work in electromagnetic induction. Remember that as a conductor moved through a magnetic field, a voltage was induced in the conductor. Further, if the conductor was a closed circuit, a current was induced, too. The induced current produced a magnetic field around the conductor in a direction that opposed the movement of the conductor. An external force was required to move the conductor.

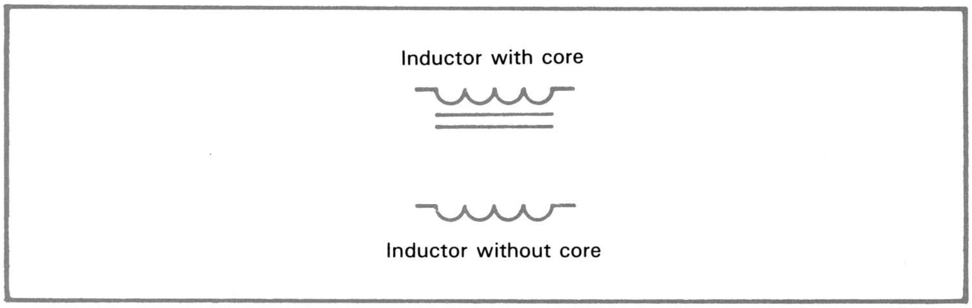

Figure 9-1. Symbols used on schematic drawings for inductors.

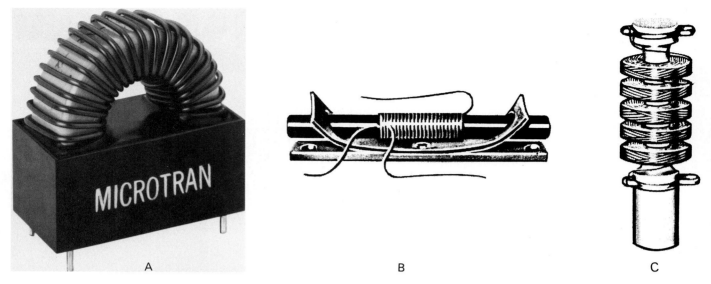

Figure 9-2. Examples of inductors. A—Toroidal power inductor for high dc current power supplies. (Microtran Company, Inc.) B—Ferrite-coil antenna for radio receivers. (J.W. Miller Co.) C—Radio frequency choke for impeding ac currents of radio frequency.   (J.W. Miller Co.)

SELF-INDUCTANCE

In Figure 9-3, an inductor, or coil, is connected to a variable dc power supply. As the circuit is switched on, a rise in current is seen. A magnetic field, starting at zero, expands to its full strength and density. As the field moves outward, it cuts across the windings of the coil.

As Faraday's experiments proved, relative motion between a magnetic field and a conductor will induce a voltage in the conductor. Either the field may be fixed and the conductor moved, or vice versa. In either case, there is relative motion. In our circuit coil, the expanding, or *moving*, field will induce a voltage in the coil. The voltage induced by the moving field is a counter emf (cemf). This effect, in which a coil induces an emf in itself, is called **self-inductance**.

Induced coil voltage will oppose the source voltage and the rise or fall of current through it in accordance with Lenz's law. In an inductor, this law

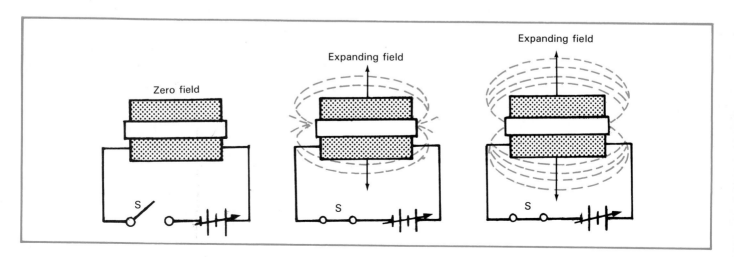

Figure 9-3. As current is increased, the expanding magnetic field cuts across the coil winding.

is concerned with the opposition to a change in current. It states that *the polarity of an induced emf is such that it sets up a current, the magnetic field of which always opposes the change in the existing field caused by the original current.*

Current in our inductive circuit builds up only gradually, and this is because of the effect of Lenz's law. After the current has risen to its full value, the magnetic field remains constant. There is no longer a change in current, and the cemf is zero.

By changing the source voltage to a lower value, the circuit current will tend to decrease. However, the decreasing magnetic field will again induce a voltage. This produces a current in a direction that aids the existing current. Thus, the change in current is opposed—tending to hold the current at its existing value. That is, the induced cemf will oppose the change in circuit current. We see, then, that an inductor in a circuit opposes *any* change of current in a circuit.

THE MAGNITUDE OF CEMF

How great will the opposition to a change in current of a specific inductor be? In the chapter on magnetism, we found that the magnitude and density of a magnetic flux will depend upon the number of turns of wire in the coil. The induced cemf will depend upon the density and magnitude of the flux and, thus, the number of turns. However, the relative motion also must be considered. A rapid movement, or a rapid change in flux, will induce a larger cemf than a slower change. The *rate of change* must be taken into account.

For a *defined interval*, the average value of cemf of an inductor may be found by the formula:

$$e_{L_{avg}} = N \frac{\Delta \Phi}{\Delta t}$$

where:

$e_{L_{avg}}$ = average induced cemf in volts
N = number of turns of the coil
$\Delta \Phi / \Delta t$ = rate of change of flux in webers/second

Note that the Greek letter Δ (delta), as used here, means a "change in." Also, it is important to note that cemf is inversely proportional to time. That is, the faster rate of change (less time) will produce a greater cemf.

THE HENRY

As mentioned, the unit of inductance is the henry. The symbol for the henry, again, is H. In electronic circuitry, inductors with values in henrys, millihenrys (1/1000 H), and microhenrys (1/1,000,000 H) will be used. An inductor is said to have an inductance of 1 henry when a changing current of 1 ampere per second induces a counter emf of 1 volt across the inductor. In other words:

$$1 \text{ H} = \frac{1 \text{ V}}{\text{A/s}} = \frac{1 \text{ V} \cdot \text{s}}{\text{A}}$$

The mathematical relationship is expressed as:

$$L = \frac{e_{L_{avg}}}{\Delta i / \Delta t}$$

where L is in henrys, $e_{L_{avg}}$ is the induced cemf in volts, and $(\Delta i / \Delta t)$ is the rate of change of current in amperes/second.

Problem:

What is the inductance of a coil that induces 100 V of cemf when the current changes 1 A in 10 ms?

Solution:

$$L = \frac{e_{L_{avg}}}{\Delta i/\Delta t} = \frac{100 \text{ V}}{1 \text{ A}/0.01 \text{ s}} = 1 \text{ H}$$

For an ac waveform, we can roughly calculate the size of an inductor from *average* counter emf, peak current, and frequency. We know that current grows from zero to maximum value in one-fourth of a cycle. The time period of one cycle is $T = 1/f$. Therefore, the time t taken for the current to grow from zero to I_{max} is:

$$t = \frac{1}{4} T = \frac{1}{4f}$$

During this time, the *average* rate of change of current in time t is:

$$\frac{\Delta i}{\Delta t} = \frac{I_{max}}{1/(4f)} = 4fI_{max}$$

Substituting into our original equation, we have a formula for figuring the approximate size of an inductor given certain ac values:

$$L = \frac{e_{L_{avg}}}{4fI_{max}}$$

Problem:

The average ac voltage induced in a certain inductor is 100 V. If the peak current is 1 A and the frequency of the waveform is 100 Hz, what is the inductance of the coil?

Solution:

$$L = \frac{e_{L_{avg}}}{4fI_{max}} = \frac{100 \text{ V}}{4(100 \text{ Hz})(1 \text{ A})} = 0.25 \text{ H}$$

Characteristics of inductors

An expression can also be derived to calculate inductance given the characteristics of the coil. It is derived from our formulas $L = e_{L_{avg}}/(\Delta i/\Delta t)$ and $e_{L_{avg}} = N(\Delta\Phi/\Delta t)$ and from certain relationships presented in the chapter on magnetism. The formula is given as:

$$L = \frac{N^2\mu A}{l}$$

where:

N = number of turns of coil
μ = permeability of core in henrys per meter
A = cross-sectional area of core in square meters
l = length of core in meters

Note that maximum inductance is obtained with a short coil having a large cross-sectional area and a large number of turns.

Problem:

A 300-turn air-core coil is wound on a hollow core that is 15 cm long. Cross-sectional area of the core is 4 cm². What is the value of the inductor?

Solution:

$$L = \frac{N^2\mu A}{l} = \frac{(300)^2(4\pi \times 10^{-7} \text{ H/m}) (0.0004 \text{ m}^2)}{0.15 \text{ m}} = 3.016 \times 10^{-4} \text{ H} = 0.302 \text{ mH}$$

MUTUAL INDUCTANCE

When the current in an inductor changes, an expanding and collapsing magnetic field is created. The field will cut across the coil windings and induce a voltage. If a second inductor is placed nearby, the changing field will also cut across its windings. Voltage will now be induced in this coil, too. See Figure 9-4. This property just described is known as **mutual inductance** (**M**). The connection between the coils is by a magnetic field only.

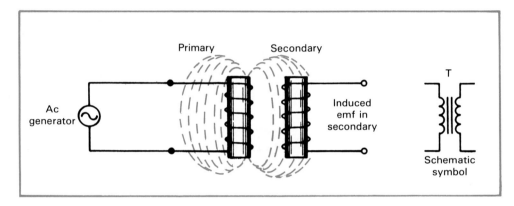

Figure 9-4. *In a transformer, energy is transferred by a moving magnetic field; voltage is induced in both coils.*

The assembly of the two coils shown in Figure 9-4 is called a *transformer*. The input coil is named the **primary winding**, or, simply, the **primary**. This coil is connected to an ac source. The output coil is named the **secondary winding**, or the **secondary**. This coil would be connected to a load resistance. By definition, a **transformer** is a component used to transfer electrical energy from one circuit to another by means of a varying magnetic field. Mutual inductance will be discussed in relationship to transformers later in this chapter.

Mutual inductance, like self-inductance, is measured in units of henrys. Two coils are said to have a mutual inductance of 1 henry when a changing current of 1 ampere per second in the primary induces an emf of 1 volt in the secondary. The mathematical relationship is expressed as:

$$M = \frac{e_{L_{\text{avg}}}}{\Delta i / \Delta t}$$

where:

M = mutual inductance in henrys
$e_{L_{\text{avg}}}$ = average emf induced in the secondary in volts
$\Delta i / \Delta t$ = rate of change of current in the primary in amperes/second

Problem:

The current in a primary coil changes 2 amperes per second and induces 20 volts in a secondary coil. What is the mutual inductance?

Solution:

$$M = \frac{e_{L\text{avg}}}{\Delta i / \Delta t} = \frac{20 \text{ V}}{2 \text{ A/s}} = 10 \text{ H}$$

COUPLING

Coils with mutual inductance are said to be *magnetically coupled*. In general, **coupling** is a means of passing energy from one circuit to another. If the coils are close to each other, many magnetic lines in the primary flux link with the secondary. Mutual inductance will be greater. On the other hand, if the coils are a distance apart, there might be very little linkage. Mutual inductance will be less. Further, the mutual inductance of two coils can be increased if a common iron core is used for both coils.

The degree to which the flux of one coil links with the other is given by its **coupling coefficient**. This number gives the fraction of flux available for mutual induction. It is assigned the letter **k**. If all the lines of force of the primary cut across the secondary, the condition is called **unity coupling**. The value of *k* is 1. If the coils are separated so that only half the flux lines cut across the secondary, then .5 is the value of the coupling coefficient.

You should be aware of the angular position of the coils, and how it affects mutual inductance. To induce a maximum emf in the secondary, the flux lines must cut across the secondary windings at right angles. This can only happen when the axes of the coils are parallel to each other. If one coil is turned so that there is an angular difference between the axes, the mutual inductance will be reduced. See Figure 9-5.

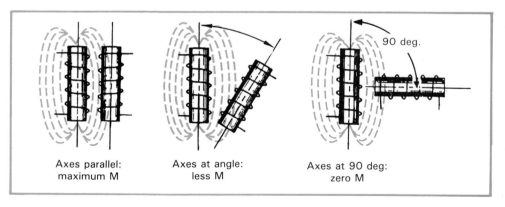

Figure 9-5. Mutual inductance (M) depends on angular relationship between coils.

Using the coefficient of coupling and the inductance of each coil, mutual inductance may be found by the formula:

$$M = k \sqrt{L_1 L_2}$$

Problem:

Two 4-H coils have a .5 coupling coefficient. What is their mutual inductance?

Solution:

$$M = k \sqrt{L_1 L_2} = .5 \sqrt{4 \text{ H} \times 4 \text{ H}} = .5 \times 4 = 2 \text{ H}$$

INDUCTORS IN SERIES

When two or more coils of any inductance are connected in series, and there is no coupling between the coils, the total inductance will be the *sum* of the individual inductances. Referring to Figure 9-6:

$$L_T = L_1 + L_2 + L_3 + \ldots$$

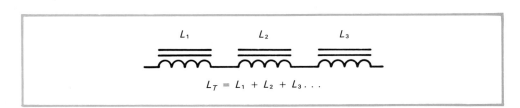

$$L_T = L_1 + L_2 + L_3 \ldots$$

Figure 9-6. Inductors in series, zero coupling.

If coupling should exist between the coils, then mutual inductance must be considered. In Figure 9-7, left, both of the coils produce magnetic fields with the *same* polarity. They combine to form a total inductance of:

$$L_T = L_1 + L_2 + 2M \qquad \textit{(series-aiding)}$$

In Figure 9-7, right, the two coils are producing opposing magnetic fields. Therefore, mutual inductance is subtracted from the total series inductance:

$$L_T = L_1 + L_2 - 2M \qquad \textit{(series-opposing)}$$

Problem:

A circuit contains a 4-henry and a 9-henry coil. They are connected series-opposing and have a .5 coupling coefficient. What is the total circuit inductance?

Solution:

Step 1. Find mutual inductance.

$$M = k \sqrt{L_1 L_2} = .5 \sqrt{4 \text{ H} \times 9 \text{ H}} = 3 \text{ H}$$

Step 2. Find total inductance.

$$L_T = L_1 + L_2 - 2M = 4 \text{ H} + 9 \text{ H} - 2(3 \text{ H}) = 7 \text{ H}$$

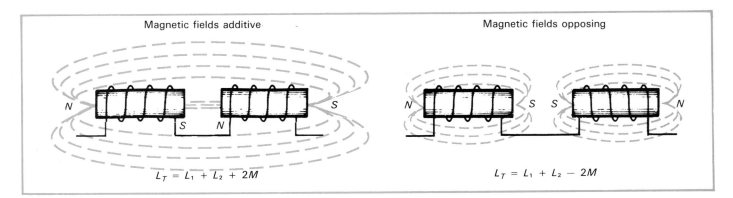

Magnetic fields additive

Magnetic fields opposing

$$L_T = L_1 + L_2 + 2M$$

$$L_T = L_1 + L_2 - 2M$$

Figure 9-7. Positions of coils to produce additive or subtractive mutual inductance.

INDUCTORS IN PARALLEL

We will consider inductors in parallel only when no coupling exists between coils. When two or more inductors are connected in parallel, Figure 9-8:

$$\frac{1}{L_T} = \frac{1}{L_1} + \frac{1}{L_2} + \frac{1}{L_3} + \ldots$$

If only two inductors are in parallel, this simplified formula may be used:

$$L_T = \frac{L_1 L_2}{L_1 + L_2}$$

Note that the formulas for series and parallel inductors are the same as series and parallel resistors (except L is used in place of R).

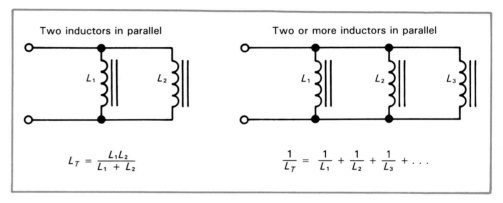

Figure 9-8. *Inductors in parallel and total inductance.*

INDUCTANCE IN DC CIRCUITS

The behavior of an inductance in a circuit is perhaps more easily understood if we first consider the *series RL circuit* with a *dc* voltage source. A circuit that contains both a resistance and an inductance is called an **RL circuit**. The voltage and current relationships in a series RL circuit are of particular interest in our studies. Look at Figure 9-9. When S_1 (switch) is moved to position A, these circuit conditions exist:

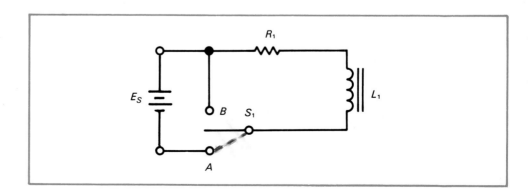

Figure 9-9. *A circuit used to demonstrate voltage and current relationships in a series RL circuit.*

- Current at the initial <u>instant</u> is zero. The voltage drop across R_1 is zero.
- Starting from zero, the actual current is very small; although the *rate* of current change is greatest. The magnetic field will change most rapidly. The cemf developed across L_1 is greatest. It will be <u>almost equal to</u>, and opposing, E_S.
- The rate of current change decreases. The cemf across L_1 decreases. Current then rises in the circuit. The voltage across R_1 increases.
- Current reaches maximum value. There is no longer a change in current, so the cemf across L_1 is zero. Assuming L_1 has no dc resistance, the voltage drop across R_1 is $e_R = i \times R_1$. This is equal to the source voltage.
- Under equilibrium conditions, with no change in current value, the circuit appears purely resistive. The effect of the inductor has disappeared.

When S_1 is moved to position B, the circuit will discharge. These circuit conditions will exist:

- Maximum cemf of opposite polarity will appear across L_1 due to rapid change from maximum current toward zero. This voltage will try to keep the current in the circuit.
- Current will decay toward zero, and the voltage across R_1 will decay toward zero.
- Voltage across L_1 will decay toward zero as the rate of change becomes less.

The rise and decay of circuit current and voltages are depicted graphically in Figure 9-10.

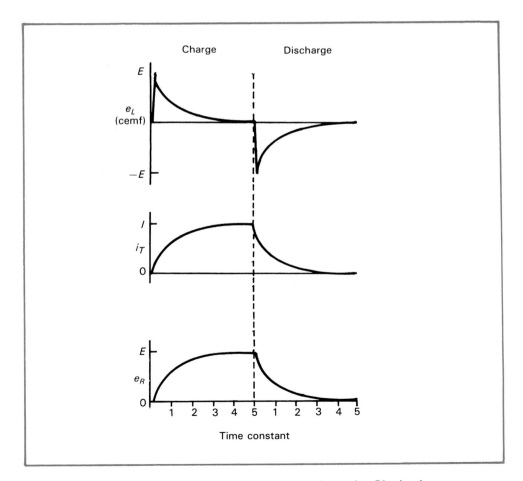

Figure 9-10. Current and voltage rise and decay in the series RL circuit.

RL TIME CONSTANT

If a resistance is added in series with an inductance, the rise and fall of current takes some interval of time. This depends upon the relationship of the inductance and resistance. The time required for current to change by 63.2% is the **time constant** of that circuit. One time constant is figured as the time required for current to *rise* to 63.2% or *fall* to 36.8% of *peak* value. It may be found by the formula:

$$t \text{ (in seconds)} = \frac{L}{R} \text{ (in seconds)}$$

where t is the time constant in seconds, L is the inductance in henrys, and R is the resistance in ohms.

It is generally accepted that after *five* time constants, the circuit will be at equilibrium state, with no further change. Refer to table of Figure 9-11 and graph of Figure 9-12. The graph is a plot of the percentages given in the table. It is called the *universal time-constant chart*. This chart is valid for RL *and* RC (resistive-capacitive) circuits. From the graph, we see that in one time constant a rising current reaches the 63.2% point. In two time constants, the current increases by 63.2% of the remaining amplitude, and so on. Notice that action of this changing state, called the *transient response*, reaches a steady state after five time constants.

Time constant	Rise: percent of maximum	Decay: percent of maximum
1	63.2	36.8
2	86.5	13.5
3	95.0	5
4	98.0	2
5	99.0	1

Figure 9-11. *Table shows percentages of current or voltage at end of each time constant. It is tabulated for both rise and decay.*

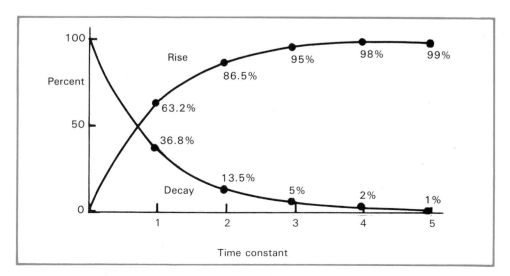

Figure 9-12. *The universal time-constant chart.*

INDUCTIVE KICK

What happens when the current through a series RL circuit is abruptly cut off — say, by opening a switch? When the switch is opened, the current will suddenly drop to zero. The magnetic field around the inductor will collapse. The collapsing field will generate a high voltage across the inductor. This voltage may be many times greater than the source voltage. Arcing occurs across the switch contacts. This reaction is called **inductive kick**. Application of inductive kick is found in the automotive ignition system, where the ignition coil, a special type of transformer, must supply 15,000 V to 20,000 V to fire the spark plugs.

Care must be taken when interrupting inductive circuits because of the inductive kick. The energy stored in the magnetic field of the coil must be dissipated. When an inductive circuit is switched open, the high induced voltage causes:

- Arcing and burning of switch contacts.
- Dissipation of heat that will break down coil insulation.
- Risk of serious injury or death.

Look at Figures 9-13 and 9-14. The NE2 is a neon glow lamp. It presents a high resistance until a firing voltage of about 65 V is applied to its terminals. Then, it will glow. The source voltage of 5 V cannot light the lamp. When switch S_1 is closed, the current rises (and cemf decays) according to the circuit time constant:

$$t = \frac{L}{R} = \frac{4\text{ H}}{100\text{ }\Omega} = 0.04\text{ s}$$

At the end of five time constants (5×0.04 s $= 0.2$ s), the circuit is fully charged. All of the voltage drop is across the resistor. Current through the circuit is:

$$I = \frac{E}{R} = \frac{5\text{ V}}{100\text{ }\Omega} = 0.05\text{ A}$$

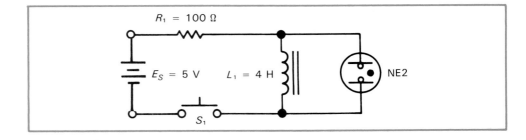

Figure 9-13. Circuit for inductive kick demonstration.

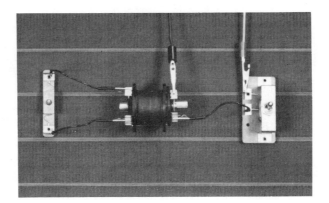

Figure 9-14. A breadboard experiment to demonstrate inductive kick. (Lab-Volt)

Recall the formula for inductance of a coil:

$$L = \frac{e_{L_{avg}}}{\Delta i / \Delta t}$$

Rearranging the formula to isolate $e_{L_{avg}}$, we get:

$$e_{L_{avg}} = L \frac{\Delta i}{\Delta t}$$

From this, it is clear that induced voltage depends directly on the rate of change of current.

In Figure 9-13, when S_1 is opened, the power source is disconnected, and the induced voltage attempts to maintain the circuit current of 0.05 A. However, the current must drop instantaneously to zero because the NE2 has a very high resistance. The rate of change will be extremely high.

Assuming the resistance of the NE2 to be 100 kΩ, then:

$$t = \frac{L}{R} = \frac{4 \text{ H}}{1 \times 10^5 \ \Omega} = 4 \times 10^{-5} \text{ s}$$

During 4×10^{-5} seconds, the current will drop 63.2% of its maximum value, or:

$$0.05 \text{ A} \times 0.632 = 0.0316 \text{ A}$$

The rate of change will be:

$$\frac{\Delta i}{\Delta t} = \frac{0.0316 \text{ A}}{4 \times 10^{-5} \text{ s}} = 790 \text{ A/s}$$

The induced voltage is calculated as:

$$e_{L_{avg}} = L \frac{\Delta i}{\Delta t} = (4 \text{ H})(790 \text{ A/s}) = 3160 \text{ V}$$

This voltage is sufficient to make the NE2 flash!

INDUCTANCE IN AC CIRCUITS

So far, we have discussed inductance in dc circuits. In an ac circuit, the applied voltage varies and reverses polarity constantly. Thus, in the inductive circuit, self-inductance develops counter emfs, which oppose the source voltage. This reduces the amount of current in a circuit much more than the resistance alone. We shall see how.

INDUCTIVE REACTANCE

In general, opposition to an alternating current is called **reactance**. When caused by an inductance, it is called **inductive reactance (X_L)**. Like resistance, it is measured in *ohms*. Figure 9-15 demonstrates inductive reactance. In this circuit, a small 6-V lamp is connected in series with a resistor and an inductor. When a dc voltage is applied, the lamp burns brightly. When the same circuit is connected to an ac voltage of the same magnitude, the lamp glows dimly. Something is limiting the current in the ac circuit. The alternating current has created a cemf in the coil that always opposes the source voltage and holds the current at a lower value.

In alternating current theory (Chapter 6), we discovered that the rate of change of the amplitude of a sine wave is determined by the angular velocity of a rotating vector. Angular velocity (ω), measured in radians per second, is the rate at which an angle changes. One complete revolution of the phasor, or one cycle of ac, represents 2π radians. Angular velocity is related to frequency, where $\omega = 2\pi f$ (f is in hertz, or cycles per second).

The rate of change of current is directly related to angular velocity and, so, to frequency. Higher frequency means that current varies at a faster rate. The cemf of an inductor is:

$$e_{L_{\text{avg}}} = L \frac{\Delta i}{\Delta t}$$

Thus, if frequency increases, $\Delta i/\Delta t$ increases, and induced voltage $e_{L_{\text{avg}}}$ increases. As induced voltage increases, so does opposition to current, or reactance. In the same manner, as frequency decreases, opposition to current decreases. The reactance is proportional to the frequency.

Likewise, induced voltage will vary directly as the size of a circuit inductor. For a given rate of current change, a larger inductor will mean a larger induced voltage; thus, opposition to current will be larger, too. Conversely, a smaller inductor will mean less opposition to current. The reactance is proportional to the inductance.

The formula for inductive reactance is:

$$X_L = \omega L = 2\pi f L = 6.28\, fL$$

where ω is in radians per second, f is in hertz, and L is in henrys. We will use this formula in an upcoming example.

THE PURE INDUCTIVE CIRCUIT

Examination of the waveforms in Figure 9-16 will reveal some interesting facts about a circuit containing only inductance. Notice the cemf, or induced voltage, waveform is 180° out of phase with the source voltage. Maximum cemf can only be induced when the current rate of change is maximum. This point must be where current crosses the zero line. At the peak of the current waveform, there is an instant where the rate of change is minimum, or zero. At this point, cemf is zero.

Upon examination, we see that the the peak current waveform occurs *later* in time than the peak voltage waveform. Since it is later, not sooner, we say

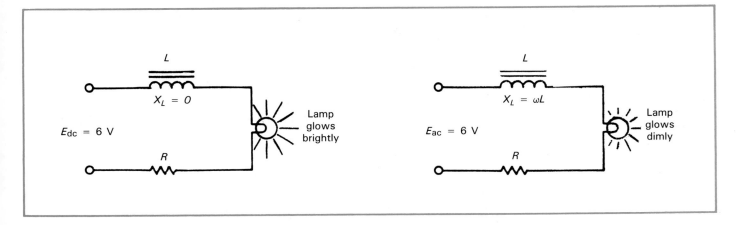

Figure 9-15. Demonstration of inductive reactance in a series RL circuit.

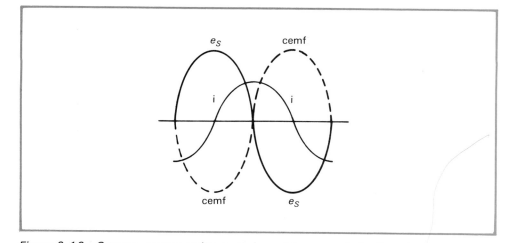

Figure 9-16. Current, source voltage, and cemf in a pure inductive circuit.

that the current *lags* behind the applied voltage. The two peaks are 90° apart. We can conclude that current *lags* behind the applied voltage by an angle of 90° in a pure inductive circuit. (In reality, a circuit will always have *some* resistance; thus, the angle will be slightly less than 90°.)

THE SERIES RL CIRCUIT

In a pure resistive circuit, the current and voltage are always in phase. In a pure inductive circuit, the current lags the voltage by an angle of 90°. In the series circuit containing both resistance and inductance, the current will be in phase with the voltage across the resistor, E_R. It will lag the voltage across the inductor, E_L, by 90°. It will lag the source voltage, E_S, by some angle between zero and 90°. The current in the series RL circuit is the same in any circuit element at any instant.

Impedance

In the series RL circuit, there are two resistive components opposing the current. These are 90° apart. One is the dc resistance in ohms; the other is the inductive reactance, X_L, in ohms. Both of these can be represented by vectors that will form two sides of a right triangle. To find the *total* opposition, or **impedance (Z)**, we find the *resultant*. It is the vector that represents the *combined* effect of the other two. It is the third side of our right triangle. Since it is opposite the right angle, impedance is equal to the hypotenuse. To find this value, we employ our resistive values and the *Pythagorean theorem*. Impedance then is:

$$Z = \sqrt{R^2 + X_L{}^2}$$

where Z is in ohms. Z is the hypotenuse of our right triangle, which is our resultant vector. It represents the *vector sum* of two vectors that are 90° out of phase. Note that the right triangle forming our *vector diagram* is called an **impedance diagram**. See Figure 9-17.

Voltage

As stated, current in the series RL circuit is in phase with E_R, and it lags E_L by 90°. Thus, we can say that E_R lags E_L by 90°. The series RL circuit, then, has two voltage components—E_R and E_L. Both of these can be represented by phasors. Since they are out of phase, we cannot simply add the two components together to figure the source voltage. We must find their *phasor sum*:

$$E_S = \sqrt{E_R{}^2 + E_L{}^2}$$

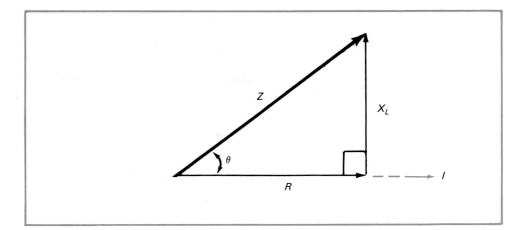

Figure 9-17. An impedance diagram, or triangle.

Figure 9-18A shows a phasor diagram. Figure 9-18B shows the phasor diagram drawn as a triangle. While an impedance diagram is sometimes referred to as a *phasor* diagram, this is not quite correct. *Z, R,* and X_L are not phasor quantities because they have fixed values. They do not change in time, unlike ac voltage and current.

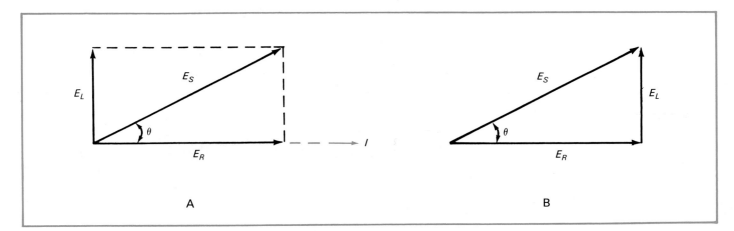

Figure 9-18. A—Voltage phasor diagram. B—Equivalent voltage triangle.

Phase angle

The **phase angle** (θ) represents the phase displacement between current and voltage that results from a reactive circuit element. The phase angle is found on the phasor diagram. We draw the diagram with a horizontal reference. The quantity that is common to all elements of a circuit is used as the reference. In the series circuit, current is common. The phase angle is the angle formed by the horizontal reference and the hypotenuse or resultant phasor.

In the impedance diagram, we do not have a current phasor; however, it is still considered the horizontal reference. Resistance is drawn as the horizontal component since there is no reactance in a resistor. In other words, current and voltage across *R* are in phase. Using this diagram, the phase angle may be found from:

$$\theta = \arctan \frac{X_L}{R}$$

In the voltage phasor diagram, we do not have a current phasor either. We use E_R as the horizontal component since it is in phase with current. Using this diagram, the phase angle may be found from:

$$\theta = \arctan \frac{E_L}{E_R}$$

Problem:

Using the circuit with assigned values in Figure 9-19, determine the phase angle between applied voltage and current. Draw the impedance triangle. Also, find the voltage drops around the circuit. Verify that voltages are correct.

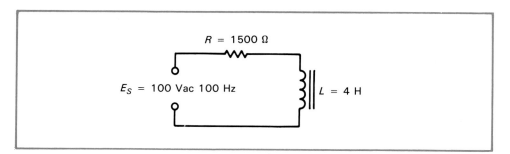

Figure 9-19. A series RL circuit.

Solution:

Step 1. Compute the value of inductive reactance, X_L.

$$X_L = 2\pi fL = 6.28 \times 100 \text{ Hz} \times 4 \text{ H} = 2512 \ \Omega$$

Step 2. Determine the phase angle.

$$\theta = \arctan \frac{X_L}{R} = \arctan \frac{2512 \ \Omega}{1500 \ \Omega}$$
$$= \arctan 1.67$$
$$= 59.1°$$

Step 3. Draw the impedance diagram. Use any convenient scale. Remember that R is drawn as the horizontal component. X_L is drawn upward at 90° from R since circuit current *lags* voltage (voltage *leads* current) across the inductor. See Figure 9-20.

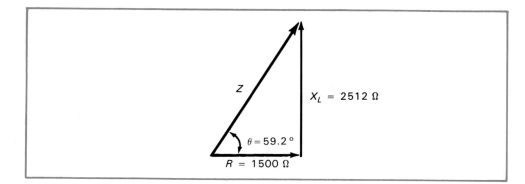

Figure 9-20. Impedance diagram.

Step 4. Compute the value of impedance, *Z.*

$$Z = \sqrt{R^2 + X_L^2} = \sqrt{(1500\ \Omega)^2 + (2512\ \Omega)^2} = 2926\ \Omega$$

Note that the value of Z could also be determined using trigonometry. Looking back at our impedance diagram:

$$\cos\theta = \frac{R}{Z}$$

Rearranging the equation to isolate *Z:*

$$Z = \frac{R}{\cos\theta} = \frac{1500\ \Omega}{\cos 59.1°} = \frac{1500\ \Omega}{.5135} = 2921\ \Omega$$

For practical purposes, the values are the same. The difference is due to rounding.

Step 5. Calculate the circuit current. Using Ohm's law:

$$I = \frac{E}{Z} = \frac{100\ V}{2926\ \Omega} = 0.034\ A$$

Step 6. Find the voltage drops around the circuit. Using Ohm's law:

$$E_R = IR = (0.034\ A)(1500\ \Omega) = 51\ V$$
$$E_L = IX_L = (0.034\ A)(2512\ \Omega) = 85\ V$$

Step 7. To verify the voltages are correct, check if their phasor sum equals the source voltage.

$$E_S = \sqrt{E_R^2 + E_L^2} = \sqrt{(51\ V)^2 + (85\ V)^2} = 99\ V$$

Again, for practical purposes, the values are the same. The difference is due to rounding.

In the foregoing problem, it is important to understand how the various quantities would be affected by a change in frequency. At a frequency of 1000 Hz:

$$X_L = 25,000\ \Omega \qquad I = 0.0039\ A$$
$$\theta = 86.6° \qquad E_R = 5.8\ V$$
$$Z = 25,300\ \Omega \qquad E_L = 97.5\ V$$

The effects of changing frequency are summarized in Figure 9-21. It shows how values change in a series RL circuit as frequency is changed and inductance is held at a constant value.

f	X_L	θ	Z	I	E_L	E_R
increases ↑	increases ↑	increases ↑	increases ↑	decreases ↓	increases ↑	decreases ↓
decreases ↓	decreases ↓	decreases ↓	decreases ↓	increases ↑	decreases ↓	increases ↑

Figure 9-21. Table shows the effect on various values in a series RL circuit as frequency is changed and the inductance value is held constant.

THE PARALLEL RL CIRCUIT

In a circuit containing a resistance and an inductance in parallel, the voltage of each circuit element will be the same as the source voltage. Further, there will be no phase difference among them. This is because they are all in parallel.

There will be a phase difference among the total and branch currents, however. Current will be in phase with voltage in the resistive branch. It will lag the voltage across the inductor by 90°. Total current will lag the source voltage by some angle between zero and 90°.

Impedance

In the parallel RL circuit, we do not find impedance through a vector sum. Instead, we apply Ohm's law:

$$Z = \frac{E_S}{I_T}$$

Current

As stated, voltage in the parallel RL circuit is in phase with I_R, and it leads I_L by 90°. Thus, we can say that I_R leads I_L by 90°. The parallel RL circuit, then, has two current components—I_R and I_L. Both of these can be represented by phasors. Since they are out of phase, we cannot simply add the two components together to figure the total circuit current. We must find their *phasor sum*:

$$I_T = \sqrt{I_R{}^2 + I_L{}^2}$$

Note that the phasor for I_L is below the horizontal reference. This is because I_L lags voltage—the horizontal reference for the parallel RL circuit. (Since a phasor, or rotating vector, rotates counterclockwise, a *lagging* phasor would be behind, or clockwise to, a *leading* phasor.)

Phase angle

The phase angle, as stated, represents the phase displacement between current and voltage resulting from a reactive circuit element. For the parallel RL circuit, it is found on the *current* phasor diagram. The horizontal reference of this circuit is voltage, since it is common to all circuit elements.

In the current phasor diagram, we do not have a voltage phasor. We use I_R as the horizontal component since it is in phase with voltage. Using this diagram, the phase angle may be found from:

$$\theta = \arctan \frac{I_L}{I_R}$$

Problem:

Using the circuit with assigned values in Figure 9-22, determine the phase angle between applied voltage and current. Draw the current phasor diagram and find the circuit impedance.

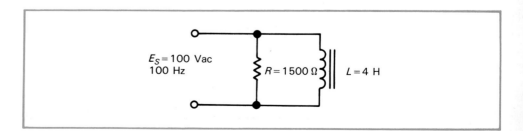

Figure 9-22. A parallel RL circuit.

Solution:

Step 1. Compute the value of inductive reactance, X_L.

$$X_L = 2\pi f L = 6.28 \times 100 \text{ Hz} \times 4 \text{ H} = 2512 \text{ }\Omega$$

Step 2. Compute branch currents. Using Ohm's law:

$$I_R = \frac{E_S}{R} = \frac{100 \text{ V}}{1500 \text{ }\Omega} = 0.067 \text{ A}$$

$$I_L = \frac{E_S}{X_L} = \frac{100 \text{ V}}{2512 \text{ }\Omega} = 0.04 \text{ A}$$

Step 3. Determine the phase angle to see by how much the circuit current lags the voltage.

$$\theta = \arctan \frac{I_L}{I_R} = \arctan \frac{0.04 \text{ A}}{0.067 \text{ A}}$$

$$= \arctan 0.597$$

$$= 30.8°$$

Step 4. Draw the current phasor diagram. Use any convenient scale. Remember that I_R is drawn as the horizontal component. I_L is drawn downward at 90° from I_R since it lags voltage — the horizontal *reference*. See Figure 9-23.

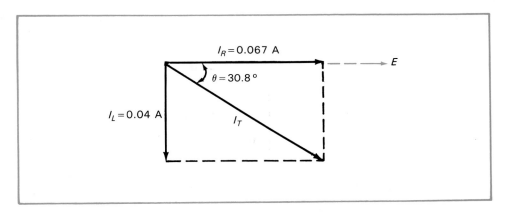

Figure 9-23. Current phasor diagram.

Step 5. Find the total circuit current.

$$I_T = \sqrt{I_R{}^2 + I_L{}^2} = \sqrt{(0.067 \text{ A})^2 + (0.04 \text{ A})^2}$$

$$= 0.078 \text{ A}$$

Step 6. Find the impedance of the circuit. Using Ohm's law:

$$Z = \frac{E_S}{I_T} = \frac{100 \text{ V}}{0.078 \text{ A}} = 1282 \text{ }\Omega$$

Once again, it is important to understand how the various quantities would be affected by a change in frequency. The effects of changing frequency are summarized in Figure 9-24. It shows how values change in a parallel RL circuit as frequency is changed and inductance is held at a constant value.

f	X_L	θ	Z	I_R	I_L	I_T
increases ↑	increases ↑	decreases ↓	increases ↑	remains constant	decreases ↓	decreases ↓
decreases ↓	decreases ↓	increases ↑	decreases ↓	remains constant	increases ↑	increases ↑

Figure 9-24. Table shows the effect on various values in a parallel RL circuit as frequency is changed and the inductance value is held constant.

POWER

Source voltage and current waveforms of a pure inductive circuit are shown in Figure 9-25. A third curve that represents consumed power is plotted over these curves. Power above the zero line is power consumed by the circuit. Power below the line is power returned by the circuit. The power curve represents the products of many instantaneous voltages and currents. When values of voltage and current are both positive, power is positive. When only one value of either voltage or current is negative, the power is negative. When values of voltage and current are both negative, power is positive. The positive half-cycles of power will equal the negative half-cycles in magnitude. Their average value is zero. Therefore, power consumed by inductance is zero. Energy required to build the magnetic field is returned to the circuit.

When speaking of power in the RL circuit, there are three types that must be considered. These are:
- *True power.*
- *Reactive power.*
- *Apparent power.*

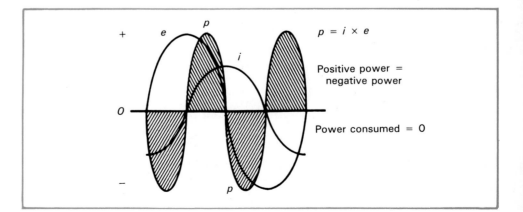

Figure 9-25. A pure inductive circuit uses zero power.

True power

In an ac circuit, only the resistive elements will dissipate power. The *reactive elements* (inductors and capacitors) simply store, and later return, power to the system. The power that is dissipated by a circuit is called **true power (P)**. Units are *watts*. True power is *resistive* power. It is also referred to as *pure power*, or *active power*. In Chapter 6, we referred to it as *average* power.

True power in the RL circuit is the power dissipated, as heat, in the resistor. It is given by:

$$P = I_R{}^2R = E_S I_T \cos \theta$$

where:

I_R = current through resistor
R = resistance of the resistor
E_S = supply voltage
I_T = supply current
θ = phase angle

Note that the factor *cos θ* in the previous formula is known as the **power factor**. Let us see where the power factor comes from. We know that power dissipated in a resistor is:

$$P = E_R \times I_R$$

Looking back to our voltage phasor diagram of Figure 9-18, we can see that:

$$\cos \theta = \frac{E_R}{E_S}$$

Therefore:

$$E_R = E_S \cos \theta$$

Now let us substitute into our formula for power dissipated in a resistor. In the series circuit, $I_R = I_T$. Therefore:

$$P = E_R \times I_T = (E_S \cos \theta) \times I_T = E_S I_T \cos \theta$$

Similarly, we can evaluate this formula using a parallel RL circuit. In the parallel circuit, we know that:

$$P = E_R \times I_R = E_S \times I_R$$

Looking back to our current phasor diagram of Figure 9-23, we can see that:

$$\cos \theta = \frac{I_R}{I_T}$$

Therefore:

$$I_R = I_T \cos \theta$$

Again, substituting into our formula for power dissipated in a resistor:

$$P = E_S \times I_R = E_S \times (I_T \cos \theta) = E_S I_T \cos \theta$$

Reactive power

An inductor does not dissipate *true power*. As mentioned, it stores and returns power to the circuit. The stored power must come from somewhere, however. It is supplied only through an increase in current drawn from the supply. It is, of course, returned, but periodically there is an increased demand on the supply that must be considered for sizing the supply. We call power that is absorbed by a reactive element, **reactive power (Q)**. The unit of reactive power is the **VAR** (for **volt-ampere reactive**). Reactive power in an inductor is given by:

$$Q = E_L I_L = I_L{}^2 X_L = \frac{E_L{}^2}{X_L}$$

A formula for reactive power can also by derived from voltage or current phasor diagrams and the formula $Q = E_L I_L$. The resulting formula is:

$$Q = E_S I_T \sin \theta$$

Apparent power

The product of source voltage and source current is **apparent power** *(S)*. This is power that at any instant *appears* to be that supplied to the circuit. The unit of apparent power is the **volt-ampere**. It is abbreviated **VA**. Apparent power is given by:

$$S = E_S I_T = I_T^2 Z = \frac{E_S^2}{Z}$$

Apparent power is the vector sum of true and reactive powers. We can draw a *power triangle* with our three power components. True power is the horizontal vector. Reactive power is drawn vertically. For a series RL circuit, it is drawn above the horizontal. For a parallel RL circuit, it is drawn below. Apparent power is the resultant phasor. It, then, is equal to:

$$S = \sqrt{P^2 + Q^2}$$

TRANSFORMERS

The basic principles of mutual inductance and the transfer of energy between circuits by means of a transformer have already been discussed. Transformers are manufactured in a large variety of shapes and sizes, with and without cores, to fill the many needs of electronic circuitry. One such need includes the *stepping up* or *stepping down* of ac voltage or ac current. Another is the *isolation* of one circuit from another. Another is for what is called *impedance matching*.

CONSTRUCTION OF THE POWER TRANSFORMER

A **power transformer** is used for raising or lowering voltage. It is designed to be used at power frequencies — usually from 30 to 400 Hz. Both the primary and secondary windings are placed on a magnetic iron core. The coefficient of coupling must be as close to unity as possible. To achieve this, various methods of winding the coils are employed. In transformers, there are two basic types of core construction:

- **Core-type transformer.** In this type, the coils surround the core. Parts of both windings are wound on each leg of the core. Primary and secondary windings may be wound one over the other (*concentric*) or side by side (*split bobbin*). The core is made up of thin plates of iron. Each plate is coated with an insulating varnish. Then, all are pressed together to form a laminated core. See Figure 9-26A.
- **Shell-type transformer.** In this type, the core surrounds the coils. The windings are wound over a center core section. Again, they may be concentric or split bobbin. The core is made up of laminations covered with varnish and pressed together. This transformer is more popular and more efficient. See Figure 9-26B.

Transformers may have one or more primary windings and/or secondary windings, all assembled on a single core. Some secondary windings have *center-tap* connections as well as taps at other points. The **center tap (CT)** is located midway between the two ends of a secondary. It is equivalent to two secondary windings with half of the total voltage across each. Relative to the center tap, one voltage will be 180° out of phase with the other. Note the symbol for the center-tapped transformer among the transformer symbols in Figure 9-27.

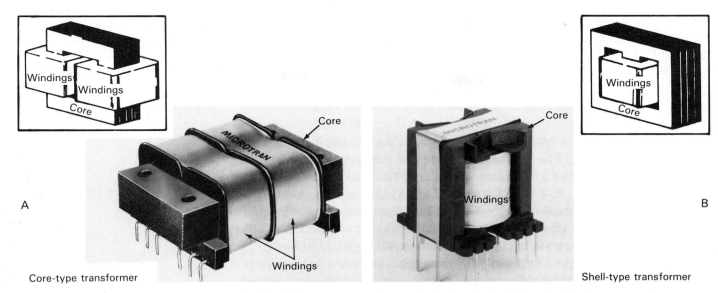

Figure 9-26. *Typical transformers. A—Split-bobbin power transformer for printed circuit boards. B—Power transformer for printed circuit boards with ferrite core, which permits use over a wide frequency range.* *(Microtran Company, Inc.)*

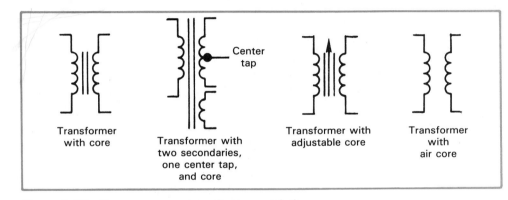

Figure 9-27. *Some common transformer symbols.*

TRANSFORMER ON NO-LOAD AND ON LOAD

When an ac voltage is applied to the primary of a transformer and the secondary is open, the transformer is said to be on *no-load*. Through self-inductance, a counter emf is induced in the primary that nearly equals the applied voltage. As a result, there will be little current in the primary. A small current, called a **no-load**, or **exciting**, **current**, still continues, however. The total of this current is made up of a current required to supply core losses and a *magnetizing current*. The **magnetizing current** creates the magnetic flux in the core. It is small and is limited by the amount of applied voltage and the reactance of the primary winding.

The expanding and collapsing magnetic fields of the primary cut across the secondary windings and induce a voltage in the secondary. When a load is connected across the secondary, a load current passes through the coil. The current sets up a magnetic field that opposes the primary field. The primary field is weakened, which reduces the cemf. Additional current flows through the primary, which strengthens the primary field. The secondary field is canceled by the extra primary field. The core is left with the original field from the magnetizing current.

DIRECTION OF WINDINGS AND MULTIPLE SECONDARIES

The polarity of the secondary voltage will depend upon the direction of the secondary winding and the external connections. Generally, a transformer is considered as inverting the signal input, but this is not always true. Waveforms are shown in Figure 9-28. The black dot at the end of each winding indicates the polarity. Note that, in any given transformer, the terminals identified by black dots will have the same polarity at every instant, with respect to the primary ac voltage. Where windings are wrapped in the same direction (with respect to the magnetic flux), input and output waveforms will be in phase; wrapped in reverse directions, they will be 180° out of phase.

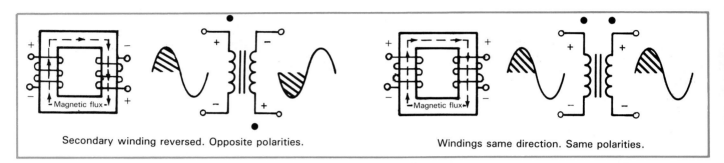

Secondary winding reversed. Opposite polarities.

Windings same direction. Same polarities.

Figure 9-28. Waveform relationships between input and output of a transformer.

As mentioned, a transformer may have multiple secondary windings. These may be connected together to achieve different voltage outputs. If they are connected such that all winding voltages are in phase, the connection is termed *series-aiding*. If not, it is termed *series-opposing*. When secondaries are connected series-aiding, voltage at the output terminals will be the sum of all secondary voltages. When connected series-opposing, voltage at the output terminals will be subtractive. See Figure 9-29.

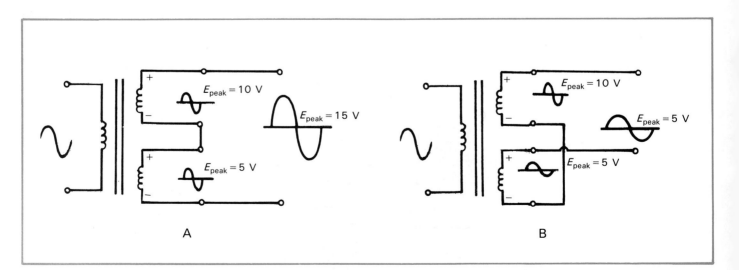

Figure 9-29.   A—Series-aiding connection: secondary voltages are added. B—Series-opposing connection: secondary voltages are subtracted.

TRANSFORMER CURRENTS AND VOLTAGES

The magnitude and density of the flux lines of the primary coil will depend upon the current and the number of turns. Since both primary and secondary are wound closely together and on a common core, the flux for the secondary must be almost the same as the primary. This is stated in a formula as:

$$I_p N_p = I_s N_s$$

where I_p and I_s are the currents in, and N_p and N_s are the number of turns in the primary and secondary. By rearranging, we get:

$$\frac{I_s}{I_p} = \frac{N_p}{N_s}$$

where N_p/N_s is called the **turns ratio**. Notice that the *current ratio*, or the ratio of *primary* to *secondary* currents (I_p/I_s), is the inverse of the turns ratio.

With unity coupling between the primary and secondary, the *voltage ratio*, or the ratio of primary to secondary voltages (E_p/E_s), is in direct proportion to the turns ratio.

$$\frac{E_p}{E_s} = \frac{N_p}{N_s}$$

The step-up transformer

If the voltage ratio is less than 1, voltage is higher in the secondary than the primary. A transformer with such a ratio is called a **step-up transformer**. The turns ratio of a step-up transformer is always less than 1. This means that a step-up transformer has more turns in the secondary than the primary. Note that current and voltage ratios are inverse of each other. Thus, current in the step-up transformer is *stepped down*. It is less in the secondary than in the primary.

The step-down transformer

If the voltage ratio is greater than 1, voltage is higher in the primary than the secondary. A transformer with such a ratio is called a **step-down transformer**. The turns ratio of a step-down transformer is always greater than 1. This means that a step-down transformer has more turns in the primary than the secondary. Current in the step-down transformer is *stepped up*. Thus, it is greater in the secondary than in the primary. A power transformer with several secondary windings, some step-up and some step-down, is shown in Figure 9-30.

Figure 9-30. A power transformer with high- and low-voltage secondaries. (Triad)

Problem:

A 500-Vac output is required of a transformer that has a 100-Vac input. What must the turns ratio be?

Solution:

$$\frac{E_p}{E_s} = \frac{N_p}{N_s} = \frac{100 \text{ V}}{500 \text{ V}} = \frac{1}{5} = 0.2$$

Problem:

A transformer with an input of 100 Vac has a turns ratio of 5. What is the output voltage?

Solution:

$$\frac{E_p}{E_s} = \frac{N_p}{N_s} = 5$$

Rearranging the proportion:

$$E_s = \frac{1}{5} E_p = \frac{1}{5} (100 \text{ V}) = 20 \text{ V}$$

PRIMARY/SECONDARY POWER RELATIONSHIP

A transformer does *not* step up or step down power. It has no way of creating electrical energy. In the transformer, assuming 100% efficiency, apparent power in the primary equals apparent power in the secondary. In other words:

$$S_{\text{in}} = S_{\text{out}}$$

Where the load is purely resistive, apparent power is equal to true power. In this case, we can add:

$$P_{\text{in}} = P_{\text{out}}$$

which states that for resistive loads, applied power equals power to the load. Since power in both cases is the product of voltage and current:

$$I_p E_p = I_s E_s$$

where I_p and I_s are the currents in, and E_p and E_s are voltages in the primary and secondary.

Problem:

Calculate the primary current in the circuit of Figure 9-31.

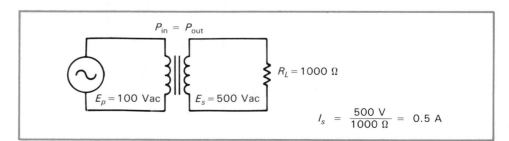

Figure 9-31. This circuit describes the power relationships between the primary and secondary of a transformer.

Solution:

$$I_p E_p = I_s E_s$$

Rearranging to isolate I_p:

$$I_p = \frac{I_s E_s}{E_p} = \frac{(0.5 \text{ A})(500 \text{ V})}{100 \text{ V}} = 2.5 \text{ A}$$

Power companies transmit electricity based on this concept. They use transformers in the transmission of electricity to reduce line losses. Power loss varies as the square of the current. Therefore, it is more economical to transmit at high voltage and low current to reduce power loss in the transmission lines. See Figure 9-32.

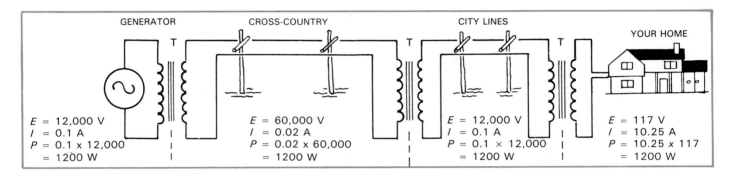

Figure 9-32. Use of transformers in the transmission of electricity.

TRANSFORMER LOSSES

Three types of losses are associated with transformer construction. All result in generation of heat. These losses were also found in generators and were discussed in Chapter 6. They include:

- Copper loss.
- Eddy current loss.
- Hysteresis loss.

Copper loss

This loss is due to the resistance of the wire that is used to wind the transformer. It is figured by the formula $P = I^2R$. As such, it is sometimes called the I^2R loss. Copper loss can be minimized by using wires of correct size to carry the expected currents. The loss increases as the square of the current as seen by our formula. Thus, when a transformer is overloaded, the heat results can seriously damage the transformer.

Eddy current loss

Since the core is a conducting material, voltages can be induced in the core as well as in the secondary winding. Currents within the core will heat up the core due to its resistance. Eddy current loss is overcome by laminating the core. The thin insulated core stampings prevent the eddy currents from flowing excessively in the core. All power transformers have laminated cores.

Hysteresis loss

This loss is due to molecular friction within the core material. As the core is magnetized, the domains are oriented in one direction. When the polarity of the magnetizing force reverses, then the domains must reverse. The constant aligning and realigning of the molecules cause heat and loss. This has been greatly reduced by using special alloys for transformer core stampings that have a minimal hysteresis loss.

TRANSFORMER EFFICIENCY AND RATINGS

No perfect machine has yet been built. There are always some losses between input and output. For a transformer:

$$\% \text{ Efficiency} = \frac{P_{\text{out}}}{P_{\text{in}}} \times 100\%$$

Efficiency is reduced by the losses just discussed. The loss of power is dissipated in the form of heat.

In selecting a transformer for a circuit, the power handling capabilities of the transformer must be considered. Required secondary voltages and currents must not exceed the specifications. Transformer ratings usually specify the primary and secondary design voltages. A power rating is also given. The rating gives apparent power ($I_p E_p$ or $I_s E_s$), rather than true power, because it is far more useful. Remember the units of apparent power are volt-amperes; whereas, the units of true power are watts.

Knowing the secondary voltage and VA ratings, the maximum allowable current draw in the secondary can be determined. For example, if the VA rating of a transformer is 12 VA and the secondary voltage is rated for 6 V, we know that a maximum 2 A can be drawn from the transformer secondary. If a power rating was given in watts, however, we may exceed maximum secondary current without exceeding wattage. If our secondary load was purely reactive, the true power delivered to the load would be zero, and we might think we were safe. Yet, we still have current in the secondary, and we may have exceeded the maximum.

The voltage handling capability of a transformer is determined by the type and thickness of insulation used in its windings. The current-carrying capacity is limited by the size of the wire used in the windings. Small wires with high resistance will produce heat and destruction. Further, the capacity to handle certain power requirements depends upon the cross-sectional area of the core and its ability to safely dissipate heat.

COUPLING TRANSFORMERS

A major use of transformers, besides current and voltage transformation, is for *coupling* between stages of electronic circuits. Transformers used for this purpose are sometimes called **coupling transformers**. These transformers are designed solely for transferring power from one circuit stage to another. The reasons for their use are:
- Isolation of one circuit from another—blocking dc signals, passing only ac.
- Matching impedances between circuits.

The isolation transformer

A common application requires the same number of turns on both primary and secondary. The turns ratio is 1:1. This type of transformer is known as an **isolation transformer**. It neither steps up nor steps down voltage. It is used to isolate one circuit from another. Since dc in a primary will not induce voltage in a secondary, a dc level in one circuit will not be passed. In a typical application, a transformer can be used to keep dc voltage from one stage from affecting the dc bias of the next. Only the ac signal is passed. You will learn more about this in later chapters.

The impedance-matching transformer

Generators and other power sources have internal resistance or impedance. In more advanced studies of electronics, the internal impedance of a source must always be considered. It has purposely been omitted in this text for simplification. To understand the concept of *impedance matching*, it must be discussed.

Look at the circuit in Figure 9-33A. It shows a generator connected to a load. The internal impedance is represented by series resistor R_G. Remember this resistance is, in fact, inside of the generator and part of it. The *ideal* voltage of the generator minus the voltage drop across the internal resistance is the actual amount of voltage at the generator output terminals.

In Figure 9-34 are tabulated current, voltage, and power values appearing in the circuit of Figure 9-33A as load R_L is changed. For obtaining these values, the ideal voltage is held a constant 100 V. The internal impedance of the generator R_G is 100 Ω.

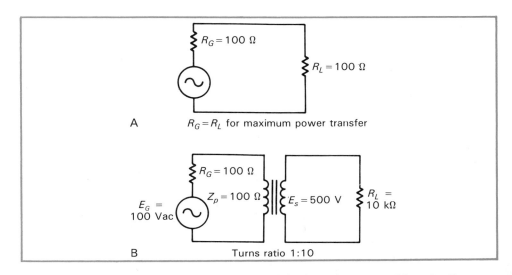

Figure 9-33. These circuits describe the need for impedance matching. A—For maximum power transfer load impedance should equal source impedance. B—An impedance-matching transformer is used here to match R_L to R_G.

Analyzing the table of Figure 9-34, we may conclude:
- For maximum transfer of power to the load, the load impedance must be equal to or *appear* to be equal to the source impedance—the internal impedance of our generator, in this case. This is called **impedance matching**. In addition, with maximum power transfer, the power developed in the load is equal to the power developed in the source.
- A low load resistance means a high current; also, poor efficiency and low power output.
- A high load resistance means a low current. A low current means less drop from the internal impedance, thus, more voltage at the generator output. It means high efficiency and low power dissipation in the internal impedance. However, it also means less power at the load.

R_L	$I = \dfrac{E}{R_G + R_L}$	IR_G	IR_L	P_{R_L}	P_{R_G}	P_{total}	Eff. P_{R_L}/P_{total}
10 Ω	0.909 A	90.9 V	9.09 V	8.26 W	82.6 W	90.86 W	9%
50 Ω	0.666 A	66.6 V	33 V	21.8 W	43.5 W	65.3 W	33%
100 Ω	0.5 A	50 V	50 V	25 W	25 W	50 W	50%
500 Ω	0.166 A	16.6 V	83 V	13.8 W	2.75 W	16.55 W	84%
1,000 Ω	0.09 A	9 V	90 V	8.1 W	0.8 W	8.9 W	91%
10,000 Ω	0.01 A	1 V	99 V	1 W	0.01 W	1.01 W	99%

Figure 9-34. Table corresponding to Figure 9-33A showing transfer of power for E_G = 100 V, R_G = 100 Ω, and selected load resistances.

We want the impedance of a generator, a signal source, or a stage of an electronic circuit to match the external load to secure a maximum power transfer. To do this, we may use a transformer to transform impedance. The *impedance ratio* of a transformer, which is the ratio of primary impedance to load impedance, will vary directly as the square of its turns ratio. In other words:

$$\frac{Z_p}{Z_L} = \left(\frac{N_p}{N_s}\right)^2 \text{ or } \frac{N_p}{N_s} = \sqrt{\frac{Z_p}{Z_L}}$$

Note that a more advanced study of electronics would include the derivation of this formula.

Problem:

A transformer is required to match a 300-ohm output impedance of a transistor power amplifier to a 3-ohm speaker coil. What will the turns ratio be?

Solution:

$$\frac{N_p}{N_s} = \sqrt{\frac{Z_p}{Z_L}} = \sqrt{\frac{300\ \Omega}{3\ \Omega}} = \sqrt{\frac{100}{1}} = \frac{10}{1}$$

Note that the power amplifier in this problem is the source impedance. You might observe that the load impedance (3 Ω) does not equal the source impedance (300 Ω) and conclude that maximum power has not been transferred. However, if we ensure a 300-Ω impedance in the transformer primary, the load will *appear* to be equal to the source impedance. In this way, maximum power will be delivered. This is a matter of selecting the correct turns ratio.

Problem:

What turns ratio of a transformer would match a generator internal impedance of 100 Ω to an R_L of 10,000 Ω? (Refer to Figure 9-33B.)

Solution:

$$\frac{N_p}{N_s} = \sqrt{\frac{Z_p}{Z_L}} = \sqrt{\frac{Z_p}{R_L}} = \sqrt{\frac{100\ \Omega}{10,000\ \Omega}} = \sqrt{\frac{1}{100}} = \frac{1}{10}$$

Note that the load impedance Z_L is a resistor, represented as R_L.

Problem:

Prove that, with the turns ratio calculated in the foregoing problem, maximum power is transferred.

Solution:

Step 1. Calculate total impedance in the primary circuit. The total impedance in the series circuit will be the sum of the individual impedance values. We know that the impedance of the transformer primary must match the generator impedance. Therefore, total impedance is:

$$Z_T = R_G + Z_p = 100\ \Omega + 100\ \Omega = 200\ \Omega$$

Step 2. Calculate current in the primary.

$$I_p = \frac{E}{Z_T} = \frac{100\ V}{200\ \Omega} = 0.5\ A$$

Step 3. Calculate current in secondary.

$$\frac{I_s}{I_p} = \frac{N_p}{N_s} = \frac{1}{10} = 0.1$$

Thus

$$I_s = 0.1(I_p) = 0.1(0.5 \text{ A}) = 0.05 \text{ A}$$

Step 4. Calculate power supplied by the generator.

$$P_{RG} = I_p{}^2 R_G = (0.5 \text{ A})^2 (100 \text{ }\Omega) = 25 \text{ W}$$

Step 5. Calculate power to the load.

$$P_{RL} = I_s{}^2 R_L = (0.05 \text{ A})^2 (10,000 \text{ }\Omega) = 25 \text{ W}$$

Power developed in the load is equal to the power in the generator, meaning we have maximum power transfer. Therefore, with 0.1 turns ratio, we have matched impedances. Note from the table of Figure 9-34 that the power in a load of 10 kΩ without impedance matching is only 1 W. In effect, as far as the generator is concerned, the transformer made 10 kΩ *appear* to look like 100 Ω.

Problem:

Calculate voltage in the primary and secondary in the transformer of the previous problem.

Solution:

Step 1. Calculate voltage in the primary.

$$E_p = I_p Z_p = (0.5 \text{ A})(100 \text{ }\Omega) = 50 \text{ V}$$

Step 2. Calculate voltage in the secondary.

$$\frac{E_p}{E_s} = \frac{N_p}{N_s} = \frac{1}{10} = 0.1$$

Thus

$$E_s = \frac{E_p}{0.1} = \frac{50 \text{ V}}{0.1} = 500 \text{ V}$$

Audio frequency (AF) and radio frequency (RF) transformers. These transformers are used as coupling devices. They are found in radio receiver amplifiers. Very often, either the primary, or secondary, or both will be tuned to a specific frequency. This will be studied in a later chapter. Figure 9-35 shows some typical AF, or *audio*, transformers.

At radio frequencies, an iron core is not used since core losses would be excessive. These transformers may have air cores or special powdered metal cores designed for low loss. *Loopstick antennas*, a type of RF transformer with adjustable cores, are shown in Figure 9-36.

HIGH INTENSITY LAMP PROJECT

A popular project that involves a working knowledge of transformer principles is the construction of a high intensity desk lamp, Figure 9-37. These lamps of a commercial variety are found in most department and appliance stores. You will enjoy making one of your own. It is an excellent study lamp.

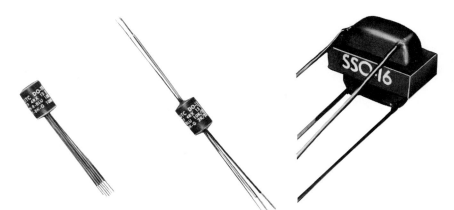

Figure 9-35. Subminiature and ultra-miniature audio transformers.   (United Transformer Co.)

Figure 9-36. Loopstick antennas used in AM radios.

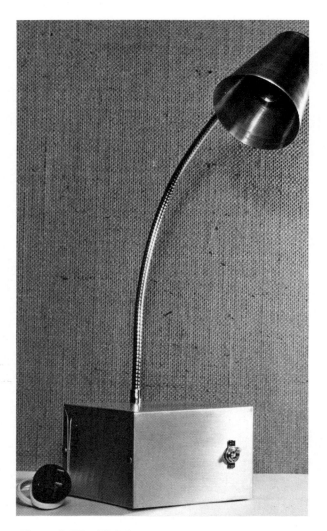

Figure 9-37. High intensity desk lamp.

The mechanical construction of the lamp can be designed to fit your individual needs and the availability of materials. You will need to purchase a socket and a No. 93, 12.8-V, 1.04-A lamp from an automotive parts store. The actual mechanical design of the lamp will depend upon your personal preference. The lamp in the photo is only a suggestion.

Enclosed within the base of our lamp is a transformer. The construction of the transformer will be the only problem for us to solve before proceeding. First, look at the specifications:

* Voltage to primary   — 120 Vac
* Voltage of secondary   — 12.8 Vac
* Current required for lamp   — 1.04 A

The size of the core required for a desired power output may be found in an electrical engineer's handbook. An abbreviated form is given in the table in Figure 9-38. Our required power output, which is applied to a purely resistive load, is:

$$P = I_s E_s = 1.04 \text{ A} \times 12.8 \text{ V} \cong 13 \text{ W}$$

Power output in watts	Core cross-sectional area in square inches
10	0.25
15	0.37
20	0.50
30	0.70
40	0.85
50	1.00
100	2.60

Figure 9-38. An abbreviated table showing size of core required for a desired power output.

To assure that we fully meet this requirement, we will select the core size for a 20-W output, which is 0.5 square inches.

We need to determine the number of turns in the primary coil. In doing so, we must see that the coil has sufficient inductive reactance at 60 Hz so that it will not create a short circuit across the 120-V line. As a rule, this demands at least 4 turns per volt. Adding a margin of safety, 6.25 turns per volt is usually used. The core area must also be considered in the desired amount of reactance. To determine the number of turns in the primary, we use the formula:

$$N_p = \frac{\text{turns per volt} \times \text{voltage}}{\text{cross-sectional area of core}}$$

Thus, for our transformer:

$$N_p = \frac{6.25 \times 120}{0.50} = 1500 \text{ turns}$$

To determine the number of turns in the secondary, we equate the turns ratio to voltage ratio and solve for N_s. In addition, we apply the factor 1.1. The factor will offset losses due to imperfect coupling between primary and secondary.

$$\frac{N_p}{N_s} = \frac{E_p}{E_s}$$

Solving for N_s:

$$N_s = \frac{E_s \times N_p}{E_p} \times 1.1 = \frac{12.8 \text{ V} \times 1500 \text{ turns}}{120 \text{ V}} \times 1.1 = 176 \text{ turns}$$

Now, we need to determine wire sizes. In doing so, we must identify the primary and secondary currents. We know the secondary current is 1.04 A. We must calculate the primary current. We know the transformer output power and primary voltage. We can use these values in the power formula to find the current. Again, we will use the factor of 1.1 to take care of losses. Current, then, in the primary is:

$$I = \frac{P}{E} = \frac{13 \text{ W}}{120 \text{ V}} \times 1.1 \cong 0.120 \text{ A}$$

Referring to the wire table of Figure 9-39, we find that the minimum size wires for these currents should be:

- For primary (0.120 A) — No. 30
- For secondary (1.04 A) — No. 21

(For added safety, we will use wire larger than the required minimum — No. 24 for the primary and No. 18 for the secondary.)

No. wire size	Current-carrying capacity in amperes
12	8.70
14	5.40
16	3.40
17	2.70
18	2.10
19	1.70
20	1.30
21	1.10
22	0.85
23	0.75
24	0.54
30	0.13

Figure 9-39. Wire table.   (Brown and Sharpe)

One final consideration in constructing a transformer is the length of the coil to be used. In Figure 9-40, standard coil lengths for various power outputs are given. Our coils will be 1.625 inches long.

Power output in watts	Coil length in inches
10	1.250
20	1.625
25	1.875
30	2.000
50	2.250
100	2.750

Figure 9-40. Coil lengths for various power outputs.

Now that the design has been completed, the coils must be wound on a temporary wooden form. Since the core will have a 0.5 inch² cross-sectional area, a core material of about 0.7 inch x 0.7 inch will be used. We make the wooden coil form, Figure 9-41, about 0.1 inch oversize (0.8 in. × 0.8 in.) so that the finished coils will fit over the core.

Use the wooden form to wind one of the coils. Remember, the primary coil will have 1500 turns of No. 24 wire; the secondary will have 176 turns of No. 18 wire. The coil is wound in layers. Wind each layer carefully and neatly. Between each layer of coil, place a single layer of cloth or paper as insulation. Leave both wire ends exposed for transformer connecting leads. When complete, remove the finished coil from the form. Repeat the entire procedure for the other coil.

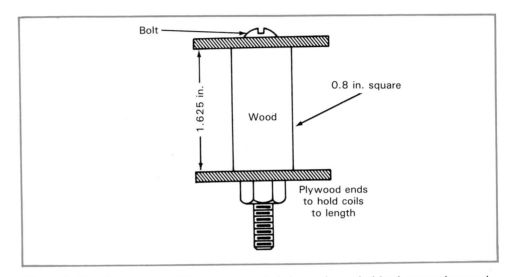

Figure 9-41. A wooden coil form, or mandrel, is used to wind both secondary and primary coils.

When finished, dip the coils in varnish, for insulation purposes, and hang them up to dry. After drying, wrap each coil with a layer of insulating tape.

Next, construct the transformer core. The core is made of laminated iron. The best source of laminated core iron is from an old transformer. However, black iron or even galvanized sheet iron from 22 to 26 gauge can be used. Cut and stack it to form the core, Figure 9-42. Cut to the dimensions shown. When stacking, stagger pieces of each layer as shown. Each lamination should be dipped in varnish and dried before stacking together.

After the core is stacked, slip out the laminations at one end and put your coils in place. Then, return the laminations to their original places. Drill holes at each corner, and bolt the core together securely with stove bolts.

With the transformer constructed, you are ready to complete your lamp project. The finished transformer is shown connected to the high intensity lamp circuit in Figure 9-43. Connect this circuit. Construct the lamp itself to suit your personal preference.

LESSON IN SAFETY: Use proper eye protection when soldering, grinding, chipping, and when working with metals, in general. Eyes cannot be replaced!

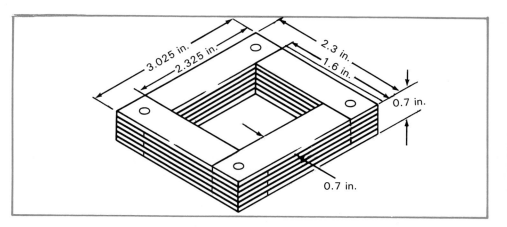

Figure 9-42. The laminations are coated with varnish and stacked with overlapping ends.

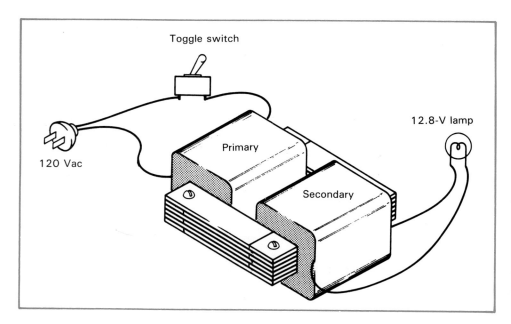

Figure 9-43. The high intensity lamp circuit showing finished transformer.

SUMMARY

- Inductance is that property of a circuit that opposes any change in current.
- The unit for inductance is the henry (H).
- Some factors that affect inductance of a coil are number of turns, diameter, and core material of the coil. Distance between windings is another factor.
- Mutual inductance is two or more coils sharing the energy of one. It is the basis for the operation of a transformer.
- The time constant of a coil is the amount of time it takes for current or voltage to rise or fall 63.2% of maximum value.
- Inductive reactance (X_L) is the opposition offered to a change in current by an inductor. It is measured in ohms.
- True power is power actually dissipated in a circuit. Reactive power is power that is alternately absorbed by a reactive component and returned to the circuit. Apparent power is power that at any instant appears to be supplied to the circuit. It is the product of effective voltage and effective current.

- Power factor is the relationship between true power and apparent power.
- The total opposition to a change of current in a reactive circuit is impedance *(Z)*. It is measured in ohms.
- Ohm's law for ac circuits substitutes Z for R as follows:

$$E = I \times Z \qquad I = \frac{E}{Z} \qquad Z = \frac{E}{I}$$

- In a transformer, energy is fed in the primary winding (input) and is taken out of the secondary winding (output).
- A step-up transformer increases voltage and decreases current.
- A step-down transformer decreases voltage and increases current.
- The relationship between voltage, current, and number of turns in transformer primary and secondary windings is:

$$\frac{E_p}{E_s} = \frac{N_p}{N_s} = \frac{I_s}{I_p}$$

- A center tap in a transformer allows the secondary winding to provide two equal voltages. Relative to the center tap, one voltage will be 180° out of phase with the other.
- The purpose of isolation in a transformer is to electrically separate one circuit from another.
- There are three basic types of transformer losses — copper, eddy current, and hysteresis losses.

KEY TERMS

Each of the following terms has been used in this chapter. Do you know their meanings?

apparent power *(S)*	inductive reactance *(X$_L$)*	RL time constant
audio frequency transformer	inductor	secondary winding (secondary)
center tap (CT)	isolation transformer	self-inductance
core-type transformer	magnetizing current	shell-type transformer
coupling	mutual inductance *(M)*	step-down transformer
coupling coefficient *(k)*	no-load current	step-up transformer
coupling transformer	phase angle *(θ)*	time constant
exciting current	power factor (cos *θ*)	transformer
henry (H)	power transformer	true power *(P)*
impedance diagram	primary winding (primary)	turns ratio
impedance matching	radio frequency transformer	unity coupling
impedance *(Z)*	reactance	volt-ampere reactive (VAR)
inductance *(L)*	reactive power *(Q)*	volt-ampere (VA)
inductive kick	RL circuit	

TEST YOUR KNOWLEDGE

Please do not write in this text. Place your answers on a separate sheet of paper.

1. Inductance is that property in a circuit which opposes a change in _____.

2. Explain the reason for increased cemf when frequency of current is increased.

3. What is the inductance of a coil that produces a cemf of 50 volts when the current changes 100 milliamps in 10 milliseconds.

4. What is total circuit inductance between two coils of 2 and 4.5 henrys when connected series-aiding with a coupling coefficient of .6?
5. The angular relationship between coils affects mutual inductance. The fields of one coil must cut across the conductors of the second coil at a _____ _____ for maximum coupling and mutual inductance.
6. What is the total inductance of 2-, 4-, and 8-henry coils in parallel without coupling?
   a. 114.0 H.
   b. 11.4 H.
   c. 1.14 H.
7. A series RL circuit has a 1000-ohm resistance and an 8-henry inductance. The applied voltage is 100 volts at a frequency of 1 kilohertz. Draw the circuit and the impedance diagram and compute impedance. By what angle does the current lag the voltage?
8. A parallel circuit having a 1000-ohm resistance and an 8-henry inductance is connected to a 100-volt source at 1 kilohertz. Find impedance. By what angle are the voltage and current out of phase?
9. A transformer has a turns ratio of 0.1, and the primary has 600 turns. How many turns are in the secondary? What is the secondary voltage when 10 volts ac are applied to the primary?
10. A power amplifier with an output Z of 2500 Ω must match the input Z of 100 Ω of another circuit. What turns ratio will the matching transformer have?
    a. 250/1
    b. 25/1
    c. 5/1
11. Should a transformer be connected to a dc source. What reason(s) can you give?
12. If a circuit has an inductance of 4 H at 10 kHz, then $X_L =$ _____.
13. If a circuit has an inductance of 2.5 mH at 100 kHz, then $X_L =$

    _____.
14. If a circuit has an inductance of 8 H at 60 Hz, then $X_L =$ _____.
15. If a circuit has an inductance of 8 H at 120 Hz, then $X_L =$ _____.
16. Draw a chart similar to the one shown in Figure 9-21, but consider frequency as a fixed value. Enter the effect (decrease or increase) in each square as inductance is decreased or increased.
17. In a parallel RL circuit, will the circuit impedance be more or less if the frequency is decreased? Why?
18. A transformer used on the 117-volt ac power line has a turns ratio of 20 to 1. What is the output voltage?
19. Assume that a transformer has a turns ratio of 1/5 and is connected to a 100-volt ac source. A load across the secondary draws 100 milliamps of current. What is the primary current?
20. Referring to the transformer and circuit arrangement described in question 19, what is the secondary voltage?
21. What is the secondary power in the circuit of question 19?
22. What is the primary power in the circuit of question 19?

Chapter 10

CAPACITANCE

Capacitance in an electronic circuit takes its place beside resistance and inductance in our studies. Only these three properties exist in any circuit. The most simple to the most complex combinations of components may be reduced to equivalent circuits composed of resistance, inductance, and capacitance. This is true even for devices like transistors and integrated circuits.

After studying this chapter, you will be able to:
☐ *Discuss the theory of capacitance in a circuit.*
☐ *Describe the construction of various types of capacitors.*
☐ *Identify common applications for capacitors in electronic equipment.*

INTRODUCTION

In our study of inductance, we found that inductors store energy in an electro*magnetic* field. In our study of capacitance, we find that certain devices, called *capacitors*, store energy in an electro*static* field. Specifically, a **capacitor** is a device that stores electrical charge. **Capacitance (C)** is the measure of the ability to store electrical charge. The basic unit of capacitance is the **farad (F)**. It is named in honor of Michael Faraday.

THE SIMPLE CAPACITOR

The most elementary capacitor is illustrated in Figure 10-1. It consists of two parallel, metal plates. The plates are separated by air or other insulative material. The insulating medium is called the **dielectric**. A voltage potential connected across these plates causes the following action:

The charge on the negative terminal of the battery instantly repels free electrons in the conductor to plate A of the capacitor. Electrons accumulate on plate A because they cannot flow through the dielectric. It takes on a negative charge. The electrons on plate B are repelled by plate A and are attracted by and to the positive terminal of the battery. Plate B takes on a positive charge.

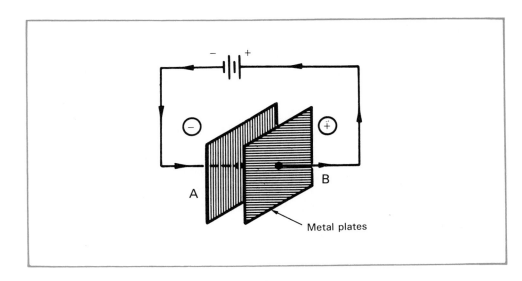

Figure 10-1. A simple parallel-plate capacitor.

This creates a potential difference between the two plates. The capacitor is now charged to a voltage equal to the source voltage but opposite in polarity. Due to the potential difference, an electrostatic field is set up. A current exists in the circuit but only during the charging and discharging of the capacitor. There is no current through the capacitor, however, because the dielectric is an insulator. There is current from one plate to the other but only through the external circuit.

Note that if you removed the capacitor from the circuit, it would remain charged and would itself be a source of voltage. In high voltage circuits, capacitors can retain charges that are dangerous. For safety, capacitors should be discharged before handling.

DIELECTRIC EFFECTS AND STORED ENERGY THEORY

In Chapter 1, you studied the properties of an electrostatic field. Demonstrations were made to prove the existence of an invisible force field surrounding a charged body. This applies to the capacitor. A strong electrostatic field exists in the dielectric between the plates of the charged capacitor. Electrons existing in this field would have a tendency to move toward the positive plate.

Figure 10-2 attempts to show one theory of how energy is stored in a dielectric. In the charged capacitor, the electrons associated with each atom of dielectric material are distorted out of their regular orbits. They assume new locations in the direction of the positive plate, while the atomic nucleus is attracted toward the negative plate. These atoms are said to be *polarized*.

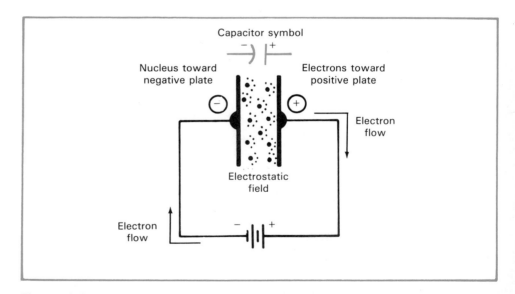

Figure 10-2. Electrons in the dielectric are distorted toward the positive plate when the capacitor is charged. (Note the symbol used for a capacitor.)

The polarized atoms give rise to charges on the dielectric surface near the capacitor plates. These charges are opposite in polarity to the charges of the plates. Their effect is to reduce the potential difference across them. Thus, for a given capacitive voltage, a dielectric will increase the amount of charge a capacitor can hold. This translates into higher capacitance, which translates into greater stored energy.

Because the dielectric is an insulator, the electrons within never detach from their atoms. Under the influence of the electrostatic field, they remain in these positions. When the capacitor is discharged, the electrons return to their original orbits. By doing so, they return energy to the circuit.

If a very strong electrostatic field, caused by too much voltage, exists between the plates, the electrons of the dielectric may be more than distorted. They may be torn from their orbits. If this happens, the material breaks down, and the insulator becomes a conductor. Sparking, or arcing, between the plates results.

All capacitors have a **working voltage direct current (WVDC) rating**. This is the specified voltage a capacitor can withstand without destruction of the dielectric. This specification must always be considered by the user when working with capacitors.

THE FARAD

As mentioned, the unit of capacitance is the farad. The symbol for the farad, again, is F. In electronics, very small values of capacitance are found. Most values will be in microfarads (1×10^{-6} F) or picofarads (1×10^{-12} F). A capacitor is said to have a capacitance of 1 farad when a changing voltage of 1 volt per second causes 1 ampere of charging current. In other words:

$$1 \text{ F} = \frac{1 \text{ A}}{\text{V/s}}$$

The mathematical relationship is expressed as:

$$C = \frac{I_{avg}}{\Delta e / \Delta t}$$

where C is in farads, I_{avg} is the average charging current in amps for a defined interval, and $(\Delta e / \Delta t)$ is the rate of change of voltage in volts/second.

Problem:

What is the capacitance of a capacitor if a charging current of 100 mA flows when the voltage changes 10 V in 10 ms?

Solution:

$$C = \frac{I_{avg}}{\Delta e / \Delta t} = \frac{0.1 \text{ A}}{10 \text{ V}/0.01 \text{ s}} = 0.0001 \text{ F} = 100 \text{ } \mu\text{F}$$

For an ac waveform, we can roughly calculate the size of a capacitor from the *average* current, E_{max}, and frequency. A formula for figuring the approximate size of a capacitor given certain ac values is:

$$C = \frac{I_{avg}}{4fE_{max}}$$

Problem:

The average ac charging current to a certain capacitor is 100 mA. If the peak voltage is 10 V and the frequency of the waveform is 100 Hz, what is the capacitance of the capacitor?

Solution:

$$C = \frac{I_{avg}}{4fE_{max}} = \frac{0.1 \text{ A}}{4(100 \text{ Hz})(10 \text{ V})} = 25 \text{ } \mu\text{F}$$

Another method of defining capacitance is by the charge, or quantity of electrons, that a capacitor will accept per volt of potential applied to it. This is expressed as:

$$C = \frac{Q}{E}$$

where C is capacitance in farads, Q is charge in coulombs, and E is voltage in volts.

By this formula, we may give another definition of the farad. A capacitor has a capacitance of 1 farad if 1 coulomb of charge is stored when 1 volt is applied to its plates. This is really just an extension of what was said before. Taking our previous definition of the farad and expanding:

$$1 \text{ F} = \frac{1 \text{ A}}{\text{V/s}} = \frac{1 \text{ A·s}}{\text{V}} = \frac{1 \text{ C}}{\text{V}}$$

Problem:

A certain capacitor will accept a charge of 0.001 C with an applied voltage of 10 V. What is its capacitance?

Solution:

$$C = \frac{Q}{E} = \frac{0.001 \text{ C}}{10 \text{ V}} = 0.0001 \text{ F} = 100 \ \mu\text{F}$$

Characteristics of capacitors

An expression can also be derived to calculate capacitance given the physical characteristics of the capacitor. Several characteristics must be considered, including:

* *Plate area.*
* *Distance between plates.*
* *Type of dielectric.*

Plate area. Capacitance is increased with an increase in plate area. The variable capacitor is a working example of this principle. (See *Air Capacitor* section towards end of chapter.) As the rotor is turned so that its plates mesh with the stator plates, a larger plate area is used. As a result, a variable capacitor in a fully meshed position is at its maximum capacitance. Larger plates can store more electrons than smaller plates, and a larger dielectric area can store a larger amount of energy.

Distance between plates. If the plates of a capacitor are very close, the field of one plate has a very pronounced effect on the field of the other plate. As the plates are separated, the ability of the fields to produce molecular distortion and store a charge within the dielectric will decrease. Therefore, with all else constant, the closer the plates are, the greater the capacitance. The further they are apart, the lower the capacitance.

Type of dielectric. When a given set of plates is separated by air, a certain value of capacitance is realized. If a piece of material, such a glass or paper, is placed between the plates, the capacitance will increase. Due to molecular formation, different materials will have different abilities to store electrical energy.

A factor ϵ, called the **permittivity** of the material, is a measure of how well a dielectric will *permit* the establishment of an electrostatic field between the plates. Capacitance varies directly with permittivity. Units are *farads per meter*. Electric permittivity is analogous to magnetic permeability.

Associated with permittivity are quantities of *permittivity of free space* and *relative permittivity*. As implied, **permittivity of free space (ϵ_0)**, is the permittivity of a vacuum. **Relative permittivity (ϵ_r)** is the ratio of the permittivity of a material to that of free space. It is also known as the **dielectric constant**. It is a unitless number. The formula for relative permittivity is:

$$\epsilon_r = \frac{\epsilon}{\epsilon_0}$$

The dielectric constants for a few materials used in capacitors are given in the table of Figure 10-3. The table tells us that a certain capacitor will have from 3 to 5 times more capacitance with waxed paper, for example, rather than air as the dielectric.

MATERIAL	DIELECTRIC CONSTANT (ϵ_r)
Vacuum	1.0000
Air	1.0006
Waxed Paper	3 to 5
Glass	5 to 10
Mica	3 to 6
Rubber	2.5 to 30
Wood	3 to 8
Pure Water	81

Figure 10-3. Dielectric constants.

We have discussed the physical factors that affect the capacitance value of a capacitor. We can combine these factors and come up with a mathematical relationship. It is given by the formula:

$$C = \frac{\epsilon_0 \epsilon_r A}{d}$$

where:

C = capacitance in farads
ϵ_0 = permittivity of free space (8.85×10^{-12} F/m)
ϵ_r = relative permittivity
A = effective area of one plate in square meters
d = distance between plates in meters

This formula shows that capacitance increases directly with the value of the dielectric constant and the plate area. It further shows that capacitance decreases as the distance between plates increases.

Capacitors in series

Connecting capacitors in series is the same as increasing the distance between the the outer two plates. Thus, total capacitance will decrease with each capacitor added. It will always be less than the lowest value in the series circuit.

All capacitors in a series circuit are part of the same current path. In the series circuit, current is the same at all points. Current is the rate of flow of charge. It follows, then, that all capacitors within a series circuit have the same charge, which is also equal to the total charge. In other words:

$$Q_T = Q_1 = Q_2 = Q_3 = \ldots$$

From Kirchhoff's voltage law, we know that the sum of the voltages across charged capacitors (voltage drop) is equal to the total voltage. In other words:

$$E_T = E_{C_1} + E_{C_2} + E_{C_3} + \ldots$$

Earlier in the chapter, we defined capacitance in terms of charge and voltage, wherein the formula $C = Q/E$ was presented. Rearranging this formula, gives $E = Q/C$. We can substitute this relationship into Kirchhoff's voltage law. We get:

$$\frac{Q_T}{C_T} = \frac{Q_1}{C_1} + \frac{Q_2}{C_2} + \frac{Q_3}{C_3} + \ldots$$

The values of charge are all equal. Thus, we can factor charge and cancel it out of the formula. From the result of this step, we can find total capacitance in a series circuit:

$$\frac{1}{C_T} = \frac{1}{C_1} + \frac{1}{C_2} + \frac{1}{C_3} + \ldots$$

If only two capacitors are in parallel, this simplified formula may be used:

$$C_T = \frac{C_1 C_2}{C_1 + C_2}$$

Capacitors in parallel

Connecting capacitors in parallel is the same as increasing the plate area. Thus, total capacitance will increase with each capacitor added in parallel.

When capacitors are in parallel, the current through each branch will depend upon the size of the capacitor. Branches with larger capacitances will have larger charging currents; smaller capacitances will have smaller charging currents. By Kirchhoff's current law, the sum of all the currents in the parallel circuit is equal to the total current. Since current is the rate of flow of charge, it follows that the sum of all charges of capacitors in parallel is equal to the total charge. In other words:

$$Q_T = Q_1 + Q_2 + Q_3 + \ldots$$

Again, rearranging a previous formula, we know that $Q = CE$. Substituting this relationship, we get:

$$C_T E_T = C_1 E_1 + C_2 E_2 + C_3 E_3 + \ldots$$

The values of voltage are equal across each branch of the parallel circuit. Thus, we can factor voltage and cancel it out of the formula. From the result of this step, we can find total capacitance in a parallel circuit:

$$C_T = C_1 + C_2 + C_3 + \ldots$$

Note that for computing capacitance, capacitors connected in parallel are treated as resistors or inductors in series. Capacitors connected in series are treated as resistors or inductors in parallel.

CAPACITANCE IN DC CIRCUITS

The behavior of a capacitance in a circuit is perhaps more easily understood if we first consider the *series RC circuit* with a dc voltage source. A circuit that contains both a resistance and a capacitance is called an **RC circuit**. The voltage and current relationships in a series RC circuit are of particular interest in our studies. Look at Figure 10-4. When S_1 (switch) contacts point A, these circuit conditions exist:

- Charge on capacitor at initial instant is zero.
- A maximum current instantaneously flows in the circuit because there is no opposing voltage from C_1. A maximum current will produce a maximum voltage drop across R_1.

- As C_1 becomes charged, the current decreases, and the drop across R_1 also decreases.
- Because the charging current has been reduced, the capacitor voltage grows at a slower rate than before.
- When C_1 is fully charged to the value of the source voltage, the current is zero, and the drop across R_1 is zero.

When S_1 contacts point B, the circuit will discharge. These circuit conditions will exist:

- Maximum discharge current will flow, decaying to zero as C_1 becomes discharged.
- The voltage across R_1 starts at maximum and decreases to zero.

The rise and decay of circuit current and voltages are depicted graphically in Figure 10-5.

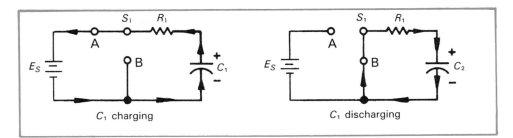

Figure 10-4. *A circuit used to demonstrate voltage and current relationships in a series RC circuit.*

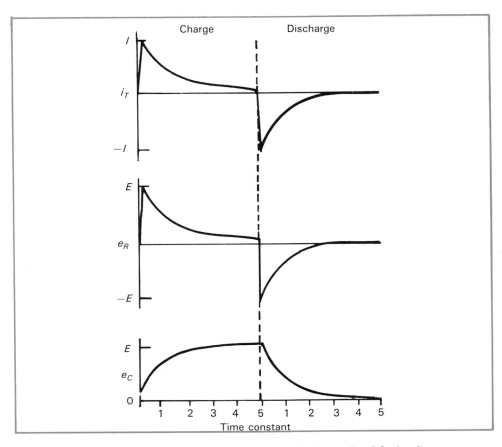

Figure 10-5. *Current and voltage rise and decay in the series RC circuit.*

RC TIME CONSTANT

If a resistance is added in series with a capacitance, the rise and fall of capacitive voltage takes some interval of time. This depends upon the relationship of the capacitance and resistance in the circuit. The time required for voltage of the capacitor to change by 63.2% is the **time constant** of that circuit. One time constant is figured as the time required for voltage to *rise* to 63.2% or *fall* to 36.8% of *peak* value. It may be found by the formula:

$$t = RC$$

where t is the time constant in seconds, C is the capacitance in farads, and R is the resistance in ohms.

It is generally accepted that after *five* time constants, the capacitor will be fully charged. Refer back to the universal time-constant chart of Figure 9-12. Remember that this chart is valid for RC as well as RL circuits. Voltages and currents at time-constant intervals may be computed from this chart. Notice that the *transient response* reaches a steady state after five time constants.

Problem:

What is the total charging time of a capacitor in a circuit containing an 8-μF capacitor in series with a 100-kΩ resistor?

Solution:

Step 1. Determine the time constant.

$$t = RC = (100 \times 10^3 \; \Omega)(8 \times 10^{-6} \; F) = 0.8 \; s$$

Step 2. Determine the total charging time.

$$\text{total charging time} = 5t = 5 \times 0.8 \; s = 4 \; s$$

The concept of *time constant* is demonstrated in the circuit of Figure 10-6. The NE2 neon glow lamp has a high resistance when not glowing. If a voltage of about 65 V is applied to the NE2, it will fire, or glow. While burning, it has a very low resistance.

As the circuit is turned on, the source voltage applies 100 V to the circuit. C_1 starts to charge. When voltage across the capacitor reaches 65 V, the lamp glows, and C_1 rapidly discharges through the NE2. The lamp goes out, and the capacitor starts to charge again.

The NE2 glow lamp will continue to flash each time the voltage reaches the firing point. You have constructed a flasher, or blinker. Technically, it is called a **relaxation oscillator** — the signal of which is a sawtooth waveform. It works by gradually charging and quickly discharging a capacitor (or inductor) through a resistor. By changing the values of either the resistor or the capacitor, the rate of flash may be controlled since the rate depends on the circuit time constant.

CAPACITANCE IN AC CIRCUITS

So far, we have discussed capacitance in dc circuits. We have seen that current in the dc circuit goes to zero after the capacitor is fully charged. In an ac circuit, the applied voltage varies and reverses polarity constantly. The capacitor alternately charges and discharges. For this reason, a capacitor allows current in an ac circuit. The current reverses directions in the circuit. It does not flow through the capacitor. Like the inductor, the capacitor reduces the amount of current in a circuit more than resistance alone. Now, we shall see how.

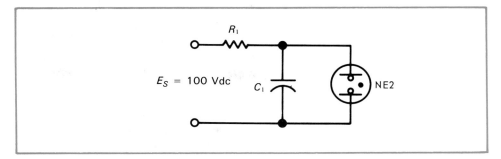

Figure 10-6. The circuit of a relaxation oscillator.

CAPACITIVE REACTANCE

When an ac voltage is applied to a capacitor, current flows by the charging and discharging of the capacitor. The voltage of the capacitor opposes the voltage applied to the circuit by the source. Thereby, a capacitor exhibits a resistance, or opposition, to charging and discharging. The amount of current depends upon this opposition to the changing current, or this **capacitive reactance** (X_C), which is measured in **ohms**.

Figure 10-7 demonstrates capacitive reactance. In Figure 10-7A, a lamp is connected in series with a large capacitor and an ac voltage source. The lamp glows brightly. Current does not flow through the capacitor; however, as the applied voltage reverses polarity, the capacitor charges and discharges. The current flows through the circuit by reversing direction periodically.

In Figure 10-7B, the large capacitor is replaced with a capacitor of a relatively small value. The lamp glows dimly. The current required to charge this small capacitor is too small to light the lamp as it should. We see that the brightness of the lamp depends upon the capacitor. In effect, the current in the ac circuit seems to be limited by the size of the capacitor. It is easy to see that a larger capacitor requires more current to charge it. Therefore, it allows more current and offers less opposition.

Frequency is another factor affecting current. As frequency is increased or decreased, a capacitor will be charged more or less times per second. An increase of frequency would cause more current, thereby, offering less opposition. In Figure 10-7C, the large capacitor is returned to the circuit. In addition, the ac source is replaced by a dc source. This is equivalent to changing the source frequency to zero. We see that the lamp does not light at all. The opposition of the capacitor is infinite.

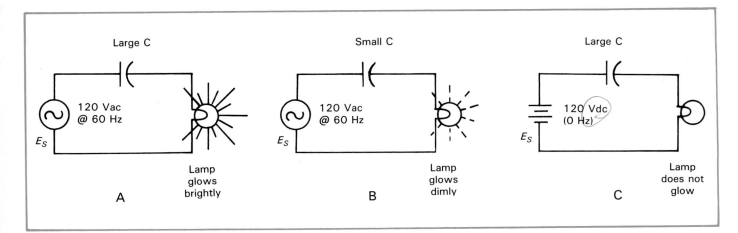

Figure 10-7. Demonstration of capacitive reactance in a series RC circuit.

The capacitor is an open circuit as far as dc ($f = 0$ Hz) is concerned. Stating it another way: a capacitor blocks dc. Capacitors are used for this purpose in many electronic circuits. In fact, a capacitor may have a very high reactance path at low frequencies, but become a very low reactance path, or an effective short circuit, at high frequencies. The technician might say, "This circuit is at ground potential for signal voltages." This means it is short circuited to ground through a capacitor. For dc voltages, however, the capacitor appears as an open circuit.

Capacitive reactance, then, is a function of the size of the capacitor and frequency. The relationship is an inverse proportion. The formula for capacitive reactance is:

$$X_C = \frac{1}{\omega C} = \frac{1}{2\pi f C} = \frac{1}{6.28 f C} = \frac{0.159}{f C}$$

where ω is in radians per second, f is in hertz, and C is in farads. Figure 10-8 shows how X_C decreases as f increases while C is held constant. Figure 10-9 shows how X_C decreases as C increases and f is held constant.

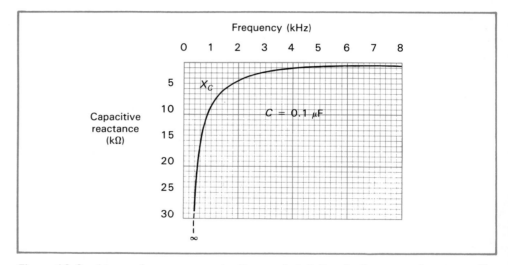

Figure 10-8. At zero frequency, or dc, X_C equals infinity. As frequency increases, X_C decreases.

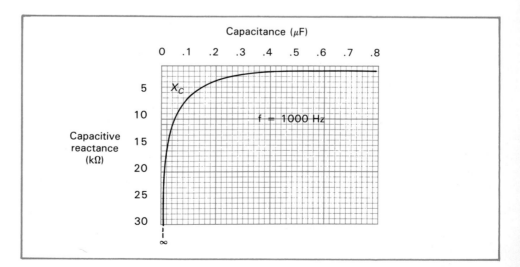

Figure 10-9. At zero capacitance, X_C equals infinity. As capacitance increases, X_C decreases.

Problem:

Compare the reactance of a 0.01 μF capacitor at 100 Hz, 1 kHz, and 100 kHz.

Solution:

Step 1. Compute reactance at 100 Hz.

$$X_C = \frac{0.159}{fC} = \frac{0.159}{(100 \text{ Hz})(0.01 \times 10^{-6} \text{ F})} = 159 \text{ k}\Omega$$

Step 2. Compute reactance at 1 kHz.

$$X_C = \frac{0.159}{fC} = \frac{0.159}{(1 \times 10^3 \text{ Hz})(0.01 \times 10^{-6} \text{ F})} = 15.9 \text{ k}\Omega$$

Step 3. Compute reactance at 10 kHz.

$$X_C = \frac{0.159}{fC} = \frac{0.159}{(100 \times 10^3 \text{ Hz})(0.01 \times 10^{-6} \text{ F})} = 159 \text{ }\Omega$$

Note that capacitive reactance at the low frequency is rather high. At the high frequency, it is rather low. This demonstrates how a low frequency in a capacitive circuit will have the same effect as an open circuit. Conversely, a high frequency will have the same effect as a short circuit.

THE PURE CAPACITIVE CIRCUIT

Examine the waveforms of Figure 10-10. They are of a pure and ideal capacitive circuit. They reveal some interesting facts about a circuit containing only capacitance. The source voltage and capacitive voltage are equal but 180° out of phase. As the source builds in a positive direction the capacitor begins to charge. The opposing capacitive voltage is small so the current is maximum. When the source voltage is equal to E_{max}, the opposing voltage of the capacitor is maximum, and the current is zero. Source voltage begins to fall; thus, capacitive voltage does, too — by discharging. The discharging current has the opposite direction of the charging current. At zero voltage, the source reverses polarity. With current continuing in the same direction,

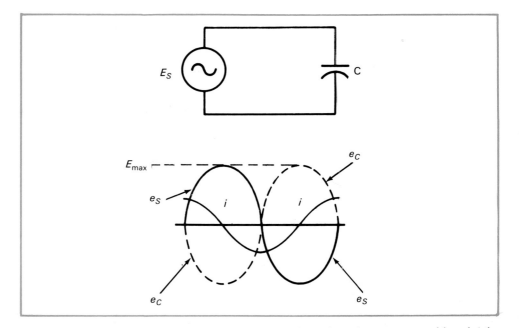

Figure 10-10. Current, source voltage, and capacitive voltage in a pure capacitive circuit.

the capacitor begins to charge to its opposite polarity. Charging current is maximum again while the capacitor is in a discharged state. This phase difference keeps going throughout each cycle.

In effect, the source voltage and capacitor voltage are alternately changing polarity and rising to maximum in opposite directions. When both reach their peak, there is no change of voltage, so current is zero. At the point where voltages cross the zero line, the change of voltage is most rapid, and the current must be at maximum to satisfy the change.

We see, then, in a purely capacitive circuit, the current leads the source voltage by an angle of 90°. This is opposite of a purely inductive circuit, where source voltage leads current. See Figure 10-11. It is a memory device. It is used to remember the effect of either inductance or capacitance in a circuit. Rather than depend upon a memory device, however, it is better that you understand these relationships.

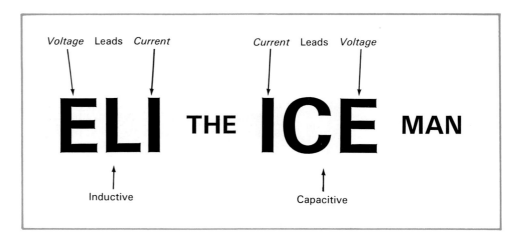

Figure 10-11. Memory aid for inductive and capacitive circuits.

THE SERIES RC CIRCUIT

In a circuit containing resistance only, the current and voltage are in phase. In a circuit of pure capacitance, the current leads the source voltage by an angle of 90°. In the series circuit containing both resistance and capacitance, the current will be in phase with the voltage across the resistor, E_R. It will lead voltage across the capacitor, E_C, by 90°. It will lead the source voltage by some angle between zero and 90°.

Impedance

In the series RC circuit, there are two resistive components opposing the current. These are 90° apart. One is the dc resistance in ohms; the other is the capacitive reactance in ohms. We can represent these components as two vectors of an impedance triangle. The hypotenuse of the triangle would represent the impedance *(Z)*. Impedance, again, is the total opposition an ac circuit offers to current. Impedance in an RC circuit is the vector sum of resistance and capacitive reactance. In other words:

$$Z = \sqrt{R^2 + X_C^2}$$

where Z is in ohms.

Voltage

As stated, current in the series RC circuit is in phase with E_R, and it leads E_C by 90°. Thus, we can say that E_R leads E_C by 90°. The series RC circuit, then, has two voltage components — E_R and E_C. Both of these can be represented by phasors. To figure source voltage, we do not simply add these two components. We must find their phasor sum:

$$E_S = \sqrt{E_R{}^2 + E_C{}^2}$$

Phase angle

Phase angle (θ), again, is the phase displacement between current and voltage that results from a reactive circuit element. In an impedance triangle, resistance is the horizontal component; capacitive reactance is the vertical. Phase angle, then, is given as:

$$\theta = \arctan \frac{X_C}{R}$$

In the voltage phasor diagram, E_R is the horizontal component since it is in phase with current. (Current, you will recall, is the common quantity of a series circuit; therefore, it is used as the horizontal reference.) E_C is the vertical component. Phase angle may be found from:

$$\theta = \arctan \frac{E_C}{E_R}$$

Problem:

Using the circuit with assigned values in Figure 10-12, determine the phase angle between applied voltage and current. Draw the impedance triangle. Also, find the voltage drops around the circuit. Verify voltages are correct.

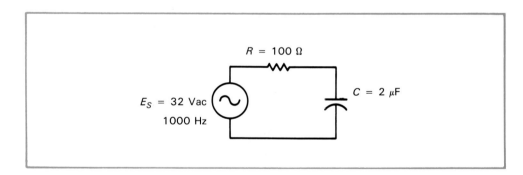

Figure 10-12. A series RC circuit.

Solution:

Step 1. Compute the value of capacitive reactance, X_C.

$$X_C = \frac{0.159}{fC} = \frac{0.159}{(1 \times 10^3 \text{ Hz})(2 \times 10^{-6} \text{ F})} \cong 80 \text{ } \Omega$$

Step 2. Determine the phase angle.

$$\theta = \arctan \frac{X_C}{R} = \arctan \frac{80 \text{ } \Omega}{100 \text{ } \Omega} = \arctan 0.8 = 38.7°$$

Step 3. Draw the impedance triangle. Use any convenient scale. Remember that, as discussed previously in Chapter 9, R is the horizontal vector since voltage is in phase with current in a resistor. (Current is the common quantity of a series circuit; thus, it is the horizontal reference.) X_C is drawn downward at 90° from R since current leads voltage across the capacitor. See Figure 10-13.

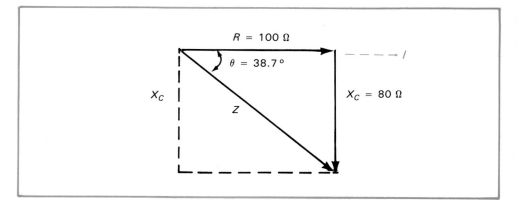

Figure 10-13. Impedance diagram.

Step 4. Compute the value of impedance, Z.

$$Z = \sqrt{R^2 + X_C^2} = \sqrt{(100\ \Omega)^2 + (80\ \Omega)^2} = 128\ \Omega$$

Note that the value of Z could also be found using trigonometry. Looking back at our impedance diagram:

$$\cos\theta = \frac{R}{Z}$$

Thus

$$Z = \frac{R}{\cos\theta} = \frac{100\ \Omega}{\cos 38.7°} = \frac{100\ \Omega}{.7804} = 128\ \Omega$$

Step 5. Calculate the circuit current. Using Ohm's law:

$$I = \frac{E}{Z} = \frac{32\ \text{V}}{128\ \Omega} = 0.25\ \text{A}$$

Step 6. Find the voltage drops around the circuit. Using Ohm's law:

$$E_R = IR = (0.25\ \text{A})(100\ \Omega) = 25\ \text{V}$$

$$E_C = IX_C = (0.25\ \text{A})(80\ \Omega) = 20\ \text{V}$$

Step 7. To verify the voltages are correct, check if their phasor sum equals the source voltage. See Figure 10-14.

$$E_S = \sqrt{E_R^2 + E_C^2} = \sqrt{(25\ \text{V})^2 + (20\ \text{V})^2} = 32\ \text{V}$$

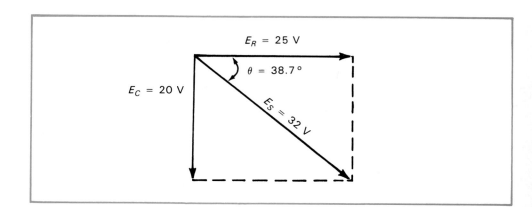

Figure 10-14. The phasor sum of the voltages will equal the supply voltage.

THE PARALLEL RC CIRCUIT

In a circuit containing a resistance and a capacitance in parallel, the voltage of each circuit element will be the same as the source voltage. Further, there will be no phase difference among them. This is because they are all in parallel.

There will be a phase difference among the total and branch currents, however. Current will be in phase with voltage in the resistive branch. It will lead the voltage across the capacitor by 90°. Total current will lead the source voltage by some angle between zero and 90°.

Impedance

In the parallel RC circuit, we do not find impedance through a vector sum. Instead, we apply Ohm's law:

$$Z = \frac{E_S}{I_T}$$

Current

As stated, voltage in the parallel RC circuit is in phase with I_R, and it lags I_C by 90°. Thus, we can say that I_R lags I_C by 90°. The parallel RC circuit, then, has two current components—I_R and I_C. Both of these can be represented by phasors. Since they are out of phase, we cannot simply add the two components together to figure the total circuit current. We must find their phasor sum:

$$I_T = \sqrt{I_R^2 + I_C^2}$$

Note that the phasor for I_C is above the horizontal reference. This is because I_C leads voltage—the horizontal reference for the parallel RC circuit.

Phase angle

For the parallel RC circuit, phase angle is found on the current phasor diagram. The horizontal reference of this circuit is voltage, since it is common to all circuit elements. On the current phasor diagram, the horizontal component is I_R since it is phase with voltage. The phase angle, then, is the angle between I_R and the total current. This is the phase displacement resulting from the reactive element. In the parallel RC circuit, phase angle is:

$$\theta = \arctan \frac{I_C}{I_R}$$

Problem:

Using the circuit with assigned values in Figure 10-15, determine the phase angle between applied voltage and current. Draw the current phasor diagram and find the circuit impedance.

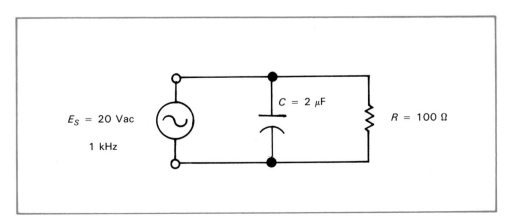

Figure 10-15. A parallel RC circuit.

Solution:

Step 1. Compute the value of capacitive reactance, X_C.

$$X_C = \frac{0.159}{fC} = \frac{0.159}{(1 \times 10^3 \text{ Hz})(2 \times 10^{-6} \text{ F})} \cong 80 \; \Omega$$

Step 2. Compute branch currents. Using Ohm's law:

$$I_R = \frac{E_S}{R} = \frac{20 \text{ V}}{100 \; \Omega} = 0.2 \text{ A}$$

$$I_C = \frac{E_S}{X_C} = \frac{20 \text{ V}}{80 \; \Omega} = 0.25 \text{ A}$$

Step 3. Determine the phase angle to see by how much the circuit current leads the voltage.

$$\theta = \arctan \frac{I_C}{I_R} = \arctan \frac{0.25 \text{ A}}{0.2 \text{ A}} = \arctan 1.25 = 51.3°$$

Step 4. Draw the current phasor diagram. Use any convenient scale. Remember that I_R is drawn as the horizontal component. I_C is drawn upward at 90° from I_R since it leads voltage — the horizontal reference. See Figure 10-16.

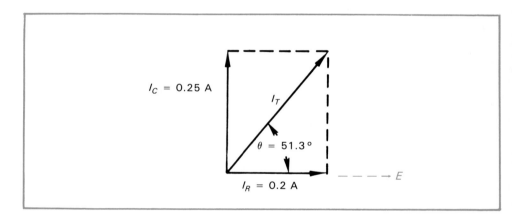

Figure 10-16. Current phasor diagram.

Step 5. Find the total circuit current.

$$I_T = \sqrt{I_R^2 + I_C^2} = \sqrt{(0.2 \text{ A})^2 + (0.25 \text{ A})^2} = 0.32 \text{ A}$$

Step 6. Find the impedance of the circuit. Using Ohm's law:

$$Z = \frac{E_S}{I_T} = \frac{20 \text{ V}}{0.32 \text{ A}} = 62.5 \; \Omega$$

POWER

Capacitive reactance uses no power. The energy stored in a capacitor is returned to the circuit. Resistance is the only element of a circuit that dissipates power. The power curve for a pure capacitive circuit is shown in Figure 10-17. The curve represents the products of many instantaneous voltages and currents. Power above the zero line is power consumed by the circuit. Power below the line is power returned by the circuit. The positive half-cycles of power will equal the negative half-cycles in magnitude. Their average value is zero. Therefore, power consumed by capacitance is zero. Energy required to build the electrostatic field is returned to the circuit.

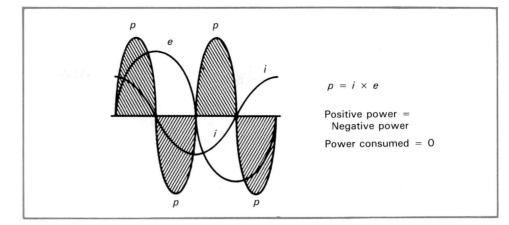

$p = i \times e$

Positive power =
Negative power

Power consumed = 0

Figure 10-17. A pure capacitive circuit uses zero power.

There are three types of power in the RC circuit. These are *true power, reactive power,* and *apparent power.* These are virtually the same to an RC circuit as they are to an RL circuit.

- True power in the RC circuit is the power dissipated, as heat, in the resistor. It is given by:

$$P = I_R{}^2R = E_SI_T \cos \theta$$

- Reactive power is power that is absorbed by a reactive element. In a capacitor, it is given by:

$$Q = E_CI_C = I_C{}^2X_C = \frac{E_C{}^2}{X_C} = E_SI_T \sin \theta$$

- Apparent power is power that at any instant *appears* to be that supplied to a circuit. It is given by:

$$S = E_SI_T = I_T{}^2Z = \frac{E_S{}^2}{Z}$$

Power factor

Power factor is the ratio of the power actually used to the power supplied. In other words, it is the ratio of true power to apparent power. On a power triangle, true power is the horizontal component; apparent power is the resultant, or hypotenuse. (Reactive power is the vertical component.) θ is the phase angle. On the power triangle, it is the angle between the true and apparent power phasors. The ratio of true power to apparent power is equal to $\cos \theta$. In other words:

$$\text{Power factor} = \cos \theta = \frac{\text{true power}}{\text{apparent power}} = \frac{P}{S}$$

We find the phase angle on our impedance triangles and voltage and current phasor diagrams, as well. Using the proper trigonometric relationships, we can also define power factor in the following terms:

$$\text{Power factor} = \frac{R}{Z} = \frac{E_R}{E_S} = \frac{I_R}{I_T}$$

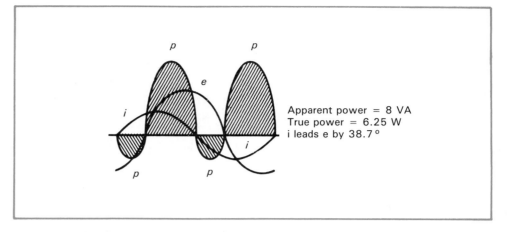

Figure 10-18. The RC circuit uses power.

Problem:

Find the power factor of the series RC circuit in Figure 10-12. (Note the power curve shown in Figure 10-18 of a typical RC circuit.)

Solution:

Step 1. Find the true power.

$$P = I_R^2 R = (0.25 \text{ A})^2 (100 \ \Omega) = 6.25 \text{ W}$$

Step 2. Find the apparent power.

$$S = I_T^2 Z = (0.25 \text{ A})^2 (128 \ \Omega) = 8 \text{ VA}$$

Step 3. Find the power factor.

$$\text{Power factor} = \frac{\text{true power}}{\text{apparent power}} = \frac{6.25 \text{ W}}{8 \text{ VA}} = .78$$

Note that the power factor of the previous problem tells us that 78% of the power supplied is actually used. The rest is returned to the source. The ideal power factor is 1. This is called the **unity power factor**. With the unity power factor, all of the power supplied is used. A power factor of less than 1 means that extra current, which is not used but returned, must be handled. Larger electrical equipment must be provided to handle the extra current. The extra current also means more power will be lost in the transmission lines.

Quality, dissipation, and capacitive power factors

Because there is no perfect insulator, some current will leak through the dielectric of a capacitor. This results in some energy loss. Other losses are due to the resistance of the capacitor leads and plates. These losses cause wasted power and heat. Capacitor checking instruments will measure losses and make compensations when measuring capacitance values.

Capacitors have ratings that give their quality in terms of loss. One such rating is indicated by its **quality factor**, or Q — the ratio of a capacitor's reactance to its resistance at a specified frequency. It should be as high as possible, since a lower ratio indicates higher power loss. **Dissipation factor** is the reciprocal of quality factor. It should be as low as possible. A high dissipation factor means a high power loss. A third rating is given by the **capacitive power factor**. This is the ratio of resistance to impedance. It represents the fraction of input power dissipated in a capacitor. Quality factor and dissipation factor are terms associated with dc capacitors. Power factor is primarily associated with ac motor start capacitors.

APPLICATIONS OF RC CIRCUITS

There are a number of standard RC circuit applications. One such application is the *low-pass filter*. A **low-pass filter network** will pass lower frequency signals through a circuit better than higher frequencies. Another application involves what is called a *high-pass filter*, which will pass higher frequency signals. One type of high-pass filter in particular is used for *coupling*. A **coupling network** is used to completely pass an ac input signal to the output and block any dc level. An RC circuit may also be connected to *bypass* an ac signal. A high-pass filter can be combined with a low-pass filter to form a **bandpass filter network**. The result here is the passing of a band of frequencies not stopped by either circuit. In contrast, a **band-reject filter network** will reject all frequencies within a certain band. In this section, we will be discussing high-pass filter and bypass networks in further detail.

The high-pass filter network

Look at Figure 10-19. These circuits have a dc power source of 10 V in series with an ac signal source of 10 V. As a general background, when ac and dc sources are connected in series, both sources supply the circuit current. Each voltage source will produce current as though the other were not there. The result is a fluctuating dc voltage or current, with ac variation superimposed on the average dc level. It is convenient to consider the fluctuating voltage in two parts. One is the steady dc component; the other is the ac component.

Consider first a circuit like one of Figure 10-19 but without an ac voltage. The capacitor will become charged to 10 V. If the voltages across the resistor and capacitor were then measured, E_C would measure 10 V; E_R would measure 0 V. The capacitor blocks the dc; thus, there is no dc output across the resistor.

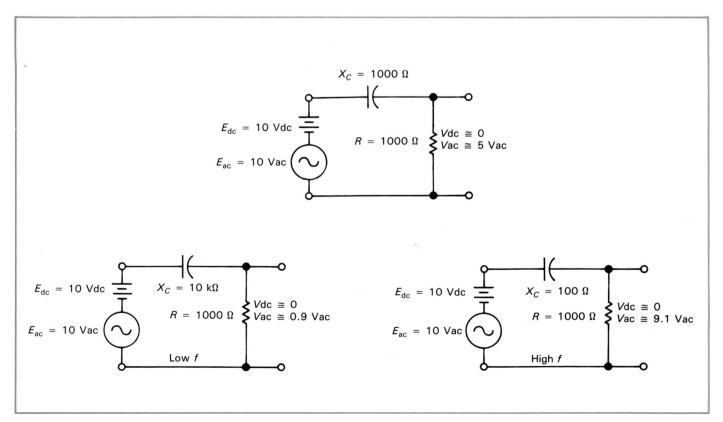

Figure 10-19. These circuits demonstrate the effect of frequency on the output of a high-pass filter.

Now, we apply the ac source in series with the dc power source. If the reactance of the capacitor is 1000 Ω, and the resistance of the resistor is 1000 Ω, the ac component of voltage across the capacitor will be 5 V_{rms}. It will also be 5 V_{rms} across the resistor.

If the frequency of the ac is *decreased* so the reactance becomes 10,000 Ω, then the ac voltage division may be approximated by:

$$E_C = E_S \times \frac{X_C}{X_C + R} = 10 \text{ V} \times \frac{10,000 \ \Omega}{10,000 \ \Omega + 1000 \ \Omega}$$

$$= 10 \text{ V} \times \frac{10,000 \ \Omega}{11,000 \ \Omega} = 9.1 \text{ Vac}$$

With 9.1 Vac across the capacitor, only 0.9 Vac of signal voltage will appear across the resistor for the circuit output. (Note that any out-of-phase condition was omitted to simplify calculations.)

If the frequency of the ac is *increased* so the reactance becomes 100 Ω, then the ac voltage division may be approximated by:

$$E_C = E_S \times \frac{X_C}{X_C + R} = 10 \text{ V} \times \frac{100 \ \Omega}{100 \ \Omega + 1000 \ \Omega}$$

$$= 10 \text{ V} \times \frac{100 \ \Omega}{1100 \ \Omega} = 0.9 \text{ Vac}$$

With 0.9 Vac across the capacitor, 9.1 Vac of signal voltage will appear across the resistor for the circuit output.

Thus, we see for higher frequencies that a high proportion of signal voltage appears across the resistor, or the output of the network. For lower frequencies, little of the signal voltage appears across the network output. This circuit passes high frequencies better than low frequencies. It is the circuit of a **high-pass filter network**.

This circuit will be used later in the RC coupling of amplifiers. To assure maximum output, the value of the resistor should be ten or more times greater than the value of X_C at operating frequency.

The bypass network

In Figure 10-20, a combination RC circuit is connected to a dc and ac source in series. First, consider the dc component only. No dc current will flow through C_1. It is practically an open circuit. Its reactance compared to the 1-kΩ R_2 is so high that X_C can be ignored as a parallel branch. Therefore, R_2 can be considered as a voltage divider in series with R_1. Since R_1 and R_2 are equal, both have 5 Vdc across them. C_1, in parallel with R_2, has 5 Vdc across it, also.

Now, consider only the ac component. The ac voltage will divide according to the resistance of R_1 and the impedance of R_2 and C_1 in parallel. In a bypass network, the ratio of R_2 to X_C must be ten or more. With this ratio, the value of the resistor is so high compared to X_C, that R_2 can be ignored as a parallel branch. Thus, we consider C_1 as a voltage divider in series with R_1. The ac voltage division may be approximated by:

$$E_{R_1} = E_{ac} \times \frac{R_1}{X_C + R_1} = 10 \text{ V} \times \frac{1000 \ \Omega}{100 \ \Omega + 1000 \ \Omega}$$

$$= 10 \text{ V} \times \frac{1000 \ \Omega}{1100 \ \Omega} = 9.1 \text{ Vac}$$

$$E_C = E_{ac} \times \frac{X_C}{X_C + R_1} = 10 \text{ V} \times \frac{100 \ \Omega}{100 \ \Omega + 1000 \ \Omega}$$

$$= 10 \text{ V} \times \frac{100 \ \Omega}{1100 \ \Omega} = 0.9 \text{ Vac}$$

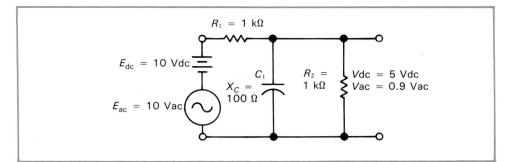

Figure 10-20. This circuit shows the effect of bypassing an ac signal.

Note that the voltage across R_2 is equal to the voltage across C_1 since they are in parallel. Little ac voltage is developed across this branch. The effect of the capacitor in this network then is to *bypass* the ac component of a pulsating dc voltage. It acts as a short circuit for ac voltage across the parallel resistor. The result is a very low ac voltage across the capacitor for any frequency at which R_2 is 10 or more times greater than X_C. This network just described is a **bypass network**.

TYPES OF CAPACITORS

Many types of capacitors are available. They are generally classified by the dielectric. Air, paper, mica, ceramic, and electrolytic are the most common. They are described in the following paragraphs.

AIR CAPACITORS

The **air capacitor** is a variable capacitor, Figure 10-21. It consists of one or more metal plates in a fixed position, called the *stator*. A second set of plates, called the *rotor*, can be rotated so that they interleave between each stator plate. The amount of capacitance is controlled by a knob that turns the rotor. Turning the rotor changes the effective plate area and, thus, the capacitance. The dielectric of this capacitor is air.

Variable capacitors are widely used in tuning circuits. The tuning knob of radios turns a capacitor of this type. Two or more of these capacitors may

Figure 10-21. A variable air capacitor. (J.W. Miller Co.)

be connected to a single shaft so that all may be rotated together. This assembly is called a **ganged capacitor**.

A typical variable capacitor will have a screw on its side, which is another small capacitor in parallel with the larger variable capacitor. This little **trimmer capacitor**, as it is called, is used to make fine adjustments on the total capacitance of the device. The trimmer capacitor will have small flexible metal plates separated by mica or some other dielectric. By turning the screw inward, the plates are compressed, and capacitance is increased.

PAPER CAPACITORS

A **paper capacitor**, Figure 10-22, is one of the more common types of fixed capacitors. Two layers of thin metal foil are separated by waxed paper or other paper dielectric. The sandwich of foil and paper is rolled into a cylindrical shape. Then it is enclosed in a paper tube or encased in a plastic or wax capsule. Leads extending from each end of the capacitor are attached to the metal foil plates. These are made in hundreds of capacitance and working voltages values. The values are usually printed on the capacitor case.

Figure 10-22. A paper capacitor. (Sprague Electric Co.)

MICA CAPACITORS

The **mica capacitor**, Figure 10-23, is a sandwich of thin metal plates separated by thin sheets of mica. Alternate plates are connected together and leads attached. The total assembly is encased in plastic. These capacitors have small capacitance values, yet high voltage ratings.

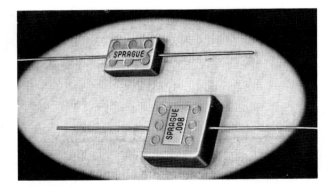

Figure 10-23. Mica capacitors. (Sprague Electric Co.)

CERAMIC CAPACITORS

Ceramic capacitors, Figure 10-24, are about the size of a dime or smaller. They have alternately deposited layers of ceramic and metal electrodes. The metal acts as the capacitor plates. The ceramic is the dielectric. Leads are attached to electrode pick-ups, and the component is encapsulated in a moistureproof coating. You will find dozens of ceramic capacitors used in electronic circuits. They have small capacity values but can be made to operate at very high voltages.

MONOLYTHIC® MULTI-LAYER CERAMIC CAPACITORS

RADIAL-LEAD RESIN-DIPPED

Voltage Ratings:
50 and 100 WVDC

Capacitance Range:
1.0 pF to 4.7 μF

Size Range:
.150" x .150" x .100" to .500" x .500" x .125"

Primary Applications:
Used where capacitors with EIA Characteristics Z5U, X7R, and C0G must be selected to meet specific requirements.

Sprague Types:
9C, 1C - 5C

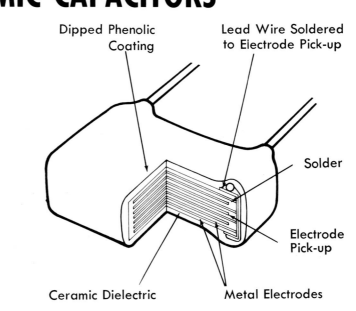

Dipped Phenolic Coating

Lead Wire Soldered to Electrode Pick-up

Solder

Electrode Pick-up

Ceramic Dielectric

Metal Electrodes

(Alternately deposited layers of ceramic dielectric material and metal electrodes fired into a rugged, solid block.)

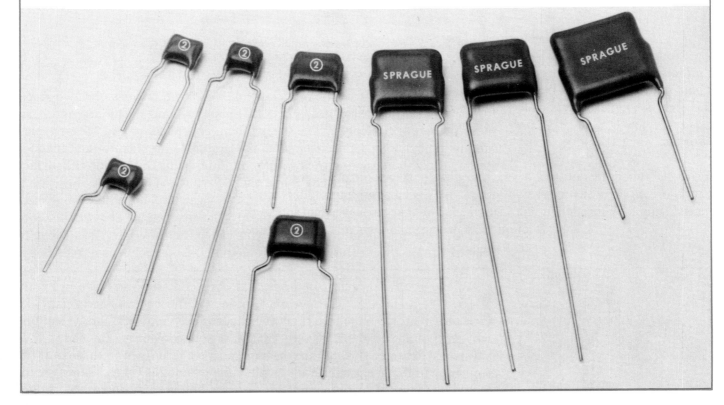

Figure 10-24. Ceramic capacitors. (Sprague Electric Co.)

ELECTROLYTIC CAPACITORS

Electrolytic capacitors are polarized; that is, one plate is positive and one is negative. These capacitors are used when large capacitance values are needed. They provide the greatest capacitance in the least space. Electrolytic capacitors are commonly made with aluminum. Tantalum and titanium are also used, however. Figure 10-25 shows a few electrolytic capacitors.

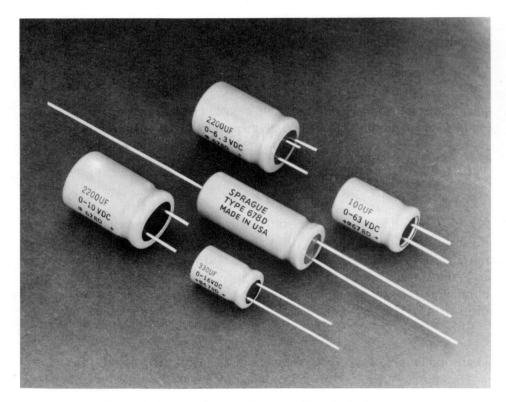

Figure 10-25. Electrolytic capacitors.   (Sprague Electric Co.)

In the general construction of an *aluminum electrolytic capacitor*, two sheets of aluminum foil are separated by a fine paper or gauze. This separator has been soaked in electrolyte. The sheets, then, are rolled up and encased in an aluminum cylinder. Once assembled, a dc voltage is applied to the capacitor terminals. This causes a thin layer of aluminum oxide to form on the surface of the positive plate, between it and the electrolyte. The aluminum oxide is the dielectric. The electrolyte makes up the negative plate. The second sheet of foil serves only as a contact surface to the electrolyte. The specific construction of the aluminum electrolytic capacitor may vary according to manufacturer. See Figure 10-26. Note that this capacitor has two paper separators.

An electrolytic capacitor in a *basic can-type* container is shown in Figure 10-27. The metal can of this capacitor is also the negative terminal. The capacitor pictured has more than one capacitance value and more than one working voltage available, as indicated on the outside of the can. The various values are obtained through the multiple positive terminals on the end of the can. Each terminal has a different symbol that denotes its value. This capacitor would be mounted on a metal plate, which is bolted to a chassis, and would stand upright. If the negative terminal is not to be connected to chassis ground, it is mounted on a fiber insulating plate.

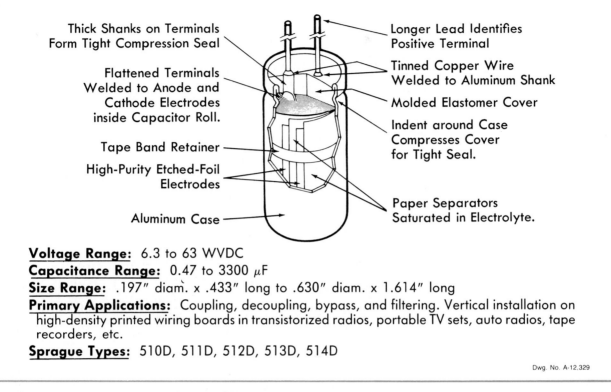

VERTI-LYTIC® MINIATURE SINGLE-ENDED ALUMINUM ELECTROLYTIC CAPACITOR

Thick Shanks on Terminals Form Tight Compression Seal

Flattened Terminals Welded to Anode and Cathode Electrodes inside Capacitor Roll.

Tape Band Retainer

High-Purity Etched-Foil Electrodes

Aluminum Case

Longer Lead Identifies Positive Terminal

Tinned Copper Wire Welded to Aluminum Shank

Molded Elastomer Cover

Indent around Case Compresses Cover for Tight Seal.

Paper Separators Saturated in Electrolyte.

Voltage Range: 6.3 to 63 WVDC
Capacitance Range: 0.47 to 3300 µF
Size Range: .197″ diam. x .433″ long to .630″ diam. x 1.614″ long
Primary Applications: Coupling, decoupling, bypass, and filtering. Vertical installation on high-density printed wiring boards in transistorized radios, portable TV sets, auto radios, tape recorders, etc.
Sprague Types: 510D, 511D, 512D, 513D, 514D

Dwg. No. A-12,329

Figure 10-26. Construction of an aluminum electrolytic capacitor. (Sprague Electric Co.)

Figure 10-27. Basic can-type electrolytic capacitor.

With the compact electronic equipment we have today, came the need for capacitors of a very small scale. We use miniature electrolytic capacitors in many projects in this text. Many of these are the tantalum type. See Figure 10-28. The pencil gives an idea of just how small these components are. Companies compete to get the most capacitance in the smallest package, with a high degree of accuracy and reliability. Since these capacitors are used in transistor circuits, the working voltages can be low. They are made in a range of 1 V – 50 V.

There is a special type of electrolytic capacitor that can be used in ac circuits. These are *nonpolarized* and specified as *NP*. These capacitors are found in motor starters and in some electronic circuits.

HERMETICALLY-SEALED SOLID-ELECTROLYTE TANTALUM CAPACITORS

AXIAL-LEAD METAL-CASE

Voltage Range:
6 to 125 WVDC

Capacitance Range:
0.0047 to 1000 µF

Size Range:
.125" diam. x .250" long to .341" diam. x .750" long ,

Primary Applications:
Industrial and military equipment where reliability, low leakage current, low dissipation factor, and stability with time and temperature are required.

Sprague Types:
150D   CSR13(MIL-C-39003)

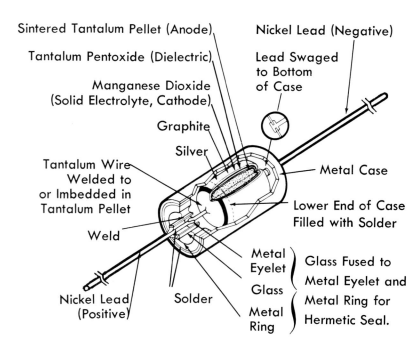

Sintered Tantalum Pellet (Anode)
Tantalum Pentoxide (Dielectric)
Manganese Dioxide (Solid Electrolyte, Cathode)
Graphite
Silver
Tantalum Wire Welded to or Imbedded in Tantalum Pellet
Weld
Nickel Lead (Positive)
Solder
Metal Eyelet
Glass
Metal Ring
Nickel Lead (Negative)
Lead Swaged to Bottom of Case
Metal Case
Lower End of Case Filled with Solder
Glass Fused to Metal Eyelet and Metal Ring for Hermetic Seal.

Dwg. No. A-13.279

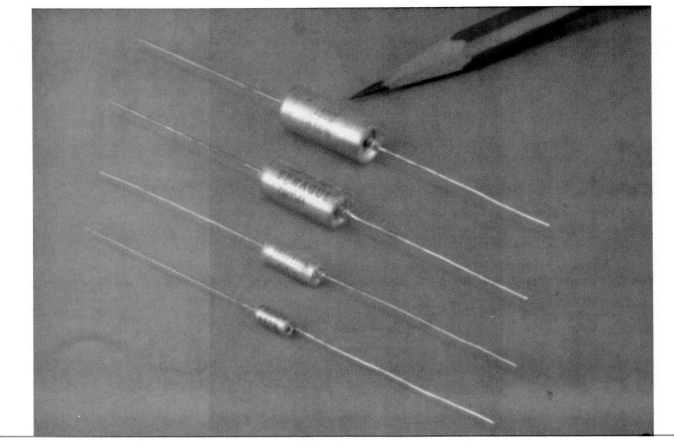

Figure 10-28. Tantalum electrolytic capacitors. (Sprague Electric Co.)

There are some disadvantages to electrolytic capacitors. They are polarity sensitive and must be installed with correct polarity in a circuit. If connected in reverse, the oxide film will dissolve and the capacitor will short circuit. A distinctive cloud of peculiar smelling white smoke, if not an explosion, will inform you of your error.

The electrolytic capacitor, further, has a low leakage resistance and, thus, a high leakage current. The resistance can be measured with an ohmmeter. Leaky capacitors are a source of trouble to the service technician, and they must be replaced. They permit unwanted dc currents in some circuits. The leakage current also reduces the shelf life.

LESSON IN SAFETY: Be on guard for electrolytic capacitors, which may remain highly charged after the equipment has been turned off. Use an insulated screwdriver to short out all capacitors to ground. Short them not only once but twice. Be certain they are discharged before touching the circuit with your bare hands.

INTEGRATED CIRCUIT CAPACITORS

Capacitors may now be formed on *integrated circuits (ICs)*. ICs are miniature packages that contain thousands of components such as semiconductors, resistors, diodes, and capacitors. Capacitors may be formed on an IC by depositing a very thin layer of metal to form a plate on a material called a *substrate*. Afterwards, an insulator and another layer of metal is deposited. Capacitors used in ICs are usually a very small value since their size is extremely limited. ICs will be covered in much more detail later in this text.

CAPACITOR COLOR CODING

Some capacitors are color coded. The code used for capacitors is basically the same as that used for resistors (Figure 3-30). Take, for example, the mica capacitor shown in Figure 10-29. Assume that the upper-left corner is black, indicating a joint army-navy (JAN) military code. Assume the next two dots are red and green, which represents 25, and the multiplier is brown. Multiply, then, by 10. The value of the capacitor is 250 pF. The tolerance value and class can also be found. However, their meanings will not be covered in this textbook.

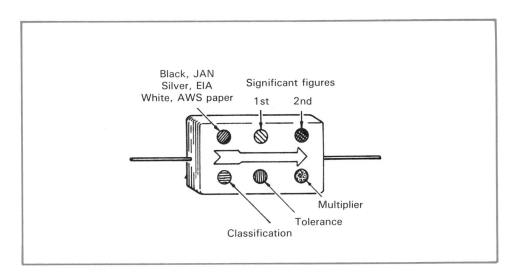

Figure 10-29. Reading a capacitor. All values are in picofarads.

SUMMARY

- Capacitance is the measure of the ability to store charge.
- A capacitor is a device that temporarily stores charge. It is made up of two plates of conductive material separated by insulation, called the dielectric.
- Capacitance is measured in farads. Commonly used units are the microfarad (μF) and picofarad (pF).
- Factors affecting capacitance include distance between plates, plate area, and dielectric material.
- The time constant of a capacitor is the amount of time it takes for the capacitor to charge or discharge to 63.2% of maximum value.
- Working voltage is the maximum voltage that can be steadily applied to a capacitor without causing an arc.
- Capacitive reactance (X_C) is opposition to alternating current due to capacitance. It is measured in ohms.
- Capacitive reactance uses no power. The energy stored in a capacitor is returned to the circuit. Resistance is the only component of a circuit that dissipates power.
- Capacitors are useful as low- and high-pass filters. They are also useful for coupling and bypassing.
- Many types of capacitors are available. Air, paper, mica, ceramic, and electrolytic are the most common.

KEY TERMS

Each of the following terms has been used in this chapter. Do you know their meanings?

air capacitor
bandpass filter
band-reject filter
bypass network
capacitance *(C)*
capacitive power factor
capacitive reactance *(X$_C$)*
capacitor
ceramic capacitor
coupling network
dielectric

dielectric constant
dissipation factor
electrolytic capacitor
farad (F)
ganged capacitor
high-pass filter network
low-pass filter network
mica capacitor
paper capacitor
permittivity *(ε)*

permittivity of free space *(ε$_0$)*
power factor
quality factor *(Q)*
RC circuit
RC time constant
relative permittivity *(ε$_r$)*
relaxation oscillator
trimmer capacitor
unity power factor
WVDC rating

TEST YOUR KNOWLEDGE

Please do not write in this text. Place your answers on a separate sheet of paper.

1. An increase in plate area of a capacitor will _____ its capacitance.
2. An increase in distance between plates of a capacitor will _____ its capacitance.
3. What is the capacitance of 0.01-μF, 0.02-μF and 0.05-μF capacitors connected in series?
4. Using the same capacitor values in Question 3, what is the total capacitance if connected in parallel?
5. Calculate the RC time constant of a series circuit of a 2.2-MΩ resistor and a 10-μF capacitor.

Compute the reactance of a 0.05-μF capacitor for the frequencies given in Questions 6 through 9.

6. At 100 Hz, X_C is _____.
7. At 1 kHz, X_C is _____.
8. At 10 kHz, X_C is _____.
9. At 100 kHz, X_C is _____.
10. A series circuit of a 10-kΩ resistor and a 0.01-μF capacitor are connected to a 50-V, 1-kHz source. Draw the circuit and impedance diagram.
11. Find Z in the circuit of Question 10.
12. Find I_T in the circuit of Question 10.
13. Find θ in the circuit of Question 10.
14. Find E_C in the circuit of Question 10.
15. Find E_R in the circuit of Question 10.
16. Find the power factor in the circuit of Question 10.
17. A 500-Ω resistor is bypassed by a capacitor. The operating frequency is 100 Hz. What is the value of the capacitor in microfarads?
18. Assume the circuit in Figure 10-6 has a 100-V source, a 1-MΩ resistor, and a 5-μF capacitor. What should the flash frequency of the NE2 lamp be? Build the circuit and prove your calculations.

Applications of Electronics Technology:
A centralized home theater such as this combines TV, VCR, and audio equipment in attractive cabinetry. (Philips)

Chapter 11

RCL CIRCUITS AND RESONANCE

In the previous two chapters, the properties and effects of both series and parallel, RL and RC circuits were presented. In many electronic circuits, inductance, capacitance, and resistance are present together. Such circuits are called **RCL circuits**. They may be series or parallel.

After studying this chapter, you will be able to:

☐ *Explain resonant frequency and how it affects various RCL circuits.*
☐ *Discuss the characteristics of a series RCL circuit at resonance.*
☐ *Explain the operation of the series RCL circuit.*
☐ *Name applications of the series RCL circuit.*
☐ *Discuss the characteristics of a parallel RCL circuit at resonance.*
☐ *Explain the operation of the parallel RCL circuit.*
☐ *Name applications of the parallel RCL circuit.*

INTRODUCTION

The principles of resistance, capacitance, and inductance should be reviewed before studying RCL circuits:

- In an ac circuit containing resistance only, the applied voltage and current are in phase. There is no reactive power. The power consumed by the circuit is equal to volts times amperes.
- In an ac circuit containing inductance only, the current lags the source voltage by an angle of 90°; they are not in phase. The power consumed by the circuit is zero.
- In an ac circuit containing resistance and inductance, the current will lag the voltage by a phase angle of less than 90°. The power consumed by the circuit is equal to volts times amperes times a power factor.
- In an ac circuit containing capacitance only, the current leads the voltage by an angle of 90°. The power consumed is zero.
- In an ac circuit containing resistance and capacitance, the current will lead the voltage by an angle of less than 90°. The power consumed by the circuit is equal to volts times amperes times a power factor.

You might think that the computations involved in RCL circuits will be rather complex. However, they are merely an extension of the calculations involved for RL and RC circuits. Certain inductive and capacitive components are 180° out of phase. In the RCL circuit, we account for this mathematically by finding their algebraic sum. This leaves us with a net value. What is meant here will become clear as you continue to read.

THE SERIES RCL CIRCUIT

When resistance, inductance, and capacitance are connected in series, the circuit behaves either as a series RL circuit or as a series RC circuit. Which one it acts as depends on which of the two reactances, X_L or X_C, is larger. (A special case exists when the two values are equal. This will be discussed shortly.) Review the following problem to see how to analyze a series RCL circuit.

Problem:

A series RCL circuit is diagramed in Figure 11-1. Calculate the circuit impedance and phase angle and draw the impedance triangle. Find the voltage drops around the circuit and verify the voltage values are correct.

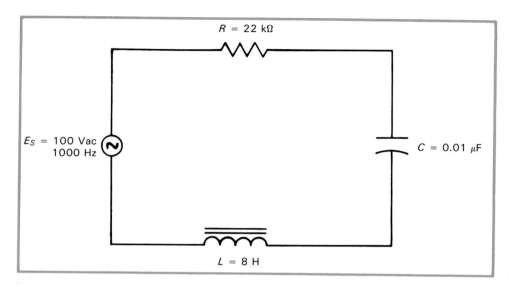

Figure 11-1. A series RCL circuit.

Solution:

Step 1. Compute the value of inductive reactance.

$$X_L = 2\pi fL = 6.28 \times 1000 \text{ Hz} \times 8 \text{ H} = 50{,}240 \text{ } \Omega$$

Step 2. Compute the value of capacitive reactance.

$$X_C = \frac{1}{2\pi fC} = \frac{0.159}{fC} = \frac{0.159}{(1 \times 10^3 \text{ Hz})(0.01 \times 10^{-6} \text{ F})} = 15{,}900 \text{ } \Omega$$

Step 3. Draw a vector diagram of R, X_L, and X_C. Use any convenient scale. Calculate net reactance X and show it on the diagram. See Figure 11-2A.

Note that R is drawn as the horizontal vector since voltage is in phase with current in a resistor. (Current is the common quantity of a series circuit; thus, it is the horizontal reference.) X_L is drawn upward at 90° from R because voltage leads current across the inductor by this angle. X_C is drawn downward at 90° because current leads voltage across the capacitor by this angle.

X_L and X_C are pointed in opposite directions. Their algebraic sum is the net reactance. Thus, the effect of one reactance is diminished by the other, and net reactance of this circuit is:

$$X = X_L + X_C = 50{,}240 \text{ } \Omega + (-15{,}900 \text{ } \Omega)$$

$$= 50{,}240 \text{ } \Omega - 15{,}900 \text{ } \Omega$$

$$= 34{,}340 \text{ } \Omega$$

Net reactance is drawn upward from the reference line because *inductive* reactance is the larger component. The circuit is inductive.

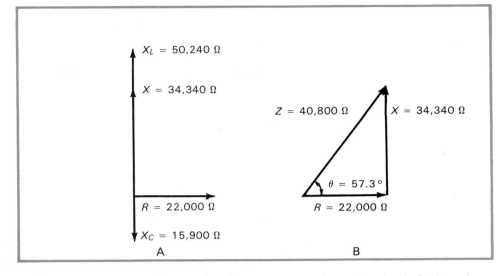

Figure 11-2. A—Vector diagram showing net reactance in a series circuit. B—Impedance triangle corresponding to vector diagram.

Step 4. Compute the value of impedance; then, draw the impedance triangle (Figure 11-2B).

The circuit impedance is the vector sum of the series resistance and the *net* reactance. Thus, the circuit impedance is:

$$Z = \sqrt{R^2 + X^2} = \sqrt{(22{,}000 \ \Omega)^2 + (34{,}340 \ \Omega)^2} \cong 40{,}800 \ \Omega$$

Step 5. Determine the phase angle. Indicate the value on Figure 11-2B.

$$\theta = \arctan \frac{X}{R} = \arctan \frac{34{,}340 \ \Omega}{22{,}000 \ \Omega}$$

$$= \arctan 1.56$$

$$= 57.3°$$

Step 6. Calculate the circuit current. Using Ohm's law:

$$I = \frac{E}{Z} = \frac{100 \ V}{40{,}800 \ \Omega} = 0.00245 \ A = 2.45 \ mA$$

Step 7. Find the voltage drops around the circuit. Using Ohm's law:

$$E_R = IR = (0.00245 \ A)(22{,}000 \ \Omega) = 53.9 \ V$$

$$E_L = IX_L = (0.00245 \ A)(50{,}240 \ \Omega) = 123.1 \ V$$

$$E_C = IX_C = (0.00245 \ A)(15{,}900 \ \Omega) = 39 \ V$$

Step 8. Verify the values of voltage are correct.

Since voltage drops across the inductor and across the capacitor are 180° out of phase, the net drop across these circuit elements is the algebraic sum of the voltage drops, or the difference between the two. Therefore, the net reactive voltage E_X is:

$$E_X = E_L + E_C = 123.1 \ V + (-39 \ V)$$

$$= 123.1 \ V - 39 \ V$$

$$= 84.1 \ V$$

If the values of voltage are correct, the phasor sum of the net reactive voltage and the resistive voltage drop will equal the applied voltage. In other words:

$$E_S = \sqrt{E_R{}^2 + E_X{}^2} = \sqrt{(53.9 \text{ V})^2 + (84.1 \text{ V})^2} \cong 100 \text{ V}$$

It is important to note that frequency is a key variable in the RCL circuit. Let us consider what effects changing frequency would have on the *series* circuit of the preceding problem.

- If frequency is increased, X_L will become larger, and X_C, smaller. As a result, the circuit becomes even more inductive. θ increases. The voltage across L increases; the voltage across C decreases. As frequency gets very large, X_L approaches ∞ Ω, and X_C approaches 0 Ω. X_L acts as an open circuit, and all voltage drop is across the inductor.
- If frequency is decreased, X_L will become smaller, and X_C, larger. The value of θ decreases toward zero. At some frequency, X_L will equal X_C. This is known as the **resonant frequency** (f_r). At this point, the value of θ is zero.
- A further decrease in frequency will make X_C larger than X_L. The circuit will become capacitive. θ will increase in a negative direction. The voltage across C increases; the voltage across L decreases. At zero frequency, or dc, X_L equals 0 Ω. X_C approaches ∞ Ω. It acts as an open circuit, and all voltage drop is across the capacitor.

Effects on circuit variables due to changes in frequency are shown for a generalized circuit in Figure 11-3. Figure 11-4 shows impedance diagrams for inductive- and capacitive-type circuits and for resonant circuits. Both figures relate specifically to series circuits.

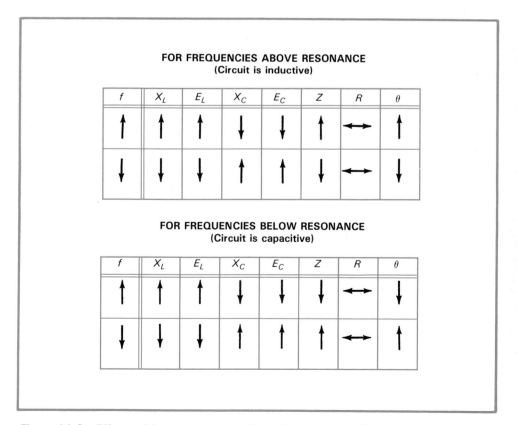

Figure 11-3. Effects of frequency on circuit variables above and below resonance (series circuits only).

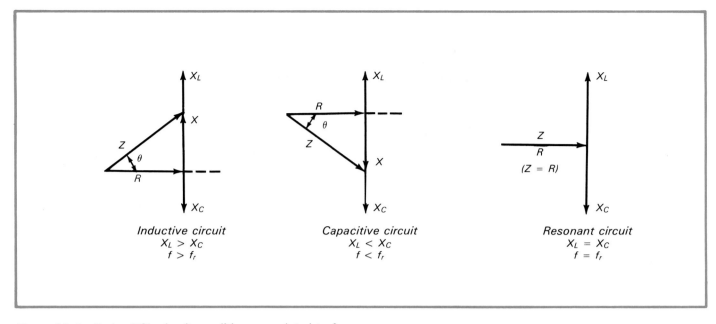

Figure 11-4. *Series RCL circuit conditions as related to frequency.*

SERIES RESONANCE

In the characteristic graph of Figure 11-5, the reactances of inductance and of capacitance are plotted as a function of frequency. Examination of the graph reveals that X_L increases as frequency increases; X_C decreases as frequency increases. The resonant frequency, f_r, occurs where the X_L plot and the X_C plot cross. At this point, X_L equals X_C. This circuit condition is called **resonance**. Such a circuit is called a **resonant circuit**. Resonant circuits — series or parallel — are sometimes called **tuned circuits**. More specifically, *tuned circuit* often implies that there is a means for adjusting, or **tuning**, the circuit for resonance.

The graph also supports the previous discussion about RCL circuits as related to frequency. That is:

- At zero frequency (dc), X_C approaches infinity, making an open circuit for dc.
- At zero frequency, X_L is zero. It is equivalent to a short circuit for dc.
- At very high frequencies, X_C approaches zero, which is a short circuit.
- At very high frequencies, X_L approaches infinity, which is an open circuit.

There is a formula for finding the resonant frequency of any RCL circuit — series or parallel. It is:

$$f_r = \frac{1}{2\pi\sqrt{LC}}$$

This formula is derived from the fact that at resonance, $X_L = X_C$. Therefore, at resonance:

$$2\pi f L = \frac{1}{2\pi f C}$$

We solve for f. To do this, we first transpose f to the left side and $2\pi L$ to the right side of the formula. This gives:

$$f^2 = \frac{1}{(2\pi)^2 LC}$$

To eliminate the square term in f, we take the square root of both sides. The end result is:

$$f = \frac{1}{2\pi\sqrt{LC}}$$

For practical purposes, this simplifies to:

$$f = \frac{0.159}{\sqrt{LC}}$$

To find the value of either L or C that will produce resonance at a given frequency, we can rearrange the resonant frequency formula. Thus, for any RCL circuit — series or parallel — at resonance:

$$C = \frac{1}{(2\pi f_r)^2 L} \text{ and } L = \frac{1}{(2\pi f_r)^2 C}$$

where C is in farads, and L is in henrys.

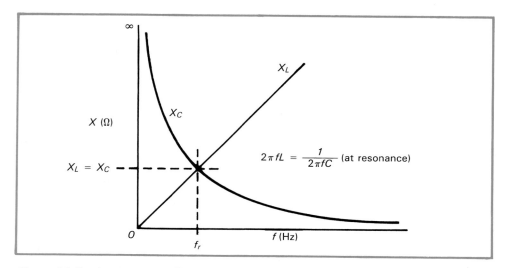

Figure 11-5. At resonance (f_r), X_L equals X_C.

Impedance and current in the series resonant circuit

Look back to the vector diagram of the resonant circuit in Figure 11-4. Since X_L and X_C vectors are equal and of opposite direction, their algebraic sum is zero. The only opposition to current in the circuit is resistance. Thus, at series resonance, Z equals R.

At frequencies above resonance, X_L is greater than X_C. The circuit becomes inductive, and Z increases. At frequencies below resonance, X_C is greater than X_L. The circuit becomes capacitive, and Z increases. The point of lowest circuit impedance is at resonance.

The series circuit is called an **acceptor circuit**. It permits maximum current at resonant frequency. Thus, it presents very small impedance to signals of frequencies close to the resonant frequency of the circuit. The *response* of a series tuned circuit appears as a bell-shaped curve, Figure 11-6. Notice that the impedance of the circuit is minimum at resonance. Note also that there is maximum current at resonance and rapidly falls off on either side of resonance. These curves showing output as a function of frequency are termed **response curves**, or, more specifically, **frequency response curves**.

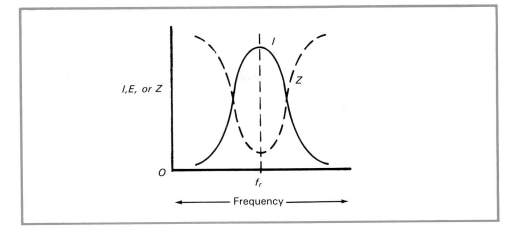

Figure 11-6. Curves showing current and impedance of a series tuned circuit.

Q of the series resonant circuit

The **quality factor**, or **figure of merit**, of a circuit is denoted by the letter *Q*. It is a measure of the relationship between stored energy and energy used by resistance. In an inductor or capacitor, it is the ratio of reactance to its effective series resistance at a given frequency.

Q is generally considered in terms of a resonant circuit, and, at resonance, *R* is the only resistance in the circuit. Further, for the resonant circuit, *Q* is usually considered in terms of X_L, not X_C, even though the reactances are equal at resonance. This is because usually the coil has the series resistance of the circuit. Thus, we give the formula for *Q* of a tuned circuit as:

$$Q = \frac{X_L}{R}$$

where *R* is the resistance in series with the inductor. It includes the winding resistance of the coil plus any added resistance. If there is no added resistance, the *Q* of the circuit will equal the *Q* of the coil. With added resistance, the *Q* of the circuit will be less than the *Q* of the coil. To secure maximum currents, *R* must be kept at a low value.

Q, the figure of merit, is an indication of the sharpness of a response curve on each side of resonance. It gives an idea of the maximum response of a circuit as well as its ability to respond within a band of frequencies. A higher value *Q* signifies a sharper response curve; a lower value signifies a flatter curve.

A circuit with a high *Q* will have a high degree of *selectivity*. This means it will have a poor response to signals of frequencies only slightly different from that to which the circuit is tuned. **Selectivity**, then, is the ability of a circuit to separate a desired signal frequency from other frequencies.

Voltage and Q. In the series tuned circuit, combined voltage across *L* and *C* is zero because the voltages across each are equal and opposite. Therefore, the drop across the series resistance is equal to the source voltage. However, there is a maximum rise in voltage across either *L* or *C* in the series resonant circuit. This is because of the maximum rise in current at resonance. This voltage rise, which can be much higher than the source voltage itself, is a function of the circuit *Q*. The relationship is derived from Ohm's law and the formula for *Q*. It is given as:

$$E_L = E_C = QE_S$$

From this, we can see how *Q* magnifies the source voltage in the series tuned circuit. In fact, *Q* is sometimes called the magnification factor.

Bandwidth and Q. Any resonant frequency has an associated band of frequencies that effectively provides resonant effects. The **bandwidth (BW)** of a tuned circuit is a range of frequencies centered about resonance, the limits of which fall where current drops to 70.7% of its peak response. The lower limit is designated f_1. The upper limit is designated f_2. The formula for bandwidth is:

$$\text{BW} = f_2 - f_1$$

The frequencies at which current drops to 70.7% of its peak value are the frequencies at which power drops to 50% of its peak value. To prove this, suppose we have a tuned circuit with a current of 1 A and a resistance of 10 Ω. Power in the circuit at resonance is:

$$P = I^2R = (1 \text{ A})^2 \times 10 \text{ } \Omega = 10 \text{ W}$$

If frequency is changed so that current drops to 70.7% of its peak, which is 0.707 A, then power at this point will be:

$$P = I^2R = (0.707 \text{ A})^2 \times 10 \text{ } \Omega = 0.5 \text{ A}^2 \times 10 \text{ } \Omega = 5 \text{ W}$$

Bandwidth, then, is the frequency above and below resonance where power drops to one-half of its peak value. Thus, the limits f_1 and f_2 are sometimes called the **half-power points**. A signal with a frequency outside of these limits is considered as *rejected* by the tuned circuit. For this reason, the frequencies f_1 and f_2 are sometimes referred to as the **cutoff frequencies**.

Bandwidth is related to Q. If Q increases, the response of the circuit increases. The *skirts*, or sides, of the curve become steeper, and the band of frequencies falling within the half-power points is greatly reduced. Conversely, if Q decreases, the bandwidth increases. A high Q circuit is sharply tuned for maximum response. It is much more selective. A low Q circuit is a *broad-band circuit*. A **broad-band circuit** has a low circuit gain and will pass a larger band of frequencies. A formula for bandwidth in a resonant circuit given in terms of Q is:

$$\text{BW} = \frac{f_r}{Q}$$

Analysis of a series resonant circuit

Review the following problem. Follow the analysis of a series RCL circuit.

Problem:

Calculate voltage drops across resistive and reactive elements in circuit of Figure 11-7 at resonance. Also, find frequencies at the half-power points.

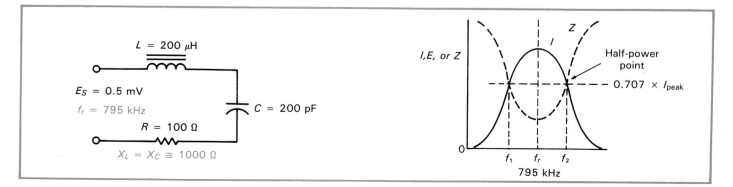

Figure 11-7. Series RCL circuit and its response curves.

Solution:

Step 1. Compute the resonant frequency of the circuit.

$$f_r = \frac{0.159}{\sqrt{LC}} = \frac{0.159}{\sqrt{(200 \times 10^{-6} \text{ H})(200 \times 10^{-12} \text{ F})}}$$

$$= \frac{0.159}{\sqrt{40,000 \times 10^{-18}}} = \frac{0.159}{\sqrt{4 \times 10^{-14}}}$$

$$= \frac{0.159}{2 \times 10^{-7}} = 0.0795 \times 10^7 = 795,000 \text{ Hz}$$

Step 2. Find reactance of L.

$$X_L = 2\pi f L = 6.28 \times 795,000 \text{ Hz} \times 200 \times 10^{-6} \text{ H}$$

$$\cong 1000 \ \Omega$$

Step 3. Find reactance of C. At resonance:

$$X_C = X_L \cong 1000 \ \Omega$$

Step 4. Find circuit current.

$$I = \frac{E_S}{Z} = \frac{0.5 \times 10^{-3} \text{ V}}{100 \ \Omega} = 5 \ \mu\text{A}$$

Step 5. Find the voltage drops around the circuit.

$$E_R = IR = (5 \times 10^{-6} \text{ A})(100 \ \Omega) = 0.5 \text{ mV}$$

$$E_L = IX_L = (5 \times 10^{-6} \text{ A})(1000 \ \Omega) = 5 \text{ mV}$$

$$E_C = IX_C = (5 \times 10^{-6} \text{ A})(1000 \ \Omega) = 5 \text{ mV}$$

Note that E_R equals E_S. E_L and E_C are 10 times higher than the source voltage. These higher voltages are a result of the exchange of energy between L and C at resonance.

Step 6. Find circuit Q.

$$Q = \frac{X_L}{R} = \frac{1000 \ \Omega}{100 \ \Omega} = 10$$

Note that we could have used Q to calculate voltages across the inductor and the capacitor. That is:

$$E_L = E_C = QE_S = 10 \times 0.5 \text{ mV} = 5 \text{ mV}$$

Step 7. Find the bandwidth.

$$\text{BW} = \frac{f_r}{Q} = \frac{795,000 \text{ Hz}}{10} = 79,500 \text{ Hz}$$

Step 8. Find frequencies at the half-power points.

$$f_1 = f_r - \frac{BW}{2} = 795,000 \text{ Hz} - \frac{79,500 \text{ Hz}}{2}$$

$$= 755,250 \text{ Hz}$$

$$f_2 = f_r - \frac{BW}{2} = 795,000 \text{ Hz} + \frac{79,500 \text{ Hz}}{2}$$

$$= 834,750 \text{ Hz}$$

Changing circuit resistance. Look at Figure 11-8. It shows how response changes by changing the circuit resistance. The broad-band circuit (flatter response curve) reflects the values of the problem just presented. (Note that f_r has been rounded to 800 kHz.) The sharper curve depicts what happens when circuit resistance is decreased from 100 Ω to 10 Ω. Notice how Q has increased and the bandwidth has decreased.

Changing circuit frequency. Figure 11-9 shows how values change as circuit frequency changes. Resonant frequency is 800 kHz. Notice that impedance is minimum at resonance. At resonance, the 10-Ω impedance is entirely resistive. Current is maximum; reactive voltages are maximum. Since Q is 10 times greater, their values are 10 times greater than those of the previous problem. Values above and below resonance change quite drastically from resonant values.

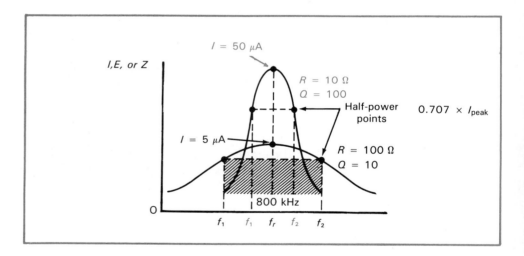

Figure 11-8. *These curves show the effect on response and bandwidth when the Q of the series circuit is changed.*

f(kHz)	$X_L(\Omega)$	$X_C(\Omega)$	$Z(\Omega)$	$I(\mu A)$	$E_L(\mu V)$	$E_C(\mu V)$
200	250	4000	3750	0.13	33	520
400	500	2000	1500	0.33	165	660
600	750	1325	575	0.87	650	1150
800	1000	1000	10	50.0	50,000	50,000
1000	1250	800	450	1.10	1375	880
1200	1500	666	833	0.60	900	400
1600	2000	500	1500	0.33	660	165

Figure 11-9. *Table corresponding to sharper response curve of previous figure shows changes that occur in the 10-Ω series tuned circuit as frequency is varied above and below resonance. (Resistance of 10 Ω is negligible. It is not included in computations, except at resonance.)*

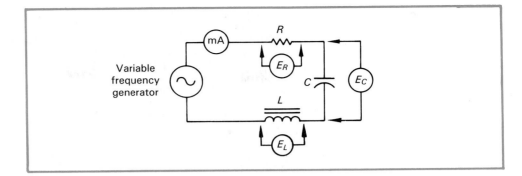

Figure 11-10. Measure voltage and current values to find resonance.

Using meters to find resonance

Resonance can be found by calculating resonant frequency for a given set of circuit components, as discussed, and adjusting the source frequency to the calculated value. It can also be determined physically through use of meters. For any RCL series circuit, measure voltages or current as shown in Figure 11-10 while varying circuit frequency and apply the previous lessons:

- Circuit current is maximum at resonance.
- E_R equals source voltage at resonance.
- E_L equals E_C at resonance.

THE PARALLEL RCL CIRCUIT

When resistance, inductance, and capacitance are connected in parallel, the circuit behaves either as a parallel RL circuit or as a parallel RC circuit. Which one it acts as depends on whether I_L or I_C is larger. (The special case of resonance exists when the two values are nearly equal. Parallel resonance will be discussed shortly.) Review the following problem to see how to analyze a parallel RCL circuit.

Problem:

A parallel RCL circuit is diagramed in Figure 11-11. Plot the current phasor diagram. Determine the phase angle between applied voltage and total current. Calculate the circuit impedance.

(Note that this is an *ideal* circuit. It is not a *practical* circuit because L will have some resistance, and C may have some leakage resistance. The circuit is simplified so that you will understand the principles involved in solving such a problem.)

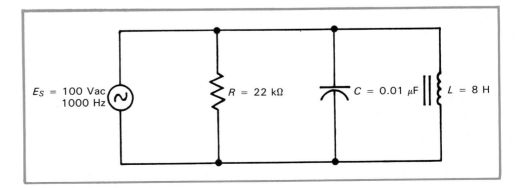

Figure 11-11. A parallel RCL circuit.

Solution:

Step 1. Compute the value of inductive reactance.

$$X_L = 2\pi f L = 6.28 \times 1000 \text{ Hz} \times 8 \text{ H} = 50,240 \ \Omega$$

Step 2. Compute the value of capacitive reactance.

$$X_C = \frac{1}{2\pi f C} = \frac{0.159}{fC} = \frac{0.159}{(1 \times 10^3 \text{ Hz})(0.01 \times 10^{-6} \text{ F})} = 15,900 \ \Omega$$

Step 3. Find branch currents. Using Ohm's law:

$$I_R = \frac{E_S}{R} = \frac{100 \text{ V}}{22,000 \ \Omega} = 4.54 \text{ mA}$$

$$I_L = \frac{E_S}{X_L} = \frac{100 \text{ V}}{50,240 \ \Omega} = 1.99 \text{ mA}$$

$$I_C = \frac{E_S}{X_C} = \frac{100 \text{ V}}{15,900 \ \Omega} = 6.29 \text{ mA}$$

Step 4. Draw the current phasor diagram. Use any convenient scale.

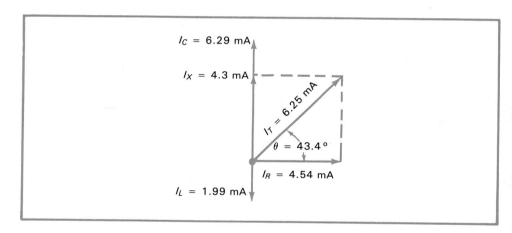

Figure 11-12. Current phasor diagram.

Note that I_R is drawn as the horizontal vector since current is in phase with voltage in a resistor. (Voltage is the common quantity of a parallel circuit; thus, it is the horizontal reference.) I_C is drawn upward at 90° from I_R because current leads voltage across the capacitor by this angle. I_L is drawn downward at 90° because voltage leads current across the inductor by this angle.

Since I_L is 180° out of phase with I_C, they are of opposite polarity. This enables the effect of one to cancel the effect of the other. The resultant current is their algebraic sum. Net reactive current, then, is:

$$I_X = I_C + I_L = 6.29 \text{ mA} + (-1.99 \text{ mA})$$

$$= 6.29 \text{ mA} - 1.99 \text{ mA}$$

$$= 4.3 \text{ mA}$$

Net reactive current is drawn upward from the reference line because *capacitive* reactance is the larger component. The circuit is capacitive.

The total current, or **line current**, is the phasor sum of the resistive current and the net reactive current. Thus, the total current is:

$$I_T = \sqrt{I_R{}^2 + I_X{}^2}$$
$$= \sqrt{(4.54 \times 10^{-3}\ \Omega)^2 + (4.3 \times 10^{-3}\ \Omega)^2}$$
$$= 6.25 \times 10^{-3}\ A = 6.25\ mA$$

Step 5. Determine the phase angle to see how much the line current leads the applied voltage.

$$\theta = \arctan \frac{I_X}{I_R} = \arctan \frac{4.3\ mA}{4.54\ mA}$$
$$= \arctan 0.947$$
$$= 43.4°$$

Step 6. Find the impedance of the circuit. Using Ohm's law:

$$Z = \frac{E_S}{I_T} = \frac{100\ V}{0.00625\ A} = 16{,}000\ \Omega$$

It is, again, important to note that frequency is a key variable in the RCL circuit. Let us consider what effects changing frequency would have on the *parallel* circuit of the preceding problem.

- If frequency is increased, X_L will become larger, and X_C, smaller. As a result, I_L will become smaller, and I_C, larger. The circuit becomes even more capacitive. θ increases. Net reactive current increases; therefore, line current increases.
- If frequency is decreased, X_L will become smaller, and X_C, larger. The value of θ decreases toward zero. At resonance, X_L will equal X_C, and I_L will nearly equal I_C. At this point, the value of θ is zero, and line current is minimum.
- A further decrease in frequency will make X_C larger than X_L. As a result, I_C will become smaller than I_L. The circuit will become inductive. θ will increase in a negative direction. Net reactive current increases; therefore, line current increases.

Effects on circuit variables due to changes in frequency are shown for a generalized circuit in Figure 11-13. This figure relates specifically to parallel circuits.

FOR FREQUENCIES ABOVE RESONANCE (Circuit is capacitive)								FOR FREQUENCIES BELOW RESONANCE (Circuit is inductive)							
f	X_L	I_L	X_C	I_C	Z	R	θ	f	X_L	I_L	X_C	I_C	Z	R	θ
↑	↑	↓	↓	↑	↑	↔	↑	↑	↑	↓	↓	↑	↑	↔	↓
↓	↓	↑	↑	↓	↓	↔	↓	↓	↓	↑	↑	↓	↓	↔	↑

Figure 11-13. Effects of frequency on circuit variables above and below resonance (parallel circuits only).

PARALLEL RESONANCE

The frequency at which X_C and X_L become equal in a parallel circuit is resonance. Resonant frequency in the parallel circuit may be computed by the same formula used for series circuits. That is:

$$f_r = \frac{1}{2\pi\sqrt{LC}} = \frac{0.159}{\sqrt{LC}}$$

Figure 11-14 illustrates a parallel LC circuit with a switch and a voltage source. (It is an LC circuit because there is no added resistor in the circuit. Of course, there will always be some resistance in the inductor.) When the switch is closed, C charges instantly to the source voltage. Then, when the switch is opened, C starts to discharge through L and the discharge current creates a magnetic field around L. As the charge on C approaches zero, the current through L starts to decrease, and the magnetic field collapses. This causes an induced voltage that charges C in the opposite polarity. Now C attempts to discharge in the opposite direction. The discharge current again builds the magnetic field, which collapses as the current approaches zero. The induced voltage charges C to its original polarity.

This action is called **oscillation**, and one cycle of oscillation has been described. If it were not for friction (in this case, resistance), the oscillations would continue indefinitely, and we would have perpetual motion. However, during each cycle of oscillation, some of the initial energy stored in C will be dissipated by the circuit resistance. Succeeding oscillations will become less and less until they die out. This is called a **damped oscillation**.

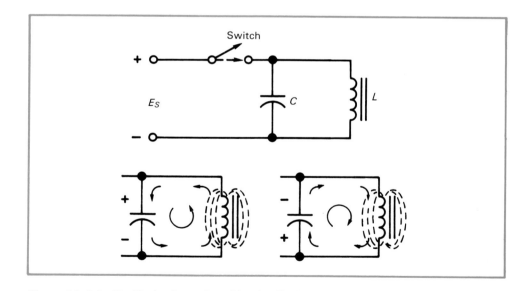

Figure 11-14. Oscillation is produced by the discharge of C through L in the tank circuit.

A damped-oscillation waveform is shown in Figure 11-15. The circulation of current in the parallel LC circuit is likened to the sloshing of water back and forth in a tank. The action is sometimes referred to as **flywheel action**, and the circuit is a **tank circuit**. The frequency of the tank circuit is found by the resonant frequency formula.

To further illustrate the point, compare the oscillations of a tank circuit to a small child being pushed on a swing. If the pushing is ended, the amplitude of the swing will decrease until it comes to rest. If not for friction, or resistance, the swing might continue forever. Now if the swing is pushed every time it reaches its maximum backward position, the added energy replaces the energy lost by friction. The tank circuit is similar. If pulses of energy are added to the oscillating tank circuit at the correct frequency, it will continue to oscillate.

Figure 11-16 illustrates a circuit by which you can observe the tank action and damped oscillations of a tuned circuit. The tank circuit is energized by pulses of voltage from across C_1, which is part of a relaxation oscillator (Chapter 10, Figure 10-6). The output from the tuned circuit may be observed on an oscilloscope. The resonant frequency of the tank circuit is much higher than the operating frequency of the relaxation oscillator. As a result, the oscillations in the tank will die out between energy pulses. A damped oscillation will be seen.

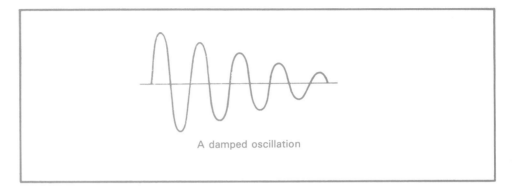

A damped oscillation

Figure 11-15. The wave decreases in amplitude due to energy used by circuit resistance.

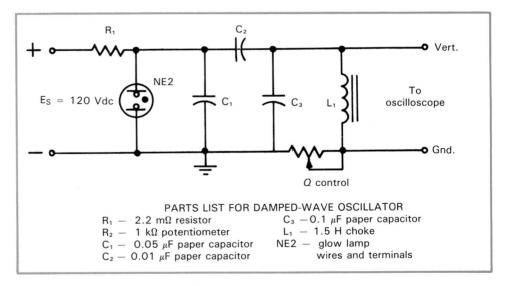

PARTS LIST FOR DAMPED-WAVE OSCILLATOR

R_1 — 2.2 mΩ resistor	C_3 —0.1 μF paper capacitor
R_2 — 1 kΩ potentiometer	L_1 — 1.5 H choke
C_1 — 0.05 μF paper capacitor	NE2 — glow lamp
C_2 — 0.01 μF paper capacitor	wires and terminals

Figure 11-16. A damped-wave oscillator circuit.
(Charles Shuler, California University of Pennsylvania, California, PA)

Current and impedance in the parallel resonant circuit

Remember that the energy required to maintain circuit oscillation depends upon the circuit resistance that uses the energy. Therefore, a parallel tuned circuit that has a very low series resistance will require very little current to sustain its oscillations at resonance. Total line current, at resonance, is minimum; thus, total impedance is maximum. Since a parallel LC circuit offers a high impedance to signals at or near its resonant frequency, it is called a **rejector circuit**. In application, a rejector circuit may be tuned to the frequency of an interfering signal in order to eliminate it.

Why does a parallel tuned circuit present maximum impedance at resonance? The current in the inductive branch is lagging the applied voltage. The current in the capacitive branch is leading the applied voltage. The currents are 180° out of phase. The total line current is the algebraic sum of the branch currents. At resonance, they are ideally equal in value; therefore, they cancel each other. Line current is close to zero. Only a small amount flows due to the resistance of the wire in the coil. By Ohm's law, since current in the circuit is minimum at resonance, impedance is maximum.

It should be noted that while I_T is minimum at resonance, the individual branch currents may be appreciable. Inside the tank circuit, I_L and I_C do not cancel because they are in separate branches. It should also be noted that as frequency moves away from resonance, the branch currents change so that they are no longer equal. The net reactive current increases; thus, the total line current increases on either side of resonance. The circuit will become either capacitive or inductive.

Q of the parallel resonant circuit

As you may suspect, the quality, or *Q*, of the tank circuit depends upon its resistance. Most of the resistance is found in the wire of the coil, which is represented by a resistor in series with the coil. Once again, the *Q* of the circuit is the relationship between inductive reactance to resistance. It is expressed as:

$$Q = \frac{X_L}{R}$$

where *R* is the winding resistance in series with the inductor.

If there is no resistance shunting the tuned circuit, the *Q of the circuit* will equal the *Q of the coil*. With a resistance in parallel with *L* and *C*, the *Q* of the circuit will be less than the *Q* of the coil. Assuming no winding resistance, *Q*, in this case, is:

$$Q = \frac{R_S}{X_L}$$

where R_S is the value of the shunt resistance. R_S is a *damping resistance* because it lowers the *Q* of the tuned circuit. If, now, we consider winding resistance, then *Q* of the circuit is figured as:

$$Q = \frac{X_L}{R + X_L{}^2/R_S}$$

More is said about this later on in this chapter.

Current and Q. In the series tuned circuit, we have *voltage* magnification. In the parallel tuned LC circuit, we have *current* magnification. This is seen in that the value of current in the tuned circuit is much larger than the total line current. The amount of rise is a function of the circuit *Q*. The relationship is given as:

$$I_L = I_C = QI_T$$

From this, we can see how *Q* magnifies the line current in the parallel tuned circuit.

Impedance and Q. The *Q* magnification factor in the parallel circuit also determines by how much impedance across the LC circuit increases at resonance. The formula is:

$$Z_T = QX_L$$

Bandwidth and Q. In Figure 11-17, the response curve of the parallel tuned circuit is plotted. Note that at resonance, its impedance is maximum, and the circuit current is negligible. At frequencies other than resonance, the impedance drops off sharply on each side. The points above and below resonant frequency at which the total circuit impedance drops to 70.7% of its maximum value sets the bandwidth of the circuit.

Again, bandwidth is related to *Q*. A higher *Q* means a narrower bandwidth. A lower *Q* means a wider bandwidth. It also means a less selective circuit and a lower circuit impedance. The bandwidth of a parallel resonant circuit is determined in exactly the same way as that for a series resonant circuit. That is:

$$\text{BW} = \frac{f_r}{Q}$$

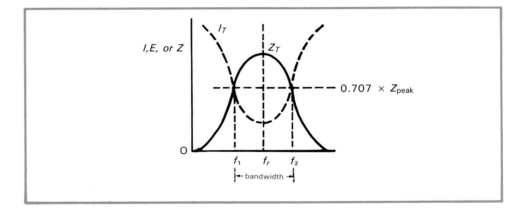

Figure 11-17. A response curve of a parallel tuned circuit showing impedance, line current, and bandwidth.

Analysis of a parallel resonant circuit

Review the following problem. Follow the analysis of a parallel RCL circuit.

Problem:

Calculate branch currents in the circuit of Figure 11-18 at resonance. Also, calculate total impedance and line current and find the cutoff frequencies.

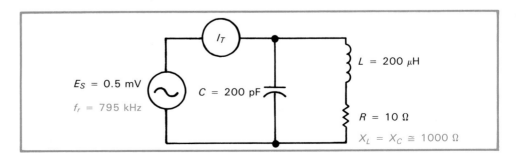

Figure 11-18. A tank circuit.

Solution:

Step 1. Compute the resonant frequency of the circuit.

$$f_r = \frac{0.159}{\sqrt{LC}} = \frac{0.159}{\sqrt{(200 \times 10^{-6} \text{ H})(200 \times 10^{-12} \text{ F})}}$$

$$= \frac{0.159}{\sqrt{40,000 \times 10^{-18}}} = \frac{0.159}{\sqrt{4 \times 10^{-14}}}$$

$$= \frac{0.159}{2 \times 10^{-7}} = 0.0795 \times 10^7 = 795,000 \text{ Hz}$$

Step 2. Find reactance of *L*.

$$X_L = 2\pi fL = 6.28 \times 795,000 \text{ Hz} \times 200 \times 10^{-6} \text{ H}$$

$$\cong 1000 \ \Omega$$

Step 3. Find reactance of *C*. At resonance:

$$X_C = X_L \cong 1000 \ \Omega$$

Step 4. Find current through *L*.

$$I_L = \frac{E_S}{X_L} = \frac{0.5 \times 10^{-3} \text{ V}}{1000 \ \Omega} = 0.5 \ \mu A$$

Note that since the value of *R* is very low in comparison to X_L, it is ignored in the computation of I_L.

Step 5. Find current through *C*. At resonance:

$$I_C \cong I_L = 0.5 \ \mu A$$

Step 6. Find circuit *Q*.

$$Q = \frac{X_L}{R} = \frac{1000 \ \Omega}{10 \ \Omega} = 100$$

Step 7. Find total circuit impedance. At resonance:

$$Z_T = QX_L = 100 \times 1000 \ \Omega = 100,000 \ \Omega$$

Step 8. Calculate the line current.

$$I_T = \frac{E_S}{Z_T} = \frac{0.5 \times 10^{-3} \text{ V}}{100,000 \ \Omega} = 5 \times 10^{-9} \text{ A} = 5 \text{ nA}$$

Note that we could have used *Q* to calculate line current. That is:

$$I_T = \frac{I_L}{Q} = \frac{0.5 \times 10^{-6} \text{ A}}{100} = 5 \text{ nA}$$

Note that net reactive current of this circuit is zero; this very small current (5 nA) is due to the resistance of the wire in the coil. It makes up the line current. This is the only current required from the source at resonance. At frequencies above and below resonance, X_L and X_C are not equal and do not cancel. The difference in current must be supplied from the source.

Step 9. Find the bandwidth.

$$\text{BW} = \frac{f_r}{Q} = \frac{795{,}000 \text{ Hz}}{100} = 7950 \text{ Hz}$$

Step 10. Find the cutoff frequencies.

$$f_1 = f_r - \frac{\text{BW}}{2} = 795{,}000 \text{ Hz} - \frac{7950 \text{ Hz}}{2}$$

$$= 791{,}025 \text{ Hz}$$

$$f_2 = f_r + \frac{\text{BW}}{2} = 795{,}000 \text{ Hz} + \frac{7950 \text{ Hz}}{2}$$

$$= 798{,}975 \text{ Hz}$$

Changing circuit resistance. Look at Figure 11-19. It shows how response changes by changing the circuit resistance. The sharper response curve reflects the values of the problem just presented. (Note that f_r has been rounded to 800 kHz.) The flatter curve depicts what happens when circuit resistance is increased from 10 Ω to 100 Ω. Increasing resistance causes a decrease in Q. This results in a decrease in total impedance and an increase in bandwidth. Changing the circuit Q has the same effect on bandwidth in the series tuned circuit. (Refer back to Figure 11-8.)

Changing circuit frequency. Figure 11-20 shows how values in the parallel LC circuit change as frequency changes. Resonant frequency is 800 kHz. Notice that impedance is maximum and line current is minimum at resonance. (Net reactive current is zero at resonance. However, a very small current exists due to the resistance of the wire in the coil.) Values above and below resonance change quite drastically from resonant values.

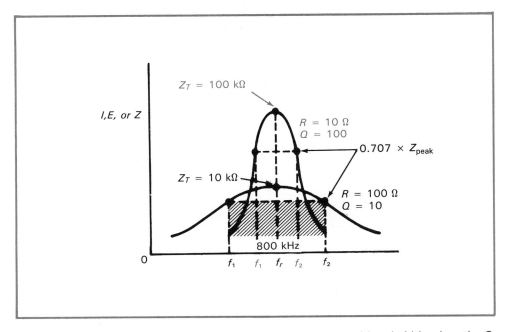

Figure 11-19. These curves show the effect on response and bandwidth when the Q of the parallel circuit is changed.

f (kHz)	X_L (Ω)	X_C (Ω)	I_L (μA)	I_C (μA)	I_T (μA)	Z_T (Ω)
200	250	4000	2.000	0.125	1.875	267
400	500	2000	1.000	0.250	0.750	667
600	750	1325	0.660	0.377	0.283	1767
800	1000	1000	0.500	0.500	0.005	100,000
1000	1250	800	0.400	0.625	0.225	2222
1200	1500	666	0.330	0.750	0.420	1190
1600	2000	500	0.250	1.000	0.750	667

Figure 11-20. Table shows changes that occur in the parallel tuned circuit as frequency is varied above and below resonance. (Values are for circuit shown in Figure 11-18.)

Using an ammeter to find resonance

An ammeter inserted in the line can be used to physically find the resonant frequency of the RCL circuit, instead of finding the value through calculation. This works because at resonance, the meter *dips*, or reads a minimum value. Current will rise sharply above or below resonance. This principle is widely used in tuning tank circuits.

Shunt damping

Earlier in the chapter, the shunt resistance and its effect on Q of a tank circuit was discussed. The effect was to lower the Q of the circuit and diminish the sharpness of resonance. Adding a parallel resistance like this to a parallel LC circuit is known as **shunt damping**. It is done, sometimes, in order to increase bandwidth. (The shunt resistance makes this an RCL circuit.)

In Figure 11-21, the current through R_S is independent of frequency. It is not canceled by the resonant conditions of the tank circuit. Line current, therefore, increases when a shunt resistor is added. This makes the dip in line current at resonance less sharp and changes the Q of the circuit.

Let us see what effect a damping resistor has on Q. Ignoring the winding resistance of L, the circuit in Figure 11-21 has a Q of:

$$Q = \frac{R_S}{X_L} = \frac{150,000\ \Omega}{1500\ \Omega} = 100$$

Further, if we consider the winding resistance, then:

$$Q = \frac{X_L}{R + X_L^2/R_S} = \frac{1500\ \Omega}{10\ \Omega + (1500\ \Omega)^2/150,000\ \Omega} = \frac{1500\ \Omega}{25\ \Omega} = 60$$

From this, we see that the winding resistance further reduces the circuit Q. Compare these two values to the circuit Q without a damping resistor. In this case:

$$Q = \frac{X_L}{R} = \frac{1500\ \Omega}{10\ \Omega} = 150$$

The damping effect of the shunt resistor is clear. Now, let us see what effect it has on bandwidth. Assume that the resonant frequency of this circuit is 800 kHz. Without R_S, bandwidth is:

$$BW = \frac{f_r}{Q} = \frac{800,000\ Hz}{150} = 5333\ Hz$$

With R_S (and the winding resistance), bandwidth is:

$$BW = \frac{800,000\ Hz}{60} = 13,333\ Hz$$

The damping resistor has clearly widened the bandwidth. Reducing the value of R_S will further increase the bandwidth. This is because reducing R_S will further increase the line current, making the response curve flatter.

We can derive formulas for BW that do not include the factor Q. Ignoring winding resistance, with shunt damping, $Q = R_S/X_L$. We know that BW = f_r/Q. Substitution of Q yields:

$$\text{BW} = \frac{f_r}{R_S/X_L} = \frac{f_r \times X_L}{R_S}$$

If winding resistance is considered, then, after substitution:

$$\text{BW} = \frac{f_r}{\dfrac{X_L}{R + X_L{}^2/R_S}} = \frac{f_r\,(R + X_L{}^2/R_S)}{X_L}$$

Now, let us see what effect lowering the value of R_S has on bandwidth. Changing the damping resistor to 100 kΩ produces:

$$\text{BW} = \frac{f_r\,(R + X_L{}^2/R_S)}{X_L}$$

$$= \frac{800{,}000\ \text{Hz}\,[10\ \Omega + (1500\ \Omega)^2/100{,}000\ \Omega]}{1500\ \Omega}$$

$$= \frac{800{,}000\ \text{Hz} \times 32.5\ \Omega}{1500\ \Omega} = 17{,}333\ \text{Hz}$$

Decreasing the shunt resistance has increased the bandwidth by 4000 Hz—from 13,333 Hz to 17,333 Hz.

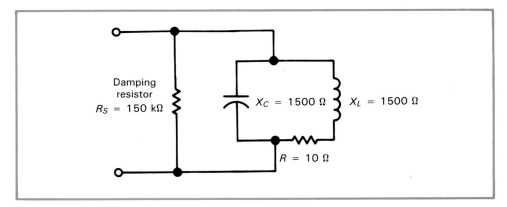

Figure 11-21. *Resistor R_S, known as a* **damping resistor,** *is connected for shunt damping the tuned circuit. The Q is lower, and the bandwidth increased.*

TUNING FOR RESONANCE

An examination of the resonant frequency formula will disclose that f_r is a function of L and C. When L and C have fixed values, we achieve resonance by adjusting the frequency of the supply to the resonant frequency of the circuit. If the frequency of the supply is fixed, however, we must adjust either L or C or both to achieve resonance. This is done by means of a variable capacitor or inductor and is how a radio receiver is tuned. Different radio stations broadcast over different frequencies. We *tune in* to the resonant frequency of our desired station. This signal is passed on to the other circuits of the receiver.

SUMMARY

- RCL networks are ac circuits that have resistors, capacitors, and inductors, which are used to pass, reject, or control current.
- Resonant frequency of a circuit occurs when the inductive reactance (X_L) is equal to the capacitive reactance (X_C).
- The formula for the frequency of resonance is:

$$f_r = \frac{1}{2\pi \sqrt{LC}}$$

- In a series resonant circuit, current is maximum and impedance is minimum.
- Tank circuits oscillate, providing an ac signal at a desired frequency.
- In a parallel resonant circuit, current is minimum and impedance is maximum.
- Parallel tuned circuits reject currents at resonant frequency.
- Series tuned circuits pass currents at resonant frequency.
- The Q of a circuit describes the relationship between X_L and R.

KEY TERMS

Each of the following terms has been used in this chapter. Do you know their meanings?

acceptor circuit	frequency response curve	resonant circuit
bandwidth (BW)	half-power point	resonant frequency *(f_r)*
broad-band circuit	line current	response curve
cutoff frequency	oscillation	selectivity
damped oscillation	quality factor *(Q)*	shunt damping
damping resistor	RCL circuit	tank circuit
figure of merit	rejector circuit	tuned circuit
flywheel action	resonance	tuning

TEST YOUR KNOWLEDGE

Please do not write in this text. Place your answers on a separate sheet of paper.
1. A series resonant circuit is called an acceptor circuit. True or False?
2. A parallel resonant circuit is called a rejector circuit. True or False?
3. At series resonance, $X_L = X_C$, and $Z =$ _____.
4. What is the Q of a tuned circuit?
5. What is a formula for finding the Q of a series resonant circuit?
6. What is the resonant frequency of a 100-mH coil and a 0.1-μF capacitor?
7. In question 6, $X_L =$ _____.
8. If the circuit in question 6 has 50 ohms of resistance, what is the circuit Q?
9. What is the effect of an increase in the frequency of input voltage in a series RCL circuit?
   a. X_L will become larger and X_C smaller.
   b. X_L will become smaller and X_C larger.
10. The point of lowest impedance of a series RCL circuit is at resonance. True or False?
11. The frequency at which X_C and X_L become equal in a parallel circuit is _____.
12. _____ is produced by the discharge of C through L in a tank circuit.
13. What determines the bandwidth of a tuned circuit?
14. What is a formula for finding the bandwidth of a tuned circuit?
15. How can the response of a parallel tuned circuit be broad-banded?

Applications of Electronics Technology:
Shown here is a telephone link for people who are deaf or hard of hearing.
(AT&T)

Section III
APPLICATIONS OF ALTERNATING CURRENT

SUMMARY

Important Points

☐ Alternating current (ac) is current flowing in two directions in a conductor, as compared to direct current (dc), which is current flowing in one direction.

☐ Inductance *(L)* is the measure of the ability of a coil to establish an induced emf as a result of a changing current. It is that property of a circuit that opposes any change in current.

☐ Capacitance *(C)* is the measure of the ability to store electrical charge.

☐ Components that give inductance to a circuit are called inductors, chokes, coils, or reactors.

☐ The time constant *(t)* of an RL circuit is the amount of time it takes for the current to rise or fall 63.2%. It is found by the formula:

$$t \text{ (in seconds)} = \frac{L \text{ (in henrys)}}{R \text{ (in ohms)}}$$

☐ The time constant *(t)* of an RC circuit is the amount of time it takes for the capacitive voltage to rise or fall 63.2%. It is found by the formula:

$$t \text{ (in seconds)} = R \text{ (in ohms)} \times C \text{ (in farads)}$$

☐ Mutual inductance, which is the basis for transformer action, occurs when two or more coils share the energy of one coil.

☐ When a tap is made on a secondary, two voltages may be obtained. With reference to the tap, the voltages are 180° out of phase.

☐ Among reasons for using a transformer are:
   • To step up or step down voltage or current.
   • To provide two voltages, 180° out of phase.
   • To provide isolation from the primary to the secondary.

☐ Transformer losses can be copper losses, hysteresis losses, or eddy current losses.

☐ Formulas for resistors, inductors, ar capacitors in series or parallel are shown in the table. All units used in the formulas must have the same prefix or basic value.

COMPONENT	IN SERIES	IN PARALLEL
RESISTORS	$R_T = R_1 + R_2 + R_3$	$\frac{1}{R_T} = \frac{1}{R_1} + \frac{1}{R_2} + \frac{1}{R_3}$
INDUCTORS	$L_T = L_1 + L_2 + L_3$	$\frac{1}{L_T} = \frac{1}{L_1} + \frac{1}{L_2} + \frac{1}{L_3}$
CAPACITORS	$\frac{1}{C_T} = \frac{1}{C_1} + \frac{1}{C_2} + \frac{1}{C_3}$	$C_T = C_1 + C_2 + C_3$

☐ Opposition to an alternating current is called reactance.

☐ Inductive reactance *(X_L)* is the opposition to change in current offered by an inductor. It is measured in ohms. The formula for inductive reactance is:

$$X_L = 2\pi f L$$

☐ Capacitive reactance *(X_C)* is the opposition to current in a capacitive circuit. It is measured in ohms. The formula for capacitive reactance is:

$$X_C = \frac{1}{2\pi f C}$$

☐ Impedance *(Z)* is the total opposition an inductive or capacitive circuit offers to alternating current. It is measured in ohms.

☐ Factors affecting resistance, inductance, and capacitance are:

Resistance (in metal conductors)
1. Length of conductor.
2. Cross-sectional area of conductor.
3. Temperature of conductor.
4. Material of conductor.

Inductance (in coils)
1. Number of windings.
2. Types of winding.
3. Type and make-up of core.
4. Diameter of coil winding.
5. Spacing between windings.

Capacitance (in capacitors)
1. Distance between plates.
2. Plate area.
3. Dielectric material.

☐ Resonance occurs when X_C equals X_L.

☐ In a series resonant circuit, the current is maximum and the impedance is minimum.

☐ In a parallel resonant circuit, the current is minimum and the impedance is maximum.

☐ *Q* of a circuit describes the relationship between X_L and R:

$$Q = \frac{X_L}{R}$$

1. Solve for the unknown quantities in the circuit shown.
   X_L _____
   Z _____
   Phase angle _____
   Power factor _____

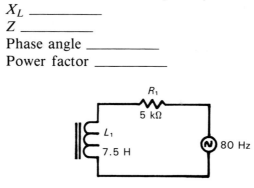

2. In the preceding circuit, I _____ (leads or lags) E.
3. Solve for the unknown quantities in the circuit shown.
   X_C _____
   Z _____
   Phase angle _____
   Power factor _____

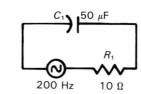

4. In the preceding circuit, I _____ (leads or lags) E.
5. The time constant of the circuit in Question 1 is _____.
6. The time constant of the circuit in Question 3 is _____.
7. Find C_T.

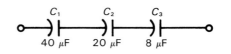

8. Find C_T.

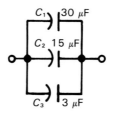

9. Find L_T.

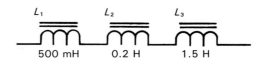

10. Solve for the unknown quantities.
    E_s _____
    I_p _____

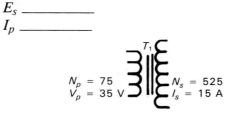

Section IV

SEMICONDUCTORS, POWER SUPPLIES, AND AMPLIFIERS

In Section IV, you will learn about the invention and development of solid-state, or semiconductor, devices. Among these devices are diodes, transistors, and integrated circuits.

Chapter 12 presents a basic explanation of the solid-state diode. The purpose and makeup of an electronic power supply is also given.

In Chapter 13, the operation and makeup of the transistor is given. This device is the basis for all semiconductor technology today. Upon its inven-

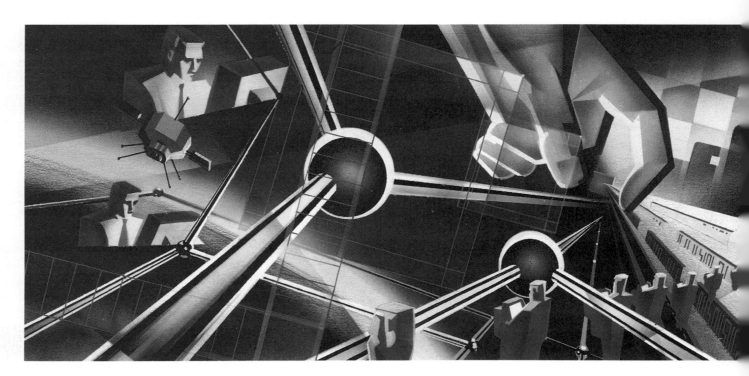

tion in 1948, the solid-state age began in electronics. Some basic transistor amplifying and switching circuits are also discussed in this chapter.

Chapter 14 goes into detail about how amplification can be achieved by using transistors. Different amplifier designs and types are presented.

Transistor power amplifiers are treated as a major area in Chapter 15. These amplifiers are used primarily to amplify power. They are used in the output stages of devices.

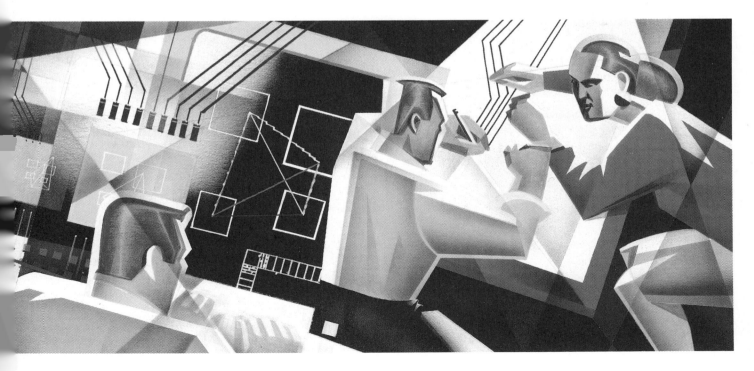

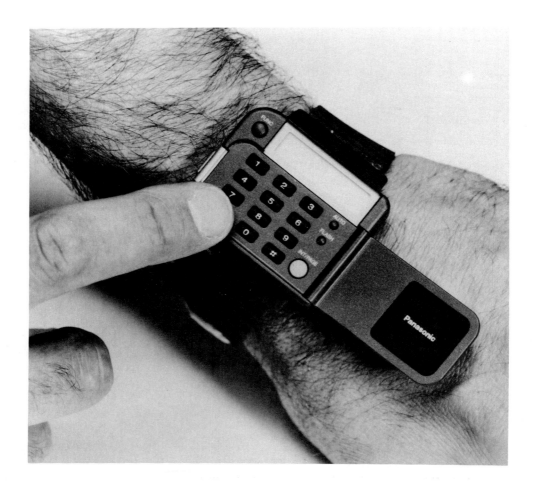

Applications of Electronics Technology:
Once considered science fiction, wrist phones, such as the one shown here, have become reality. (Panasonic)

Chapter 12

INTRODUCTION TO SEMICONDUCTOR DEVICES AND POWER SUPPLIES

The basic semiconductor device is the *diode*. It is the simplest component of the family of *semiconductors*—materials having current conductive properties falling between insulators and conductors. Diodes perform a very important role in electronics today.

After studying this chapter, you will be able to:
☐ *Explain the theory and characteristics of semiconductor diodes.*
☐ *Describe methods of rectification using diodes.*
☐ *Discuss the principles and application of filtering circuits.*
☐ *Cite the requirements of a low-ripple, well-regulated power supply.*
☐ *Give examples of some of the more common special diode devices.*

INTRODUCTION

A **diode** is a two-element, unilateral conductor. This means that a diode has two connections—*anode* and *cathode*, Figure 12-1. It also means that current will flow through this component in only one direction. It will have a high resistance to a current in its reverse direction. The theory of semiconductor conduction in Chapters 1 and 3 should be reviewed at this time. Remember that conduction through an *n-type* material is by electrons; conduction through a *p-type* material is by holes. Also, remember that both n- and p-type crystals will have a small quantity of minority carriers. In the n-type crystal, these minority carriers are holes; in the p-type, they are electrons.

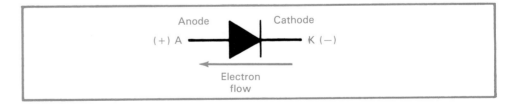

Figure 12-1. Schematic symbol for a semiconductor diode.

BASIC JUNCTION DIODES

A **junction diode** is a semiconductor diode consisting of n- and p-type semiconductor crystals. The diode has a transition region, or surface, where the two crystals join. This surface is called a **pn junction**; hence, the name junction diode. See Figure 12-2. This type of diode is the main focus of this chapter.

Forming a junction diode

Diodes are made by alloying pure silicon or germanium with trivalent or pentavalent impurity atoms. These are atoms from such elements as aluminum and boron. There are several methods of forming a junction diode. They are

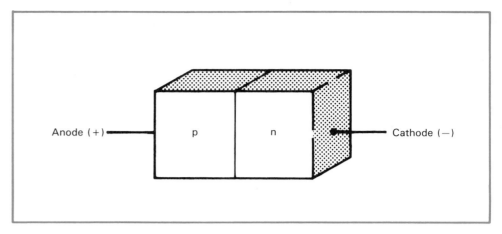

Figure 12-2. Block diagram of a pn-junction diode.

used individually or, sometimes, in combination. The desired result is to produce a surface that separates the n-type and p-type crystals. Methods of forming a junction diode include the:

- *Rate-growth method.*
- *Grown-junction method.*
- *Diffusion method.*

Rate-growth method. In this method, both n-type, or pentavalent, impurities and p-type, or trivalent, impurities are placed in a molten semiconductor material. The temperature is suddenly raised and lowered to produce alternate p-type and n-type layers.

Grown-junction method. In this method, n- and p-type impurities are alternately added to molten semiconductor material during the growth of the crystal. This results in a pn junction.

Diffusion method. This process of forming junction diodes can employ either *solid* or *gaseous diffusion*. In **gaseous diffusion**, one surface of an n-type semiconductor is exposed to a high-temperature atmosphere containing a p-type impurity. The gas diffuses gradually into the crystal. In **solid diffusion**, a p-type impurity is *painted* on the surface of an n-type semiconductor. The two are then heated until the impurity diffuses into the n-type material. (In each case, the roles of the p- and n-type materials can be reversed.) By masking certain areas, precise pn junctions can be formed. For this reason, the diffusion method is used extensively in the making of integrated circuits.

Mechanisms of current

Two mechanisms are found in the study of semiconductors for producing current. One is an electric potential. A **drift current** is caused by a potential difference. It is established as free electrons are drawn toward a positive potential. The resulting movement of these electrons is called **drift**. Whenever a voltage is applied to a conductor, the current that exists is a drift current.

The other mechanism for producing current is an imbalance of charges, or **concentration gradient**. A **diffusion current** is caused by this condition. It is established as charge carriers repel each other from areas of higher to lower concentration. The resulting movement of these charges is called **diffusion**. Current can flow due to diffusion even in the absence of an applied voltage.

Keep in mind that the only charge that moves is that carried by electrons. Even the movement of holes (which are *thought* of as positive charges) is really the jumping of valence electrons from hole to hole. The appearance of holes moving from right to left is, in reality, the flow of electrons from left to right.

Barrier potential

Diffusion is the movement of carriers from a higher concentration to a lower concentration. Diffusion of a junction diode is shown in Figure 12-3. Majority carriers are attracted by diffusion across the junction. Here at the junction, holes and electrons meet and recombine. In the process, the n-type material loses electrons, and *uncovered* bound positive charges, or ions, are left behind at its junction. Conversely, the p-type material loses holes, and uncovered bound negative charges are left behind at its junction.

It becomes more and more difficult for the majority carriers to cross the junction. This is because as the charges build up, the electrons must overcome both the attraction of the positive ions and the repulsion of the negative ions at the junction. Holes must overcome the attraction of the negative ions and repulsion of positive ions. The regions near each side of the junction become void of current carriers. This region of fixed ionic charges and no current carriers is called the **depletion region**, **transition layer**, or **space-charge region**.

Due to diffusion, a potential difference develops. This is shown in Figure 12-3 by a small battery in dotted lines across the junction. This voltage is called the **barrier potential *(V_B)***, or **contact potential *(V_0)***. Its polarity is indicated in the illustration. Since positive ions now reside at the n side of the junction, the n side is positive. Likewise, the p side of the junction is now negative due to the negative ions.

Note that, while it is not pictured in the diode representation, impurity ions exist throughout both n and p regions. These regions are electrically neutral, however. This is because there are about the same number of impurity ions as oppositely charged current carriers. Only the ions at the junction are pictured. Here, there is a depletion of charge carriers, and the charges of these ions are not compensated for. These unneutralized ions are said to be **uncovered charges**.

In Figure 12-3, the *charge-distribution* graph below the junction shows the polarity of the barrier potential. It also shows approximately the concentration of charge at, as well as the relative width of, the depletion region. Net

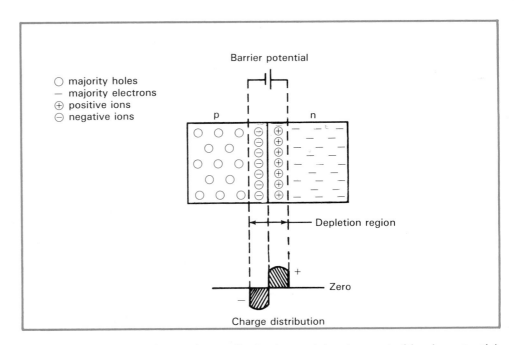

Figure 12-3. Diagram shows charge distribution and development of barrier potential by diffusion across a pn junction.

charge density is approximately zero everywhere, with the exception of the depletion region. The barrier potential for germanium is typically about 0.3 V; for silicon, about 0.7 V.

Note that the net *flow* of charge in the pn diode does not change due to the barrier potential. With no applied voltage, the net flow is zero. Also, the net *charge* on the diode does not change due to the formation of the barrier potential. For every mobile charge carrier, electron or hole, there is a corresponding atom carrying an equal charge of opposite sign. The total component neither gains nor loses electrons because of diffusion across the junction; net charge is zero. The barrier potential cannot be used to cause external flow.

Forward bias

A dc potential applied or developed between two electrodes, which is used to control the width of the depletion region, is referred to as **bias voltage**, or **bias**. Figure 12-4 shows a diode with a **forward bias** applied—positive battery terminal to p-type material; negative to n-type. In this case, electron minority carriers in the p crystal are attracted to the positive source. For each electron leaving the p crystal, a hole is injected at that exit site into the crystal. The holes are repelled toward the junction by the positive battery terminal. Electrons in the n crystal are repelled toward the junction by the negative terminal. The depletion region decreases in width. Ions at the junction begin to neutralize. The potential across the junction decreases.

As the source voltage is increased and the *junction potential* decreases, the chance of majority carriers having enough energy to cross the junction greatly increases. When the source voltage becomes greater than the barrier potential, current flows almost unimpeded. In this way, the barrier potential is overcome by the source voltage. The majority electrons and holes will then recombine at the junction, and conduction will occur. As electrons leave the n crystal, more are supplied from the negative battery terminal. Electron flow is through the external circuit as indicated by the arrows.

It should be noted that the forward bias of a junction diode can be achieved in one of two ways. One way is to apply a potential to the anode that is more positive than the cathode. The other way is to apply a potential to the cathode that is more negative than the anode.

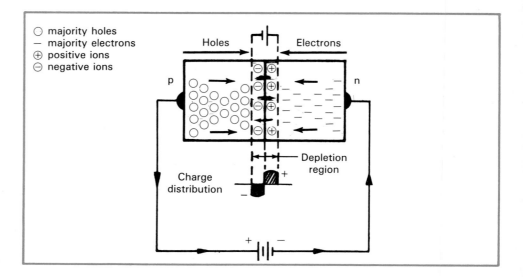

Figure 12-4. The diode is forward biased, which reduces the potential at the junction. Conduction in p crystal is by holes; conduction in n crystal is by electrons.

Reverse bias

A diode is connected in a circuit diagramed in Figure 12-5. In this case, the negative terminal of the battery is connected to the p crystal and the positive terminal to the n crystal. The holes, being positive, are attracted to the negative source; the electrons to the positive source. As current carriers move away, more fixed ionic charges are left behind at the junction. The depletion region widens. Buildup of charges at the junction is indicated by the plot of charge distribution. The potential across the junction increases. It is now equal to the sum of the barrier and source potentials; the two join together in preventing majority carriers from crossing the junction. The diode has a **reverse bias** applied. In the operating range of the diode, except for a small **leakage current**, which is due to minority carriers, no current flows.

As **reverse voltage (V_R)** is applied to the diode, a minute current will flow due to minority carriers. This current is referred to as leakage current, or **reverse current (I_R)**. It is made up of two separate currents — *reverse saturation current* and *surface-leakage current*. **Reverse saturation current (I_S)** is the maximum current that will flow through the *volume* of the diode. It is caused by thermal activity in the semiconductor material. It is affected primarily by temperature and accounts for most of the reverse current. **Surface-leakage current (I_{SL})** is the current along the surface of the diode. This current increases with an increase in reverse bias.

Minority carriers are thermally produced. At normal temperatures, they are being created continually, even in the depletion region. The increase in junction potential due to reverse bias has the opposite effect on the *minority* carriers present *in the depletion region*. The junction potential, which acted to oppose diffusion of majority carriers, aids *drift* of minority carriers across the junction. Those carriers in the depletion region are drawn across the junction to the ionic charges of opposite polarity. (Remember that drift current is current due to an electric field.) In this way, a small current is sustained. This reverse current also travels negative to positive, in the external circuit.

Drift current increases with reverse voltage until all of the available minority carriers have crossed the junction. At this point, reverse saturation current cannot increase even though the junction potential does. The current, then, is limited by the rate of thermal generation. It is largely independent of the

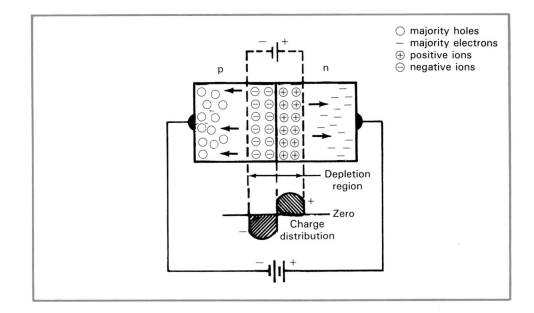

Figure 12-5. The diode is reverse biased. The depletion region increases in width.

junction potential; except that the larger the reverse voltage, the larger the depletion region. This means there are more minority carriers available to drift across the reverse-biased junction.

Interelement capacitance

In the junction diode, each crystal acts as a plate of a capacitor. The depletion region, which is depleted of charge carriers, appears as the insulating dielectric. This effect is called **interelement capacitance**. When forward biased, the junction capacitance increases as voltage increases. This is partly the result of the decrease in the width of the depletion region. This is, in effect, a reduction of the thickness of the dielectric. Therefore, capacitance will increase until the barrier voltage is reached and the depletion region is neutralized. Forward-biased capacitance is called **diffusion capacitance**.

When reverse biased, the opposite holds true. The reverse voltage increases the width of the depletion region, which increases the effective thickness of the dielectric. Therefore, capacitance decreases with reverse bias. Reverse-biased capacitance is called **transition capacitance**. When diodes are used in the higher frequency ranges, the capacitive reactance of the interelement capacitance will be an important consideration.

Volt-ampere characteristics

The typical operation of a diode is displayed graphically in Figure 12-6. As forward bias voltage is applied to the diode, the barrier potential must first be overcome. At this **knee voltage**, the plot turns upward sharply; the diode starts conduction. Beyond about 0.7 V (silicon diode), the current increases greatly for small increases in **forward voltage (V_F)**, the voltage drop across the diode. Here, there is an approximate linear relationship between voltage and current. At some point, not shown on this curve, *saturation* will occur, and current will level out (provided the diode is not destroyed). **Saturation** is the point where any further increase in voltage does *not* result in an increase in current.

Refer now to the reverse-biased region of the graph. A very minute reverse current flows until, at a certain reverse voltage, conduction suddenly increases. The reverse voltage at this point is called the **breakdown**, or **reverse-breakdown, voltage (V_{BR})**, or **zener voltage (V_Z)**. The current at this point is called the **avalanche current**, or **zener knee current (I_{ZK})**. A very small increase in voltage will cause a large current increase.

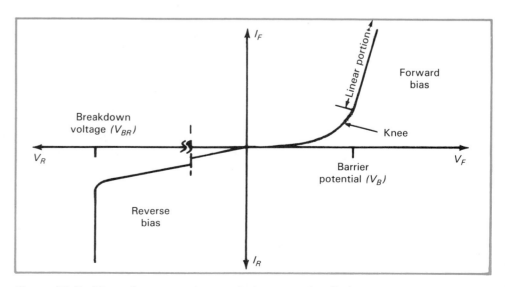

Figure 12-6. The volt-ampere characteristic curve of a diode.

The breakdown of the diode appears to be the result of ionization due to the strong electric field and collisions between current carriers. The high reverse electric field pulls the carriers out of the atomic structure. The fact that the diode has a varying current range with a rather constant voltage would make it very useful in precise voltage regulation. However, excessive heat caused by friction will ruin the standard diode. The *zener diode* made for this purpose will be discussed later in this chapter.

The dc resistance of a diode can be computed from the volt-ampere curves by using Ohm's law:

$$R_{dc} = \frac{V}{I}$$

Remember that resistance of a semiconductor is not a constant value; it varies with the applied voltage. For resistance during standard operation, select a point on the linear portion of the forward-bias curve. (Note that in this book *V* is used for voltage in electronic circuits, where *E* was used in ac/dc electrical circuits.)

Specification of diodes

In order to select a diode for a specific application, you must be aware of the many specifications supplied by the manufacturer. These specifications usually include the absolute maximum ratings of the diode. They include the typical operating conditions and curves showing typical operation. They also include mechanical data for leads and mounting. Note that many diode specification numbers have a 1N prefix, for example: 1N4001 and 1N5400. A data sheet giving typical diode characteristics is found in the appendix of this text. Diodes ratings include the following:

- **Peak inverse voltage *(PIV)*.** This is the maximum voltage that can be applied to the diode in a reverse direction without destruction.
- **Average rectified forward current *(I_O)*.** The maximum average forward current the diode can handle.
- **Peak rectified current.** The maximum current the device can conduct for a partial cycle of operation.
- **Peak surge current *(I_{FSM})*.** The maximum *surge* value of forward current a diode can handle.
- **Forward voltage drop *(V_F)*.** The resultant voltage drop when current flows through a diode in the forward direction. This voltage remains approximately equal to the barrier potential. Also called forward voltage.
- **Maximum reverse current *(I_R)*.** The maximum current that will flow when a reverse voltage is applied to a diode.
- **Power dissipation rating *(P_D)*.** The maximum average power that can be continuously dissipated at 25°C (77°F). *Derating* specifications and curves, Figure 12-7, are also given.

Diode derating. This means operating a diode at less than maximum current. A diode must be derated when ambient temperature increases above 25°C to compensate for reduced heat dissipation. Refer again to Figure 12-7. The curve shows that power dissipation of a certain diode is 500 mW. The curve is derated linearly to zero at 150°C (302°F). During a change in temperature of 125°C (225°F), power dissipation changes 500 mW. The diode is derated at:

$$\frac{\Delta P}{\Delta T} = \frac{500 \text{ mW}}{125°C} = 4 \text{ mW/°C}$$

To prevent excessive derating of a diode due to temperature, a *heat sink* is generally used. A **heat sink** is a mass of metal used to draw heat away from a component. The semiconductor device is mounted in contact with a larger mass of metal, usually aluminum. Commercial heat sinks have fins for air circulation and greater heat dissipation. See Figure 12-8. Almost all high-power diodes are mounted in contact with a heat sink. In addition, sometimes a silicone grease is used to aid the heat transfer from the device to the heat sink. In some applications, the chassis in which the device is housed serves as a heat sink.

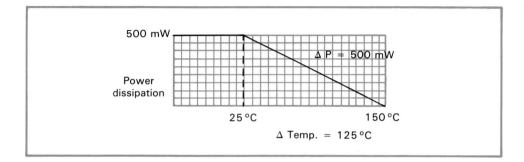

Figure 12-7. Curve used in derating a diode due to an increase in temperature.

Figure 12-8. A heat sink. Note fins, which are for air circulation and greater heat dissipation.

Testing diodes

It is important to remember that a diode presents a high resistance when reverse biased and a low resistance when forward biased. We use this fact to check the quality of an unknown diode. Use an ohmmeter (or multimeter) and set it to its highest resistance range ($R \times 10k$ or higher). Remove the power if the diode is part of a circuit and disconnect one of the diode leads. Connect the two meter leads across the diode. Resistance will be low in the forward-biased direction. It will be high in the reverse-biased direction. A low reading in both directions or high reading in both directions indicates a damaged diode.

Using a digital multimeter, you may test diodes using the built-in diode test function. This function is found on multimeters having low-powered ohm-meters, which cannot be used for diode testing. Remove power from your circuit. Set the meter for the diode test function. Connect the two test probes to the diode in question. If it is good, the display should show about 600 mV − 700 mV with probes in one direction (forward-biased silicon diode). It should read a larger voltage — say about 1.5 V — with probes in other direction (reverse biased). If the lower reading is indicated in both directions, the diode is shorted. If the higher reading (1.5 V) is indicated in both directions, the diode is open. With some multimeters, instead of a voltage, a "1," meaning continuity, and a "0," meaning an open condition, are used. Also, with some, the test is performed internal to the meter and switching of leads is not required. The meter beeps to let you know a diode is good.

The forward-biased and reverse-biased direction of a diode can be determined by testing; however, generally, the direction of conduction is indicated on the device, Figure 12-9. Diodes come in many different cases depending on the type. Figure 12-10 shows some typical diodes.

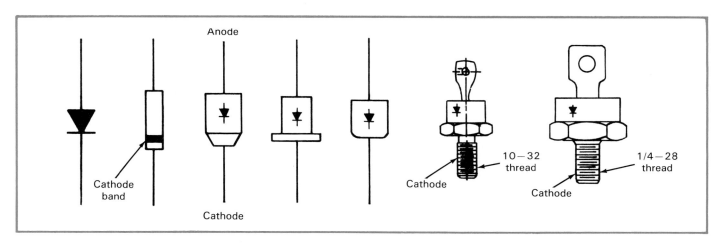

Figure 12-9. *Typical diode case outlines. Note conduction direction indicated on case.*

Figure 12-10. *A comparison of diode sizes and shapes.*

POWER SUPPLIES

A **power supply** may be defined as an electronic circuit or unit that is designed to provide ac and/or dc voltages, at specified currents, for equipment operation. Sometimes such devices as batteries or generators are considered power supplies. This chapter will discuss the electronic circuit that may be used as a power supply.

The circuit of a power supply can contain a:
- *Transformer.*
- *Rectifier.*
- *Filter.*
- *Voltage regulator.*

Figure 12-11 is a block diagram of a typical power supply. Note that not *all* power supplies have transformers or regulators. However, in general, most electronic equipment does require a well-regulated dc power supply. Many electronic units have built-in power supplies, Figure 12-12A. Instrumentation, telecommunications, process control, and robotics are just a few examples of where they are used. In addition, laboratory power supplies can be used for circuit testing or troubleshooting to provide proper voltages and currents, Figure 12-12B.

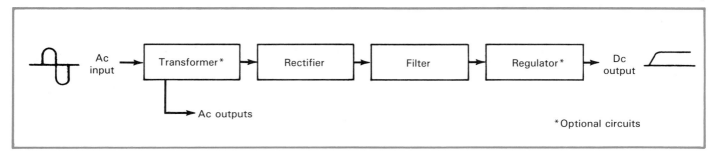

Figure 12-11. Block diagram of a typical power supply.

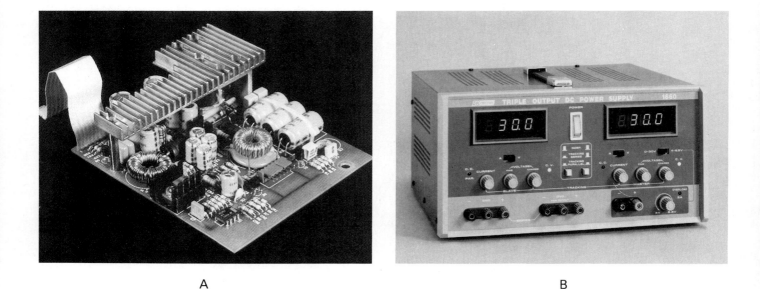

A   B

Figure 12-12. Examples of power supplies. A—Power supply for a CRT display. (General Motors Corp.) B—Laboratory power supply. (B&K Precision)

Transformers

In Chapter 9, the transformer was presented as a device that operated from the principle of mutual induction. Transformers are used in power supplies to step up or step down voltage. They are also used for isolation. In a regular transformer, there is no physical connection between the primary and secondary windings. Thus, the secondary is isolated from the primary. This is a safety feature to look for in power supplies.

Transformers may also be used to provide two voltages, one 180° out of phase with the other. These voltages are provided by a center-tapped transformer. This concept will be discussed later in this chapter in the section on "full-wave rectifiers."

As mentioned, not all power supplies have transformers; they may, instead, use input voltage directly from the line, or **line voltage** (120 Vac, for example). Without a transformer, voltage or current will not be stepped up or stepped down. Such power supplies are called **line-operated power supplies**.

Rectifiers

Rectifiers were briefly discussed in Chapter 8. These circuits are found within multimeters having the capability of measuring ac voltages and currents. They are found in power supplies as well. **Rectification** is a process that converts an alternating current into a unidirectional current. The circuit or device used to do this is a **rectifier**. There are three basic types of rectifiers used in dc power supplies. These are the *half-wave rectifier*, the *full-wave rectifier*, and the *full-wave bridge rectifier*. Another type of circuit combines rectification with voltage *multiplying*. One such circuit, in particular, is called a *voltage doubler*.

Half-wave rectifier. Figure 12-13 illustrates the operation of the *half-wave rectifier* circuit. This circuit employs a transformer, single diode, and resistor. An ac voltage is supplied to the circuit. On the positive half-cycle, point A is positive; point B is negative. Diode D_1 conducts. With current through the circuit, a voltage appears across load resistor R_1. On the negative half-cycle, point B is positive; point A is negative. D_1 is reverse biased and does not conduct. No voltage appears across R_1. This is shown by a straight line in the output waveform (points 2 to 3). The cycles repeat. Output is a pulsating dc wave with a frequency the same as the input frequency. Since only half of the input voltage is used, this circuit is called a **half-wave rectifier**.

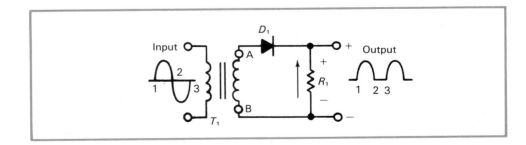

Figure 12-13. A half-wave rectifier showing input and output waveforms.

Rectification changes alternating current to direct current. Rectifier peak output voltage, $V_{\text{peak out}}$, will be the same as the peak input (secondary) voltage, *less the voltage drop across the diode*. For the half-wave rectifier, the *average* output voltage is one-half the average that would arise from the full wave. This is because the diode only conducts on half-cycles; thus:

$$V_{\text{avg}} = 0.318 \, V_{\text{peak out}} = \frac{V_{\text{peak out}}'}{\pi}$$

V_{avg} is shown graphically in Figure 12-14. The output waveform of a half-wave rectifier may be observed by an oscilloscope connected across R_1.

Full-wave rectifier. The basic *full-wave rectifier* circuit is shown in Figure 12-15. This circuit employs two diodes, center-tapped transformer, and resistor. The center tap of the transformer is taken as a voltage reference point of zero. The voltage from one end of a center-tapped transformer to the center tap is always one-half of the total secondary voltage.

When point A in Figure 12-15 is positive with respect to the center tap, diode D_1 is forward biased and conducts as indicated by the arrows. A voltage appears across resistor R_1. Its peak value is roughly equal to the peak voltage from the secondary. (Remember that voltage obtained from the secondary is *one-half* the value across the entire secondary. This is because this rectifier is connected from the center tap.) Point B is negative with respect to the center tap; D_2 is reversed biased. It does not conduct.

During the next half-cycle of the input voltage, point B in Figure 12-15 is positive and point A is negative with respect to the center tap. D_2 conducts. Follow the arrows. Observe that current is in the same direction across R_1 as in the first half-cycle. A voltage again appears across R_1. It has the same polarity as that of the first half-cycle. Its peak value is again roughly one-half of the peak voltage appearing across the entire secondary. It is slightly less due to the voltage drop across the diode.

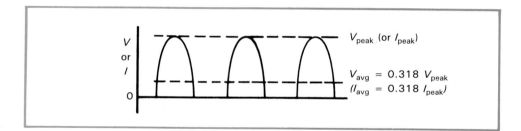

Figure 12-14. Average value of a half-wave rectifier output.

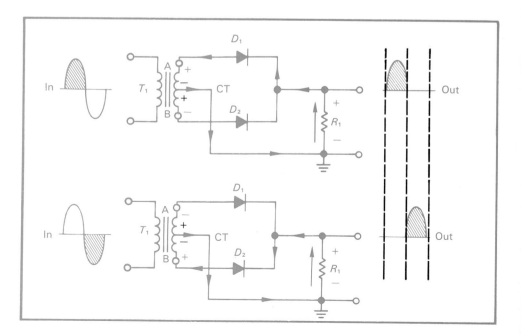

Figure 12-15. A full-wave rectifier showing input and output waveforms.

The circuit of a **full-wave rectifier** has just been described. Both half-cycles of ac input voltage produce a dc output voltage. The frequency of this waveform is twice the frequency of the input waveform. Again, the peak output voltage will be the same as the peak input voltage (the value across one-half of the entire secondary), *less the voltage drop across the diode*. The *average* output voltage is equal to the average full-wave value because the diode now conducts on both half-cycles, thus:

$$V_{avg} = 0.637 \ V_{peak \ out} = \frac{2 \ V_{peak \ out}}{\pi}$$

This is shown graphically in Figure 12-16. The output waveform of this rectifier may also be observed by an oscilloscope connected across R_1.

Full-wave bridge rectifier. A full-wave rectifier that uses the total secondary is called a **full-wave bridge rectifier**, or, commonly, a **bridge rectifier**. See Figure 12-17. The bridge rectifier does not require a center-tapped transformer. It does, however, require four diodes instead of two.

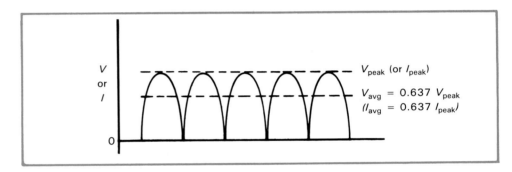

Figure 12-16. Average value of a full-wave rectifier circuit.

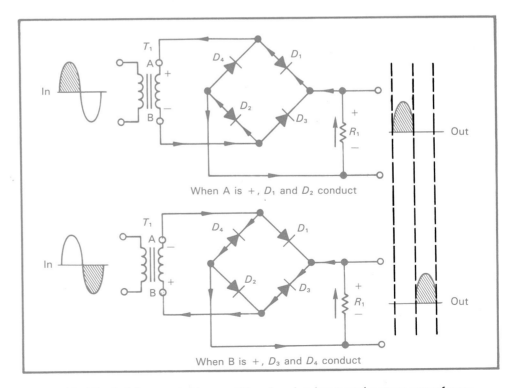

Figure 12-17. A full-wave bridge rectifier showing input and output waveforms.

When point A on transformer T_1 secondary is positive, both diodes D_1 and D_2 are forward biased, and both conduct as indicated by arrows. A voltage across resistor R_1 results from this current. Its value is roughly that of the input voltage, or the voltage across the *entire* transformer secondary. It is less due to the voltage drop across two diodes. When point B is positive and point B is negative, diodes D_3 and D_4 conduct as indicated by arrows. Voltage across R_1 results. Its value is the same as that of the first half-cycle.

Polarity of the output waveform does not change; thus, output current is unidirectional. Since both half-cycles of input voltage produce a dc waveform, it is a type of full-wave rectifier. Output is pulsating dc with a frequency that is twice input frequency. The *peak* output voltage will be the same as the peak input (secondary) voltage, less the voltage drop across the *two* diodes. Average value is figured the same as for the other full-wave rectifier.

Many times bridge rectifiers come as four separate diodes. Other times they come as a molded assembly with all four diodes in one unit. See Figure 12-18 for a drawing of this unit.

Voltage doubler. It is possible to *step up* rectifier output voltage by adding to the circuit. This type of a rectifier circuit is called a **voltage multiplier**. In Figure 12-19, a half-wave *voltage doubler* is shown. The **voltage doubler** is simply a type of voltage multiplier. Specifically, it is a rectifier circuit having a dc output that is about twice the peak value of the applied ac voltage.

During the first half-cycle of input, Figure 12-19, diode D_1 will conduct. Capacitor C_1 will charge to about the peak input voltage. Diode D_2 is reverse biased; no current flows through it. No charge accumulates on C_2. During

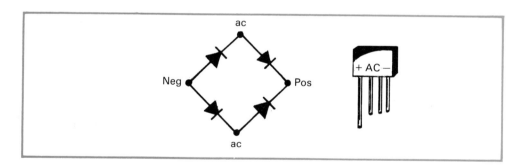

Figure 12-18. Bridge-rectifier assembly.

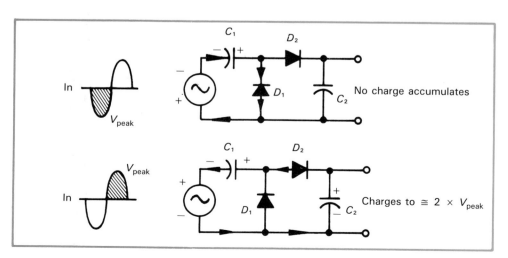

Figure 12-19. Half-wave voltage doubler operation. Arrows show current direction.

the second half-cycle, D_1 is reverse biased. D_2 conducts, and capacitor C_2 charges. C_1 is in series with the input voltage, and their voltages are additive. Therefore, C_2 charges to about the peak value of the input wave *plus* the charge on C_1, or about twice the peak input voltage.

Under no load, C_2 stays charged to about $2V_{peak}$. If a load resistance is connected across the output, C_2 discharges through the load on the next half-cycle. It then recharges on the next positive half-cycle. The output that results is a half-wave, capacitor-filtered voltage.

Another type of voltage doubler circuit is shown in Figure 12-20. This is the circuit of a *full-wave* voltage doubler. During the first half of input voltage, diode D_1 conducts. Capacitor C_1 charges to about peak input voltage. During the second half-cycle, diode D_2 conducts. Capacitor C_2 charges to about peak input voltage. The output is taken across C_1 and C_2, which are in series and additive. Output, then, is equal to about two times the peak input voltage.

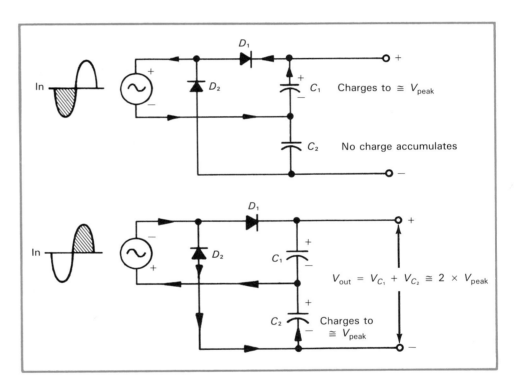

Figure 12-20. Full-wave voltage doubler operation. Arrows show current direction.

Filters

The output of a rectifier has a largely fluctuating signal. In order to smooth this output, a **power-supply filter** is used. This type of filter is used to reduce the amount of **ripple voltage**. This is an alternating component that is found in the rectified signal. Filtering is accomplished by connecting series resistors or inductors and parallel capacitors to the rectifier output.

Ripple voltage. Examine the output wave of a rectifier in Figure 12-21. This pulsating dc wave is made up of the combination of a dc voltage, which is the average voltage, and an ac wave, or ripple.

It is very important when working with transistor power supplies that the *percent ripple* be kept at a very low value ($\le 3\%$). **Percent ripple** is the ratio of the rms value of the ripple voltage to the average value of the total voltage. The formula for percent ripple is:

$$\% \text{ Ripple} = \frac{V_{rms} \text{ of ripple voltage}}{V_{avg} \text{ of total voltage}} \times 100\%$$

The rms value of the signal may be measured with an oscilloscope by eliminating the dc component of the rectified waveform. Also, a dc and an ac voltmeter may be used to obtain the necessary values. The dc voltmeter will read the average, or dc, level of the output voltage. The ac voltmeter will read only the rms value of the ac component of the output voltage (provided the signal is coupled to the meter through a capacitor to block out the dc level.)

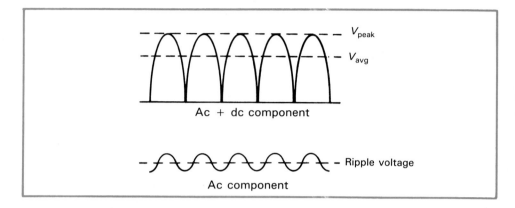

Figure 12-21. This output wave of a full-wave rectifier consists of a dc component plus an ac component. The dc component is equal to the average value of the waveform. The ac component is due to the ripple voltage.

Problem:

The peak value of the output of a full-wave rectifier is 9 V. The ripple is measured as 0.1 $V_{\text{p-p}}$. What is the percent ripple?

Solution:

Step 1. Convert peak value of the total voltage to average value.

$$V_{\text{avg}} = 0.637 \times V_{\text{peak}} = 0.637 \times 9 \text{ V} = 5.73 \text{ V}$$

Step 2. Find peak value of the ripple voltage.

$$V_{\text{peak}} = \frac{V_{\text{p-p}}}{2} = \frac{0.1 \text{ V}}{2} = 0.05 \text{ V}$$

Step 3. Find the rms value of the ripple voltage.

$$V_{\text{rms}} = 0.707 \times V_{\text{peak}} = 0.707 \times 0.05 \text{ V} = 0.035 \text{ V}$$

Step 4. Find the percent ripple.

$$\% \text{ Ripple} = \frac{V_{\text{rms}} \text{ of ripple voltage}}{V_{\text{avg}} \text{ of total voltage}} \times 100\% = \frac{0.035 \text{ V}}{5.73 \text{ V}} \times 100\% = 0.61\%$$

Since the power supply is common to most circuits within a piece of equipment (amplifier and oscillator circuits, for example), an excessive ripple will introduce hum and distortion in the final output. For this reason, a very low percent ripple is desired.

Voltage regulation. Some power supplies are designed to maintain a constant voltage. Some are designed to maintain a constant current. When a power supply is designed to produce a definite voltage output, it is desirable that the output voltage remains relatively constant when the load is applied to the supply. The **voltage regulation** is a specification of quality of a power supply. As discussed in Chapter 6, it is defined as the percentage of voltage drop between no load ($I_{R_L} = 0$) and full load. It is expressed as:

$$\% \text{ Regulation} = \frac{V_{\text{no load}} - V_{\text{full load}}}{V_{\text{full load}}} \times 100\%$$

Problem:

A power supply has a no-load voltage of 12 V. The output drops to 11.5 V when connected to a load. What is the voltage regulation?

Solution:

$$\% \text{ Regulation} = \frac{V_{\text{no load}} - V_{\text{full load}}}{V_{\text{full load}}} \times 100\% = \frac{12 \text{ V} - 11.5 \text{ V}}{11.5 \text{ V}} \times 100\% = 4.3\%$$

Capacitor-input filter. Since ripple and poor voltage regulation are undesirable, what can be done about them? In Figure 12-22, a single capacitor is connected to a rectifier. This circuit is that of a **capacitor-input filter**. During the first half-cycle, capacitor C charges up to some voltage, and during the next half-cycle, C becomes fully charged. It remains charged at the polarity shown. (If resistance of the rectifier circuit was low, C might charge during one half-cycle.) Current only flows during the charging interval of C. When C is charged, no more current is required from the rectifier.

In Figure 12-23, load R_L has now been connected across C. Now C will discharge through R_L; it will continually need current from the rectifier to maintain its voltage. The output voltage across R_L is shown by the waveform in Figure 12-23. Observe that a capacitor raises the average voltage level of the output. It also removes the ripple to some extent.

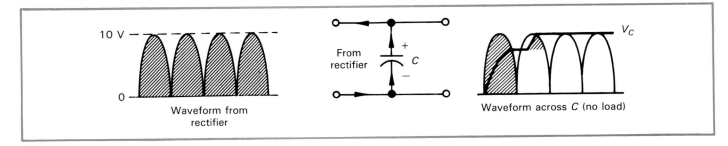

Figure 12-22. *The capacitor charges to peak rectifier voltage and remains charged.*

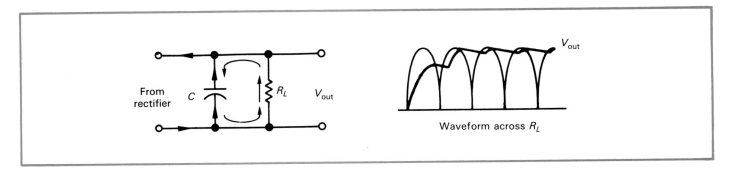

Figure 12-23. *The charge on C tends to keep the output voltage at a constant value. Rectifier current recharges C as needed.*

The same circuit is redrawn in Figure 12-24. This time, assume a load that requires more current is connected across *C*. Note the greater discharge of *C*, seen by the more pronounced voltage fluctuation. The rectifier must supply more current as a result. The average dc voltage has dropped. The ripple is more severe.

Before making improvements to this filter, consider the initial **surge current** required to charge *C*. A discharged capacitor, for an instant, appears like a short circuit across a rectifier. The initial pulse of current to charge a filter capacitor could damage a diode if it had an insufficient surge current rating. Consult the specifications. Frequently, a resistor of a few ohms is placed in series with each diode to limit the initial surge of current. After *C* has become charged, then the current drawn by the R_L should not exceed the average rectifier current specification.

Choke-input filter. The ability of a coil to resist a change in current has already been studied. In the case of the inductor, energy is stored in a magnetic field. In Figure 12-25, a rectifier is connected to an inductor in series with a load. This reduces the current variation a great deal, but some sacrifice is made in voltage. Nevertheless, the regulation is better in circuits that require larger currents. Inductors used in this manner are called **chokes**, Figure 12-26. A power-supply filter in which a choke is used in series with rectifier input is called a **choke-input filter**.

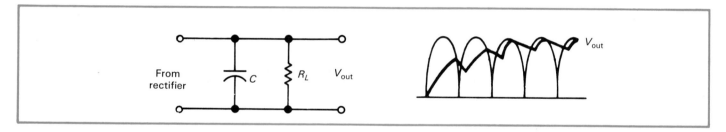

Figure 12-24. If R_L draws more current, the rectifier must supply more current. The ripple is more severe, and the average voltage drops.

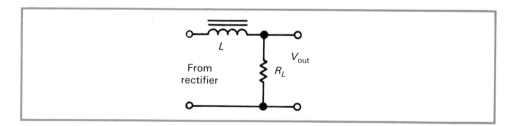

Figure 12-25. A choke-input filter. The choke limits a change in current and improves filtering.

LC filter. The combination of an inductor and a capacitor in a filter circuit, Figure 12-27, combines the advantages of each. The filter shown is called an **LC filter**. The output of this filter approaches a dc voltage with only a little ripple and with good regulation.

It is interesting to see what effect an LC filter has on a ripple voltage. In Figure 12-28, an LC filter is drawn with a load resistor across the capacitor. (Compare this filter with that of Figure 12-27 and assure yourself that they are the same.) At a ripple frequency of 120 Hz, the reactance of inductor

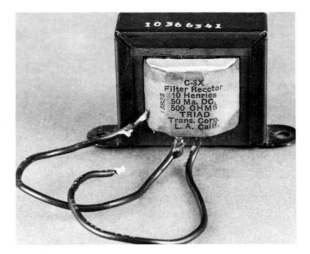

Figure 12-26. A choke used in a power-supply filter.

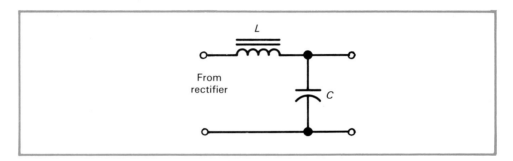

Figure 12-27. An LC filter consisting of choke and capacitor.

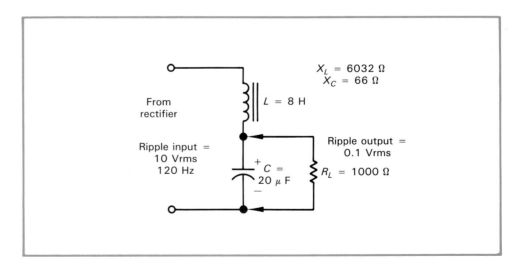

Figure 12-28. Effect LC filter has on ripple voltage.

L is 6032 Ω. The reactance of capacitor *C* is 66 Ω. The net reactance *(X)* of X_L and X_C in series is their algebraic sum, which is $(6032 - 66)$ Ω, or 5966 Ω. A ripple voltage with an effective value of 10 Vrms is fed to the circuit.

Voltage drops are proportional to resistance values in series circuits. Each resistance provides a voltage drop equal to its proportional part of the applied voltage. Thus, we can use our voltage-divider formula from Chapter 4 to determine the effective voltage across *C*:

$$V_{\text{rms out}} = \frac{X_C}{X_C + X_L} \times V_{\text{rms in}} = \frac{X_C}{X} \times V_{\text{rms in}}$$

In our circuit, then:

$$V_{\text{rms out}} = \frac{66 \ \Omega}{5966 \ \Omega} \times 10 \text{ Vrms} = 0.113 \text{ Vrms}$$

This is a fine improvement in the ripple voltage. By the filter network, the ripple voltage across the load has been reduced substantially.

You may have noticed that we ignored the loading effect of the load resistance across the capacitor in the calculation. Since the load is parallel to the capacitor, equivalent resistance R_{eq} would be less than X_C alone. Thus, you might have thought that the equivalent resistance of the parallel branch should be used to figure the voltage across *C*. However, R_L was large enough compared to the value of X_C that the difference between R_{eq} and X_C could be ignored. As a rule of thumb, we can neglect the loading by the load resistor on the capacitive impedance if R_L is at least five times as large as X_C.

Pi filter. Further improvement can be made by connecting a second capacitor to make a **pi filter**. See Figure 12-29. From the rectifier, the capacitor C_1 is selected to have a low reactance to the ripple frequency. As a result, a major part of the filtering is achieved by C_1. Most of the remaining ripple is filtered out by the LC network that follows.

Although the filtering is better with the pi filter, C_1 is still connected directly across the rectifier, and high pulses of current are needed to keep it charged, if high currents are needed for the load. These high currents may damage the rectifier diodes. It is, therefore, only used with low-current equipment. The inductor of the pi filter and others makes for better regulation, but it detracts from the voltage output. The capacitor-input filter will produce a higher output voltage.

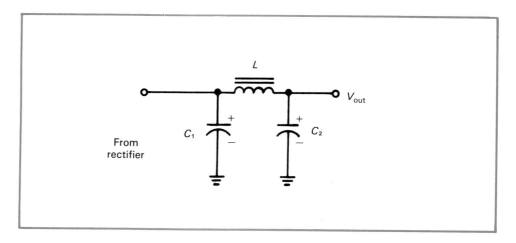

Figure 12-29. A pi filter. The circuit resembles the Greek letter π.

RC filter. In any of the previously discussed filters using chokes, a resistance may be substituted for the choke to save expense. However, the filtering will be impaired, and there will be some loss of dc voltage because of the voltage drop across the resistor. When using a choke, the dc loss is due to the resistance of the choke winding, which is relatively low. A schematic of the **RC filter** is found in Figure 12-30. The resistor R is usually a power resistor with sufficient wattage rating to carry the load current.

Bleeder resistor. Very often, a resistor is placed across the output of a power-supply filter. See Figure 12-31. This resistor serves several purposes. As related to filters, these include:

- The resistor provides a path to discharge filter capacitors after the power supply is turned off. The resistor is called a **bleeder resistor** for this reason. Without a bleeder, a power supply may retain, for some time, a high voltage, which can be dangerous to anyone working on the equipment.
- A bleeder maintains a current through the filter choke and stores energy in its magnetic field. The filter action of the choke is greatly improved. The resistance selected for this application should conduct 10% of the total load current. It should have sufficient wattage rating to carry the current.

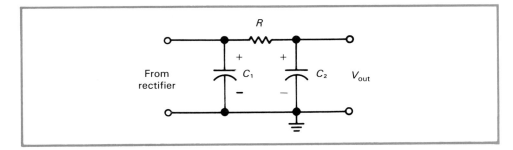

Figure 12-30. A power resistor is used in this filter in place of a choke. It is cheaper but less effective.

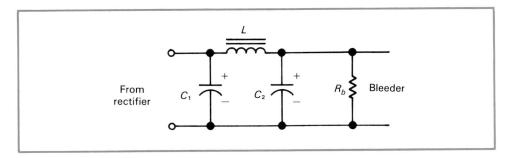

Figure 12-31. A bleeder resistor across a filter circuit.

Voltage regulators

The input voltage and the output load can vary the output of a power supply. For this reason, a *voltage regulator* can be used to maintain a more consistent output. **A voltage regulator** is a circuit designed to provide a constant output voltage from a changing input voltage and with changing load conditions.

A circuit leading from a power supply filter is diagramed in Figure 12-32. The filter output voltage is designated as 25 V. The circuit shown consists of a load R_L and a variable resistance R_{var} in series. The total input voltage, V_{IN}, is divided between R_{var} and R_L: 5 V and 20 V, respectively. Consider the following circumstances:

CASE 1. If the input voltage happened to increase to 26 V, the voltage drop across R_{var} would have to increase to 6 V to maintain a voltage across R_L of 20 V. This would mean a change in resistance value of R_{var}.

CASE 2. A decrease in the resistance of R_L would increase circuit current and increase voltage drop across R_{var}. It would be necessary to decrease the value of R_{var} to maintain 20 V across R_L.

CASE 3. An increase in R_L would cause a decrease in current. R_{var} would have to be increased to hold voltage across R_L at 20 V.

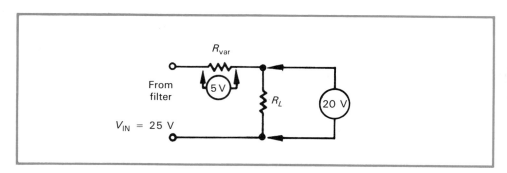

Figure 12-32. The output voltage can be held constant under varying voltage inputs by varying regulation resistor R_{var}.

We see that by varying the regulator resistance R_{var}, a constant voltage can be maintained across R_L. However, it would be impossible to vary this resistance manually and follow any sudden changes in load current. It can be done, but it must be done electronically. A special type of diode is used for this purpose, as you will learn.

Zener-diode regulation. At a certain point in a reverse-biased diode, zener voltage is reached. Beyond this point, voltage across the diode remains relatively constant and diode current increases drastically. In other words, regardless of the amount of current through it, voltage across a diode operated in this region will be relatively constant. (Zener voltage was discussed earlier in this chapter. Refer to the section covering volt-ampere characteristics and Figure 12-6.)

The constant-voltage characteristic of the diode near the zener voltage would make it a good voltage regulator when operated in the breakdown region; however, a normal diode cannot take the heat from the high currents. Special diodes called **zener diodes** are manufactured purposely to operate in a reverse direction. These diodes are used for dc voltage regulation. Zener diodes can tolerate high currents that normal diodes cannot. They operate with desired reverse-breakdown voltages from about 3 V − 250 V with minimum impedance. They are available in power ratings ranging from about 0.15 W − 50 W.

In Figure 12-33, zener diode D_1 is selected to have an operating voltage of 9 V at 7.5 mA. Any variation of input voltage within operating limits of D_1 will appear across resistor R_1, and the voltage across D_1 will remain 9 V. For example, if V_{IN} is increased to 26 V, V_{R_1} will increase to 17 V.

In Figure 12-34, a 5-kΩ load is attached to the circuit. To compute the value of regulator resistor R_1 for this load, we first find load current I_{R_L}:

$$I_{R_L} = \frac{V_{R_L}}{R_L} = \frac{9 \text{ V}}{5 \text{ k}\Omega} = 0.0018 \text{ A}$$

Knowing the load current, we can find total current I_T:

$$I_T = I_{R_L} + I_{D_1} = 0.0018 \text{ A} + 0.0075 \text{ A} = 0.0093 \text{ A}$$

We now have the values needed to determine R_1:

$$R_1 = \frac{V_{R_1}}{I_T} = \frac{16 \text{ V}}{0.0093 \text{ A}} = \cong 1700 \text{ } \Omega$$

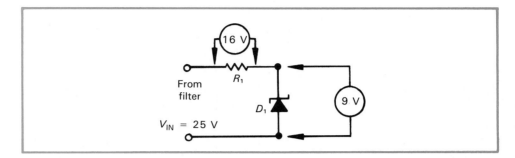

Figure 12-33. Basic zener-diode voltage regulation circuit. Note the component symbol used for a zener diode (D_1).

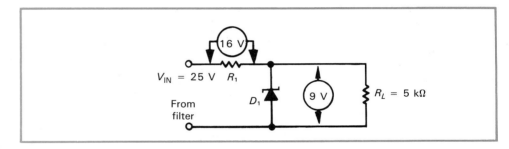

Figure 12-34. Zener-diode circuit with 5-kΩ load.

In the operating region of D_1, V_{D_1} is relatively constant. With a constant voltage input, V_{R_1} will be constant and so will I_T. If the load resistance should change, the load current will change. The current through D_1 will also change. It will change in a direction that will hold the total current at a constant value. If load current decreases, zener current increases by the same amount. If load current increases, zener current decreases by that amount. Voltage across the diode, and so to the load, stays the same.

If the load resistance stays the same but the input voltage changes, total current will change. Current through D_1 will change, but V_{D_1} will remain relatively constant. With V_{R_L} and R_L constant, current through the load remains constant. Any increase or decrease in total current is reflected in the current through the diode. The current through the diode varies and our desired output voltage is maintained, as long as we are within the operating region of D_1.

Zener limiting. In some *ac* applications, it is necessary to limit a voltage from exceeding a specified level. This is accomplished with a zener diode. See Figure 12-35. In the positive half-cycle, the diode is forward biased, and the voltage developed across it is limited to its barrier potential. In the negative half-cycle, the diode is reverse biased; the voltage developed across it is limited to the selected zener voltage. Notice that the peaks of the applied signal have been *clipped off*. This application is known as **limiting**, or **clipping**. If the zener diode is reversed in the circuit, the negative peak is limited to the barrier potential, and the positive peak to the zener voltage.

Two zener diodes may be connected back-to-back to limit both peaks to the zener voltage plus the barrier voltage. See Figure 12-36. A 4-V zener is used. Its barrier potential is 0.7 V. A 6-V_{peak} ac wave is clipped off at 4.7 V in both positive and negative directions.

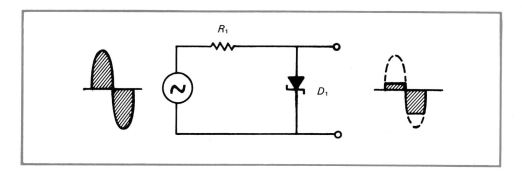

Figure 12-35. Zener clipping.

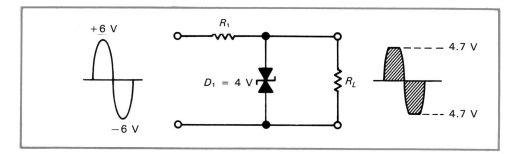

Figure 12-36. Back-to-back zener diodes are used to limit voltage in both positive and negative directions. Note that this circuit uses one double-anode, symmetrical zener diode in place of two separate zeners.

Fixed-voltage, three-terminal regulator. This device is very popular today. These regulators operate extremely well. They are simple and self-contained; there is no need for external components. Most of these regulators use integrated circuitry, have built-in short-circuit protection, and have automatic thermal shutdown. Schematic and connection diagrams for a typical regulator are shown in Figure 12-37. More on these devices will be presented in Chapter 19, *Linear Integrated Circuits*.

Bleeders as voltage regulators and dividers. Bleeders are used to improve filtering, as mentioned earlier. They also have a regulating function. The bleeder acts to ensure that rectifier current flows at all times. This keeps filter capacitors from charging to full voltage at no load ($I_{R_L} = 0$). In other words, it reduces the no-load voltage output. Since the difference between no-load and full-load voltage has been decreased, the regulation of the filter output is much improved. The load will see less voltage variation. A rule of thumb is the bleeder resistor R_b should be ten times the value of the load resistance R_L at full load. For example, if R_L is 1200 Ω at full load, R_b should be about 12 kΩ.

schematic and connection diagrams

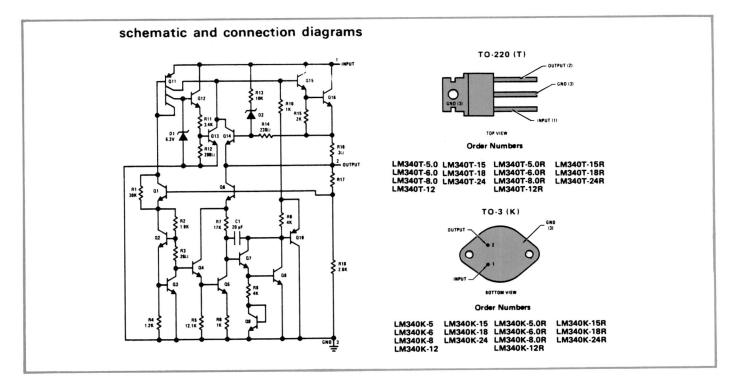

Figure 12-37. A three-terminal, fixed-voltage regulator.

A bleeder may be used as a **voltage divider** — that is, it can provide a way of obtaining several voltages from the supply. Replacing the single resistor with a resistor with taps or a slider adjustment will allow voltages other than full voltage to be secured. A bleeder of two or more resistors in series can also serve as a voltage divider. See Figure 12-38.

In the illustration, two 100-Ω resistors are in series across the output of a 20-V supply. The total divider resistance is:

$$R_T = R_1 + R_2 = 100 \ \Omega + 100 \ \Omega = 200 \ \Omega$$

The divider current is:

$$I_T = \frac{V_T}{R_T} = \frac{20 \ \text{V}}{200 \ \Omega} = 0.1 \ \text{A}$$

In reference to ground, the voltage at point A is:

$$V_{R_1} = I_T R_1 = 0.1 \ \text{A} \times 100 \ \Omega = 10 \ \text{V}$$

This seems simple enough. Now, our circuit has a bleeder resistance and two output voltages are available: 10 V at the tap and 20-V full supply voltage. (Note that we could have used the voltage-divider formula to find V_{R_1}.)

We have determined that 10 V is available at the tap. However, if a 100-Ω load is connected between ground and the tap, the voltage division is upset. Refer to Figure 12-39. The load is in parallel with R_2 of the divider, and it is equal to R_2. The equivalent resistance of the parallel circuit is:

$$R_{\text{eq}} = R/2 = 100 \ \Omega/2 = 50 \ \Omega$$

Total resistance of the circuit is:

$$R_T = R_1 + R_{\text{eq}} = 100 \ \Omega + 50 \ \Omega = 150 \ \Omega$$

Now, voltage available at the tap is:

$$V_{R_1} = \frac{R_{\text{eq}}}{R_T} \times V_T = \frac{50 \ \Omega}{150 \ \Omega} \times 20 \ \text{V} = 6.67 \ \text{V}$$

This remarkable change is caused by loading the divider circuit. Changes of this magnitude are not acceptable. If 10 V are required at the tap under load, the original divider resistance must be designed to provide this voltage under load.

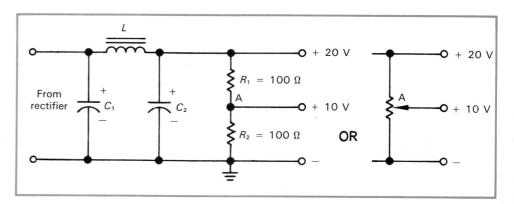

Figure 12-38. *The bleeder resistor may be used as a voltage divider.*

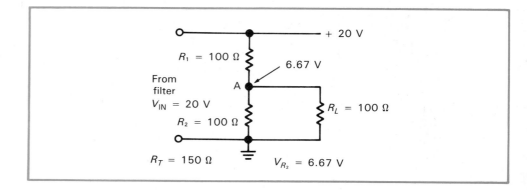

Figure 12-39. *This circuit shows the effect of loading the divider. Corrections must be made.*

OTHER JUNCTION DIODES

Up until now, this chapter has been devoted to the common rectifier diode and to the zener diode. Two other junction diodes, optimized for specific applications, are the *light-emitting diode* and the *point-contact diode*.

Light-emitting diodes

In just a few short years, *light-emitting diodes* became a major indicator device in transistor electronics. In Chapter 8, a project was presented that used these diodes in making a voltage and continuity tester. Perhaps you built one.

A **light-emitting diode (LED)** is a type of diode that is made up of a pn junction. Like regular diodes, these devices have a low forward voltage threshold. When the threshold is overcome, the junction has a low opposition, and current flows easily. Usually, an external resistor in the circuit limits this current. LEDs emit light when forward biased. The amount of light is directly proportional to the forward current.

In operation, an LED receives energy from a dc power source. When forward biased, the electrons and holes combine at the pn junction of the diode. This creates photons. Each photon is from one hole and one electron and is a particle of light that can be seen. See Figure 12-40. A large exposed surface on one layer of the semiconductor material allows the photons to be emitted as visible light. Silicon and germanium are not used as the semiconductor material because they are poor at producing light. Other compounds, such as gallium phosphide (GaP), are used instead.

LEDs come in all sizes and colors, Figure 12-41. They are smaller in size than a regular incandescent lamp. Typical colors are red, yellow, and green. LEDs consume very little power and have an extremely long life (100,000 hours or more).

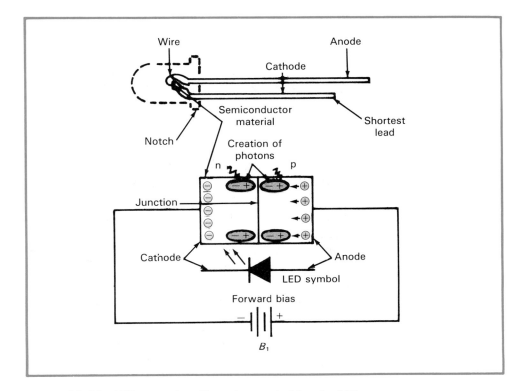

Figure 12-40. LED operation. Note the symbol for the LED.

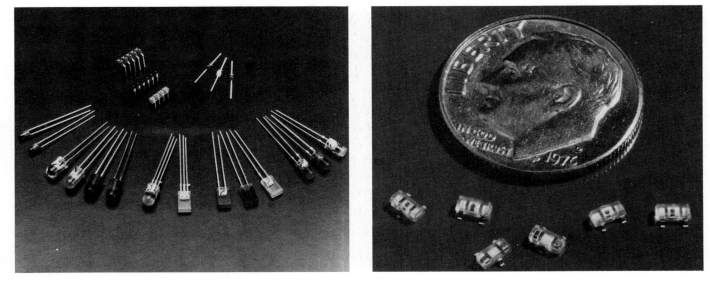

A

B

Figure 12-41. LEDs come in all shapes and sizes. A—Various types of LEDs. B—Subminiature LEDs for surface mounting. (Siemens)

Point-contact diodes

A type of diode used in the detection of radio signals is the **point-contact diode**, Figure 12-42. It consists of a very small piece of n- or p-type crystal against which a sharply pointed wire (catwhisker) is pressed. During manufacture, a high current is run through this combination. This forms a region of the opposite type (n-type in the case of a p-type crystal, and vice versa) around the contact point. The advantage of the small contact area is it reduces undesirable capacitance. The point-contact diode can be made extremely small, and it adapts itself well to miniaturized and portable equipment. Other names for this device are *crystal rectifier, crystal diode,* and *point-contact crystal diode.*

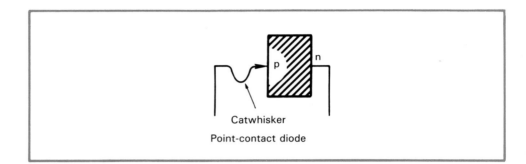

Catwhisker
Point-contact diode

Figure 12-42. A point-contact diode.

THYRISTORS

A group of semiconductor devices that act as open or closed switches are **thyristors**. These devices are constructed of semiconductive material of three or more junctions. They act as open circuits until triggered. When triggered on, they stay on, acting as low-resistance current paths. They are turned off when the current falls below a certain level, or until they are triggered off. Thyristors are found in dimmer switches, toys, power hand tools, appliances, and motor speed controls, to name a few. Among devices included in this group are the *silicon controlled rectifier* and the *triac.*

Silicon controlled rectifiers

One type of thyristor device is the **silicon controlled rectifier (SCR)**. This device is a four-layer pnpn semiconductor. See Figure 12-43. When in its normal state, the SCR blocks a voltage applied in either direction. It is enabled to conduct in the forward direction when an appropriate signal is applied to its *gate* electrode.

Overall, the SCR in the figure is forward biased; the positive lead of the voltage source is connected to the anode and the negative lead, to the cathode. Note, however, that junctions 1 and 3 are forward biased and junction 2 is reverse biased. Also, note that a third connection is made to the device. This connection, which is called the **gate**, is the control point for the SCR.

An SCR may be viewed very simply as two diodes formed back to back. An SCR that is reverse biased will not conduct. A forward-biased SCR will not conduct in its normal state since one of the diode junctions will be reverse biased. However, the forward-biased SCR will conduct if a voltage or current pulse is applied to the gate in a direction to forward bias junction 2 (Figure 12-43). Once the SCR has been *triggered*, or *fired*, the gate signal is of no consequence. The circuit continues to conduct regardless of the presence of, or lack of, the gate signal.

The SCR will remain in conduction until the forward current drops below a certain level. The current at this level is called the **forward holding current**

(I$_H$). Current can be made to fall below the value of I_H by reducing or eliminating the difference in potential between anode and cathode. If the SCR is being used in an ac application, once it fires, conduction will continue until signal polarity changes and the device becomes reverse biased.

It should be mentioned that an SCR can be made to conduct without a signal to the gate. Without a gate signal, the device will block current up to a point. This point is the **forward breakover voltage** *(V$_{(BR)F}$)*. Once this potential is reached, the SCR will break down and conduct even without a gate signal.

In the circuit of Figure 12-44, the action of an SCR can be observed. In its present state, with switch S_1 open, the SCR does not conduct, and the lamp is out. When S_1 is momentarily closed, a positive voltage applied to the gate forward biases the center pn junction. The SCR is pulsed into conduction and remains so until forward voltage is removed or reversed.

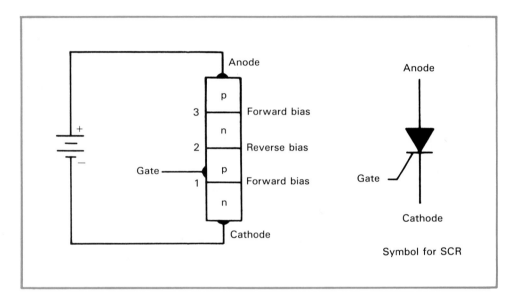

Figure 12-43. A block diagram and symbol for a silicon controlled rectifier.

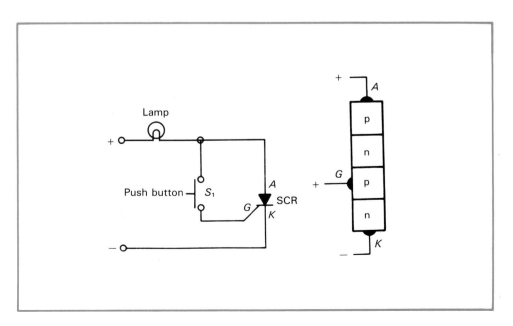

Figure 12-44. A simple SCR switching circuit.

Note that the explanation just offered views the SCR as two back-to-back diodes. The SCR's construction may also be viewed as a two-transistor arrangement. However, an explanation of the SCR's operation based on the latter view is not appropriate at this point in the text.

In recent years, hundreds of devices have been designed around the SCR. These devices include motor controls, light dimmers, automotive ignition systems, and battery chargers. They are made in current ranges from a few milliamperes to greater than 100 amperes. A typical example is shown in Figure 12-45.

Figure 12-45. A high-power SCR with rating to 1200 peak inverse volts and a current rating of 275 amperes rms. (IRC, Inc., Semiconductor Div.)

Testing SCRs. Identifying terminal leads can be a problem. Often the parts package or appropriate data book is missing. Without this information, you will have difficulty placing the device correctly into a circuit. Further, even if the terminals are known, you still need to know whether the device is good or defective. SCR testing allows the user to identify terminals and test for conduction. You will need a VOM to test an SCR.

The following procedure can be used to identify terminals:

1. Randomly assign each of the three terminals the number 1, 2, or 3.
2. Set the meter at the $R \times 1$ position. Make up a table like that shown in Figure 12-46. Obtain resistance measurements across the terminals using the combinations shown in the table. If the SCR is good, only one reading will be low. The other readings should be close to infinity. More than one low reading means there is a possible short, and the terminals cannot be identified by the VOM. In the example for Figure 12-46, we determine that the SCR is good and proceed to step 3.

Positive VOM lead	Negative VOM lead	Resistance (Ω)
1	2	∞
1	3	∞
2	1	∞
2	3	100 Ω
3	1	∞
3	2	∞

Figure 12-46. A table like this should be set up to determine the terminals of an SCR. Resistance values are measured and recorded, as shown in this example.

3. Use the low-reading combination to identify the terminals. The gate is given by the number of the terminal under the positive lead column; the cathode, the negative lead column. The remaining terminal is the anode. Refer back to Figure 12-46. We find that the terminal that has been assigned the number 2 is the gate. The cathode is terminal number 3. The number 1 terminal is the remaining terminal; therefore, it is the anode.

The following procedure can be used to test for conduction:

1. Set the VOM at the $R \times 1$ position. Put the negative lead on the cathode; the positive, on the anode. Resistance reading should be close to infinity. This means the SCR is off unless it is triggered by the gate. A low resistance reading indicates either a short or an excessive leakage current between the anode and cathode, in which case, the device should not be used.
2. Jumper a 10-Ω resistor across the gate and anode. This connects the gate to the positive VOM lead, which is connected to the anode. Keep the negative lead on the cathode. The SCR should show a large decrease in resistance. This means the SCR will fire.
3. Remove the jumper. The VOM should still show the same low resistance reading. This means that the SCR is still in conduction, indicating the SCR can be properly turned on.

Triacs

The SCR was a unidirectional device, meaning that current through it was permitted in only one direction (from anode to cathode). **Triacs** are *bi*directional, four-layer devices that conduct current in either direction. They are a type of ac switch. The direction of current through the triac is controlled by the gate voltage.

Triacs, like SCRs, have three leads, or electrodes. Refer to Figure 12-47. Note that the triac has no designation of anode and cathode. The three leads are designated: Terminal No. 1, Terminal No. 2, and Gate. Figure 12-47 also shows a block diagram of the triac junction.

A comparison can be drawn from a triac and two SCRs connected in parallel and turned in opposite directions. See Figure 12-48. The output waveforms of the SCR and the triac are compared in Figure 12-49. Note how the SCR only conducts when forward biased. The triac, however, conducts in either direction after receiving the trigger signal.

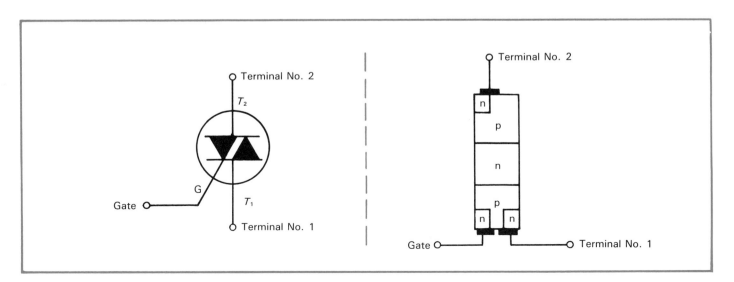

Figure 12-47. Triac symbol and junction diagram.

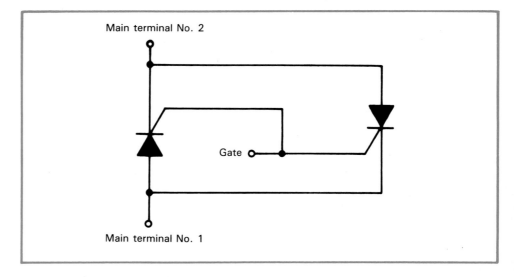

Figure 12-48. Triac comparison as two SCRs connected in parallel.

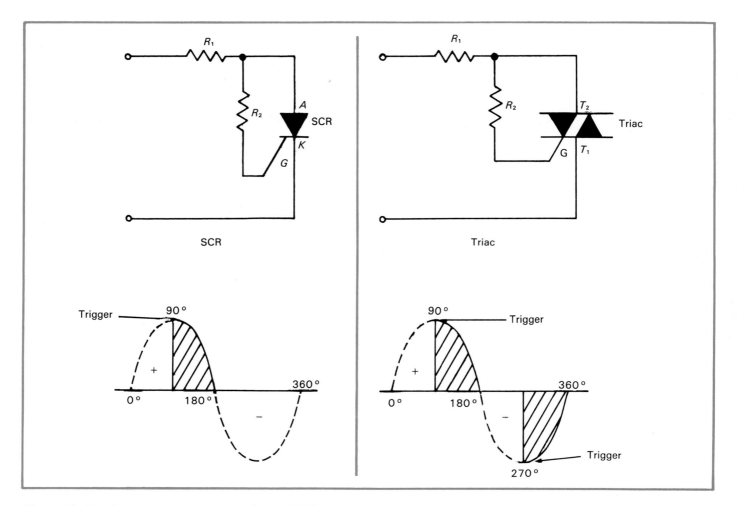

Figure 12-49. Output waveform comparison of SCR and triac.

Testing triacs. Triac testing involves terminal identification and testing for conduction. A VOM is needed.

The following procedure can be used to identify terminals:

1. Terminal No. 2 is connected to the case of the package for most triacs. To identify the other two terminals, set the VOM at $R \times 1$. Connect the meter leads randomly to the two terminals. A low resistance reading will confirm these two terminals are the gate and Terminal No. 1.
2. Now, put the positive lead to Terminal No. 2. Put the negative meter lead to either of the remaining two terminals. Use a jumper wire to short Terminal No. 2 to the last remaining terminal. Record the measured resistance as R_1.
3. With the positive lead still on Terminal No. 2, repeat procedure step 2; but interchange the jumper and negative lead connections. (Keep one end of jumper connected to Terminal No. 2.) Record the resistance as R_2. If R_1 is less than R_2, the terminal presently jumpered is Terminal No. 1, and the other is the gate. If R_1 is greater than R_2, then the jumpered terminal is the gate, and the other is Terminal No. 1.

The following procedure can be used to test for conduction:

1. With VOM in $R \times 1$ position, put the positive lead to Terminal No. 2. Put the negative lead to Terminal No. 1. The resistance reading should be close to infinity. This means the triac is off unless it is triggered by the gate.
2. Jumper a 10-Ω resistor across the gate and Terminal No. 2. This connects the gate to the positive VOM lead. Keep the negative lead on Terminal No. 1. The triac should show a large decrease in resistance. This means it can be fired on the positive half-cycle.
3. Remove the jumper. The VOM should still show the same low resistance reading. This means that the triac is still in conduction.
4. Reverse the meter leads, putting the positive lead to Terminal No. 1 and the negative lead to Terminal No. 2. The resistance reading should be close to infinity.
5. Again, jumper the 10-Ω resistor across the gate and Terminal No. 2. This connects the gate to the negative lead. The triac should show a large decrease in resistance. This means it can be fired in the negative ac half-cycle.
6. Remove the jumper. The VOM should still show the same low resistance reading. This means that the triac is still in conduction.

POWER SUPPLY PROJECT

Build a power supply that produces a 9-Vdc output. This project can be used to power any electronic device including a transistor radio that operates from 9-V batteries. This circuit uses many of the concepts that you have learned in this chapter.

A printed circuit layout for the 9-V power supply is shown in Figure 12-50. The schematic and parts list is given in Figure 12-51. Figure 12-52 shows the chassis layout for this power supply project.

LESSON IN SAFETY: In performing service work, check and recheck replacement components to be sure they are the correct wattage or power rating. Be certain that components such as electrolytic capacitors are connected with the correct polarity. Errors in connection can produce smoke, fumes, short circuits, and burns.

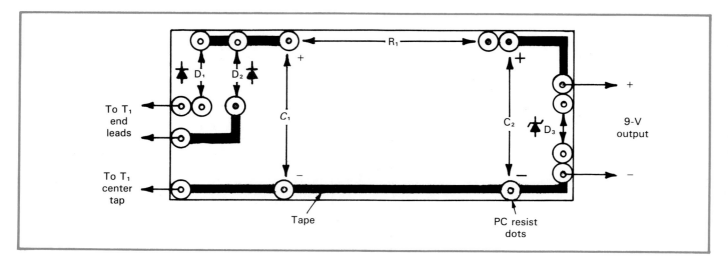

Figure 12-50. Printed circuit layout for 9-V power supply (actual size).

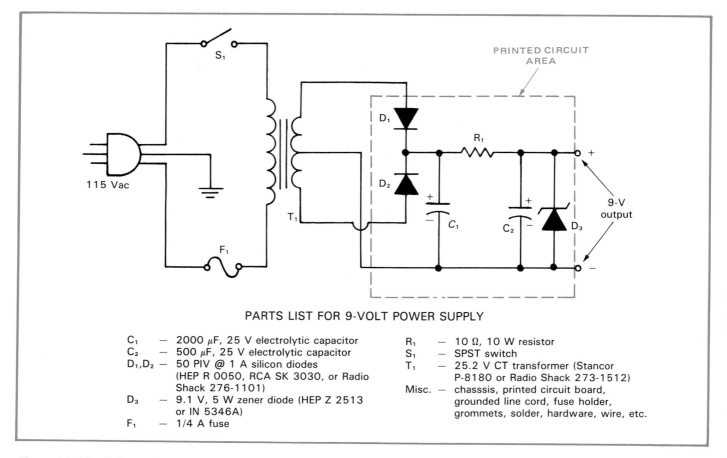

PARTS LIST FOR 9-VOLT POWER SUPPLY

C_1 — 2000 μF, 25 V electrolytic capacitor
C_2 — 500 μF, 25 V electrolytic capacitor
D_1, D_2 — 50 PIV @ 1 A silicon diodes
(HEP R 0050, RCA SK 3030, or Radio Shack 276-1101)
D_3 — 9.1 V, 5 W zener diode (HEP Z 2513 or IN 5346A)
F_1 — 1/4 A fuse

R_1 — 10 Ω, 10 W resistor
S_1 — SPST switch
T_1 — 25.2 V CT transformer (Stancor P-8180 or Radio Shack 273-1512)
Misc. — chasssis, printed circuit board, grounded line cord, fuse holder, grommets, solder, hardware, wire, etc.

Figure 12-51. Schematic and parts list for 9-V power supply.

Figure 12-52. A 9-V power supply.

SUMMARY

- Diodes are made by alloying pure silicon or germanium with trivalent or pentavalent impurity atoms.
- A junction diode is a semiconductor diode consisting of n- and p-type semiconductor crystal.
- Forward bias provides low resistance in a diode. Reverse bias provides high resistance in a diode.
- Diodes can be tested with an ohmmeter or with the diode test function of a digital meter.
- The barrier potential for germanium is 0.3 V; for silicon, it is 0.7 V.
- Rectification is the process of changing ac to pulsating dc. Diodes are used for this process.
- A power supply may be defined as an electronic unit that is designed to provide ac and/or dc voltages for equipment operation.
- The three basic types of rectifiers used in dc power supplies are the half-wave rectifier, the full-wave rectifier, and the full-wave bridge rectifier.
- A filter is used to reduce the amount of ripple voltage.
- Zener diodes are used for voltage regulation.
- Light-emitting diodes are special function diodes that, when connected in a forward-biased direction, give off light.
- An SCR is enabled to conduct in the forward direction when an appropriate signal is applied to its gate electrode.

KEY TERMS

Each of the following terms has been used in this chapter. Do you know their meanings?

avalanche current
average rectified forward current (I_0)
barrier potential *(V_B)*
bias voltage (bias)
bleeder resistor
breakdown voltage (V_{BR})
bridge rectifier
capacitor-input filter
choke
choke-input filter
clipping
concentration gradient
contact potential (V_0)
depletion region
diffusion
diffusion capacitance
diffusion current
diffusion method
diode
diode derating
drift
drift current
forward bias
forward breakover voltage ($V_{(BR)F}$)
forward holding current *(I_H)*
forward voltage *(V_F)*
forward voltage drop
full-wave bridge rectifier

full-wave rectifier
gaseous diffusion
gate
grown-junction method
half-wave rectifier
heat sink
interelement capacitance
junction diode
knee voltage
LC filter
leakage current
light-emitting diode (LED)
limiting
line-operated power supply
line voltage
maximum reverse current *(I_R)*
peak inverse voltage (PIV)
peak rectified current
peak surge current (I_{FSM})
percent ripple
pi filter
pn junction
point-contact diode
power dissipation rating (P_D)
power supply
power-supply filter
rate-growth method
RC filter

rectification
rectifier
reverse bias
reverse current (I_R)
reverse saturation current (I_S)
reverse-breakdown voltage (I_{BR})
reverse voltage (V_R)
ripple voltage
saturation
silicon controlled rectifier (SCR)
solid diffusion
space-charge region
surface-leakage current (I_{SL})
surge current
thyristor
transition capacitance
transition layer
triac
uncovered charge
voltage divider
voltage doubler
voltage multiplier
voltage regulation
voltage regulator
zener diode
zener knee current (I_{ZK})
zener voltage (V_Z)

TEST YOUR KNOWLEDGE

Please do not write in this text. Place your answers on a separate sheet of paper.

1. Why must forward voltage reach a certain level before conduction starts?
2. _____ are majority carriers in an n-type crystal.
3. _____ are the minority carriers in a p-type crystal.
4. What is meant by the depletion region, or space-charge region, of a diode?
5. In a forward-bias condition, the _____ terminal of the source is connected to the _____-type crystal, and the _____ terminal of the source is connected to the _____-type crystal of the diode.
6. What is meant by the PIV of a diode?
7. Why are heat sinks used with some diodes?
8. Rectification is the process of changing _____ to _____.
9. The output frequency of a full-wave rectifier is four times input frequency. True or False?
10. An inductor used in series with a rectifier and load is called a _____.
    a. Choke.
    b. Triac.
    c. Voltage divider.
11. What is the purpose of placing a small resistor in series with a diode and a capacitive-input filter?
12. How does a choke act to improve filtering?
13. A 12-volt, 10-milliamp zener diode is used in a regulation circuit. The power supply has a 25-volt output. Compute the value of the regulation resistance when using a 5-kilohm load.
14. Referring to Question 13, what power must be dissipated in both zener diode (P_Z) and regulation resistor (P_R)?
15. The SCR gate signal must be present for conduction through the device to be maintained. True or False?
16. What are two types of thyristors?
17. What is the purpose of a power supply?
18. In power supplies, _____ are used to step up or step down voltages.
19. A light-emitting diode is made up of a/an _____ junction.
    a. Npn.
    b. Pnp.
    c. Pn.
20. A bleeder resistor provides a path to discharge the filter capacitors after the power supply is turned off. True or False?

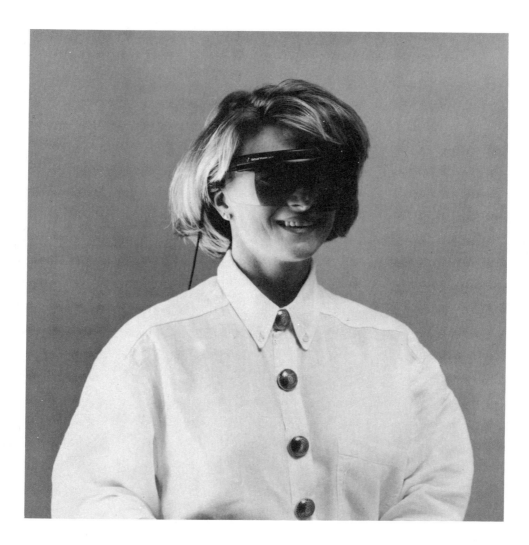

Applications of Electronics Technology:
Viewers can see big-screen television images through these 5-oz. eyeglasses.
The glasses also contain earphones that provide stereo-quality sound.
(Virtual Vision)

Chapter 13

BIPOLAR JUNCTION TRANSISTORS

The *transistor* replaced the vacuum tube as a basic amplifier and revolutionized the electronics industry. The device was conceived at Bell Laboratories; its invention was announced in 1948. The advent of the transistor opened up endless possibilities for design engineers.

After studying this chapter, you will be able to:

☐ *Discuss bipolar junction transistor theory.*
☐ *Describe the action of a transistor.*
☐ *Cite transistor characteristics.*
☐ *Identify the common configurations of transistor circuits and their phase relationships.*

INTRODUCTION

Transistors are key devices in electronics for several reasons. They are able to amplify signals. They can create ac signals at desired frequencies *(oscillation)*. They can be used as switching devices. Some features of transistors inherent in their design are:

* Extremely long life. This contributes substantially to the dependability and maximum life of electronic equipment.
* Low power required for operation. This saves electrical energy.
* No warm-up required. The transistor is ready to go to work at once. This also saves electrical energy. (The vacuum tube, forerunner of the transistor, had to heat up before it was ready to function.)
* Cool operation. Low power consumption assures cooler operation. Also, the device is not required to heat up in order to operate.
* Operates on low voltage. Since the current drain is low, small portable equipment may be supplied by low-voltage dry cells.
* Small physical size. Size has been a major factor in the development of light and portable equipment. Transistors are now found in integrated circuits (ICs), which have made possible the endless array of electronic devices used in our space programs and orbiting satellites. ICs have decreased the size of complex computers, which once filled large rooms, to personal- and laptop-sized units. The transistors of these integrated circuits are so small that they must be observed under a powerful magnifying glass. (Integrated circuits will be discussed in more detail in Chapter 19.)
* Extremely durable. The transistor is a rugged component. It will withstand excessive vibration and shock.
* Low cost. Transistors are very inexpensive.

Is it any wonder that the transistor has assumed such a place of importance in all industries concerned with the design and fabrication of electronic equipment? A major disadvantage, as you will discover, is the inability of the transistor to operate satisfactorily in high surrounding temperatures. The characteristics of a transistor are very sensitive to changes in ambient temperatures.

BIPOLAR JUNCTION TRANSISTORS

There are two basic types of transistors—the *bipolar junction transistor* and the *field-effect transistor*. The broad applications of both devices are similar, but they differ greatly in theory of operation and in construction. In this chapter, we will be concerned only with the former device.

The **bipolar junction transistor (BJT)** is a three-element device made up of semiconductor materials. BJTs are made and the junctions between elements are formed by methods similar to the making of a basic junction diode. (Review Chapter 12.) Manufacture usually begins by growing a crystal from a single crystal *seed*. The seed is the foundation of the crystal-lattice structure of the semiconductor. Junctions are formed in one of several ways. In one popular method, n- and p-type impurities are alternately added to molten semiconductor material. Most frequently used, however, is the diffusion process. In this process, impurities of one type (n or p) are diffused into a substrate of another type. Both processes produce sandwiches of semiconductors of specific types.

Figure 13-1 shows the two types of bipolar junction transistors—the *pnp* type and the *npn* type. The **pnp transistor** consists of two p-type regions separated by an n-type region. The **npn transistor** consists of two n-type regions separated by a p-type region. Note that the three regions of any transistor are named the **emitter (E)**, the **base (B)**, and the **collector (C)**.

The transistor has two junctions. The pn junction where the emitter and the base regions join is called the **emitter-base (EB) junction**. The pn junction where the collector and base regions join is called the **collector-base (CB) junction**.

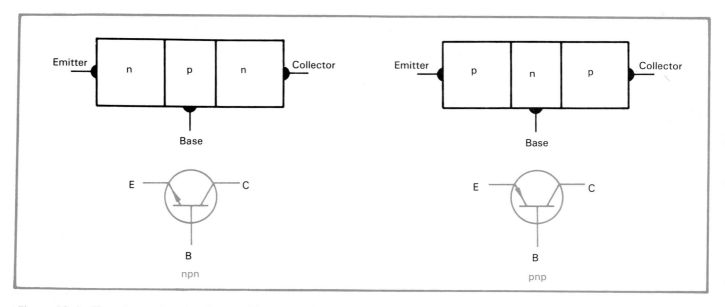

Figure 13-1. Transistors are manufactured in two basic types. Notice the symbols used in schematic diagrams for each type. Remember that the arrow in the schematic of the **npn** transistor **never** points in.

After the transistor crystals are formed and leads attached, the entire assembly is welded in a metal enclosure. This protects the transistor from heat, light, dust, and moisture, which might change its characteristics. Transistor case outlines and lead placements can be found in the *JEDEC* diagrams of any transistor manual. A data sheet giving typical transistor characteristics is found in the appendix at the back of this text.

Transistor biasing

Transistor operation is based upon the theory behind the junction diode. To understand the transistor, you must first understand the junction diode. If you need to, you should review this section in Chapter 12 before proceeding further.

In order for the transistor to operate properly, the two junctions must be correctly *biased*. This means the correct bias voltages must be applied or developed across the transistor terminals. Figure 13-2 shows the correct bias arrangement for both pnp and npn transistors. It is of special importance to note that the EB junction of each transistor is connected in a forward-biased manner. The EB junction, then, offers low resistance or, more accurately, low impedance to current. Current will flow rather freely through it. The CB junction, however, is connected in a reverse-biased manner. The CB junction offers high impedance.

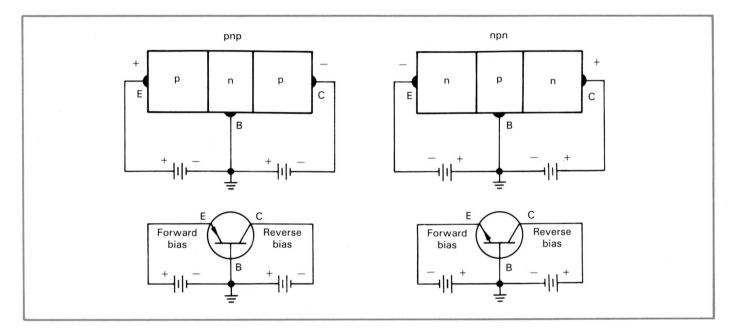

Figure 13-2. These circuits show the forward and reverse bias of transistor junctions. Both block and schematic diagrams are shown.

Direct current in the collector of a transistor is denoted I_C; in the emitter, I_E; in the base, I_B. The emitter current in a transistor is equal to the sum of the base and collector currents. In other words:

$$I_E = I_B + I_C$$

Only a very small portion of emitter current flows through the base; thus, I_E and I_C are approximately equal.

A number of voltages are normally involved in the discussion of transistors. *Collector biasing voltage (V_{CC}), emitter biasing voltage (V_{EE}), and base biasing voltage (V_{BB})* are dc supply voltages that are directly or indirectly applied to the respective terminals. V_C, V_E, and V_B are dc voltages measured from the respective terminals to ground. V_{CE}, V_{EB}, and V_{CB} are the dc voltages measured across the terminals indicated by the subscripts, Figure 13-3. Note that in electronics uppercase letters are used for dc signals; lowercase letters are used for ac signals. A composite signal uses both lowercase and uppercase letters, with the subscript in uppercase.

In measuring transistor voltages, it is recommended that you use an *electronic voltmeter (EVM)* or a *digital voltmeter (DVM)*. The high internal impedance of these meters will not load the circuit, which would otherwise produce inaccurate readings. (Note that the original EVMs used vacuum tubes, which have been replaced by semiconductor devices, for the most part. These instruments were called vacuum-tube voltmeters, or VTVMs.)

It is important to understand that while two elements of a transistor may be positive (or negative) with respect to ground, one may be more positive or more negative than the other. Thus, you need not have one element positive and one negative (with respect to ground) for biasing. What is needed, rather, is a potential difference between two elements.

Finally, you will learn later how ac signals *ride* on a dc level to keep the transistor operating. Voltage waveforms with three different dc levels are drawn in Figure 13-4.

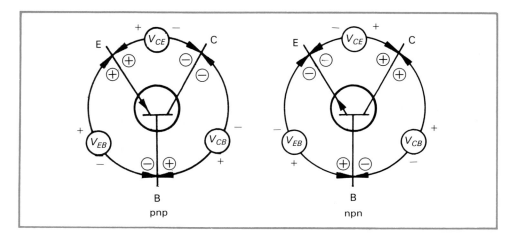

Figure 13-3. These diagrams summarize the polarities of voltages at transistor terminals and indicate correct meter connections.

Zero-biased transistor. Figure 13-5 shows a pnp transistor that is *zero-biased*. This means there is no potential applied to either of the junctions. Notice the depletion regions at the junctions. The establishment of these regions is a result of diffusion and recombination of electron-hole pairs along both the EB and CB junctions. As a result, negative ions build up at the junctions in the p regions; positive ions build up at the junctions in the n region. The potential developed at the junctions acts to oppose further diffusion of majority carriers. Without external biasing, the depletion regions remain in a static, or equilibrium, state. The equilibrium, or barrier, potential at each junction is roughly 0.7 V for silicon; 0.3 V for germanium. Net charge of the transistor is zero. The same discussion can be applied to an npn transistor.

Active region of operation. Biasing supply voltages are applied in the proper directions in Figure 13-6. The transistor is operating in the **active region** when the EB junction is forward biased and the CB junction is reverse biased. Consider the depletion regions and junction potentials. Like the junction diode, a forward bias reduces the width of the depletion region and reduces the junction potential. The barrier potential is overcome if the bias voltage is greater than it is. In the pnp transistor shown, the holes in the emitter will then recombine with electrons in the base section, and conduction across this junction will occur.

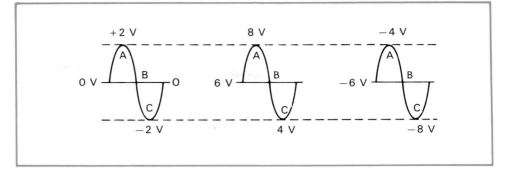

Figure 13-4. *Waveforms show the voltage variation of a 2-V_{peak} signal around three different reference levels. A is always more positive than B. C is always more negative than B.*

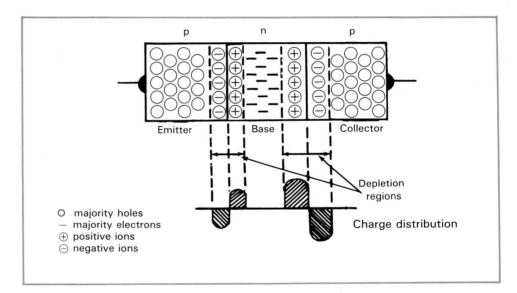

Figure 13-5. *Like the diode, depletion regions are formed at pn junctions. Generally, the CB depletion region is larger than the EB region due to the amount of dopants used in the collector crystal.*

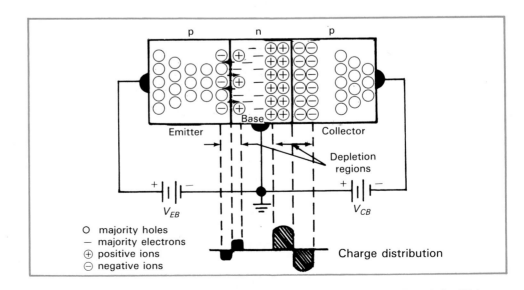

Figure 13-6. *Internal effects of forward/reverse bias. The forward bias of the EB junction decreases its depletion region. The reverse bias of the CB junction increases its depletion region.*

In the case of the reverse-biased CB junction, majority carriers are drawn away from the junction. More positive and negative ions are left behind. The junction potential increases. Very little current will flow between the collector and base leads. However, a large amount will flow between the collector and the emitter leads. A detailed explanation of this follows.

● **Pnp transistor operation.** Study the circuit of the pnp transistor in Figure 13-7. Current flows through the EB junction since it is forward biased. Conduction in the p-type emitter is by holes; conduction in the n-type base is by electrons. The majority carriers in the emitter (holes) and base (electrons) are repelled by V_{EB} to the junction. The holes and electrons recombine upon diffusion across the junction. The flow of electrons from this diffusion constitutes the very small base current I_B. The electrons flow from the source and enter at the base terminal. The recombined electrons flow out of the emitter lead as valence electrons and return to the source.

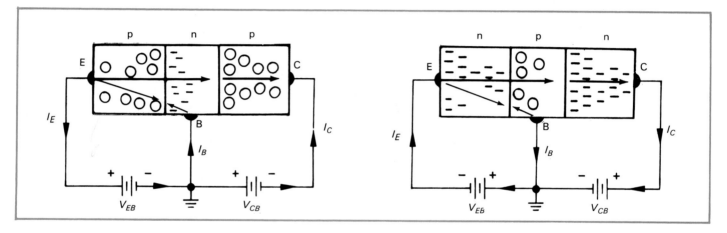

Figure 13-7. Arrows indicate the current paths of the majority carriers in both types of transistors. Note that most of the majority carriers flow right through the base and on to the collector. This is shown by the heavier arrow.

The base material is very thin and lightly doped, and far more holes than electrons converge at the junction. Thus, as the holes diffuse across the junction, only a few are able to recombine with electrons. This makes for the very small base current.

Most of the holes diffuse across the base into the CB depletion region despite the positive ions on the base side of the CB junction. (The repelling force on the holes due to the positive ions at the junction is overcome by the overwhelming concentration of holes injected into the base region.) The holes are minority carriers in the n-type base. In the CB depletion region, they readily drift across the reverse-biased junction. They are drawn across by the negative ions in the collector depletion region. Once in the collector, they are attracted to the negative terminal of V_{CB}. Electrons from V_{CB} fill the holes in the p-type collector, and this constitutes collector current I_C.

Emitter current I_E, then, is the combination of the base and collector currents, with base current being very much less than collector current. Increasing the forward bias voltage will increase the diffusion current, and this will increase the output current I_C. This is the principle that makes the transistor useful as an amplifier—a small input current effectively controlling a much larger current from the emitter to the collector. It is important to note that a BJT is a current-controlled device. The amount of output current depends directly on the amount of input current.

Note that, regarding emitter current, holes are replaced at the emitter terminal. For each electron leaving this p crystal, a hole is injected at the exit site. Note also that regardless of the direction of conduction in the transistor, the electron flow in the external circuit is from negative to positive.

- **Npn transistor operation.** The action within the npn transistor is also described in Figure 13-7. The polarities of the supply voltages, V_{EB} and V_{CB}, have been reversed for proper biasing. The direction of current is also reversed, and the roles of electrons and holes are reversed; otherwise, operation is the same as for the pnp transistor. Electrons from V_{EB} flow through the n-type emitter and recombine with holes in the p-type base. The flow of electrons from this diffusion constitutes base current I_B. The electrons flow from the source and enter at the emitter lead. The recombined electrons flow out of the base lead and return to the source as valence electrons.

The base section is very thin and lightly doped. Therefore, many more electrons than holes are drawn to the EB junction, and few holes are available to recombine with the electrons. Most of the electrons diffuse into the CB depletion region. From here, the electrons, which are minority carriers in the p-type base, readily drift across the reverse-biased junction and on to the positive terminal of V_{CB}. This flow of electrons is the collector current I_C.

Saturation and cutoff regions of operation. The two junctions of a transistor are normally operated in one of three regions of bias. One of these, the active region, has already been discussed. Two others are *saturation* and *cutoff* regions.

Saturation is the condition where collector current is maximum. It is where further increases in base current will not cause further increases in collector current. When a transistor operates in the **saturation region**, both junctions are forward biased. This condition would arise as increasing base current caused increasing collector current. As collector current increased, so would the drop across a **bias resistor** tied to the collector terminal. Its purpose would be to produce the voltage drop needed for a desired bias. A point would be reached when the EB junction switches from a reverse-biased to a forward-biased condition.

When both junctions of a transistor are reverse biased, it is said to be in **cutoff**. In the **cutoff region** of operation, there is no collector current except for a very small leakage current due to thermally produced carriers. A zero-biased transistor is reverse biased in both directions; thus, it is also in a cutoff condition.

TRANSISTORS AS AMPLIFIERS

As mentioned, transistors can be used as amplifiers. An **amplifier** is an electronic circuit that uses a small input signal to control a larger output signal. This can be compared to a phenomenon of physics—the amplification of a weight using a lever. A small weight is used to lift a larger weight—*a small input controlling a larger output*. The amount of amplification in a circuit is know as **gain**. There are three types of *gain* in an amplifier—*current, voltage,* and *power*.

Current gain

Overall **current gain (A_i)** of an amplifier is the ratio of output current to input current. For the transistor itself, there are two general measures of current gain. These are the *current-transfer ratios* of *alpha (α)* and *beta (β)*.

Alpha. This is used in reference to *common-base* circuit configurations. (These will be discussed later in this chapter.) **Dc alpha (α_{dc})** is the ratio of collector to emitter dc currents:

$$\alpha_{dc} = \frac{I_C}{I_E} \bigg| V_{CB} \text{ constant}$$

Note that leakage, which affects I_C, is not accounted for here. Note, too, that α_{dc} is also expressed as h_{FB}. The subscript F means *forward* direction, and B refers to a common-*base* circuit.

Values of α_{dc} typically range from 0.95 to 0.99. In the common-base circuit, the input signal is applied to the emitter; the output is taken from the collector. Since the emitter current is always greater than the collector current, the current gain is always less than one.

Figure 13-8 relates circuit current values to α_{dc}. These relationships, assuming no leakage current, include:

* I_C equals the difference between I_E and I_B:

$$I_C = I_E - I_B$$

* I_C equals the product of I_E and α_{dc}:

$$I_C = \alpha_{dc}I_E$$

* I_E is the sum of base current I_B and collector current I_C:

$$I_E = I_B + I_C$$

Since I_C is equal to $\alpha_{dc}I_E$, then:

$$I_E = I_B + \alpha_{dc}I_E$$

* I_B is the difference between emitter current I_E and collector current I_C:

$$I_B = I_E - I_C$$

Since I_C is equal to $\alpha_{dc}I_E$, then:

$$I_B = I_E - \alpha_{dc}I_E$$

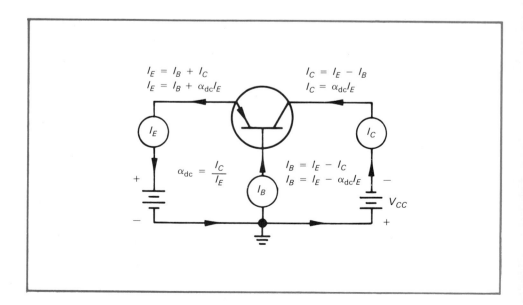

Figure 13-8. Relationships between I_C, I_B, and I_E are illustrated in this diagram.

Values of ac and dc current gain are not quite the same. This is because I_C and I_E (and I_B) are constant for dc, so the value of α_{dc} stays the same. For ac, the values of the collector and emitter (and base) currents change, and the value of alpha changes from one value to the next. Thus, an average value is used. **Ac alpha (α)**, also expressed as h_{fb}, is the ratio of the *change* in collector current to the *change* in emitter current:

$$\alpha = \frac{\Delta i_C}{\Delta i_E} = \frac{i_{c_{p\text{-}p}}}{i_{e_{p\text{-}p}}} \bigg| v_{CB} \text{ constant}$$

Values of α and α_{dc} are generally close for any given transistor.

Beta. This is used in reference to *common-emitter* circuit configurations. (These, too, will be discussed later in this chapter.) **Dc beta (β_{dc})** is the ratio of collector current to base current:

$$\beta_{dc} = \frac{I_C}{I_B} \bigg| V_{CE} \text{ constant}$$

Note that leakage, which affects I_C, is not accounted for here. Note, too, that *dc beta* is also expressed as h_{FE}. The subscript F, again, means *forward* direction, and E refers to a common-emitter circuit.

Values of β_{dc} typically range from 50 to 400. In the common-emitter circuit, the input signal is applied to the base; the output is taken from the collector. Since the collector current is always very much greater than the base current, the current gain is always significantly greater than one.

Conversion of β_{dc} to α_{dc} and vice versa can be done by combining formulas. From the formula $I_C = I_E - I_B$ and $I_C = \alpha_{dc}I_E$, the following relationship for a given transistor is found:

$$\beta_{dc} = \frac{\alpha_{dc}I_E}{I_E - \alpha_{dc}I_E}$$

Factoring and canceling yields:

$$\beta_{dc} = \frac{\alpha_{dc}}{1 - \alpha_{dc}}$$

Further, by complementary relationships:

$$\alpha_{dc} = \frac{\beta_{dc}}{\beta_{dc} + 1}$$

As is true for alpha, beta also changes from one value to the next with ac. Thus, again, an average value is used. **Ac beta (β)**, also expressed as h_{fe}, is the ratio of the *change* in collector current to the *change* in base current:

$$\beta = \frac{\Delta i_C}{\Delta i_B} = \frac{i_{c_{p\text{-}p}}}{i_{b_{p\text{-}p}}} \bigg| v_{CE} \text{ constant}$$

Values of β and β_{dc} are generally close for any given transistor. You may also hear β (or α) referred to as the *small-signal current gain* or even the *small-signal, short-circuit, forward current transfer ratio*. (*Small signal* refers to the condition that the ac current values are small enough that linear relationships hold between them.)

Voltage gain

Some transistor circuit configurations amplify voltage. In the common-base circuit, current amplification is less than one; even so, the transistor will amplify the voltage signal. This is possible due to the difference in the resistance of the output circuit, or **output impedance**, and the resistance of the input circuit, or **input impedance**. The ratio of these two values, output to input, is the *resistance ratio*. This "transfer" of resistance is the basis upon which the transistor (*trans*fer res*istor*) was named. It has already been explained that due to forward bias, the emitter-base resistance is low. Due to reverse bias, the collector-base resistance is high.

Study Figure 13-9. It shows an approximate equivalent circuit of a transistor (pnp) with bias voltages applied. The current i_E flows in the emitter circuit. It is relatively high due to the low resistance R_{EB}. Current αi_E flows in the high-resistance, collector-base circuit. It is somewhat less than i_E by the factor alpha. The approximate *voltage gain* may be computed by comparing the voltage drops across the resistances. **Voltage gain (A_v)** is the factor by which an ac signal voltage increases from the input of an amplifier to the output. It is the ratio of ac output voltage to ac input voltage. By formula:

$$A_v = \frac{v_{\text{out}}}{v_{\text{in}}}$$

An approximate formula for transistor voltage gain in a common-base circuit is:

$$A_v = \frac{\alpha i_E R_{CB}}{i_E R_{EB}} = \frac{\text{voltage across } R_{CB}}{\text{voltage across } R_{EB}}$$

The term i_E cancels out, and the formula is rewritten to:

$$A_v = \alpha\left(\frac{R_{CB}}{R_{EB}}\right) = \alpha \text{ (resistance ratio)}$$

Problem:

From a transistor data sheet, emitter-base input resistance is found to be 50 Ω and collector-base output resistance is found to be 500 kΩ. With $\alpha = .99$, compute voltage gain.

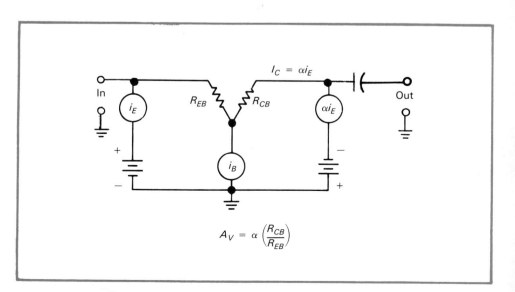

$$A_V = \alpha\left(\frac{R_{CB}}{R_{EB}}\right)$$

Figure 13-9. Voltage gain in a transistor is realized by a high resistance ratio.

Solution:

$$A_v = \alpha \left(\frac{R_{CB}}{R_{EB}}\right) = .99 \left(\frac{500 \times 10^3 \ \Omega}{50 \ \Omega}\right) = .99 \times 10,000 = 9900$$

The computed voltage gain will be *considerably* reduced by external resistors that are needed in a practical circuit. This explanation has purposely omitted several contributing factors to voltage gain in order to explain the resistance ratio concept of voltage amplification.

Power gain

The factor by which an ac signal power increases from the input of an amplifier to the output is the **power gain** *(A_p)* of the amplifier. It is the ratio of ac output power to ac input power. By formula:

$$A_p = \frac{P_{\text{out}}}{P_{\text{in}}}$$

Power gain, further, is equivalent to the product of current gain and voltage gain. By formula:

$$A_p = A_i A_v$$

Since β (or h_{fe}) is current gain in the common-emitter circuit, the power gain for this circuit may be approximated by:

$$A_p = \beta A_v$$

Since α (or h_{fb}) is current gain in the common-base circuit, the power gain for this circuit may be approximated by:

$$A_p = \alpha A_v$$

Consideration of the resistance ratio may be used to approximate power gain. By substitution, since $A_v = \alpha(R_{CB}/R_{EB})$, for the common-base circuit:

$$A_p = \alpha \ (\alpha) \left(\frac{R_{CB}}{R_{EB}}\right) = \alpha^2 \left(\frac{R_{CB}}{R_{EB}}\right) = \alpha^2 (\text{resistance ratio})$$

Note that these formulas given for power gain are approximations for the transistor only. They do not consider external circuit components.

LEAKAGE CURRENT

Up to this point, we have primarily discussed the current in a transistor by *majority* carriers, or current due to transistor action. There are always some *minority* carriers in each crystal produced by impurities and thermal or light energy. In Figure 13-10, a very small leakage current flows across the CB junction. It is due to the *thermally generated* minority carriers in the CB

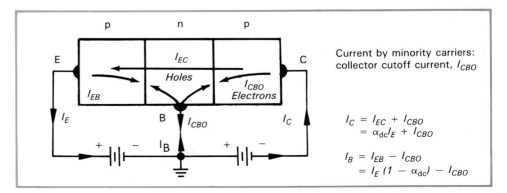

Figure 13-10. Diagram shows direction of I_{CBO} in a pnp transistor.

depletion region. Those carriers in the depletion region are drawn across the junction to the ionic charges of opposite polarity. The result is a flow of electrons moving in the same direction as the collector current and adding to it. (Remember that hole flow in one direction has the same effect as electron flow in the opposite direction.)

The *theoretical* transistor leakage current is called the **reverse saturation current (I_{CO})**. Values for small silicon transistors might be in the picoampere range at room temperature; for germanium, the microampere range. The *actual* current due to this leakage, called the **collector cutoff current (I_{CBO})**, exceeds I_{CO}. For silicon transistors, values might be in the nanoampere range. The designation I_{CBO} means: Current from C to B with emitter circuit open (O).

In the properly biased transistor, the total collector current is the sum of two currents:
- Current due to transistor action ($\alpha_{dc}I_E$).
- Collector cutoff current (I_{CBO}).

Therefore:

$$I_C = \alpha_{dc}I_E + I_{CBO}$$

In a similar diagram for the npn transistor, Figure 13-11, the bias battery polarities have been reversed. Minority carriers in the n collector are holes; minority carriers in the p base are electrons. Compare the current paths and carriers in the npn and pnp transistor to satisfy yourself that the same equations apply.

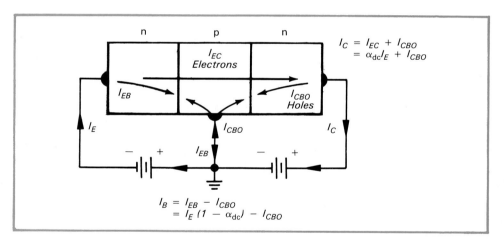

Figure 13-11. *Diagram shows direction of I_{CBO} in an npn transistor.*

TRANSISTOR CIRCUIT CONFIGURATIONS

The transistor can be connected in three circuit configurations. The type of circuit is identified by observing which element is common to both input and output circuits. Usually, the common element is at ground potential. The three types of circuit configurations are the:
- *Common-base circuit.*
- *Common-emitter circuit.*
- *Common-collector circuit.*

The common-base circuit

When a transistor is connected so that the *base* is common to both the input and the output circuits, it is a **common-base**, or **CB**, **circuit**. See Figure 13-12. You will recognize this circuit as one we have already briefly discussed.

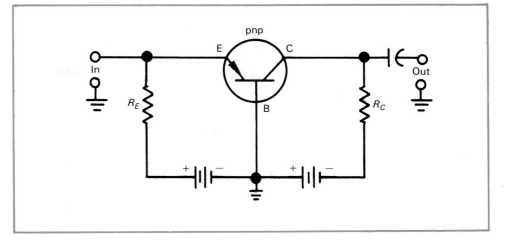

Figure 13-12. The common-base circuit. Note that it is common practice to omit the ground symbol on the signal inputs and output; in which case, the symbol is implied.

The input signal is applied across the EB junction. The output signal is taken from across the CB output circuit. The base is common to both signals and is usually grounded. The circuit is sometimes called a **grounded-base circuit**. Note that, in this circuit, I_E is always the input current; I_C is always the output current. Do not be confused by the fact that in the pnp transistor I_C flows *into* the collector terminal and I_E flows *out* of the emitter terminal.

CB collector characteristics. The circuit in Figure 13-13 is used to measure the static *collector characteristics* of the transistor connected in this configuration. Setting emitter current I_E at a fixed value, the collector voltage V_{CB} will be increased in steps to determine its effect on collector current I_C. For each fixed value of emitter current, the measurements are repeated. From the data obtained, a **collector characteristic curve** may be plotted. Figure 13-14 is a typical graph of collector characteristics for the CB circuit. The graph shows the effect collector voltage has on collector current for each constant value of emitter current. It makes up what is called a *family* of collector curves for the transistor. On transistor data sheets, they are generally referred to as *output characteristic curves*. They would be found as part of a set of curves called *performance curves*.

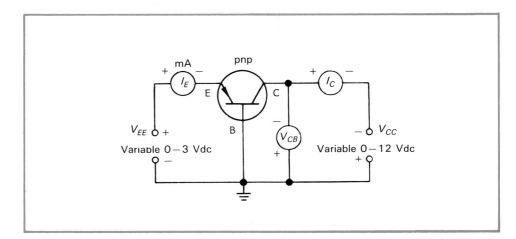

Figure 13-13. This circuit is used to determine the static collector characteristics of a transistor in a common-base configuration.

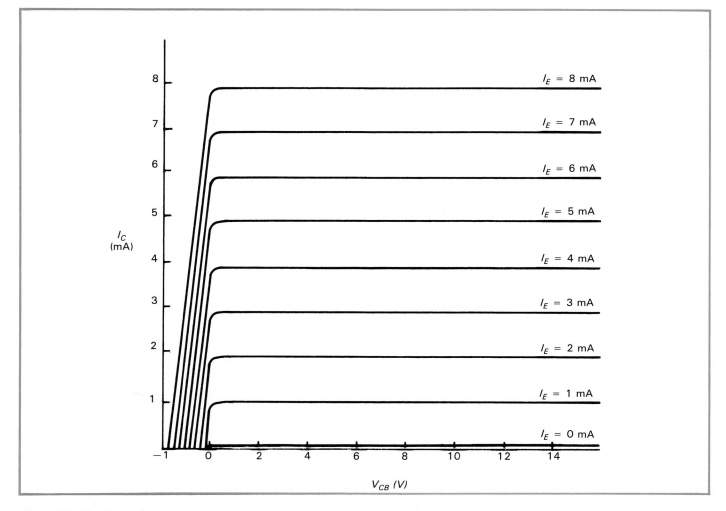

Figure 13-14. The collector characteristic curves of a transistor connected in a CB circuit.

The uniform spacing between each emitter current curve in Figure 13-14 would indicate a linear relationship between input and output currents. Also, the fact that the curves are very flat indicates a relatively high output impedance. Ohm's law will verify this. For example, if V_{CB} were to change from 8 V to 10 V, with a 200-Ω output impedance, I_C would change by 10 mA. However, if this 2-V change were seen with a 200-kΩ output impedance, I_C would change by only 10 μA. Such a small current change is depicted by a very flat curve.

Current amplification (gain) in the CB circuit is given by alpha. Dc gain is $\alpha_{dc} = I_C/I_E$ where V_{CB} is constant, as previously discussed. At any given value of voltage V_{CB}, the current amplification factor may be found from the collector characteristic curve and this relationship. For small applied signals, the dynamic (ac) value of alpha is given as $\alpha = \Delta i_C/\Delta i_E$; again, where v_{CB} is constant. This was also discussed earlier. The value of α can also be found from the curves.

CB phase relationships. For every type of amplifier, input and output *currents* are in phase. Figure 13-15 shows input and output voltage waveforms. The input signal appears across emitter resistor R_E, and the output signal appears across collector load resistor R_C. Before going on to describe the waveforms in Figure 13-15, it should be helpful to describe the transistor at two extreme conditions—saturation and cutoff.

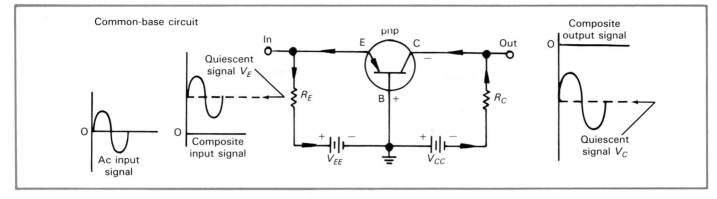

Figure 13-15. The common-base circuit does not invert the voltage signal. Input and output are in phase. Note the ac input and output signals superimposed on the quiescent (dc-bias) signal.

First, consider the transistor at cutoff and not conducting. Since there is no collector current, the voltage drop across R_C equals zero. In the nonconducting state, the reverse-biased CB junction appears as an open circuit. All of the voltage drop in the collector circuit loop is across it. The same holds true for the emitter voltage V_E, since at cutoff *both* junctions are reverse-biased. Thus, collector voltage V_C is maximum and equal to V_{CC} when V_E is maximum.

Next, consider the transistor near saturation and in full conduction. Collector current is maximum. Collector voltage V_C is equal to V_{CC} less the drop across R_C $(V_C = V_{CC} - I_C R_C)$. The same holds true for V_E, since at saturation, both I_C and I_E are maximum. With maximum collector current, voltage drop across R_C is maximum, and V_C is minimum as is V_E. The value of V_{R_C} approaches V_{CC}, and the value of V_C approaches zero.

In Figure 13-15 for the pnp transistor, an operating current I_C is considered to be flowing, which produces some voltage drop across R_C. V_C is less negative than supply voltage V_{CC} by the amount of the voltage drop across R_C. *Quiescent signals* are established with the bias voltage sources. (A **quiescent signal** is strictly the dc level. A quiescent amplifier is one that has no ac signal applied.) An ac signal is now applied to the emitter circuit loop. Follow the action:

1. As the input signal voltage moves in a positive direction, the composite voltage increases, and, thus, the emitter current increases. This increases collector current i_C.
2. An increase in i_C produces a larger voltage drop across R_C.
3. Voltage v_C, which is also the output voltage, becomes even less negative and moves in a positive direction.
4. As the input signal moves in a negative direction, the emitter current is reduced. This will cause i_C to decrease.
5. A decrease in i_C will produce less of a voltage drop across R_C.
6. Voltage v_C becomes more negative than it was at its quiescent state. At the peak negative input signal, the cycle is repeated.

CONCLUSION: The CB amplifier does not cause a phase reversal of the voltage signal. The input and output signals are the same.

In Figure 13-16, the same circuit is drawn for the npn transistor. All polarities are reversed; yet, the same theory of operation will apply. Study the circuit and be able to explain the in-phase signal relationship with either type of transistor.

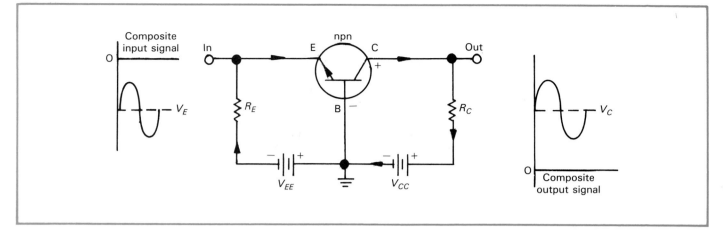

Figure 13-16. *A positive input decreases forward bias of EB junction and decreases* i_C *and the voltage drop across* R_C. *Therefore, C becomes more positive.*

The common-emitter circuit

When a transistor is connected so that the *emitter* is common to both the input and the output circuits, it is a **common-emitter**, or **CE, circuit**. See Figure 13-17. The input signal is applied across the EB junction. The output signal is taken from across the CE output circuit. The emitter is common to both signals and is usually grounded. The circuit is sometimes called a **grounded-emitter circuit**. It is the most common configuration used. It has decided advantages over the other two.

CE collector characteristics. The circuit in Figure 13-18 is used to measure the static *collector characteristics* of the transistor connected in this configura-

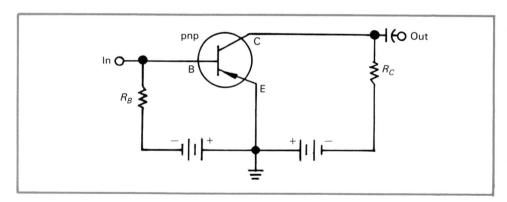

Figure 13-17. *The common-emitter circuit.*

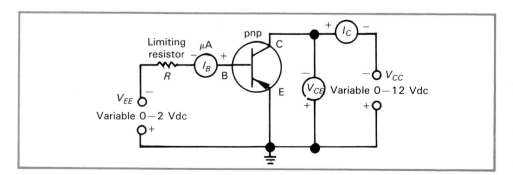

Figure 13-18. *This circuit is used to determine the static collector characteristics of a transistor in a common-emitter configuration.*

tion. Resistor R is a current limiting resistor used to protect the transistor from excessive currents while the test is being made. Base current I_B is first set at zero by setting V_{EE} at zero. Collector voltage V_{CE} is increased in steps, and collector current I_C is recorded for each value of V_{CE}. Ideally, when I_B equals 0 A, I_C will equal 0 A, regardless of the value of V_{CE}. In reality, a very small current will flow due to leakage.

I_B is adjusted to another fixed value by adjusting V_{EE}, and V_{CE} and I_C are recorded in steps as before. A family of curves for a typical transistor is shown in Figure 13-19. There is a separate curve for each fixed value of I_B. (Note that I_B is in microamperes.) Usually, transistor manufacturers will supply specifications and common-emitter characteristic curves for any specified transistor.

Let us examine the CE collector characteristic curves and see what information they contain:

- At a certain value of V_{CE} and I_B, I_C may be found. By increasing V_{CE} and holding the value of I_B fixed, very little increase in I_C is realized.
  CONCLUSION: A change in collector voltage has only a minimal effect on collector current.
- If, at a certain value of V_{CE}, the value of I_B is increased, a relatively large increase in I_C will be observed.
  CONCLUSION: A small change in base current will produce a large change in collector current.

Compare these characteristics to those of the CB circuit. Observe that changing collector voltage had little, if any, effect on collector current in either circuit. For the common-base circuit, changing input current I_E produced roughly the same current change in the output current I_C. The CB circuit exhibits no current gain.

Current amplification (gain) in the CE circuit is given by beta. Dc gain is $\beta_{dc} = I_C/I_B$ where V_{CE} is constant, as previously discussed. At any given value of voltage V_{CE}, the current amplification factor may be found from the collector characteristic curve and this relationship. For small applied signals, the dynamic (ac) value of beta is given as $\beta = \Delta i_C/\Delta i_B$; again, where v_{CE} is constant. This was also discussed earlier. The value of β can also be found from the curves.

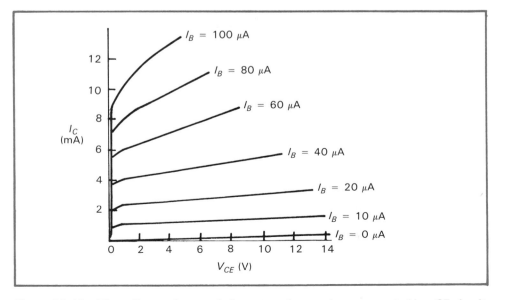

Figure 13-19. The collector characteristic curves of a transistor connected in a CE circuit.

CE phase relationships. In Figure 13-20 for the pnp transistor, an operating current I_C is considered to be flowing. This produces some voltage drop across R_C. Voltage V_C is less negative than supply voltage V_{CC} by the amount of the voltage drop across R_C. A quiescent signal, established with the bias voltage sources, is assumed. An ac signal is now applied to the base circuit loop. Follow the action:

1. Composite voltage is the algebraic sum of the dc bias voltage and the signal voltage. In the circuit, the *magnitude* of the composite voltage v_B is *reduced* by a positive-going signal voltage since the bias voltage at B is negative. When this happens, a reduction is seen in i_B.
2. A decrease in i_B results in a decrease in i_C, which results in a lesser voltage drop across R_C.
3. Voltage v_C, which is also the output voltage, becomes more negative and moves toward the value of V_{CC}.
4. With a negative-going input signal voltage, the magnitude of the composite voltage is increased. When this happens, an increase is seen in i_B.
5. Consequently, i_C will increase, and the voltage drop across R_C will increase.
6. Voltage v_C becomes less negative, or moves in a positive direction. At the peak negative input signal, the cycle is repeated.

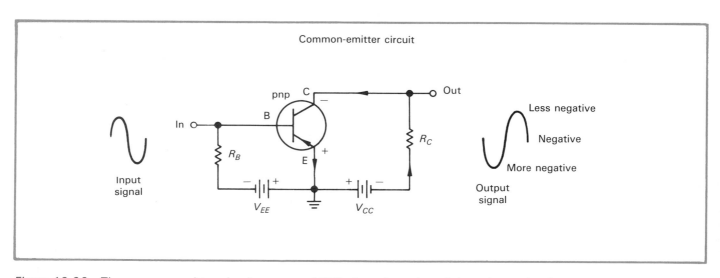

Figure 13-20. *The common-emitter circuit causes a 180° phase inversion of the voltage signal.*

CONCLUSION: The CE amplifier causes a 180° phase shift of the voltage signal. It *inverts* the signal. (Input current and voltage signals are in phase; input and output voltage signals are 180° out of phase.)

In Figure 13-21, the same circuit is drawn using the npn transistor. Study the circuit and understand how phase inversion is accomplished.

The common-collector circuit

When a transistor is connected so that the *collector* is common to both the input and the output circuits, you have a **common-collector**, or **CC**, **circuit**. Figure 13-22 shows the common-collector circuit as it is normally drawn. It does not clearly show why it is called a common *collector*, however. The circuit can be redrawn in a way that would show the collector common to both the input and output signals. Can you figure out how?

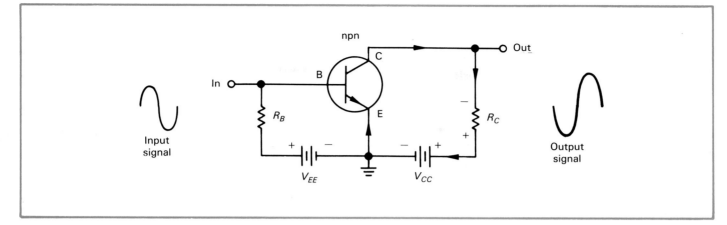

Figure 13-21. *A positive input signal makes B more positive and increases* v_{EB}*. As a result,* i_C *increases, causing a voltage drop across* R_C*. C then becomes less positive, or more negative. The voltage signal is inverted.*

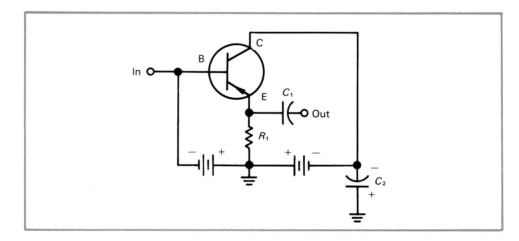

Figure 13-22. *The common-collector circuit.*

The value of resistor R_1 shown in the circuit is quite large. It effectively isolates the transistor emitter from ground. The value of capacitor C_2 is selected for a low-reactance path for signal frequencies to ground. The collector is at ground potential for signal frequencies. (Review Chapter 10; section covering capacitive reactance and Figure 10-7.) The circuit is sometimes called a **grounded-collector**, or **emitter-follower**, **circuit**. The input signal is applied across the EB junction. The output signal is taken from across R_1, or between the emitter lead and ground.

CC collector characteristics. This particular circuit has a high input impedance and a low output impedance. It is used as an impedance matching circuit, performing the same basic function as an impedance-matching transformer. Current and power gain are seen with this circuit; however, its voltage gain is less than one.

CC phase relationships. In the CC circuit, Figure 13-23, the output signal is taken from the emitter end of R_E. Capacitor C_2 is a low reactance path to ground for the signal. The collector is at signal ground potential. Again, a definite quiescent current, which produces a voltage drop across R_E, is assumed. V_E is more negative than ground by the amount of voltage drop V_{R_E}. An ac signal is now applied to the base circuit loop. Follow the action:

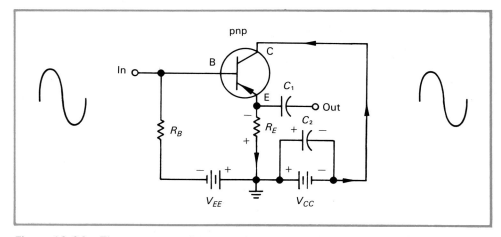

Figure 13-23. The common-collector circuit does not invert the signal.

1. In the circuit, the magnitude of the composite voltage is reduced by a positive-going signal voltage. When this happens, a reduction is seen in i_B.
2. This will cause a decrease in i_C and also i_E. The voltage drop across R_E will decrease.
3. Voltage v_E, which is also the output voltage, will become less negative, or will move in a positive direction.
4. A negative-going input signal will increase the magnitude of the composite voltage. When this happens, an increase is seen in i_B.
5. Currents i_C and i_E will both increase as a result.
6. Voltage v_E becomes more negative. At the peak negative input signal, the cycle is repeated.

CONCLUSION: The input and output voltage signals of a CC amplifier are in phase. (Input and output currents are in phase.)

Follow the diagram of the npn transistor in Figure 13-24. Assure yourself that the conclusion holds true for it as well.

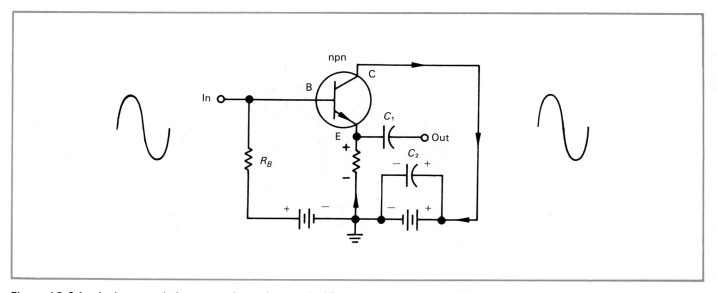

Figure 13-24. An increase in i_E causes the emitter end of R_E to become more positive. The output voltage signal is not inverted.

Figure 13-25 gives phase relationships of waveforms in transistor amplifiers. It also gives typical input and output impedances and gain values for a transistor in CB, CE, and CC configurations. You should learn this table. (Note that power gain is expressed in unit of *decibels*. Decibels will be discussed in the next chapter.)

Type of Circuit	Z_{in}	Z_{out}	A_v	A_i	A_p	Phase relationship	
						Input voltage	Output voltage
Common base—CB	Low 50—150 Ω	High 300—500 kΩ	High 500—1500	Less than one	Medium 20—30 db	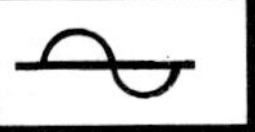 ⌒⌄	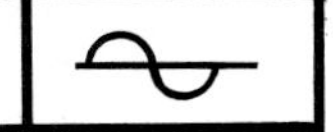 ⌒⌄
Common emitter—CE	Medium 500—1500 Ω	Medium 30—50 kΩ	Medium 300—1000	Medium 25—50	High 25—40 db	 ⌒⌄	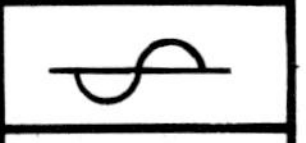 ⌄⌒
Common collector—CC	High 20—500 kΩ	Low 50—1000 Ω	Less than one	Medium 25—50	Medium 10—20 db	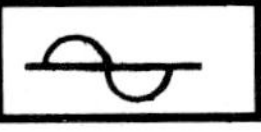 ⌒⌄	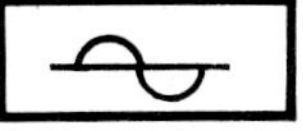 ⌒⌄

Figure 13-25. Typical characteristics of a transistor in CB, CE, and CC configurations.

TRANSISTOR TESTER PROJECT

This chapter dealt with the fundamental principles of the transistor. A transistor tester is an excellent related project. It is something that a person who works with them might find very useful. This inexpensive tester will test almost any type of conventional transistor. The circuit uses an oscillator, or ac signal generator, and the transistor to be tested becomes a part of the oscillator circuit. A good transistor will give an audio tone in the speaker. If the transistor is bad, no tone will be heard.

The tester is pictured in Figure 13-26. The current gain (beta) is controlled by variable resistor R_3, which is adjusted by the dial found on the face of

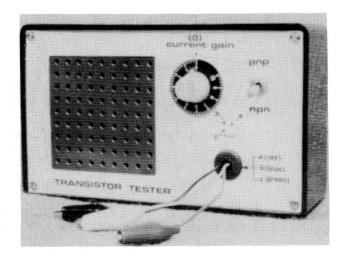

Figure 13-26. Transistor tester.

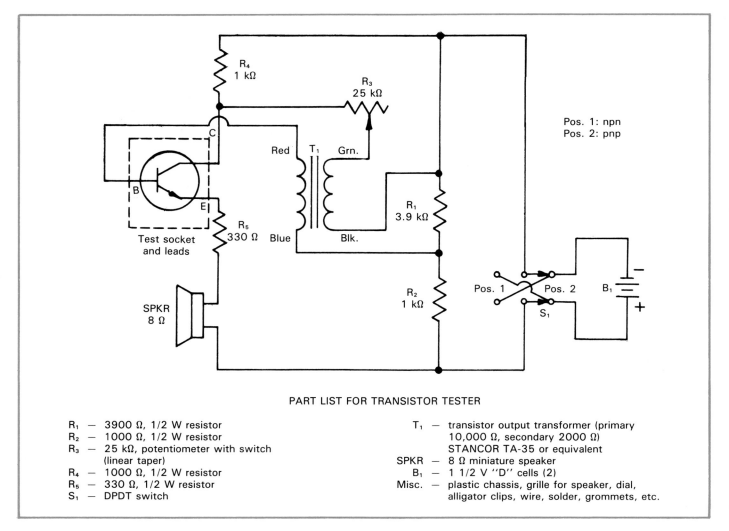

PART LIST FOR TRANSISTOR TESTER

R$_1$ — 3900 Ω, 1/2 W resistor	T$_1$ — transistor output transformer (primary
R$_2$ — 1000 Ω, 1/2 W resistor	10,000 Ω, secondary 2000 Ω)
R$_3$ — 25 kΩ, potentiometer with switch	STANCOR TA-35 or equivalent
(linear taper)	SPKR — 8 Ω miniature speaker
R$_4$ — 1000 Ω, 1/2 W resistor	B$_1$ — 1 1/2 V "D" cells (2)
R$_5$ — 330 Ω, 1/2 W resistor	Misc. — plastic chassis, grille for speaker, dial,
S$_1$ — DPDT switch	alligator clips, wire, solder, grommets, etc.

Figure 13-27. Schematic and parts list for transistor tester.

the tester. The transistor may be plugged into a test socket, also on the face, or test leads may be used to connect the transistor. The type of transistor (pnp or npn) needs to be known. The switch (S_1) must be set to the proper position before operation. A schematic and parts list are provided in Figure 13-27.

NOTE: Always remove the power from a transistor circuit before making any disconnects or changes in a circuit. Transient voltages and current surges may be sufficient to destroy a transistor.

SUMMARY

- The transistor is a semiconductor device capable of amplification, oscillation, and switching.
- The bipolar junction transistor is a three element device made up of semiconductor materials. The two types are the npn and the pnp.
- The three leads of a BJT are the emitter, base, and collector.
- Amplifiers are electronic circuits that control output signals with input signals.
- Bias is the voltage difference between the two input elements of a transistor.

- In the active region of a transistor amplifier, the input elements are forward biased while the output elements are reverse biased.
- The amount of amplification in a circuit is know as gain.
- Alpha is the current transfer ratio used in reference to common-base circuits. Beta is the current transfer ratio used in reference to common-emitter circuits.
- Three transistor circuit configurations are the common-base, common-emitter, and common-collector circuits.
- CB circuits are used for voltage amplification. CE circuits are used for current amplification. CC circuits are used for impedance matching.
- Input and output voltages of CB and CC circuits are in phase. In CE circuits, they are 180° out of phase.

KEY TERMS

Each of the following terms has been used in this chapter. Do you know their meanings?

ac alpha (α or h_{fb})
ac beta (β or h_{fe})
active region
amplifier
base (B)
bias resistor
bipolar junction transistor (BJT)
collector (C)
collector characteristic curve
collector cutoff current (I_{CBO})
collector-base (CB) junction
common-base circuit
common-collector circuit

common-emitter circuit
current gain (A_i)
cutoff
cutoff region
dc alpha (α_{dc} or h_{FB})
dc beta (β_{dc} or h_{FE})
emitter (E)
emitter-base (EB) junction
emitter-follower circuit
gain
grounded-base circuit
grounded-collector circuit

grounded-emitter circuit
input impedance
npn transistor
output impedance
pnp transistor
power gain (A_p)
quiescent signal
reverse-saturation current (I_{CO})
saturation
saturation region
transistor
voltage gain (A_v)

TEST YOUR KNOWLEDGE

Please do not write in this text. Place your answers on a separate sheet of paper.

1. List four transistor design features.
2. The three regions and the terminals of the transistor are called _____, _____, and _____.
3. What causes depletion regions to form at the transistor junctions?
4. In a transistor, the emitter-base junction is always _____ biased, and the collector-base junction is always _____ biased.
5. What is the cause of collector cutoff current?
6. A certain CB-configured transistor has an I_E of 10 mA and I_B is 50 μA. What is its current gain?
7. A certain transistor has a dc alpha of .98. I_E is 10 mA. I_C is _____.
8. A certain CB-configured transistor has a dc alpha of .99 and a resistance ratio of 100,000/50. What is its voltage gain?
9. What is the power gain of the transistor used in Question 8?
10. Draw a circuit for a common-base transistor configuration.
11. Draw a circuit for a common-emitter transistor configuration.
12. Draw a circuit for a common-collector transistor configuration.
13. The designation I_{CBO} stands for current from C to B with _____ circuit open.

14. Generally, about 95-99% of emitter current I_E flows directly to the collector and constitutes collector current I_C. True or False?

15. In measuring transistor voltages, it is recommended that you use an _____ or a _____ .

16. What is the phase relationship between input and output voltages of a common-base amplifier circuit?

17. What is the phase relationship between input and output voltages of a common-emitter amplifier circuit?

18. In a common-emitter transistor circuit, a change in i_B of 40 μA produces a change in i_C of 4 mA. What is the beta?

Chapter 14

TRANSISTOR AMPLIFIERS

A transistor is designed to have certain characteristics. These are only achieved if the transistor is connected into a circuit that has the required voltage and current values. Such a circuit contains the necessary resistors and other passive devices to obtain these desired values.

After studying this chapter, you will be able to:

☐ *Describe various methods of biasing transistor circuits.*
☐ *Design a transistor graphically by using load lines.*
☐ *Apply maximum power-dissipation specifications in the design of a transistor amplifier.*
☐ *Identify different types of amplifiers.*

INTRODUCTION

In the previous chapter, you learned that a major function of the transistor is to amplify. You were also introduced to basic amplifier principles. By now, these should be firmly established in your mind. In this chapter, amplifier basics are extended to the point where practical, working circuits are presented. This includes a more detailed explanation of biasing the bipolar junction transistor. In addition, you will learn about *field-effect* and *unijunction transistors*.

GENERAL BIAS CIRCUITS

The general circuit for biasing a transistor is shown in Figure 14-1. This circuit can be applied to any of the three circuit configurations — CB, CE, or CC — by selecting the required input and output points. V_{CC} is the collector supply voltage that places a reverse bias on the collector-base junction. V_{EE} is the emitter supply voltage that places a forward bias on the emitter-base junction. *Collector resistor (R_C), emitter resistor (R_E), base resistor (R_B),*

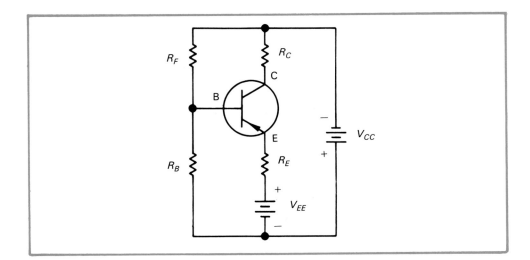

Figure 14-1. The general bias circuit for transistors. Pnp transistor is shown here. For npn transistor, polarities of V_{EE} and V_{CC} are reversed.

and *feedback resistor (R_F)*, are the circuit bias resistors.

The general bias circuit is converted to the common-base circuit in Figure 14-2. In this circuit, the input signal is applied across the emitter and base; the output is taken across the collector and base. The base is common to both input and output and is connected directly to ground. R_B is short circuited. R_F is omitted. It is like an open circuit and is considered to have infinite resistance.

The general bias circuit is converted to the common-emitter circuit in Figure 14-3. In this circuit, the input signal is applied across the base and emitter; the output is taken across the collector and emitter. The emitter is common to both input and output and is connected directly to ground. R_E is short circuited. R_F is omitted and considered to have infinite resistance.

The general bias circuit is converted to the common-collector circuit in Figure 14-4. In this circuit, the input signal is applied across the base and collector. The output signal is taken across the emitter and collector. The collector is common to both input and output and is connected directly to ground. R_C is short circuited. Again, R_F is omitted and considered to have infinite resistance. In practice, C is grounded for signal voltages only through a low-reactance capacitor.

Up to this point, R_F seems to have always been omitted, or considered as an infinite resistance. In practical circuits, R_F will assume a value. This will be explained later in this chapter.

METHODS OF BIAS

A transistor must be properly biased to operate as an amplifier. This means applying the proper *forward-* and *reverse-bias voltages*, which are, respectively, the dc potentials across the transistor EB and CB junctions. The reason for the dc bias is to set up a steady level of transistor voltage and current. Without proper dc bias, transistor amplifiers will function improperly or not at all. There are several methods of establishing dc bias, including:

* *Base bias.*
* *Voltage-divider bias.*
* *Emitter bias.*
* *Collector-feedback bias.*
* *Combination bias.*

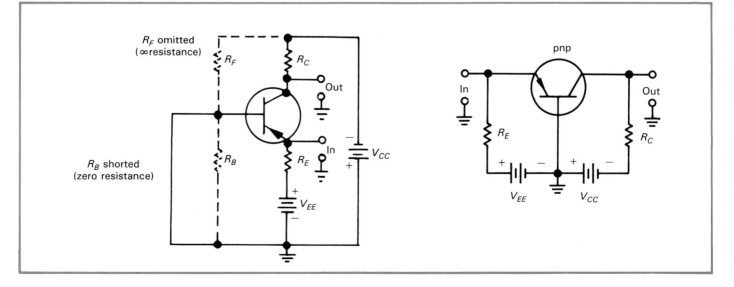

Figure 14-2. General bias circuit converted to a CB amplifier. Compare these two schematics.

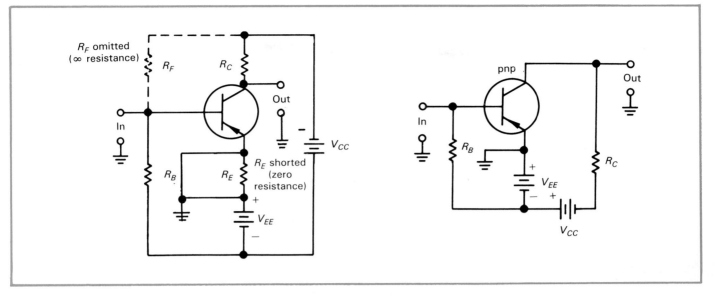

Figure 14-3. General bias circuit converted to a CE amplifier. Compare these two schematics.

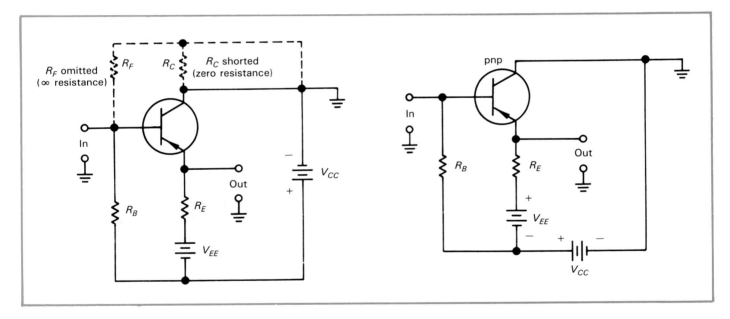

Figure 14-4. General bias circuit converted to a CC amplifier. Compare these two schematics.

Base bias

For simplification, two voltage sources have been used in circuits discussed so far. One source is used for the forward biasing of the EB junction; the other is used for the reverse biasing of the CB junction. Typically, however, a single voltage source is used to bias both junctions. In the CE amplifier of Figure 14-5, only one source is used — V_{CC}.

Figure 14-5 illustrates a **base-bias**, or **fixed-bias**, **circuit**. This is the simplest type of biasing. This circuit sets essentially a constant base current. It is very sensitive to variations in the circuit; as such, the circuit makes a poor amplifier and is primarily used in switching applications. By proper selection of R_B, the required forward bias voltage and base current may be set up.

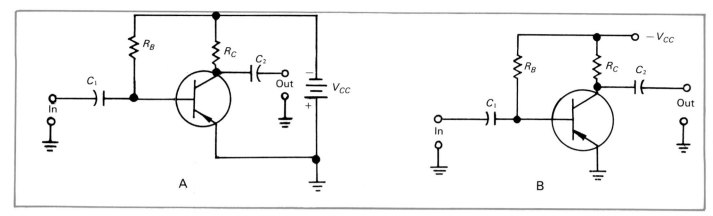

Figure 14-5. Base bias. A—Circuit with battery symbol. B—Same circuit but with battery replaced by a line termination and voltage indicator, for simplicity.

Analysis of this circuit for the linear region is given by the following relationships:

(a)
$$V_{R_B} = V_{CC} - V_{EB}$$

(b)
$$I_B = \frac{V_{R_B}}{R_B} = \frac{V_{CC} - V_{EB}}{R_B}$$

(c)
$$I_C = \beta_{dc}I_B$$

(d)
$$V_{CE} = V_{CC} - I_C R_C$$

Voltage-divider bias

Transistor characteristics are very temperature dependent. Leakage current nearly doubles every 8°C in germanium transistors and every 10°C in silicon. An increase in temperature results in an increase in collector current. This makes transistors somewhat unstable in environments of fluctuating temperature. What is needed is a method of compensation, where an increase in collector current would cause a decrease in base current. After all, base current controls collector current. Likewise, a decrease in collector current should be compensated for by an increase in base current.

The base bias circuit sets a constant base current. It does not compensate for temperature variations. It is an unstable circuit and would be practical only in a constant temperature environment. One bias scheme that changes base current to compensate for leakage is the **voltage-divider-bias circuit**. This scheme is the most popular method of biasing a transistor. It is a stable circuit and requires only a single power source. Figure 14-6 shows a typical circuit. Note that the voltage divider $R_F R_B$ is called a *bias stick*.

If an increase in temperature causes collector current to increase due to leakage, V_E, the voltage at E to ground, becomes more negative. This pulls down V_B, the voltage at B to ground. This causes more voltage across R_B and less across R_F, which means an increased current through R_B but decreased current through R_F. By Kirchhoff's current law, current through B is equal to current through R_F minus current through R_B:

$$I_B = I_{R_F} - I_{R_B}$$

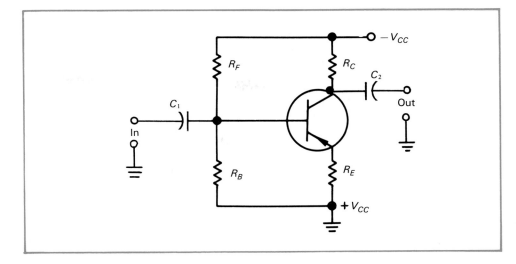

Figure 14-6. Voltage-divider bias.

Therefore, under these conditions, the base current must decrease. In this circuit, a decrease in I_B caused by an increase in I_C, limits the increase in I_C.

In general, good temperature stability is achieved if the value of R_E is between the values of R_C and 10% of R_C. Further, the value of R_B should be between the value of R_E and ten times R_E. By selecting the proper values for R_F and R_B, the desired forward bias voltage and current can be established. The series combination of R_F and R_B must be large enough so that current drain from the supply battery will be small and long life will be assured.

Analysis of this circuit for the linear region is aided by the following relationships:

(a)

$$V_B \cong \left(\frac{R_B}{R_B + R_F}\right)V_{CC}$$

(b)

$$V_E = V_B - V_{EB}$$

(c)

$$I_E = \frac{V_E}{R_E}$$

(d)

$$I_C \cong I_E$$

(e)

$$V_{CE} = V_{CC} - I_C R_C - I_E R_E = V_{CC} - I_E(R_C + R_E)$$

Emitter bias

Figure 14-7 illustrates an **emitter-bias circuit**. R_E is selected to provide the proper forward bias. R_B is small compared to R_E. Most of the voltage drop is across R_E. B is close to ground potential. Dc emitter-base voltage V_{EB} is a *nearly* constant 0.7 V for a silicon transistor. (V_{EB} is subject to slight variations in any given circuit.) Thus, V_E is more positive than V_B and the EB junction of this pnp transistor is forward biased. Electron flow is from the negative collector supply, V_{CC}, through the collector and out of the emitter to the positive emitter supply, V_{EE}. As in voltage-divider bias, an increase in collector current is compensated for by a decrease in base current, which limits the collector current.

A signal applied to the amplifier will produce an ac component in the collector current. However, this does not upset the emitter bias, because a low-reactance path around R_E is provided by bypass capacitor C_3.

Typically, for this circuit, R_B values between 10 kΩ but less than ten times R_E are chosen. This circuit almost guarantees a stable operating point for the transistor. However, this advantage is overshadowed by the need for two dc voltage sources.

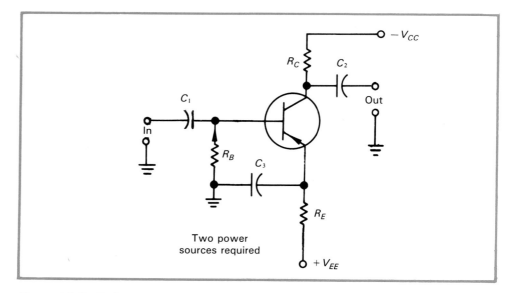

Figure 14-7. Emitter bias.

Analysis of this circuit for the linear region is aided by the following relationships:

(a)
$$V_B \cong 0$$

(b)
$$V_E \cong -V_{EB}$$

(c)
$$I_E = \frac{V_E - V_{EE}}{R_E}$$

(d)
$$I_C \cong I_E$$

(e)
$$V_C = V_{CC} - I_C R_C$$

(f)
$$V_{CE} = V_C - V_E$$

Collector-feedback bias

Figure 14-8 shows a **collector-feedback-bias circuit**. This circuit differs from the base-bias circuit in that bias resistor R_B is connected to the collector rather than V_{CC}. This scheme provides a more stable quiescent level than the base bias does. It, too, requires only one power source.

Dc collector voltage, V_C, provides the bias for the EB junction. Its value varies, so dc base voltage, V_B, varies; whereas, V_B was constant in the base-bias circuit. An increase in temperature and leakage causes collector current to increase. An increase in I_C causes an increased voltage drop across R_C. This lowers the magnitude of the collector voltage. This, in turn, reduces I_B, which limits the increase in I_C.

Since B is connected to C (through R_B), a certain amount of **negative feedback** will result. This is a process by which a part of the output signal of a CE amplifier is fed back to the input circuit. The signal fed back is 180° out of phase with the input signal. The two voltages are opposing; thus, the input voltage is *reduced* by the amount of the feedback voltage. This arrangement reduces **distortion**, an undesired change in the waveform of the original signal, by stabilizing I_C. However, it also reduces amplifier gain. From this standpoint, negative feedback can be a disadvantage. In this context, negative feedback is sometimes referred to as **degeneration**.

Analysis of this circuit for the linear region is aided by the following relationships:

(a)

$$I_B = \frac{V_C - V_{EB}}{R_B}$$

(b)

$$V_C \cong V_{CC} - I_C R_C$$

(c)

$$I_B = \frac{I_C}{\beta_{dc}}$$

(d)

$$I_C = \frac{V_{CC} - V_{EB}}{R_C + R_B/\beta_{dc}}$$

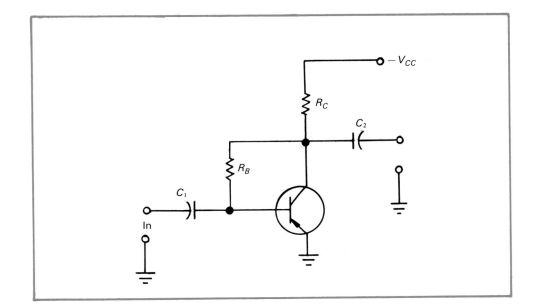

Figure 14-8. Collector-feedback bias.

Combination bias

This type of circuit is shown in Figure 14-9. It is a combination of emitter-bias and collector-feedback-bias circuits. The combination makes it a stable bias circuit.

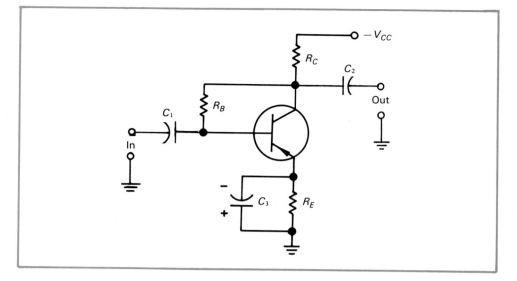

Figure 14-9. Combination bias.

LOAD LINES

A **load line** graphically depicts the performance of an amplifier circuit under loaded conditions. It shows how output current changes with input voltage when a specific load resistance is used. The line is drawn on a plot of collector characteristics between two extremes of operation—cutoff and saturation. The **dc load line** is a plot of dc values. For a CB amplifier, this plot is of I_C versus V_{CB} when I_E varies and V_{CC} is held constant. For a CE amplifier, it is of I_C versus V_{CE} when I_B varies and V_{CC} is held constant. Figure 14-10 is an example for a typical CE-configured transistor.

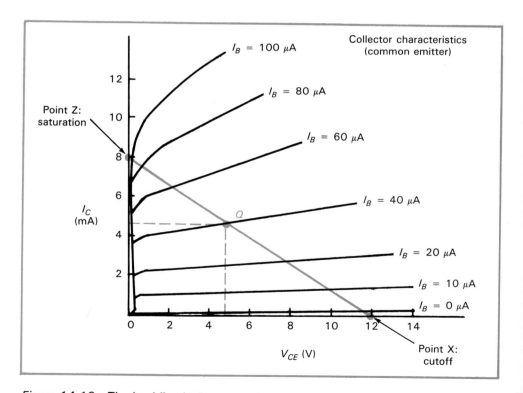

Figure 14-10. The load line is drawn on the family of collector characteristic curves.

Figure 14-11 shows a CE amplifier circuit. At cutoff, except for a very small leakage current, no current flows through the collector. An open circuit, in effect, exists across the collector and emitter terminals. Most of the voltage drop in this circuit loop, then, is across these two terminals. Thus, V_{CE} is close to the value of V_{CC}.

At saturation, the opposite is true. There is, in effect, a short circuit across the collector and emitter terminals. Most of the voltage drop at saturation is across the collector resistor R_C. Voltage across collector and emitter terminals is close to zero. Current through the collector, then, is $I_C = V_{CC}/R_C$. Knowing these relationships, the endpoints of the load line can be determined.

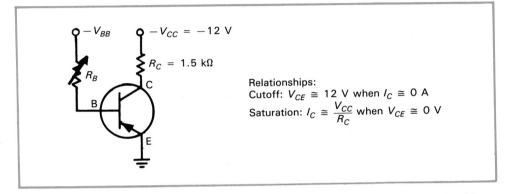

Figure 14-11. The CE circuit used with characteristic curves to establish dc load line.

The data for the load line in Figure 14-10 is based on the CE amplifier circuit shown in Figure 14-11. Follow the computations to see how endpoints X and Z were found.

1. At the point of zero collector current, the voltage drop across R_C is equal to zero, and the collector-emitter voltage V_{CE} is close to V_{CC} (12 V). This cutoff point, where $I_C = 0$ A and $V_{CE} = 12$ V, is plotted in Figure 14-10 as point X.

2. Using collector resistance $R_C = 1500 \ \Omega$, the opposite end of the load line may be found. At this point, there is maximum collector current. The assumption is made that the transistor has zero resistance in full conduction, and the current is limited only by R_C. With maximum current, the voltage drop across R_C equals V_{CC}, and V_{CE} equals zero. The current, then, is:

$$I_C = \frac{V_{RC}}{R_C} = \frac{V_{CC}}{R_C} = \frac{12 \text{ V}}{1500 \ \Omega} = 0.008 \text{ A} = 8 \text{ mA}$$

This saturation point is plotted as point Z on Figure 14-10, using $V_{CE} = 0$ V and $I_C = 8$ mA.

3. A straight line is drawn connecting the two points (X and Z). This is the dc load line. The slope of the line is determined by the load resistance. At any point along this line, values of I_B, I_C, and V_{CE} can quickly be determined.

4. Once the load line is drawn on the family of curves, a suitable *operating point* can be determined. The **dc operating point**, or **Q-point (quiescent**

point), is the point representing the condition existing when there is no input signal. For the CE amplifier, it is defined by I_C and V_{CE}. In most cases, the selected Q-point should fall on or near the midpoint of the load line. At this position, an amplifier is said to be **midpoint biased** and is set up for optimum ac operation. At any rate, the Q-point is selected with the purpose of the amplifier in mind and by reference to the transistor specifications.

In Figure 14-10, the operating point is selected at $I_B = 40 \ \mu A$. Q-point, then, is the point where I_B and the load line cross. A vertical line dropped from this point to the axis will show V_{CE}. A horizontal line extended left to the current axis will show the quiescent current at this Q-point. From the graph, we see, then, that V_{CE} is about 5 V and I_C about 4.75 mA.

In Figure 14-12, the circuit is redrawn to include emitter resistor R_E. Now, the total resistance would be used to plot the dc load line. The total resistance would also be used for the *ac load line*, unless it is bypassed by a low reactance capacitor, shown as C_1 in Figure 14-12. In this case, only the resistance R_C would be considered; R_E would be ignored. (An **ac load line** is a plot of ac values. For the CE amplifier, it is of i_c versus v_{ce}. It differs from the plot of the dc load line.)

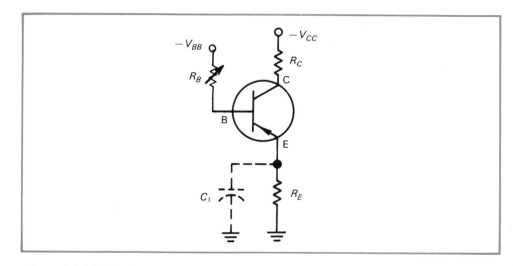

Figure 14-12. R_E is not considered part of circuit resistance for ac if bypassed by C.

Distortion of the output

It was mentioned that the Q-point should be selected to fall at about the midpoint of the load line. This is done when the output signal is to be maintained as an amplified replica of the input signal. This relationship defines a category of amplifiers known as **linear amplifiers**. Stereo amplifiers, which are linear amplifiers, do this so that music will be duplicated with minimal distortion. Selecting an operating point too near the saturation or cutoff points of the load line can result in *limiting* of the signal, which is a form of output distortion. An excessively large input signal can also result in distortion. Let us see how.

Figures 14-13, 14-14, and 14-15 show what happens to transistor outputs for different Q-points. The graphs show the transistor load line and ac input and output signals. The signals are plotted off the load line. Quiescent levels are projected from the Q-point. Maximum and minimum levels are also projected. Maximum output signals cannot exceed saturation or cutoff levels.

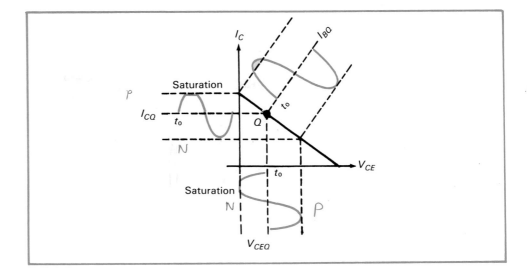

Figure 14-13. Transistor driven into saturation.

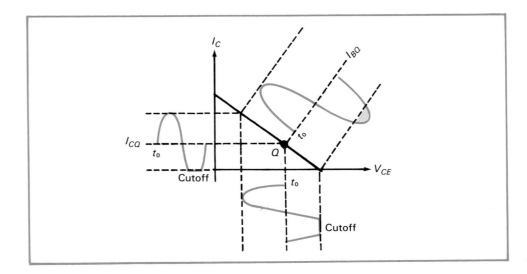

Figure 14-14. Transistor driven into cutoff.

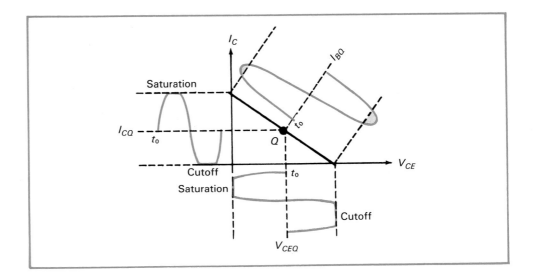

Figure 14-15. Transistor driven into both saturation and cutoff.

Figure 14-13 shows the graph of a transistor driven into saturation. Note that the positive peak of the output current signal and the negative peak of the output voltage signal have been clipped off. At this point, the transistor has reached its maximum output even though the input signal continues to increase. The selected operating point is too close to the saturation point.

Figure 14-14 shows the graph of a transistor driven into cutoff. Note that the negative peak of the output current signal and the positive peak of the output voltage signal have been clipped off. Again, the transistor has reached a maximum output before the input signal has peaked. The selected Q-point is too near the cutoff point.

Figure 14-15 shows the graph of a transistor driven into saturation *and* cutoff. Both negative and positive peaks are clipped off before the input signal peaks. This transistor is midpoint biased; however, the input signal is too large. All three transistor outputs have been distorted.

Output distortion can also occur by operating an amplifier in a region that is *nonlinear*. See Figure 14-16. A different type of graph is shown in Figure 14-17. It shows a selected operating point at 100 μA of base current. At this base current and a constant collector-emitter voltage of 6 V, the collector current is 3 mA. A 10-μA change in signal input causes a 1-mA change in signal output on either side of quiescence. This curve shows a linear relationship between I_B and I_C.

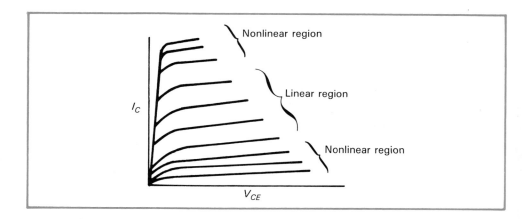

Figure 14-16. Linearity of output curves. Note in nonlinear regions that I_B lines are not equally spaced.

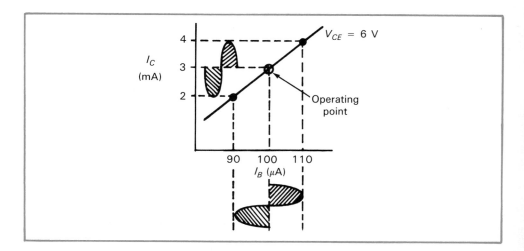

Figure 14-17. The operating point is selected on the linear portion of the curve.

This is not always the case. Observe Figure 14-18 in which part of the curve is nonlinear. If the operating point were moved to this portion of the curve, the signal output, i_C, would become very distorted. Obviously, if reliable operation of a transistor is to be expected, the operating point must be stable and must not stray into the nonlinear region of operation.

HEAT AND THE TRANSISTOR

The worst enemy of a transistor is heat. Heat can cause an undesired shift in Q-point. It can also cause *thermal runaway*. Let us look at both cases.

Heat and Q-point

Heat energy, when added to a semiconductor, raises the energy level of electrons and produces many electron-hole pairs. The p crystal is affected because heat energy increases the number of free electrons, or minority carriers, in it. In the n crystal, many holes, or minority carriers, are created.

Refer to a formula from Chapter 13:

$$I_C = \alpha_{\text{dc}}I_E + I_{CBO}$$

I_{CBO}, as you know, represents the current due to minority carriers across the reverse-biased CB junction. Now, the value of α_{dc} remains rather constant for a given transistor. Most variations are the result of the manufacturing process. There is little we can do about that. Heat will raise the collector current by raising I_{CBO}. We conclude that heat can change the design Q-point of a transistor amplifier.

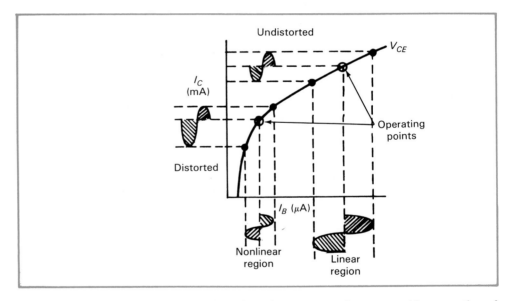

Figure 14-18. This graph shows distortion of output waveform caused by operation of amplifier in nonlinear region.

Heat and thermal runaway

Consider a transistor in a common-base circuit. Semiconductors have a negative temperature coefficient; thus, with a rise in temperature, resistance decreases, and leakage current I_{CBO} increases. I_C increases. The increase in I_C creates more heat, further reducing resistance and further raising I_{CBO}. This, again, raises I_C, which creates more heat, and the cycle repeats until the transistor is destroyed by this **thermal runaway**.

Increasing temperatures will also cause the emitter current to increase. The net effect may be dampened by placing a relatively large **swamping resistor** before the emitter lead. See Figure 14-19. With the swamping resistor, the effect on I_E due to decreasing resistance and, also, on I_C ($I_C = \alpha_{dc}I_E$) will be minimized. In other words, the swamping resistor masks the effect of changing semiconductor resistance caused by changing temperature.

STABILITY FACTOR

The previous discussions give rise to a new term called the *stability factor* of a transistor circuit. The reasons that a circuit should be stable, both from the voltage and current points of view, have been illustrated. The **stability factor (S)** is a measure of the bias stability of an amplifier. It is defined as the ratio between a change in collector current to a change in collector cutoff current. In equation form:

$$S = \frac{\Delta I_C}{\Delta I_{CBO}}$$

S is a unitless number. The ideal circuit would have a stability factor of zero, which would mean that any change in I_{CBO} would have no effect on collector current. In other words:

$$S = \frac{0}{\Delta I_{CBO}} = 0$$

In practice, the design engineer attempts to keep the stability factor under 5, except in highly sophisticated types of circuits. In such circuits, stability factors of much better than 5 are required.

The formula for S can be derived mathematically to give a very useful approximate formula for the value of S:

$$S = \frac{R_F R_B}{R_E(R_F + R_B)} + 1$$

where R_F, R_B, and R_E are values of bias resistors. Note that the derivation of this formula is beyond the scope of this text.

If the value of R_F is ten times greater than R_B, then the equation may be further simplified to:

$$S = \frac{R_B}{R_E} + 1$$

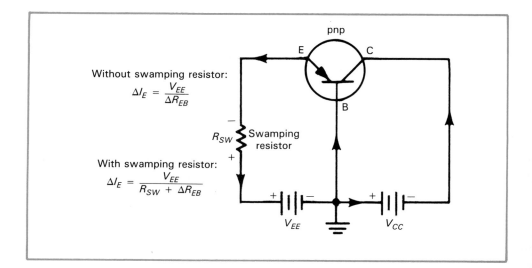

Figure 14-19. A swamping resistor helps to stabilize operation.

This approximate equation will always give a slightly worse stability factor than actually exists, so you may assume your circuit will be better.

AMPLIFIER ANALYSIS

In this section, we will analyze the characteristic curves and load line to graphically obtain amplifier ac operation data. Then, we will analyze a circuit and see how to determine bias resistors needed for proper dc biasing of an amplifier. Our analysis will be based on the load line of Figure 14-10.

Common-emitter amplifier graphical analysis

Earlier in this chapter, we plotted a load line for a typical transistor. Now, by continuation of this problem, assume that an input signal is applied such that the base current is caused to vary between a 30-μA minimum and a 50-μA maximum. These points are marked A and B in Figure 14-20.

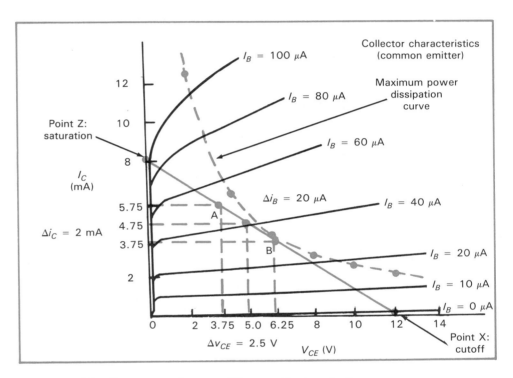

Figure 14-20. Common-emitter amplifier graphical analysis.

By projecting down to the voltage scale, V_{CE} at point A is about 3.75 V. At point B, V_{CE} is about 6.25 V. The change in collector-emitter voltage due to signal input is:

$$\Delta V_{CE} = \Delta v_{CE} = 6.25 \text{ V} - 3.75 \text{ V} = 2.5 \text{ V}$$

By projecting horizontally to the current scale, I_C at point A is about 5.75 mA. At point B, I_C is about 3.75 mA. The change in collector current due to signal input is:

$$\Delta I_C = \Delta i_C = 5.75 \text{ mA} - 3.75 \text{ mA} = 2 \text{ mA}$$

Also, the input signal has caused a change in base current:

$$\Delta I_B = \Delta i_B = 50 \text{ }\mu\text{A} - 30 \text{ }\mu\text{A} = 20 \text{ }\mu\text{A}$$

So, an input signal of 20 μA$_{\text{p-p}}$ at the base produces an output signal of 2 mA$_{\text{p-p}}$ at the collector. Stated another way, a change in base current i_B of 20 μA produced a change in collector current i_C of 2 mA.

Current gain. Now that we have found both input and output currents, the gain may be calculated. Current gain, you may recall, is the ratio of output current to input current. Current gain for the transistor in the CE circuit is approximated by β—the ratio of changes in collector to base currents:

$$A_i \cong \beta = \frac{\Delta i_C}{\Delta i_B} = \frac{2 \text{ mA}}{20 \text{ }\mu\text{A}} = \frac{2 \times 10^{-3} \text{ A}}{20 \times 10^{-6} \text{ A}} = 100$$

Voltage gain. Assuming a typical input impedance of 500 Ω, the change in voltage required to produce a change of base current can be found by Ohm's law:

$$\Delta v_{EB} = \Delta i_B \times R_E = 20 \times 10^{-6} \text{ A} \times 500 \text{ }\Omega = 0.01 \text{ V}$$

The ac input voltage of $v_{eb} = 0.01 \text{ } V_{\text{p-p}}$ produces the output voltage of $v_{ce} = 2.5 \text{ } V_{\text{p-p}}$. Stated another way, a change in the input voltage v_{EB} of 0.01 V produced a change in the output voltage v_{CE} of 2.5 V. Voltage gain can be stated as:

$$A_v = \frac{v_{\text{out}}}{v_{\text{in}}} = \frac{v_{ce}}{v_{eb}} = \frac{2.5 \text{ } V_{\text{p-p}}}{0.01 \text{ } V_{\text{p-p}}} = 250$$

NOTE: Until now, we have seen voltage across the EB junction to be about 0.7 V (for silicon). However, this is the *dc* voltage drop across the junction, V_{EB}. The *ac* input signal voltage is v_{eb}. Its value is whatever the value of the input signal is. Generally, it is much smaller than 0.7 V. Finally, v_{EB} is the composite ac and dc emitter-base voltage. See Figure 14-21.

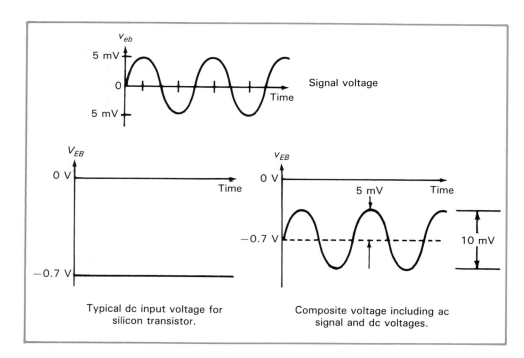

Figure 14-21. Voltages at emitter-base junction.

Power gain. Knowing the current and voltage gains, the power gain may be found:

$$A_p = A_i A_v = 100 \times 250 = 25,000$$

Experiments have shown that sound intensity must change by at least a factor of 2 before a change in level can be detected by the human ear. For example, if an amplifier puts out 1 W of power, the change in volume intensity could be distinguished at 0.5 W or 2 W. The relationship is a logarithmic one. It is more convenient to discuss levels of intensity in relative terms rather than actual levels, especially when a power level or intensity varies over a wide range. For this reason, gain in a circuit, whether it is current, voltage, or power, is commonly expressed in **decibels (dB)**. These units are defined by a logarithmic relationship between output and input. The intensity of sound is commonly expressed in decibels. Power, in decibels, is given by:

$$A_p \text{ (in dB)} = 10 \log \left(\frac{P_{\text{out}}}{P_{\text{in}}} \right) = 10 \log A_p$$

Expressing our power gain of 25,000, then:

$$A_p = 10 \log 25{,}000 = 10 \times 4.4 = 44 \text{ dB gain}$$

Note that 44 dB gives us a *ratio* of output power to input power. We cannot tell by this what our actual output power is when we do not know what the input power is. For this reason, on some specification sheets, you will see units of *dBm* (called **dBm reference**) giving the power values. This is a standard way of specifying output power. It gives power relative to a reference level (P_{in}) of 1 mW. When units of dBm are given, actual output power in watts may be determined. Thus, these units have more meaning than either power gain or output power (in watts) by themselves would express. For example, we could tell that an amplifier listed as having an output power of 44 dBm would have an actual power output of 25.1 W. The mathematics is as follows:

$$A_p \text{ (in dB)} = 10 \log \left(\frac{P_{\text{out}}}{P_{\text{in}}} \right)$$

Substituting values:

$$44 \text{ dBm} = 10 \log \frac{P_{\text{out}}}{1 \text{ mW}}$$

Transposing and simplifying the equation:

$$\frac{44 \text{ dBm}}{10} = \log \frac{P_{\text{out}}}{1 \text{ mW}}$$

$$4.4 = \log \frac{P_{\text{out}}}{1 \text{ mW}}$$

To find antilog of 4.4:

$$10^{4.4} = \frac{P_{\text{out}}}{1 \text{ mW}}$$

With a calculator, we easily find that $10^{4.4} = 25{,}118$. Therefore, solving for P_{out} yields:

$$P_{\text{out}} = (25{,}118.86)(0.001 \text{ W}) = 25.1 \text{ W}$$

Maximum power-dissipation curve. All transistors are rated according to their ability to dissipate power at a given ambient temperature. If this maximum power rating is exceeded, the transistor may be destroyed. Remember, too, that if a transistor is to be operated above the standard temperature of 25°C (77°F), its ability to dissipate power is derated according to a scale supplied by the manufacturer.

Power dissipation (P_D) of a transistor may be calculated at any given point on an output characteristic curve. It is found using the power formula: $P = IV$; more specifically for the CE amplifier, $P_D = I_C V_{CE}$. The **maximum power-dissipation curve** (Figure 14-20) delimits the region in which a transistor may be operated. The curve is a plot of I_C versus V_{CE} for a given *maximum power-dissipation rating*, $P_{D_{max}}$. P_D cannot exceed $P_{D_{max}}$.

For example, the maximum power-dissipation curve of Figure 14-20 is plotted for values of I_C and V_{CE} assuming a $P_{D_{max}} = 25$ mW. Given this rating, we find the permissible current for any value of V_{CE}. To do this, the power formula is transposed to:

$$I_C = \frac{P_{D_{max}}}{V_{CE}}$$

Substituting values from our example:

$$I_C \text{ (mA)} = \frac{25 \text{ mW}}{V_{CE}}$$

Selecting different values of V_{CE} yields the following:

For $V_{CE} = 12$ V, $I_C = 2.1$ mA
$V_{CE} = 10$ V, $I_C = 2.5$ mA
$V_{CE} = 8$ V, $I_C = 3.1$ mA
$V_{CE} = 6$ V, $I_C = 4.2$ mA
$V_{CE} = 4$ V, $I_C = 6.3$ mA
$V_{CE} = 2$ V, $I_C = 12.5$ mA

These points of maximum power dissipation are plotted in Figure 14-20, and a smooth curve is drawn through them. The transistor can be damaged if the coordinates given by the current I_C and the voltage V_{CE} yield a point falling above this line. At such a point, P_D, the product of current and voltage will exceed $P_{D_{max}}$.

If maximum power is to be realized from a transistor, the load line should be tangent to the maximum power-dissipation curve. The resistance load R_L that will produce this line is found by drawing a load line from the cutoff voltage tangent to the curve. (This is 12 V in our example.) It will intersect the current scale at some value of I_C. (This would be about 8.25 mA in our example.) This is the current at saturation. In the circuit (refer to Figure 14-11):

$$V_{RC} = V_{R_L} = V_{CC} - V_{CE}$$

At saturation:

$$V_{CE} = 0 \text{ V}$$

Therefore:

$$V_{R_L} = V_{CC}$$

The load resistance then will be:

$$R_L = \frac{V_{CC}}{I_{C_{sat}}}$$

In our example, if the load line were tangent to the maximum power-dissipation curve:

$$R_L = \frac{12 \text{ V}}{8.25 \text{ mA}} = 1455 \ \Omega$$

Common-emitter amplifier circuit biasing

The previous characteristic curves and computations will be used as the basis for a voltage-divider-bias circuit. The circuit uses a silicon transistor and has a stability factor of 5. Examine Figure 14-22 and be sure you understand all voltages around the circuit.

The collector current at our selected Q-point is 4.75 mA. In general, emitter current is approximately equal to collector current. If we assume base current is zero, then:

$$I_E = I_C = 4.75 \text{ mA}$$

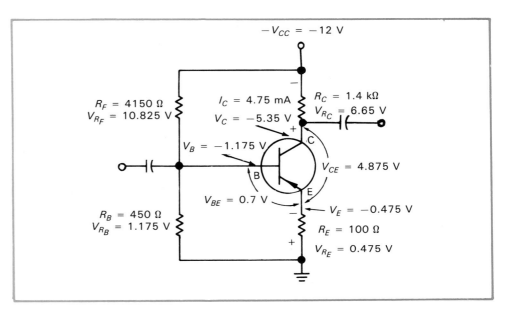

Figure 14-22. Common-emitter stabilized amplifier.

In the collector-emitter circuit, the load has been divided between R_E and R_C. We need R_E to assure better stability. Voltage drop across R_E is:

$$V_{R_E} = I_E R_E = 0.00475 \text{ A} \times 100 \text{ } \Omega = 0.475 \text{ V}$$

E is more negative than ground by the amount of voltage drop across R_E. V_E, the potential at E, then, is -0.475 V. Voltage drop across R_C is:

$$V_{R_C} = I_C R_C = 0.00475 \text{ A} \times 1400 \text{ } \Omega = 6.65 \text{ V}$$

C is less negative than $-V_{CC}$ by the amount of voltage drop across R_C. V_C, the potential at C, then, is:

$$V_C = -V_{CC} + V_{R_C} = -12 \text{ V} + 6.65 \text{ V} = -5.35 \text{ V}$$

Voltage drop across the collector and emitter leads, V_{CE}, is:

$$V_{CE} = V_E - V_C = -0.475 \text{ V} - (-5.35 \text{ V}) = -0.475 \text{ V} + 5.35 \text{ V} = 4.875 \text{ V}$$

We can check this by applying Kirchhoff's voltage law:

$$V_{R_C} + V_{CE} + V_{R_E} = V_{CC}$$

Transposing:

$$-V_{CC} + V_{R_C} + V_{CE} + V_{R_E} = 0 \text{ V}$$

Substituting:

$$-12 \text{ V} + 6.65 \text{ V} + 4.875 \text{ V} + 0.475 \text{ V} = 0 \text{ V}$$

Simplifying:

$$0 \text{ V} = 0 \text{ V}$$

The left side of the equation adds up to zero as it is supposed to. We have figured the correct values.

The emitter-base junction must be forward biased. For the silicon transistor, the voltage drop across the EB junction is typically about 0.7 V. Therefore, V_B must be 0.7 V more negative than V_E, or at a potential of -1.175 V with respect to ground. V_{RB}, the voltage drop across R_B, then, is 1.175 V. By applying Kirchhoff's voltage law:

$$-V_{CC} + V_{RF} + V_{RB} = 0 \text{ V}$$

Transposing, substituting, and solving:

$$V_{RF} = V_{CC} - V_{RB} = 12 \text{ V} - 1.175 \text{ V} = 10.825 \text{ V}$$

It is convenient now to find the voltage ratio between V_{RF} and V_{RB}:

$$\text{voltage ratio} = \frac{V_{RF}}{V_{RB}} = \frac{10.825 \text{ V}}{1.175 \text{ V}} = 9.21$$

Applying the formula for stability factor, we can find the resistance of R_B:

$$S = \frac{R_F R_B}{R_E(R_F + R_B)} + 1$$

Rearranging the formula yields:

$$S - 1 = \frac{R_F R_B}{R_E(R_F + R_B)}$$

$$R_E(S - 1) = \frac{R_F R_B}{R_F + R_B}$$

From our voltage ratio, we know that:

$$R_F = 9.21 R_B$$

Therefore, substituting values into the stability formula yields:

$$100 \ \Omega \ (5 - 1) = \frac{(9.21 R_B) R_B}{9.21 R_B + R_B}$$

Simplifying:

$$400 \ \Omega = \frac{9.21 R_B^2}{R_B(9.21 + 1)} = \frac{9.21 R_B}{9.21 + 1} = \frac{9.21 R_B}{10.21}$$

Solving for R_B:

$$R_B = \frac{10.21(400 \ \Omega)}{9.21} \cong 450 \ \Omega$$

From our voltage ratio, we know that R_F is 9.21 times greater than R_B. Thus:

$$R_F = 9.21 \times 450 \ \Omega \cong 4150 \ \Omega$$

As further proof of the voltage division:

$$V_{RF} = \left(\frac{R_F}{R_B + R_F}\right) V_{CC} = \left(\frac{4150\ \Omega}{450\ \Omega\ +\ 4150\ \Omega}\right) \times 12\ \text{V}$$

$$= \left(\frac{4150\ \Omega}{4600\ \Omega}\right) \times 12\ \text{V} = 10.83\ \text{V}$$

$$V_{RB} = \left(\frac{R_B}{R_B + R_F}\right) V_{CC} = \left(\frac{450\ \Omega}{4600\ \Omega}\right) \times 12\ \text{V} = 1.17\ \text{V}$$

CLASSES OF AMPLIFIERS

Amplifiers may be classified in a variety of ways: by range of operating frequency; by area of application; as voltage or current amplifier; as small-signal or large-signal (power) amplifier. They may also be classified according to their bias current, or quiescent level. This determines the portion of the input signal for which there is an output current. *Power amplifiers*, in particular, are classified this way. You will learn more about these in the next chapter.

Up until now, discussion has focused on the amplifier whose output is an amplified replica of the input signal. However, it is sometimes desirable to use only a portion of the input signal. This is a matter of Q-point selection. Amplifiers are sometimes classified accordingly to Q-point. They may belong to one of the following categories:

- *Class-A amplifier.*
- *Class-B amplifier.*
- *Class-AB amplifier.*
- *Class-C amplifier.*

Class-A amplifier

The **class-A amplifier** is biased so that an output signal is produced for 360° of the input signal. This applies to CB, CE, and CC circuit configurations. See Figure 14-23. The output of a class-A amplifier is an amplified reproduction of the input signal. I_C flows all of the time. Amplifier **collector efficiency**

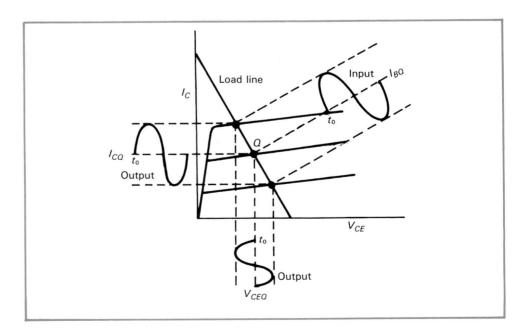

Figure 14-23. A class-A amplifier has a 360° signal output for a 360° signal input. Note that the voltage signal is 180° out of phase with the current signals, typical of a CE amplifier.

(η_C), which is the ratio of ac output power to dc input power, is low—25% at most (or 50% if transformer coupled). This class of amplifier produces a high quality of amplification. It is widely used in high-fidelity sound systems.

Class-B amplifier

The **class-B amplifier** is biased so that an output signal is produced during 180° of signal input and is in cutoff for 180°. See Figure 14-24. The class-B amplifier is biased at cutoff. Its output is distorted; I_C flows only one-half of the time. It is similar to a half-wave rectifier.

A two-transistor configuration, called a **push-pull amplifier**, can be used to get a sufficiently good reproduction of the input waveform. The push-pull circuit (discussed in Chapter 15 in greater detail) more or less restores both halves of the signal output. Improved efficiency and greater ac power output result. The amplifier has a much higher efficiency than the class A; ideally, as much as 78.5%. Efficiency is high because the quiescent current is small. A push-pull amplifier is the final output stage in many audio amplifiers.

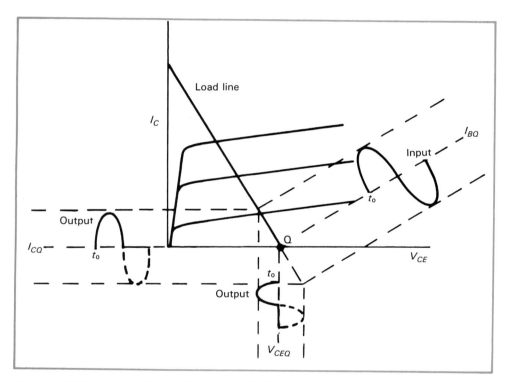

Figure 14-24. A class-B amplifier has a 180° signal output for a 360° signal input.

Class-AB amplifier

A class-AB amplifier is biased between the class-A operating point and the class-B operating point. Current flows more than one-half of the time but not all the time. A phenomenon called *crossover distortion* is reduced at the expense of a small decrease in efficiency. (Crossover distortion is discussed in Chapter 15.)

Class-C amplifier

A class-C amplifier conducts only during a small portion of the input signal. It is biased so that less than 180° of output signal is produced for 360° of input signal. This amplifier is of no use in linear applications because the output does not replicate the input. It is used mostly as a radio-frequency amplifier and for providing energy to oscillators and switching circuits. Maximum collector efficiency of this amplifier is about 99%.

OTHER AMPLIFIER DEVICES

Discussion up until now has been restricted to the bipolar junction transistor. There are other devices besides the BJT that are used for amplifying including the *field-effect transistor* and the *unijunction transistor*. These are explained in this section.

FIELD-EFFECT TRANSISTORS

There are two basic types of transistors. The BJT, which you are now familiar with, is one type. The other is the **field-effect transistor (FET)**. It, like the BJT, is a three-element semiconductor device. The three elements of the FET are the *source*, the *drain*, and the *gate*. The **source (S)** is the input terminal. It compares to the BJT emitter. The **drain (D)** is the output terminal. It compares to the BJT collector. Voltage at the **gate (G)** controls the amount of current through the device. This voltage sets up an electrical field within the device. (FETs are voltage-driven devices; whereas, BJTs are current-driven.) The gate compares to the BJT base.

Unlike the BJT, the FET is a **unipolar device**. This means that either electrons or holes, but not both, act as current carriers within the device; whereas, with the BJT, a **bipolar device**, current is carried by *both* electrons *and* holes. FETs use both n-type and p-type materials. Depending on their makeup, they are classified as either *n-channel* or *p-channel*. (This will be discussed shortly.)

The FET may serve as an amplifier or a switch. A major advantage of the FET over the BJT is the high input impedance of the FET, which translates into better power amplification. For this reason, the FET can operate on low power. There are two main types of FETs: the *junction field-effect transistor*, and the *metal oxide semiconductor field-effect transistor*.

Junction field-effect transistors

The basic construction of the **junction field-effect transistor (JFET)** is shown in Figure 14-25. The drain and source are connected by a common material that constitutes the **channel**. It serves as the conductive pathway between these two electrodes. A JFET having a channel made of n-type material is an **n-channel JFET**. One having a channel made of a p-type material is a **p-channel JFET**. Two regions of doped material are fused to opposite sides of the channel. Both are connected to the gate electrode. In an n-channel JFET, these

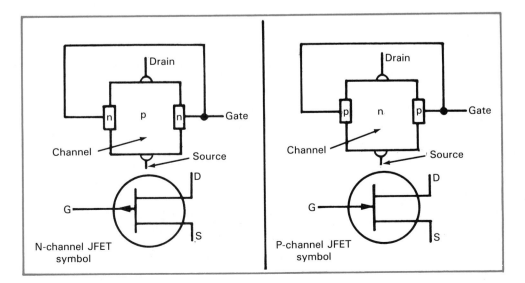

Figure 14-25. Construction and symbols for JFETs.

regions are of a p-type material. In a p-channel JFET, they are of an n-type material.

To understand the operation of the JFET, refer to Figure 14-26. It features an n-channel JFET. P-channel operation is the same except the voltage polarities are reversed. The drain is connected through resistor R_D to the positive lead of the *drain-supply voltage (V_{DD})*. The source is connected to the negative lead of V_{DD}. Reverse, *never* forward, bias is provided to the pn junctions between the gate and the source through the *gate-supply voltage (V_{GG})*. Depletion regions are formed at these junctions. Current will not flow through the depletion regions. As the current flows from source to drain *(drain current, I_D)*, it is restricted here. As V_{GG} is increased and the regions get larger (for constant V_{DD}), the current decreases and vice versa. At some point, reverse bias voltage will almost completely cut off current from source to drain.

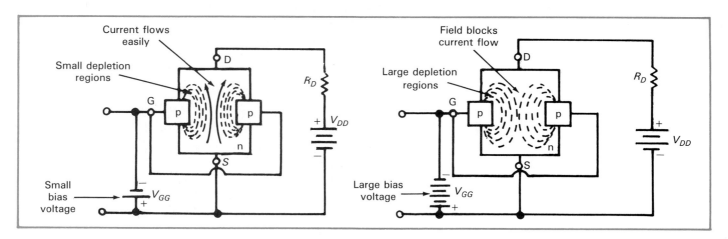

Figure 14-26. Operation of an n-channel JFET.

The voltage applied across the gate-source leads at this point is called the **cutoff voltage,** or $V_{GS(off)}$.

A graph showing the output characteristics of a typical JFET is shown in Figure 14-27. The family of curves show *drain current I_D* as a function of *drain-source voltage V_{DS}* for fixed values of reverse-biased *gate-source voltage V_{GS}*. Figure 14-28 shows the circuit used to obtain the data for plotting the curve. Compare it to the circuit of Figure 14-26 and see that they are essentially the same.

From the graph, we see I_D increasing linearly in the *ohmic region*. In this region, V_{DS} and I_D are related by Ohm's law. The channel resistance is essentially constant because the depletion region is not large enough to have much effect.

In the *pinch-off region*, I_D becomes relatively constant. Here, the reverse-biased *gate-source voltage V_{GS}* produces a depletion region to effectively narrow the channel. Channel resistance begins to increase in proportion with V_{DS}, keeping current constant. The reason for this is that electron flow, in itself, causes depletion regions at the pn junctions. The size of these regions varies directly with the amount of flow. This is why we have a pinch-off region, even for the zero-biased device, and why the resistance increases with V_{DS}.

The voltage marking the beginning of the pinch-off region is the **pinch-off voltage (V_p)**. Specifically, it is the value at which I_D reaches a relatively constant value when V_{GS} = 0 V. The pinch-off region is the operating region

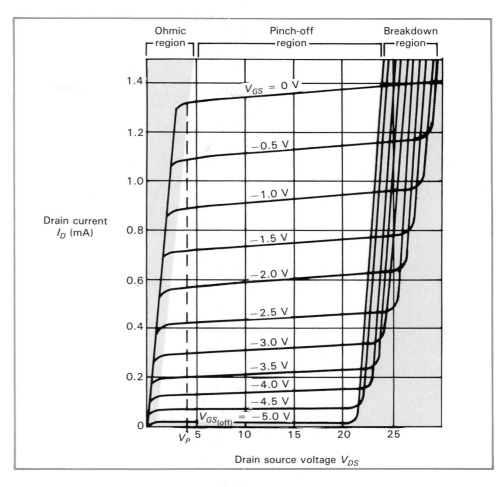

*Figure 14-27. JFET **drain characteristic curves**.*

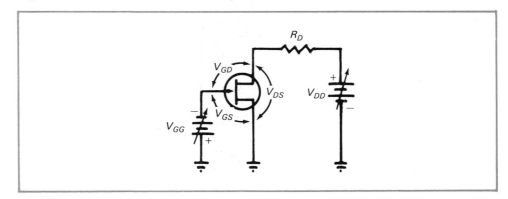

Figure 14-28. Circuit used to obtain drain characteristic curves.

of the device. By controlling the value of V_{GS}, the drain current is controlled.

In the *breakdown region*, I_D increases rapidly. In this region, damage to the device is likely.

Metal oxide semiconductor field-effect transistors

The **metal oxide semiconductor field-effect transistor (MOSFET)** is a three- or four-terminal device. The leads have the same designations as the JFET. In addition, the four-terminal device has a *substrate* lead. Like the JFET, the MOSFET has a channel whose effective width is controlled by the gate potential. Most FETs presently in use are MOSFETs. They have great application in digital computers.

There are two basic types of MOSFETs. One type is the **depletion-enhancement MOSFET**, or **DE MOSFET**. The other type is the **enhancement-only MOSFET**, or **E MOSFET**. Figure 14-29 shows their basic structure. The major difference is the presence of a physical channel connecting the source and drain in the DE MOSFET.

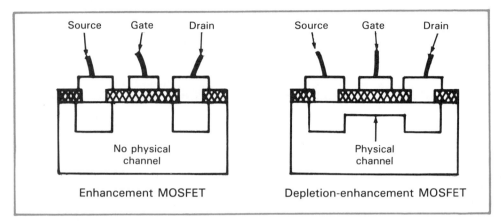

Figure 14-29. Difference between E MOSFET and DE MOSFET.

The MOSFET is constructed of three layers—metal, oxide, and semiconductor. See Figure 14-30. The foundation, usually referred to as **substrate**, is either a p-type or an n-type semiconductor material. The channel is a semiconductor material of the opposite type. The channel allows current from the source to the drain. The gate is the controlling element. It controls the flow of the electrons or holes from the source to the drain region. The source, drain, and gate are insulated from each other by a thin layer of silicon dioxide (SiO_2).

The DE MOSFET operates in one of two modes. Figure 14-30 shows it operating in the **depletion mode**. In this mode, a negative voltage is applied to the gate. This causes a large negative field, which, by repelling conduction electrons, restricts current from the negative source to the positive drain. Increasing the negative gate voltage increases the field. This further decreases the source-to-drain current. At some point, current will be cut off.

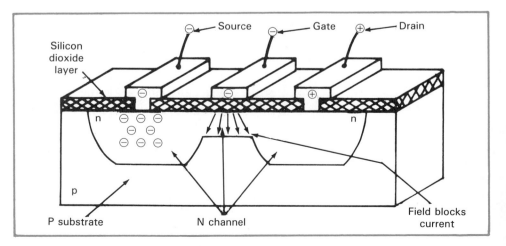

Figure 14-30. N-channel DE MOSFET shown blocking current from source to drain.

Figure 14-31 shows the DE MOSFET operating in the **enhancement mode**. In this mode, positive voltage is applied to the gate. This effectively widens the channel. Current from source to drain increases even beyond what it is with no voltage applied to the gate. This happens as electrons in the p-type substrate are attracted to the positive gate. They are drawn up toward the channel. This buildup of electrons near the channel effectively widens it, and more current is allowed. Thus, by varying the signal on the gate, the current from source to drain can be modulated or switched on and off.

The E MOSFET has no depletion mode. It operates only in the enhancement mode. It has no physical channel. For an n-channel device, a positive gate voltage above some threshold *induces* a channel. Increasing the voltage draws more electrons into the channel and increases source-to-drain current. Below the threshold value, there is no channel; thus, there is no current. Figure 14-32 shows symbols for n- and p-channel DE MOSFETs and E MOSFETs.

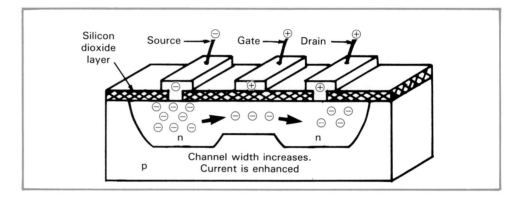

Figure 14-31. N-channel DE MOSFET shown allowing current from source to drain.

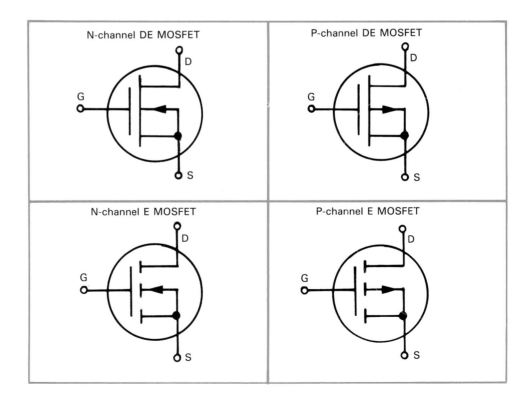

Figure 14-32. MOSFET symbols.

UNIJUNCTION TRANSISTORS

The **unijunction transistor (UJT)**, Figure 14-33, is a type of controlled diode with some amplifying capabilities. A small bar of lightly doped n-type material has a region of heavily doped p-type material within. The n-type material serves as the base. It has two leads — base 1 (B_1) and base 2 (B_2). The lead connected to the p-type material is the emitter. It is located closer to B_2 than B_1. A sufficient bias voltage, V_E, applied to it can trigger current in the emitter circuit.

The n-type material is uniformly doped. Its resistance is uniform. The voltage at any point between B_1 and B_2 is proportional to the length of crystal involved. We can think of the base section as a voltage divider made up of two resistances. One is R_{B_1}, the resistance between B_1 and E. The other is R_{B_2}, the resistance between B_2 and E. R_{B_1} is about 60% of the total resistance; R_{B_2} is about 40%. (Remember E is nearer B_2 than B_1.)

As a simplified example, assume 10 V are applied between B_1 and B_2. The base voltage at E (measured from B_1) would be $+6$ V. If a bias voltage of $+5$ V is applied at the emitter, E would be 1 V less positive than the base voltage at E. The pn junction would be reverse biased. Current in the emitter circuit would be almost zero.

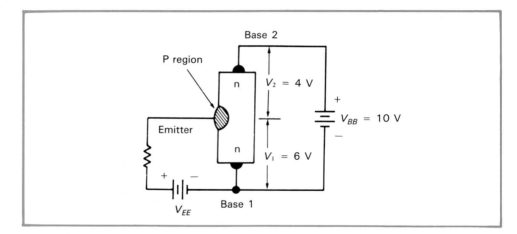

Figure 14-33. This circuit is used to explain the theory of a unijunction transistor.

If bias voltage V_E is increased to about $+7$ V, E would be about 1 V more positive than the base voltage at E. The junction would be forward biased. Under these conditions, holes are injected into the B_1 region of the base. The base is lightly doped, and the holes have little chance to recombine. The resistance of the B_1 region drops way down, as it is now full of current carriers (holes). The current increases as the resistance goes down, which results in the injection of more holes, which further decreases the resistance and further increases the current, and so on. Meanwhile, the voltage at E in the base material decreases as a result of the decrease in R_{B_1}, and a corresponding decrease is seen in the bias voltage V_E. This cycle continues until the device reaches a saturation point.

The unijunction transistor is classified as a *negative-resistance device*. Why is this so? Under the previous circumstances, the base resistance was effectively reduced between B_1 and E because of forward bias. This upset the voltage division. The base voltage at E became less. In addition, bias voltage V_E became less. The emitter current increased even though the applied emitter voltage decreased. This is **negative resistance**. The symbol for the unijunction transistor is shown in Figure 14-34.

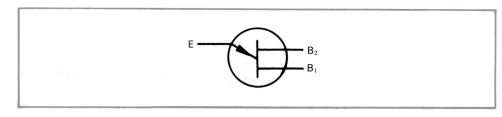

Figure 14-34. Symbol for the unijunction transistor.

APPLAUSE METER PROJECT

Perhaps you have seen applause meters used on television. This meter measures the loudness of noise or applause. The circuit and parts list for building an applause meter are shown in Figure 14-35.

In this circuit, the speaker is used as a microphone. It is matched to the amplifier circuit by transformer T_1. The applause gets converted to an electrical signal and enters the amplifier through this circuit. Potentiometer R_8 is the signal **attenuator**. It is used to adjust the amplitude of the input signal. (The most basic attenuator is a simple voltage divider.) The signal is amplified by three stages of amplification. The output of Q_3 is rectified by the bridge rectifier. The rectified signal registers on meter μA. (The bridge rectifier is not necessary if your meter has an ac current function.) The deflection of the meter will increase by the loudness of the applause. The circuit is easy to build using printed circuit techniques or a breadboard.

NOTE: A major cause of circuit failure in transistor circuits built by students is using incorrect voltage or polarity, or both. Check and recheck the polarity of your voltage source before connecting it to the circuit. A little flashlight cell can destroy a transistor if connected with the wrong polarity.

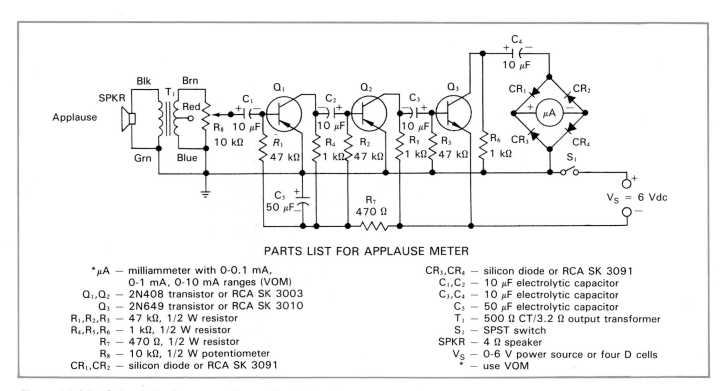

PARTS LIST FOR APPLAUSE METER

*μA — milliammeter with 0-0.1 mA, 0-1 mA, 0-10 mA ranges (VOM)	CR_3, CR_4 — silicon diode or RCA SK 3091
Q_1, Q_2 — 2N408 transistor or RCA SK 3003	C_1, C_2 — 10 μF electrolytic capacitor
Q_3 — 2N649 transistor or RCA SK 3010	C_3, C_4 — 10 μF electrolytic capacitor
R_1, R_2, R_3 — 47 kΩ, 1/2 W resistor	C_5 — 50 μF electrolytic capacitor
R_4, R_5, R_6 — 1 kΩ, 1/2 W resistor	T_1 — 500 Ω CT/3.2 Ω output transformer
R_7 — 470 Ω, 1/2 W resistor	S_1 — SPST switch
R_8 — 10 kΩ, 1/2 W potentiometer	SPKR — 4 Ω speaker
CR_1, CR_2 — silicon diode or RCA SK 3091	V_S — 0-6 V power source or four D cells
	* — use VOM

Figure 14-35. Schematic diagram and parts list for applause meter.

BASIC TRANSISTOR AMPLIFIER PROJECT

A transistor amplifier can be an excellent piece of equipment for the hobbyist or experimenter. See Figure 14-36. This amplifier has a simple design. The output is about 1.5 W. The schematic and parts list are shown in Figure 14-37. The layout for a printed-circuit board is shown in Figures 14-38 and 14-39.

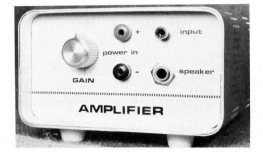

Figure 14-36. Basic transistor amplifier.

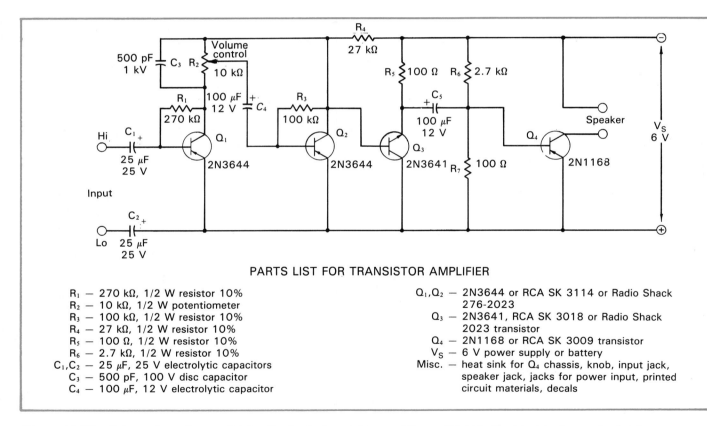

PARTS LIST FOR TRANSISTOR AMPLIFIER

R_1 — 270 kΩ, 1/2 W resistor 10%
R_2 — 10 kΩ, 1/2 W potentiometer
R_3 — 100 kΩ, 1/2 W resistor 10%
R_4 — 27 kΩ, 1/2 W resistor 10%
R_5 — 100 Ω, 1/2 W resistor 10%
R_6 — 2.7 kΩ, 1/2 W resistor 10%
C_1, C_2 — 25 μF, 25 V electrolytic capacitors
C_3 — 500 pF, 100 V disc capacitor
C_4 — 100 μF, 12 V electrolytic capacitor

Q_1, Q_2 — 2N3644 or RCA SK 3114 or Radio Shack
276-2023
Q_3 — 2N3641, RCA SK 3018 or Radio Shack
2023 transistor
Q_4 — 2N1168 or RCA SK 3009 transistor
V_S — 6 V power supply or battery
Misc. — heat sink for Q_4 chassis, knob, input jack,
speaker jack, jacks for power input, printed
circuit materials, decals

Figure 14-37. Schematic and parts list for the basic transistor amplifier. (Hickok Electrical Instruments, Inc.)

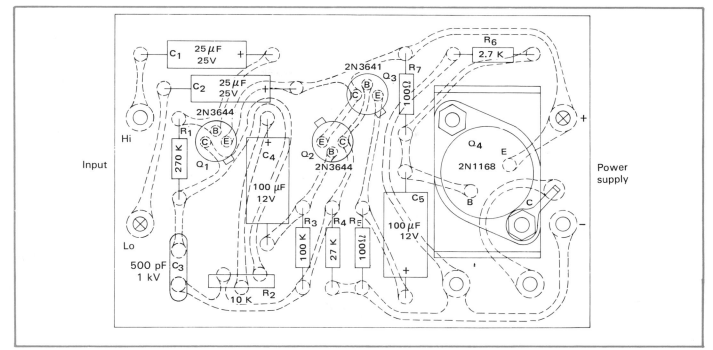

Figure 14-38. Printed-circuit parts layout for amplifier.   (Hickok Electrical Instruments, Inc.)

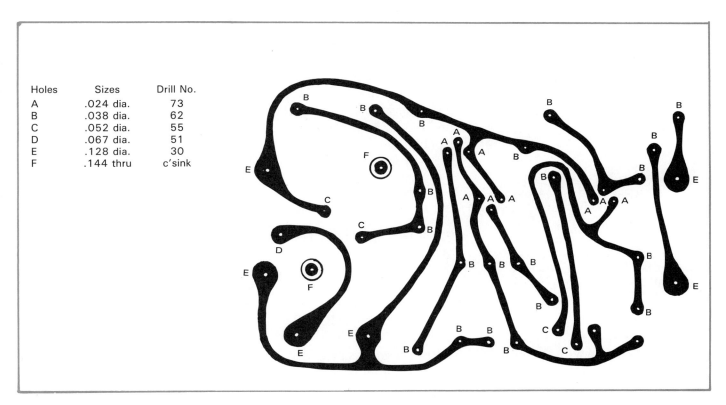

Holes	Sizes	Drill No.
A	.024 dia.	73
B	.038 dia.	62
C	.052 dia.	55
D	.067 dia.	51
E	.128 dia.	30
F	.144 thru	c'sink

Figure 14-39. Layout and hole sizes of printed circuit for transistor amplifier.

SUMMARY

- Amplifiers are electronic circuits that control output signals with input signals.
- Bipolar junction, field-effect, and unijunction transistors are all amplifier devices.
- Base-bias is the simplest biasing circuit. This circuit sets essentially a constant base current. It is very sensitive to variations in temperature.
- Voltage-divider bias is the most popular biasing circuit. It is a stable circuit and requires only a single power source.
- Emitter biasing almost guarantees a stable operating point for the transistor. However, this advantage is overshadowed by the need for two power supplies or batteries.
- In collector-feedback biasing, the base resistor is connected to the collector rather than V_{CC}. This scheme provides a stable quiescent level and requires only one power source.
- A dc load line is drawn on a plot of collector characteristics between two extremes of operation—cutoff and saturation.
- The dc operating point is the point representing the condition existing when there is no input signal.
- Selecting an operating point too near the saturation or cutoff points of the load line can result in distortion of the output. An excessively large input can also result in distortion.
- The field-effect transistor is a three-element semiconductor device. The three elements are the source, the drain, and the gate.
- There are two main types of FETs. The JFET is one type. The MOSFET is another type.
- The two main types of MOSFETs are DE MOSFETs and E MOSFETs. The former has a physical channel. The latter does not; its channel is induced.
- The unijunction transistor is a type of controlled diode with some amplifying capabilities.

KEY TERMS

Each of the following terms has been used in this chapter. Do you know their meanings?

ac load line
attenuator
base-bias circuit
bipolar device
channel
class-A amplifier
class-B amplifier
collector efficiency (η_C)
collector-feedback-bias circuit
cutoff voltage ($V_{GS[off]}$)
dBm reference
dc load line
dc operating point
decibel (dB)
degeneration
depletion mode
depletion-enhancement MOSFET (DE MOSFET)

distortion
drain (D)
drain characteristic curve
emitter-bias circuit
enhancement mode
enhancement-only MOSFET (E MOSFET)
field-effect transistor (FET)
fixed-bias circuit
gate (G)
junction field-effect transistor (JFET)
linear amplifier
load line
maximum power-dissipation curve
metal oxide semiconductor field-effect transistor (MOSFET)

midpoint biased
n-channel JFET
negative feedback
negative resistance
p-channel JFET
pinch-off voltage (V_p)
push-pull amplifier
Q-point
quiescent point
source (S)
stability factor (S)
substrate
swamping resistor
thermal runaway
unijunction transistor (UJT)
unipolar device
voltage-divider-bias circuit

TEST YOUR KNOWLEDGE

Please do not write in this text. Place your answers on a separate sheet of paper.

1. A capacitor can be used in an amplifier circuit to provide a low-reactance path to ground for ac signals. True or False?
2. Name four bias methods used in transistor circuits.
3. What three CE circuit values can be identified from the Q-point?
4. Consideration must be given to the purpose of an amplifier when selecting Q-point. True or False?
5. What is the standard operating temperature of a transistor?
   a. 25°C (77°F).
   b. 35°C (95°F).
   c. 45°C (113°F).
6. A fixed-bias circuit is very stable because base current is fixed. True or False?
7. Name two factors that contribute to thermal runaway.
8. What principle underlies the swamping resistor?
9. The unijunction transistor is classified as a _____ resistance device.
   a. Positive.
   b. Negative.
10. A unijunction transistor is a type of controlled diode with some amplifying capabilities. True or False?
11. A field-effect transistor is considered to be a _____-controlled amplifier.
12. The main advantage of the FET over the typical transistor is its high input _____.
   a. Voltage.
   b. Current.
   c. Impedance.

Indicate whether the following relationships are *true* or *false*.

13. $I_C = I_E - I_B$
14. $\alpha_{dc} = \dfrac{I_C}{I_B}$
15. $\beta_{dc} = \dfrac{I_C}{I_B}$
16. $I_B = I_E + I_C$
17. $\alpha_{dc} I_E = I_C$
18. $\beta_{dc} I_B = I_C$
19. $I_E = I_C - I_B$
20. $A_p = A_i A_v$
21. $A_p = \alpha^2 \left(\dfrac{Z_{out}}{Z_{in}} \right)$
22. $A_p = \beta^2 \left(\dfrac{Z_{out}}{Z_{in}} \right)$
23. $\alpha = \dfrac{1}{\beta + 1}$
24. $\beta = \dfrac{\alpha}{1 - \alpha}$
25. $P_D = I_C \times V_{CE}$

Applications of Electronics Technology:
This stereo system features a CD player, cassette player, and wireless remote.
(Aiwa America)

Chapter 15 POWER AMPLIFIERS

Communications electronics is an important branch of solid-state technology. It involves the circuits and systems that are used to transmit information. Television and radio transmitters and receivers, two-way radios, and radar systems are all examples. Underlying all of these systems is the *power amplifier*.

After studying this chapter, you will be able to:
☐ *Discuss several methods of coupling transistor stages.*
☐ *Explain the major difference between small- and large-signal amplifiers.*
☐ *Describe operation of phase-inversion circuits.*
☐ *Select components and predict performance of a push-pull amplifier.*
☐ *Describe operation of complementary-symmetry circuits.*
☐ *Identify speaker types and cite their advantages.*

INTRODUCTION

A **power amplifier** is designed to deliver *power* to a load. Power amplifiers are **large-signal amplifiers**. As such, they operate over a large portion of the load line. These amplifiers are normally used as the final stage of a receiving or transmitting system to drive a power device such as a loudspeaker or transmitting antenna.

In truth, all junction transistors are *current* amplifiers. There are major differences, however, between a tiny transistor used to increase the voltage level of a microsignal and the larger power transistor used to drive a speaker or output device. Power transistors are designed with larger collector surfaces and with provisions for heat dissipation. More often than not, they are mounted on heat sinks with cooling fins. A power transistor is shown in Figure 15-1.

Figure 15-1. A typical power transistor. The case is the collector terminal. When mounted in a heat sink, the collector will have an increased cooling surface. (Westinghouse)

MULTISTAGE COUPLING

As mentioned, a power amplifier is the final *stage*. This suggests that more than a single transistor is required to amplify a signal to a useful value, which in fact, is true. A power amplifier is preceded by a *preamplifier*. The **preamplifier** amplifies weak signals so they are strong enough to drive the power amplifier. In other words, its function is to increase overall gain (current, voltage, and power). It consists of one or more amplifiers *coupled* together in **cascade**, which means they are connected in series. Each amplifier in the cascaded arrangement is known as a **stage**. As there are more than one of these, the overall arrangement may be referred to as a **multistage amplifier**.

There are several methods of coupling the stages together, including:
- *RC coupling.*
- *Transformer coupling.*
- *Direct coupling.*

RC COUPLING

A simple method of coupling transistor amplifier stages is by using **RC coupling**. Refer to Figure 15-2 in which a step-by-step explanation is given. (A review of RC networks in Chapter 10 may be necessary before proceeding further.) When the ac signal is added, we find that the voltage appearing across R rises and falls at the same frequency. We also find that where the input signal varies around a dc level of 10 V, the output signal varies around the zero level, Figure 15-3. The dc component has been removed. The capacitor blocks dc. In this respect, it is called a **blocking capacitor**.

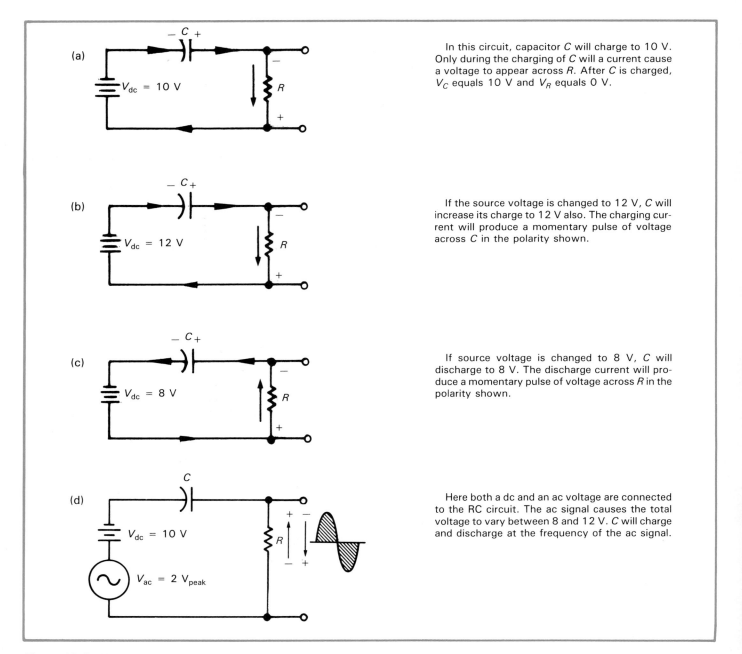

(a) In this circuit, capacitor C will charge to 10 V. Only during the charging of C will a current cause a voltage to appear across R. After C is charged, V_C equals 10 V and V_R equals 0 V.

(b) If the source voltage is changed to 12 V, C will increase its charge to 12 V also. The charging current will produce a momentary pulse of voltage across C in the polarity shown.

(c) If source voltage is changed to 8 V, C will discharge to 8 V. The discharge current will produce a momentary pulse of voltage across R in the polarity shown.

(d) Here both a dc and an ac voltage are connected to the RC circuit. The ac signal causes the total voltage to vary between 8 and 12 V. C will charge and discharge at the frequency of the ac signal.

Figure 15-2. Explanation of RC coupling.

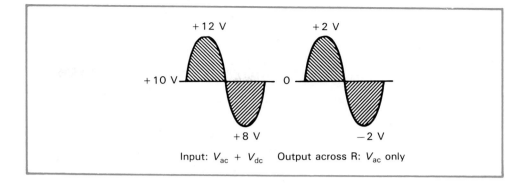

+12 V   +2 V

+10 V —   0

+8 V   −2 V

Input: $V_{ac} + V_{dc}$ Output across R: V_{ac} only

Figure 15-3. The capacitor blocks the dc voltage but permits the ac signal to pass.

Now, examine this circuit from a mathematical point of view; C and R in series form an ac voltage divider. Refer to Figure 15-4. The voltage division will depend upon the reactance of X_C at the frequency of the signal. The voltage output across R is the important consideration, and we want as much of the ac voltage to appear across R as possible. Assuming a frequency of 1000 Hz and a value of C equal to 0.01 μF, then:

$$X_C = \frac{1}{2\pi fC} = \frac{0.159}{(1 \times 10^3 \text{ Hz})(0.01 \times 10^{-6} \text{ F})} \cong 16,000 \ \Omega$$

If the value of R is ten or more times greater than X_C, then most of the voltage will appear across R. Assuming R equals 160 kΩ, then:

$$V_R = \left(\frac{R}{R + X_C}\right) V_{ac}$$

$$= \left(\frac{160 \times 10^3 \ \Omega}{160 \times 10^3 \ \Omega + 16 \times 10^3 \ \Omega}\right)(2 \text{ V}_{peak})$$

$$= (0.91)(2 \text{ V}_{peak}) = 1.82 \text{ V}_{peak}$$

(From the practical point of view, phase shift is disregarded.)

It seems that almost all the voltage does appear across R. Only a fraction has been lost as signal output. If R were made larger, then even more of the total output would be developed across R.

Look at the two-stage amplifier using transistors in Figures 15-5 and 15-6. Note the use of blocking capacitors C_1, C_2, and C_5. These **coupling capacitors**, as they are also called, are used to join circuits. They block dc current between stages for isolation purposes. Yet, they allow ac signals to pass with little opposition.

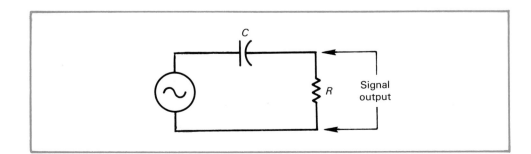

C

R   Signal output

Figure 15-4. R and C form a voltage divider for the ac signal.

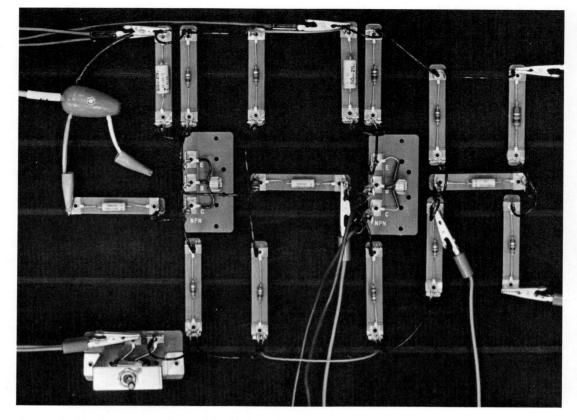

Figure 15-5. RC coupling is used in this breadboard circuit. It is a popular method of coupling transistor stages.

In the final stage of this circuit, the input impedance of the amplifier, R_{in}, is used for R. R_{in} is the equivalent resistance of R_5, R_6, and R_{be} in parallel. The basis for this is as follows: R_5 is tied to ac ground through V_S. In ac analysis, ac ground and actual ground are treated as the same point. R_{be} is the ac resistance between base and emitter terminals of Q_2. R_E is effectively bypassed by C_4 for signal voltages. Thus, for ac purposes, R_E does not exist. (In general, the ac resistance R_{be} is termed h_i. Specifically, for a CE-circuit configuration, it is termed h_{ie}, where e stands for common emitter: $h_{ie} = v_b/i_b$.)

Neglecting the internal resistance of the signal source, output voltage is divided between X_{C_2} and R_{in}. As stated, the value of R (R_{in}, in this case) should be at least ten times the value of X_C. CE amplifiers do not have a high input impedance; thus, the reactance of C_2 must be quite low to ensure most of the voltage is dropped across R_{in}. If not, it will rob a large part of the signal from Q_2.

Also, consider the signal at the collector of Q_1. It has a choice of paths to follow and will take the easiest path. It can go through R_3 or through the RC coupling network, formed by C_2 and R_{in}, which may be considered in parallel. If the impedance of the network is higher than R_3, then signal current will go through R_3 instead of to the next stage.

Looking at Figure 15-6, let us find out what minimum frequency we should apply to our circuit. Assume the amplifier input impedance R_{in} is 500 Ω. This is a typical input impedance value for a CE circuit. (Refer back to Figure 13-25.) With R_{in} = 500 Ω, X_C would need to be 50 Ω or less. Transposing our equation for capacitive reactance yields:

$$f = \frac{0.159}{CX_C} = \frac{0.159}{(10 \times 10^{-6}\text{ F})(50\text{ }\Omega)} = 318\text{ Hz}$$

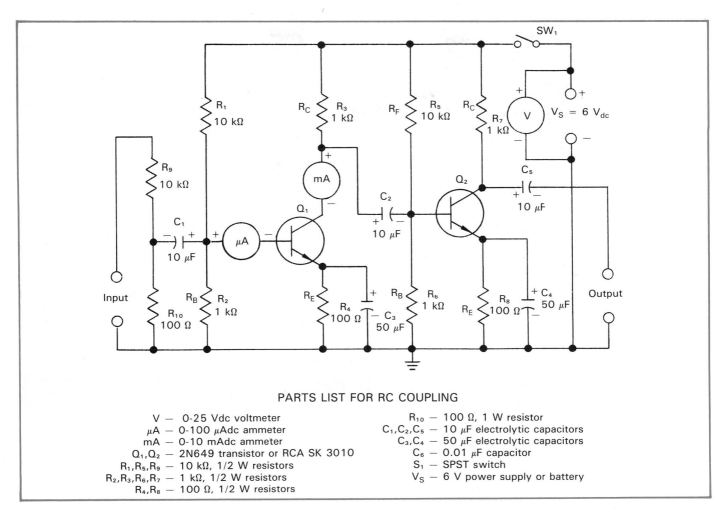

PARTS LIST FOR RC COUPLING

V — 0-25 Vdc voltmeter	R_{10} — 100 Ω, 1 W resistor
μA — 0-100 μAdc ammeter	C_1, C_2, C_5 — 10 μF electrolytic capacitors
mA — 0-10 mAdc ammeter	C_3, C_4 — 50 μF electrolytic capacitors
Q_1, Q_2 — 2N649 transistor or RCA SK 3010	C_6 — 0.01 μF capacitor
R_1, R_5, R_9 — 10 kΩ, 1/2 W resistors	S_1 — SPST switch
R_2, R_3, R_6, R_7 — 1 kΩ, 1/2 W resistors	V_S — 6 V power supply or battery
R_4, R_8 — 100 Ω, 1/2 W resistors	

Figure 15-6. Schematic and parts list for circuit with RC coupling.

We would not want to apply a frequency less than 318 Hz to our circuit. Remember that reactance of a capacitor increases as the frequency applied decreases. If our frequency goes below 318 Hz, X_C will go above 50 Ω, and we will begin to drop too much voltage across the capacitor.

As for the values of R_5 and R_6, they were determined by the required bias and stability of the circuit. Note that these resistors are in parallel with the emitter-base circuit. Their values should be sufficiently high so that the ac signal will not be bypassed around the EB junction. This will also help reduce current drain on the dc source.

Circuit design with transistors can become quite complex, and it is a matter of give-and-take. To further illustrate this point, compare the output and input impedances of the CE amplifier. The output can be in the range of 30 kΩ – 50 kΩ, and the input impedance, in the range of 0.5 kΩ – 1500 kΩ. This is a severe mismatch between the output of one stage and the input of the next. With the RC coupling, the mismatch must be tolerated with its accompanying reduction of power gain. However, when cost is a factor, it may be cheaper to add another transistor stage than to purchase a transformer for interstage matching. This will help offset the loss due to mismatch.

TRANSFORMER COUPLING

Transformer coupling is shown in Figure 15-7. The circuit shows two stages of transistor amplifiers coupled with a transformer. Assume that transistor

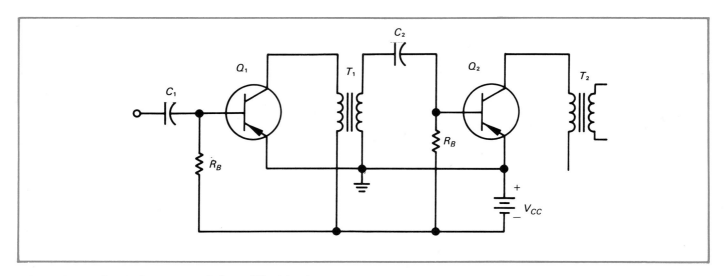

Figure 15-7. A transformer-coupled amplifier circuit.

Q_1 has an output impedance of 20 kΩ, and Q_2 has an input impedance of 1 kΩ. This translates to a severe mismatch and loss of gain between stages. The transformer is a convenient device to match these impedances and correct this problem. A step-down transformer is required. A low secondary voltage means a higher secondary current. This is fine for junction transistors because they are current operated. See Figure 15-8.

The purpose of C_2 in the transformer-coupled circuit of Figure 15-7 is to block the dc bias voltage of the transistor from ground. Notice that with C_2 omitted, the transistor base would be grounded directly through the secondary of transformer T_1.

Figure 15-8. These subminiature and ultra-miniature transformers may be used for transformer coupling. A typical application is seen in power amplifiers in coupling the high resistance output of a CE amplifier to a low impedance speaker.   (Triad)

A significant drawback to transformer coupling is presented by the transformer. Aside from its undesirable size, cost, and weight, a transformer has a poor frequency response. This means that transformer-coupled circuits are limited in the range of frequencies that they can handle. Typically, the use of transformer-coupled amplifiers is limited to *audio* frequencies between about 500 Hz—20 kHz. Outside of this range, amplifier gain quickly falls off due to the transformer. (Note that the range of frequencies from 20 Hz— 20 kHz is considered the *audio frequency range*.) Compare this to the RC-coupling arrangement, which can handle frequencies in the approximate range of 100 Hz—1 MHz.

As the higher audio frequencies are approached, transformers tend to *saturate*. At this point, all the magnetic alignment that can be established within the transformer core has been attained, and the magnetic flux has peaked. Thus, changes in primary current (collector output) will not be reflected in the secondary, or the next stage input. At radio frequencies, inductive reactance and winding capacitance will present problems.

A variation of transformer coupling you will see in many circuits uses a tapped transformer. These taps can be at low, medium, and high impedance points. With this version of a transformer, good impedance matching can be attained as well as good coupling and gain. See Figure 15-9. For radio frequency amplifier circuits, both the transformer primary and secondary windings can be tuned by variable capacitors for frequency selectivity. This will be discussed further in Chapter 17.

Another point about the transformer is that its primary impedance acts as a collector load for the prior-stage transistor. For example, T_2 primary acts as the load for the output of Q_1. Since this impedance only appears under signal conditions, the load is the inductive reactance of the primary. From the dc point of view, the only load is the ohmic resistance of the wire used to wind the transformer primary. It is not involved in the dc analysis of the amplifier because the dc resistance is too small to consider. You will use this information later in the design of a power amplifier, which will require an understanding of both ac and dc load lines.

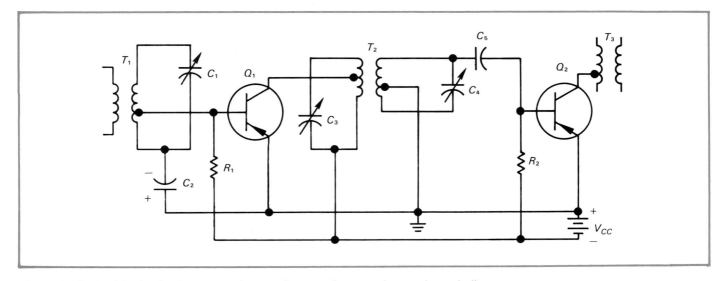

Figure 15-9. In this circuit, the taps on the transformer primary and secondary windings provide a convenient matching point. The transformer can be designed for good overall gain.

DIRECT COUPLING

In many industrial circuits used today, it is necessary to amplify either very low-frequency signals, or it is required to retain the dc value as well as the ac value of a signal. RC coupling introduces capacitive reactance. At low frequencies, it becomes excessive, which results in signal loss and gain reduction. Further, RC and transformer coupling will block out the dc component. **Direct coupling**, Figure 15-10, is an answer.

In this circuit, the collector of Q_1 is connected directly to the base of Q_2. The collector load resistor R_2 also acts as a bias resistor for Q_2. In general, any change of bias current is amplified by each successive stage of the direct-coupled circuit. Therefore, it is very sensitive to temperature changes. This disadvantage can be overcome with stabilizing circuits. Another disadvantage is that each stage requires different bias voltages for proper operation. This was not a problem with RC and transformer coupling where the dc bias of one stage did not affect other stages.

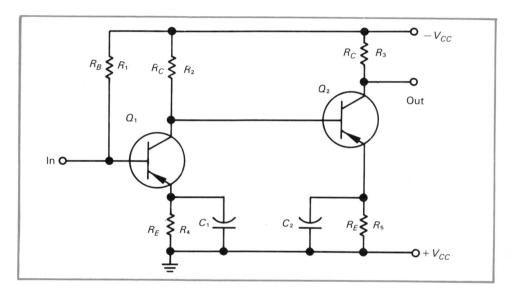

Figure 15-10. The circuit of a direct-coupled amplifier.

THE FINAL STAGE

The final stage in multistage amplifiers designed to deliver *power* to the load is the power amplifier. Voltage and current gain are not important considerations. However, the transistor must be capable of carrying relatively large currents. Provisions must be made for heat dissipation. This large-signal amplifier is usually mounted on a heat sink.

Figure 15-11 shows an exploded view of a mounting assembly. This drawing shows a chassis serving as a heat sink. Note that the transistor case is the collector terminal of the power transistor. This provides greater surface area for heat dissipation in the collector circuit. The collector is connected to the external circuit by means of a screw and solder lug.

Power amplifiers are commonly categorized by class. In this section, we will specifically discuss the class-A and class-B power amplifier.

CLASS-A POWER AMPLIFIER

The power amplifier circuit shown in Figure 15-12 is classified as a **single-ended amplifier**. In this instance, it means that the stage has only one transistor. We will analyze a similar circuit; however, before doing so, you should

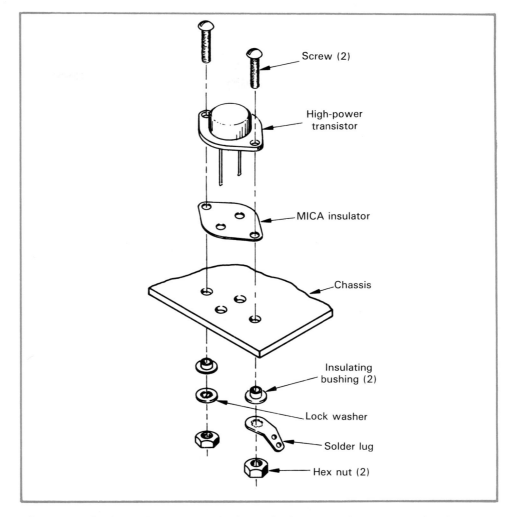

Figure 15-11. *A special mounting kit is used when mounting a power transistor on a heat sink. The kit contains the required hardware.*

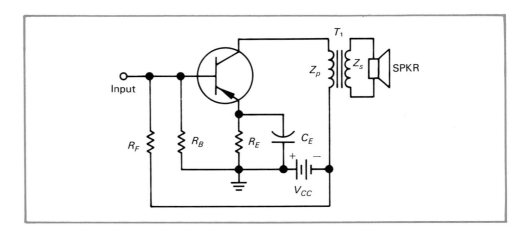

Figure 15-12. *Circuit of a single-ended power amplifier.*

review the function of the components in this circuit.

T_1 serves as a matching transformer between the high output impedance of the transistor and the low impedance of the *speaker voice coil.* (In a *moving-coil speaker*, the **speaker voice coil** is moved back and forth by electric impulses.

Sound waves are produced by the speaker cone which is fastened to it.) Primary impedance Z_p is the collector load.

R_E is the emitter resistance to improve stabilization. C_E is the emitter resistor bypass for signal currents to keep the emitter terminal at a constant voltage and prevent degeneration. R_B and R_F form a voltage divider to bias the base terminal.

At this point, we should take time to review relevant formulas from earlier lessons:

* The peak value of a sine wave is one-half of the peak-to-peak value.

$$V_{\text{peak}} = \frac{V_{\text{p-p}}}{2} \qquad I_{\text{peak}} = \frac{I_{\text{p-p}}}{2}$$

* The effective (rms) value of a sine wave is 0.707 (or $1/\sqrt{2}$) of the peak value.

$$V_{\text{rms}} = \frac{V_{\text{peak}}}{\sqrt{2}} \qquad I_{\text{rms}} = \frac{I_{\text{peak}}}{\sqrt{2}}$$

* The relationship between rms and peak-to-peak current and voltage values is:

$$V_{\text{rms}} = \frac{V_{\text{p-p}}}{2\sqrt{2}} \qquad I_{\text{rms}} = \frac{I_{\text{p-p}}}{2\sqrt{2}}$$

* The impedance ratio of a transformer varies directly as the square of its turns ratio.

$$\frac{Z_p}{Z_L} = \left(\frac{N_p}{N_s}\right)^2 \text{ or } \frac{N_p}{N_s} = \sqrt{\frac{Z_p}{Z_L}}$$

* Transformer efficiency is the ratio of power out to power in, or power in the secondary *(P_s)* to power in the primary *(P_p)*.

$$\% \text{ Efficiency} = \frac{P_s}{P_p} \times 100\%$$

* Collector efficiency is the amount of power drawn from the supply that is actually delivered to the load. It is the ratio of ac output power to dc input power.

$$\% \ \eta_C = \frac{P_{\text{out}}}{P_{\text{dc}}} \times 100\%$$

Note that for an RC-coupled, class-A amplifier, maximum collector efficiency is 25%. Due to its operating characteristics, maximum efficiency of a transformer-coupled, class-A amplifier is 50%.

* The dc input power for RC- or transformer-coupled amplifiers is the dc supply voltage times the current drawn from the supply. Since most of the supply current flows through the collector, we can approximate supply current to equal collector current. Thus, dc input power may be approximated by the formula:

$$P_{\text{dc}} = V_{CC}I_{C_Q}$$

* In general, ac power is average power. (Review Chapter 6: section entitled *Sine wave power values.*) Assuming voltage and current are in phase:

$$P = \frac{1}{2} V_{\text{peak}}I_{\text{peak}} = \frac{1}{8} V_{\text{p-p}}I_{\text{p-p}} = V_{\text{rms}}I_{\text{rms}} = I_{\text{rms}}^2 R = \frac{V_{\text{p-p}}^2}{8R}$$

• The maximum ac output power of a class-A amplifier is:

$$P_{out} = \frac{1}{2} V_{CEQ} I_{CQ}$$

Problem:

A power amplifier that will supply 50 mW of power to a 4-Ω speaker is desired. A matching transformer with an efficiency of 85% will be used. A battery of 9 V is available. With a maximum signal applied, collector efficiency of 30% is desired. The circuit is shown in Figure 15-13. Characteristic curves for the silicon transistor are shown in Figure 15-14. The transistor has a maximum power dissipation rating of 200 mW. Determine the following values for the amplifier:

(a) Q-point.
(b) Transformer turns ratio.
(c) Value of base resistor.

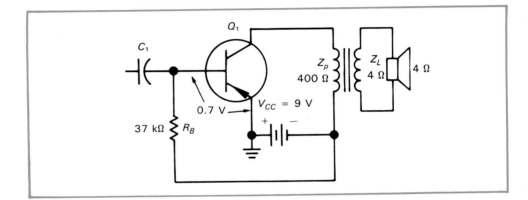

Figure 15-13. Circuit design for a basic power amplifier.

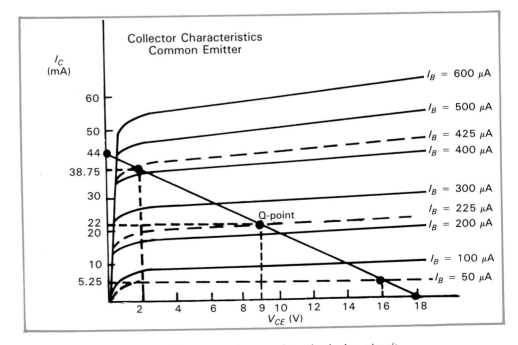

Figure 15-14. Characteristic curves for transistor in design circuit.

Solution:

Step 1. The output transformer has an efficiency of 85%, and 50 mW of power are required at the speaker. We must determine what the power in the primary must be to achieve this.

$$\text{efficiency} = \frac{P_s}{P_p}$$

Transposing to solve for P_p and substituting yields:

$$P_p = \frac{P_s}{\text{efficiency}} = \frac{50 \text{ mW}}{0.85} = 58.8 \text{ mW}$$

Step 2. The primary must receive 58.8 mW of power from the transistor output. We must determine what dc power input will yield 58.8 mW ac output power. We use the formula for collector efficiency, which is:

$$\eta_C = \frac{P_{\text{out}}}{P_{\text{dc}}}$$

Transposing to solve for P_{dc} and substituting yields:

$$P_{\text{dc}} = \frac{P_{\text{out}}}{\eta_C} = \frac{58.8 \text{ mW}}{0.30} = 196 \text{ mW}$$

Note that maximum efficiency for transformer coupling is 50%, ideally. To attain this, the maximum swing, or **excursion**, of a signal must cover the entire length of an ac load line. The ac load line, for the CE amplifier, gives all possible combinations of i_c and v_{ce}. It tells you the maximum possible output voltage swing for a given amplifer. For our circuit, this would be 18 $V_{\text{p-p}}$. However, at the extreme ends of the load line, distortion occurs. For this reason, it is advisable to design an amplifier at an efficiency of less than 50%.

Step 3. With a power source $V_{CC} = 9$ V, compute the current required for this power.

$$P_{\text{dc}} = V_{CC} I_{CQ}$$

Transposing to solve for I_{CQ} and substituting yields:

$$I_{CQ} = \frac{P_{\text{dc}}}{V_{CC}} = \frac{196 \text{ mW}}{9 \text{ V}} \cong 22 \text{ mA}$$

For the ideal transformer-coupled amplifier, $V_{CEQ} = V_{CC}$ and $V_{CE\text{max}} = 2V_{CC}$. The reason we can have a value of $V_{CE\text{max}}$ that is greater than the value of V_{CC} for the circuit has to do with the counter emf produced by the transformer. In short, when the collector current is minimum, the *change* in current is maximum. Thus, the counter emf is also maximum, and its polarity is such that it adds to the value of V_{CC}. For the ideal RC-coupled circuit, however, $V_{CE\text{max}} = V_{CC}$. This is why efficiency of the transformer-coupled amplifier is double that of the RC-coupled amplifier.

Step 4. We can now establish our Q-point for the load line. Refer to Figure 15-14. We plot the point at $V_{CEQ} = 9$ V and $I_{CQ} = 22$ mA.

Step 5. We must check the maximum power dissipation of the transistor to be used. Note that quiescent power (P_{dc}) is the maximum power that the class-A transistor must handle; therefore, its maximum power dissipation rating should exceed this value. Our quiescent power is 196 mW. The specifications state 200 mW, so we are safe and ready to continue.

Step 6. With a maximum signal excursion about the Q-point, the limits would occur at saturation and cutoff at equal and opposite distances along the load line. One end of the load line can be established at the cutoff point, which is at $I_C \cong 0$ A and $V_{CE} = V_{CE_{max}}$. Ideally, in this case, $V_{CE_{max}} = 18$ V. The load line must pass through the Q-point, so the current at zero voltage can be found graphically. This is roughly 44 mA.

Step 7. The slope of the load line is the negative inverse of output impedance Z, since slope $= \Delta I_C / \Delta V_{CE} = 1/-Z$. (Thus, the steeper the slope, the lower the impedance and vice versa.) For impedance matching, this should be primary load impedance Z_p. Therefore:

$$Z_p = \frac{v_{CE}}{i_C} = \frac{18 \text{ V}_{\text{p-p}}}{44 \text{ mA}_{\text{p-p}}} \cong 400 \text{ } \Omega$$

Step 8. This impedance is matched to the 4-Ω secondary, and the transformer turns ratio must be:

$$\frac{N_p}{N_s} = \sqrt{\frac{Z_p}{Z_L}} = \sqrt{\frac{400 \text{ } \Omega}{4 \text{ } \Omega}} = \frac{10}{1}$$

Step 9. The Q-point falls just above the 200-μA I_B curve at about 225 μA. We need to determine what value of R_B will give us this value of I_B. Base-emitter voltage V_{EB} is assumed to be 0.7 V. Thus:

$$R_B = \frac{V_{CC} - V_{EB}}{I_B} = \frac{9 \text{ V} - 0.7 \text{ V}}{225 \times 10^{-6} \text{ A}} \cong 37 \text{ k}\Omega$$

Having performed these calculations, we now have a circuit that will work. Now, let us analyze this circuit for current and voltage swing and to see what we can expect for power, voltage, and current gains.

We have already discussed that with a full signal excursion between 0 V $-$ 18 V, distortion would occur at both ends of the load line. In these regions, operation becomes nonlinear. So we plan to operate at only 30% efficiency. What will the voltage swing ΔV_{CE} be for this efficiency? It is given by the formula:

$$\Delta V_{CE} = \sqrt{\eta_C 8 V_{CC}^2}$$

Follow the derivation to learn from where this comes:

$$P = \frac{1}{2} V_{\text{peak}} I_{\text{peak}} = \frac{1}{2} \left(\frac{V_{\text{p-p}}}{2}\right)\left(\frac{I_{\text{p-p}}}{2}\right) = \frac{1}{8} V_{\text{p-p}} I_{\text{p-p}}$$

From Ohm's law:

$$I_{\text{p-p}} = \frac{V_{\text{p-p}}}{R}$$

Substituting this into our power formula gives:

$$P = \frac{1}{8} V_{\text{p-p}} \left(\frac{V_{\text{p-p}}}{R}\right) = \frac{V_{\text{p-p}}^2}{8R}$$

We know that efficiency (η_C) is the ratio of ac output power to dc input power. Using this ratio and substituting in values for ac and dc power:

$$\eta_C = \frac{P_{\text{out}}}{P_{\text{dc}}} = \frac{\dfrac{V_{\text{p-p}}^2}{8R}}{V_{CC} I_{C_Q}} = \frac{V_{\text{p-p}}^2}{8R V_{CC} I_{C_Q}} = \frac{\Delta V_{CE}^2}{8R V_{CC} I_{C_Q}}$$

(Note that $V_{p\text{-}p}$ is the peak-to-peak value of the ac voltage. It is equal to v_{CE}, or ΔV_{CE}.) Transposing, to solve for ΔV_{CE} gives:

$$\Delta V_{CE} = \sqrt{\eta_C 8 R V_{CC} I_{CQ}}$$

Again, from Ohm's law:

$$I_{CQ} = \frac{V_{CEQ}}{R} = \frac{V_{CC}}{R}$$

Substituting this into our formula yields:

$$\Delta V_{CE} = \sqrt{\eta_C 8 R V_{CC}\left(\frac{V_{CC}}{R}\right)} = \sqrt{\frac{\eta_C 8 R V_{CC}^2}{R}} = \sqrt{\eta_C 8 V_{CC}^2}$$

Now, we are looking for the voltage swing of V_{CE}, which is ΔV_{CE}. It must produce an output at a specified collector efficiency. Applying our formula:

$$\Delta V_{CE} = \sqrt{\eta_C 8 V_{CC}^2} = \sqrt{0.3(8)(9 \text{ V})^2} = \sqrt{194.4 \text{ V}^2} \cong 14 \text{ V}_{p\text{-}p}$$

One-half of this voltage swing will be above the Q-point and the other half, below. We can locate the points along our load line.

$$V_{CE_{min}} = V_{CEQ} - \frac{1}{2}\Delta V_{CE} \qquad V_{CE_{max}} = V_{CEQ} + \frac{1}{2}\Delta V_{CE}$$

V_{CEQ} is ideally equal to V_{CC} for the transformer-coupled circuit. Substituting values into our formula:

$$V_{CE_{min}} = 9 \text{ V} - 7 \text{ V} = 2 \text{ V} \qquad V_{CE_{max}} = 9 \text{ V} + 7 \text{ V} = 16 \text{ V}$$

Projecting from these points over to the I_C axis will give the current swing. See Figure 15-14. From the graph, we find that $I_{C_{min}} \cong 5.25$ mA, and $I_{C_{max}} \cong 38.75$ mA. Current swing ΔI_C is the difference of these two values; we find that $\Delta I_C = 33.5$ mA$_{p\text{-}p}$.

Proof of the currents swing plot can be found in these calculations:

$$P_{out} = \frac{1}{8}\Delta V_{CE}\Delta I_C$$

Transposing to solve for ΔI_C and substituting gives:

$$\Delta I_C = \frac{8 P_{out}}{\Delta V_{CE}} = \frac{8(58.8 \text{ mW})}{14 \text{ V}_{p\text{-}p}} = 33.6 \text{ mA}_{p\text{-}p}$$

(For all practical purposes, the calculated and plotted values of ΔI_C are the same.) Now, half of the current swing will be above the Q-point, and the other half, below. As a result:

$$I_{C_{min}} = I_{CQ} - \frac{1}{2}\Delta I_C \qquad I_{C_{max}} = I_{CQ} + \frac{1}{2}\Delta I_C$$

Substituting values into our formula:

$$I_{C_{min}} = 22 \text{ mA} - \frac{1}{2}(33.5 \text{ mA}) = 5.25 \text{ mA}$$

and

$$I_{C_{max}} = 22 \text{ mA} + \frac{1}{2}(33.5 \text{ mA}) = 38.75 \text{ mA}$$

The numbers check out. These plotted points now represent the maximum peak-to-peak current and voltage swings with maximum signal.

Looking at Figure 15-14 again, the change in base current ΔI_B may be approximated. Base current curves do not run through the maximum excursion points on the load line, but the values of I_B can be estimated.

$$\Delta I_B = 425 \ \mu A - 50 \ \mu A = 375 \ \mu A_{p\text{-}p}$$

Note that $I_{B_{\min}}$ and $I_{B_{\max}}$ should be an equal distance from I_{B_Q}. The points in the figure are. However, since we are picking points off the graph, we are working with approximate values. This is the reason that $I_{B_{\max}} - I_{B_Q} = 200$ μA, and $I_{B_Q} - I_{B_{\min}} = 175 \ \mu A$. Since we have approximated, their differences do not figure to be exactly the same.

In addition to output characteristic curves, there are **input characteristic curves**. For the CE circuit, this is a plot of base current versus base-emitter voltage for different values of V_{CE}. See Figure 15-15. Notice that the base current does not begin until V_{EB} is about 0.6 V.

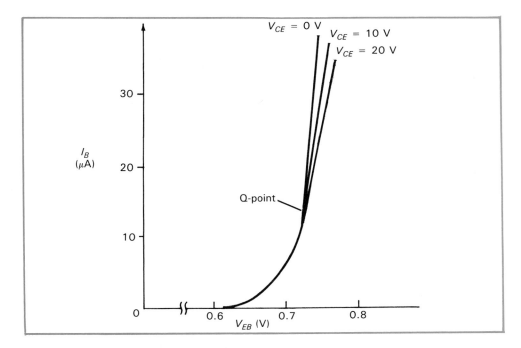

Figure 15-15. Typical input characteristic curves for CE circuit.

Refer now to Figure 15-16. It is a single curve showing a portion of the input characteristic of our power amplifier circuit with V_{CE} equal to 9 V. We plot our points where base current values intersect the 9-V curve. Projecting from these points down to the V_{EB} axis will give the base-emitter voltage excursions. We find the following values:

$$V_{EB} \text{ at Q-point} = 0.70 \text{ V}$$
$$V_{EB} \text{ at } I_{B_{\min}} = 0.60 \text{ V}$$
$$V_{EB} \text{ at } I_{B_{\max}} = 0.77 \text{ V}$$

The input resistance is found by Ohm's law:

$$R_{\text{in}} = \frac{\Delta V_{EB}}{\Delta I_B} = \frac{0.77 \text{ V} - 0.60 \text{ V}}{425 \ \mu A - 50 \ \mu A} = \frac{0.17 \text{ V}}{375 \ \mu A} = \frac{0.17 \text{ V}_{p\text{-}p}}{375 \times 10^{-6} \text{ A}_{p\text{-}p}} = 453 \ \Omega$$

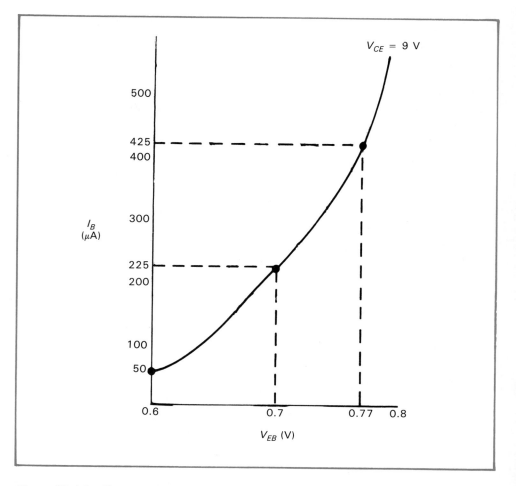

Figure 15-16. Characteristic input curve for transistor in design circuit.

Using the power formula for ac input power:

$$P_{in} = \frac{V_{p-p}I_{p-p}}{8} = \frac{\Delta V_{EB}\Delta I_B}{8}$$

$$= \frac{(0.17\ V_{p-p})(375 \times 10^{-6}\ A_{p-p})}{8} = 7.97 \times 10^{-6}\ W = 7.97\ \mu W$$

We can now calculate the circuit gain values. The power gain A_p is calculated:

$$A_p = \frac{P_{out}}{P_{in}} = \frac{58.7\ mW}{7.97\ \mu W} = \frac{58.8 \times 10^{-3}\ W}{7.97 \times 10^{-6}\ W} = 7378$$

In decibels, this is:

$$A_p\ (dB) = 10 \log A_p = 10 \log 7378 = 38.7\ dB$$

The voltage gain A_v is computed as:

$$A_v = \frac{\Delta V_{CE}}{\Delta V_{EB}} = \frac{14\ V_{p-p}}{0.17\ V_{p-p}} = 82.4$$

The current gain A_i is computed as:

$$A_i = \frac{\Delta I_C}{\Delta I_B} = \frac{33.5\ mA_{p-p}}{375\ \mu A_{p-p}} = \frac{33.5 \times 10^{-3}\ A_{p-p}}{375 \times 10^{-6}\ A_{p-p}} = 89.3$$

This completes the analysis of this single-ended, class-A power amplifier. Do not proceed further until you have mastered the procedure. It may seem difficult at first, but, with practice, it does get easier.

CLASS-B POWER AMPLIFIER

The class-B amplifier obtains greater efficiency by moving the bias point to the cutoff point. Thus, no current is drawn and no power is used until an input signal drives the Q-point back up the load line. With the class-B amplifier, there is an output current only for one-half of a cycle. Two transistors can be connected in parallel to allow an output current for both positive and negative halves of a signal. This is referred to as **push-pull operation**. The output waveform is a reproduction of the original input wave as both halves of the input are combined. Arrangements that employ push-pull operation include standard *push-pull* and *complementary-symmetry amplifiers*.

Push-pull amplifier

In order to attain the maximum power output and efficiency from a power amplifier, two transistors may be operated in push-pull. Figure 15-17 shows a standard **push-pull amplifier**. Note that the circuit in this figure is operating class B, or at zero bias. No dc voltage is applied across the EB junction, so the current I_B is zero. Only the signal acts to bias the transistor terminals.

When point A of the input transformer becomes negative as a result of the input signal v_{in}, Figure 15-17A, then base B_1 of transistor Q_1 is more negative than E_1. This is a forward bias on this pnp transistor, and Q_1 conducts. At the same time, base B_2 of transistor Q_2 is being driven positive. E_2 is more negative than B_2. The EB junction of Q_2 is reverse biased. Q_2 does not conduct.

On the second, or positive, half of the input signal to T_1 primary, Figure 5-17B, the reverse is true. Q_2 is driven into conduction when a negative signal is applied to its base. Q_1 is cut off, as a reverse bias is applied by a positive signal to its base. So, first Q_1 conducts, then Q_2 conducts. On one half-cycle, current i_{C_1} flows in Q_1; on the second half-cycle, current i_{C_2} flows in Q_2.

These transistors are operating class B, which has a much greater efficiency. Half the time each transistor is resting and cooling. A single transistor would run hotter. A transistor must not exceed its rated heat dissipation; so, a single transistor is much more limited in the amount of power it can put out. Two transistors in class B will each run cooler, so they may be driven harder. They can put out nearly six times the power as one similar transistor in class A.

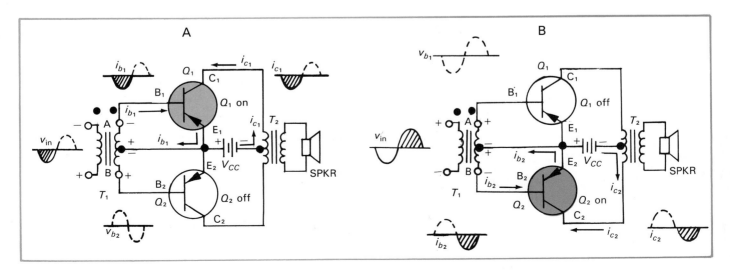

Figure 15-17. The basic push-pull amplifier circuit. A—Negative half-cycle of v_{in}. B—Positive half-cycle of v_{in}.

In addition to the ac power output, the dc power input is less. So, efficiency is much improved. In fact, the ideal maximum efficiency is 78.5%.

The two half-waves of current from the transistors are nearly restored to their original input form by transformer action in the output transformer. This requires an explanation. In Figure 15-18, when Q_1 conducts, current flows as indicated by arrows and creates a magnetic field of one polarity. This produces a half wave in the secondary of the transformer. When Q_2 conducts, the polarity of the primary reverses and induces a wave of the other polarity in the secondary. Thus, the complete wave is nearly restored. The wave is not an exact replica of the original input v_{in} because an effect called *crossover distortion* is introduced in the process.

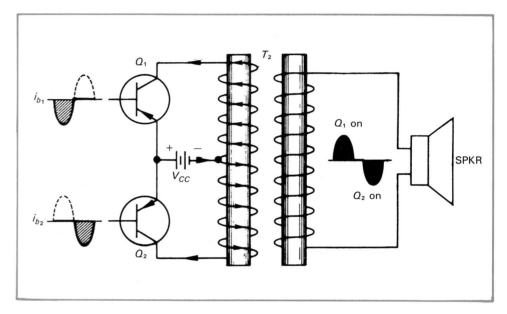

Figure 15-18. Each half of the signal is joined in the output transformer to restore the original wave.

Crossover distortion. Due to the biasing arrangement, operation of the push-pull amplifier as class B does cause some distortion of the waveform. At zero bias, the input voltage signal must be larger than the barrier potential before a transistor conducts. As a result, there is a time interval between positive and negative half-cycles of the input when neither transistor is conducting. See Figure 15-19. This distortion of the output waveform is called **crossover distortion**.

To remedy this distortion, a small bias voltage is applied to each transistor, and the push-pull amplifier is operated class AB. This is done in the usual manner with a voltage divider. See Figure 15-20. Resistors R_1 and R_2 act as a voltage divider, dividing the voltage V_{CC}. We only desire a small bias voltage, so most of the voltage is dropped across R_1. Without R_1, all V_{CC} would be dropped across R_2, which would be too much bias.

R_2, then, sets up the bias for both transistors. The base terminals are connected to R_2 through T_1. This side of R_2 is more negative. Now the bases of both transistors are slightly negative with respect to their emitters. Therefore, both are forward biased. This small bias effectively cancels out the crossover distortion.

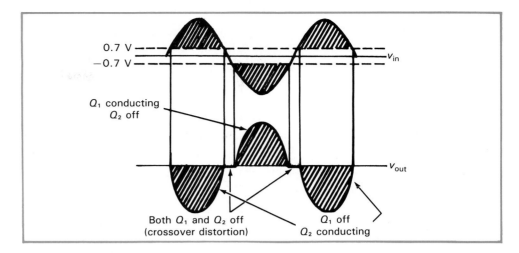

Figure 15-19. Crossover distortion in class-B, push-pull amplifier. Note that the transistors conduct only during portions of the input indicated by shaded areas.

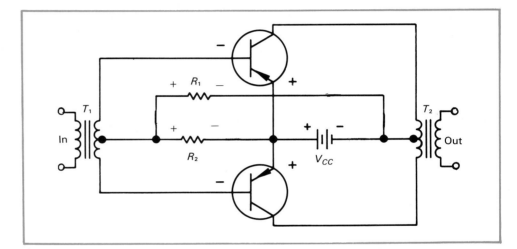

Figure 15-20. A voltage divider provides bias for the transistors.

Phase splitters. For the standard push-pull amplifier, a transformer is used to provide the 180° out-of-phase signals. Transformers can be expensive, however. They also have a rather poor frequency response. A simple **phase splitter**, which will supply two outputs in *antiphase* (180° out of phase), can be used to drive a push-pull amplifier. See Figure 15-21.

A review of transistor operation will tell you $I_C = I_E - I_B$ and $I_C \cong I_E$. An increase in I_E also produces an increase in I_C. If R_1 and R_2 are equal, then the voltage drops across them will be about equal. An increase in collector current will make C more positive and, at the same time, make E more negative. With a decrease in collector current, the opposite is true. Thus, equal but out-of-phase signals may be taken from emitter and collector terminals. This circuit does not provide any amplification of the signal.

Another circuit that produces two out-of-phase signals but also amplifies the signal is drawn in Figure 15-22. It is a simple circuit to understand when you realize that a CE amplifier always inverts the signal. In this circuit, Q_1 acts as a **phase inverter**. This is a stage that functions chiefly to change the phase of a signal 180°. The output for one signal is taken from the collector of the Q_1 stage. The output at this point is inverted. A small part of this inverted

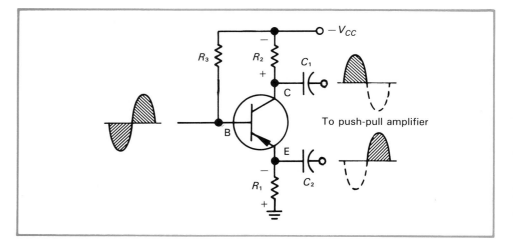

Figure 15-21. *A phase-splitter circuit. Transistors of push-pull circuit (not shown) conduct during shaded portions of waveforms.*

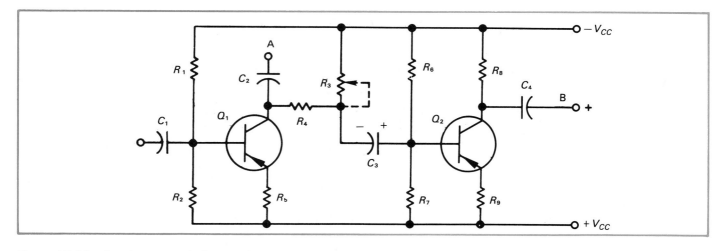

Figure 15-22. *Another type of phase-splitter circuit.*

output is fed to stage Q_2 where it is amplified and once again inverted. The second output is taken from the collector of Q_2.

Note that the load for Q_1 consists of R_3 and R_4 in series. The reason for dividing this load resistance is to secure a small part of the output of stage Q_1 to drive stage Q_2. If R_3 is made variable, then the exact amount of signal may be obtained which, after amplification by Q_2, will produce the same amplitude signal as found at the output of Q_1. In other words, the signals at points A and B should be equal but opposite in phase.

Other resistors in the circuit include networks R_1R_2 and R_6R_7. These are voltage dividers to fix operating bias. R_5 and R_9 are emitter resistors.

Complementary-symmetry amplifier

A circuit in which an npn and a pnp transistor with matched characteristics are connected in parallel is a **complementary-symmetry amplifier**. The transistors operate on half-cycles of the input signal. The npn conducts during one half-cycle, and the pnp, during the other half-cycle.

Study Figure 15-23 where two transistors, Q_1 and Q_2, are fed with a single input signal. Q_1 is an npn transistor. Q_2 is a pnp transistor. A positive-going signal causes the base of Q_1 to become positive, and Q_1 will conduct. At the same time, the same signal causes the base of Q_2 to become positive, which cuts it off. The waveforms and their phase relationships are shown. The signals

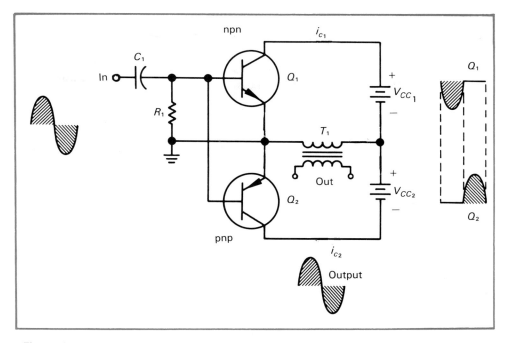

Figure 15-23. The circuit of a complementary-symmetry amplifier.

are joined in output transformer T_1 for the full-wave output. T_1 is used for impedance matching. It qualifies as an output load in this application, but it is not a requirement of the complementary-symmetry amplifier. In a different application, the output could be connected directly to a resistor, for example.

The complementary-symmetry circuit shown has two supply voltages. If two supply voltages are used, the dc level of the output is near 0 V. In this case, the output may be connected directly to a load resistance. It is not necessary to have two supplies, however. If a single supply voltage is used, the dc level of the output will be more than 0 V. In this case, a capacitor (or transformer) must be used to couple the output signal to the resistive load. This will block the dc level at the output from getting to the load.

The complementary-symmetry circuit is by far the most commonly used. One advantage to this circuit is that it does not require a transformer or inverter circuit to provide out-of-phase signals for the input. Transformers can be heavy and expensive. A disadvantage of this circuit is it, too, is subject to crossover distortion.

SPEAKERS

A *speaker* is a type of **transducer**; that is, it converts energy from one form into another. Specifically, a **speaker** converts electrical energy into sound energy. The sound energy is delivered in the form of a wave. A physicist would say that a **sound wave** is alternate **rarefactions** (expansions) and **condensations** (compressions) of air that produce the sensation of sound when impressed upon the ear. The effect sound waves have on air is depicted in Figure 15-24. Electrically, these sound waves may be set in motion by vibrating a flexible membrane known as a **diaphragm**.

Sound waves are sinusoidal in nature. They have a definite frequency, amplitude, and wavelength. Audible sound waves are at **audio frequency**, which ranges roughly from 15 Hz − 20,000 Hz. Amplitude of a sound wave is a measure of air pressure, not electrical pressure, which mostly we are more familiar with. The peak value of the sine wave is the maximum departure of pressure from the normal static value. Peak positive pressure coincides with maximum compression. This is where air pressure is higher than normal. Peak negative pressure coincides with maximum rarefaction. This is where air pressure is less than normal static pressure.

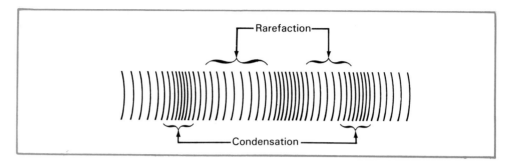

Figure 15-24. A representation of sound waves as alternate rarefactions and condensations of air.

PERMANENT-MAGNET SPEAKERS

A **permanent-magnet (PM) speaker** is sketched in Figure 15-25. This is by far the most commonly used speaker in audio systems today. It basically consists of a permanent magnet, a coil, and a speaker cone. The coil is suspended in the center of the magnet in the magnetic field. The speaker cone, which is made of flexible paper, is attached to the coil. A varying current through the voice coil will produce its own magnetic field. This reacts with the field of the permanent magnet and causes motion.

This action is similar to the motor action described in Chapter 7. A current of one polarity will cause the coil to move in one direction; the opposite polarity will cause the coil to move in the opposite direction. As a result, the coil and cone move inward and outward at the same frequency and amplitude as the incoming electrical signal fed to the voice coil. The cone movement sets up sound waves, which the human ear will sense as sound.

The voice coil is wound with only a few turns of low resistance wire. The common types have an impedance of around 4 Ω at signal frequencies. It is for this reason that an output transformer is used to match this low impedance.

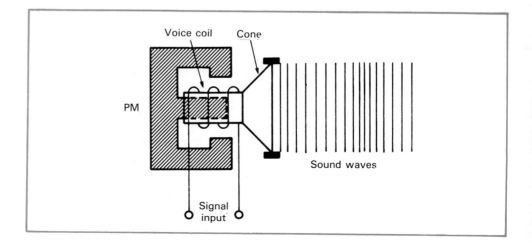

Figure 15-25. A sketch of a permanent-magnet speaker.

You will find speakers with an impedance of 4 Ω, 8 Ω, or 16 Ω commonly in use with audio equipment. The better quality audio amplifiers may use a **universal output transformer**, which has a number of taps on its winding.

ELECTRODYNAMIC SPEAKERS

The action of the **electrodynamic speaker**, Figure 15-26, is the same as the PM speaker. The major difference is the use of an electromagnet rather than a permanent magnet. The electromagnetic field is established in a field coil and supplied by a source of dc taken from the equipment power supply. Again, a varying current through the voice coil produces a magnetic field that reacts with the electromagnetic field and causes motion.

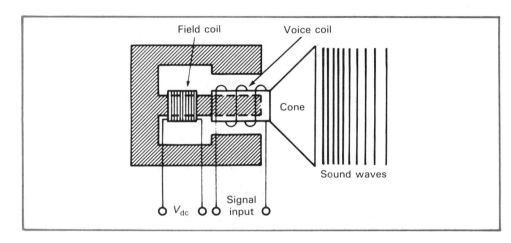

Figure 15-26. A sketch of an electrodynamic speaker.

ELECTROSTATIC SPEAKERS

An **electrostatic speaker** usually is used for high frequencies. It depends upon the characteristics of capacitance for its operation. In Figure 15-27, a high voltage is applied to the two capacitor plates in the speaker. One of these plates will flex inward and outward slightly. The speaker cone is attached to the flexible plate. A signal now applied in series with the high voltage will cause a varying high voltage, which, in turn, varies the electrostatic field between the capacitor plates. The flexible plate moves according to the applied signal. The cone converts the movement to sound waves. An electrostatic speaker is usually called a **tweeter**.

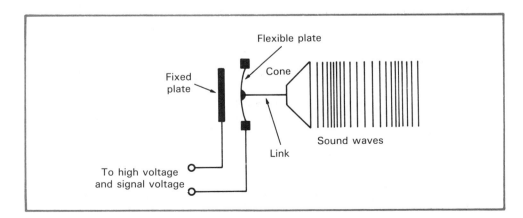

Figure 15-27. A sketch of an electrostatic speaker.

CROSSOVER NETWORKS

A good-quality speaker system will have three speakers. One of them is the tweeter just mentioned. The other two are the *woofer* and the *midrange*. A speaker should be of the proper size and design to best produce the audio sounds in a given range. Generally, the larger speakers are necessary to produce the low bass notes. These speakers are called **woofers**. There are intermediate range speakers for the great majority of audio sounds. Such a speaker is called a **midrange**

The various kinds of speakers suggest some circuitry that will direct a specified range of frequencies to the speaker that will best reproduce them. Such is the **crossover network** in Figure 15-28. The network is based on the characteristics of X_L and X_C. With higher frequencies, X_L increases in ohms. Therefore, it will pass the lower frequency currents better. With higher frequencies, X_C decreases in ohms. Therefore, it offers less impedance to higher frequencies.

The more sophisticated circuit in Figure 15-29 takes advantage of both changing values of X_L and X_C to form a high-pass and low-pass filter. Apply your previous knowledge of inductive and capacitive circuits. Be sure you understand the principles of these crossover circuits found in most *high fidelity* systems.

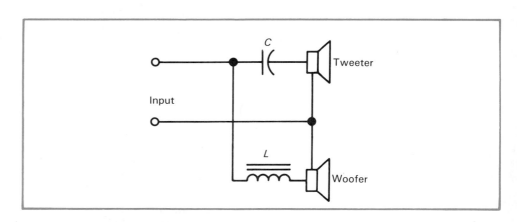

Figure 15-28. Crossover network.

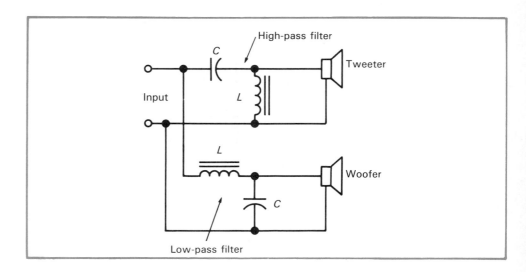

Figure 15-29. Crossover network using filter circuits.

Figures 15-30 through 15-33 are typical components found in a stereo system. The cassette deck and compact disc player function as signal inputs for the amplifier. The speaker system is, of course, an output component for the amplifier.

Figure 15-30. Double cassette deck.   (Marantz)

Figure 15-31. Compact disc player.   (Marantz)

Figure 15-32. Stereo amplifier.   (Marantz)

Figure 15-33. A 3-way speaker system showing woofer, midrange, and tweeter. (Cerwin-Vega)

PUSH-PULL AMPLIFIER PROJECT

Here is a practical amplifier that may be used for many purposes. You can use it to increase the *volume* of oscillators and sirens. You can connect a record player or a microphone to its input. The complete push-pull amplifier is shown in Figure 15-34. A schematic and parts list are given in Figure 15-35. The circuit is easy to build using printed circuit techniques, Figure 15-36, or a breadboard.

Figure 15-34. A practical push-pull amplifier.

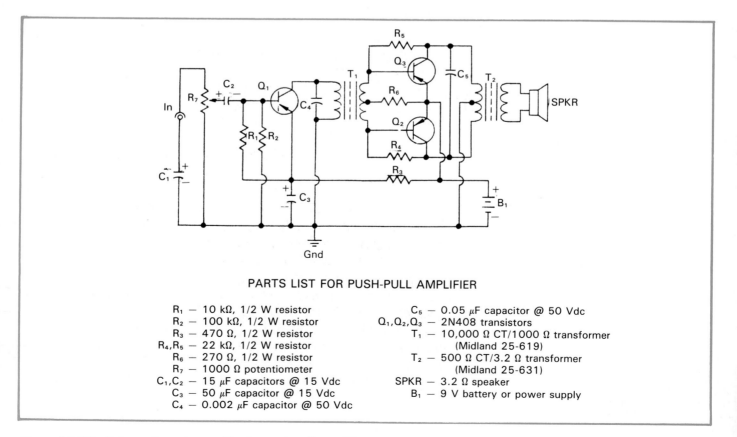

PARTS LIST FOR PUSH-PULL AMPLIFIER

R₁ — 10 kΩ, 1/2 W resistor
R₂ — 100 kΩ, 1/2 W resistor
R₃ — 470 Ω, 1/2 W resistor
R₄,R₅ — 22 kΩ, 1/2 W resistor
R₆ — 270 Ω, 1/2 W resistor
R₇ — 1000 Ω potentiometer
C₁,C₂ — 15 μF capacitors @ 15 Vdc
C₃ — 50 μF capacitor @ 15 Vdc
C₄ — 0.002 μF capacitor @ 50 Vdc

C₅ — 0.05 μF capacitor @ 50 Vdc
Q₁,Q₂,Q₃ — 2N408 transistors
T₁ — 10,000 Ω CT/1000 Ω transformer
(Midland 25-619)
T₂ — 500 Ω CT/3.2 Ω transformer
(Midland 25-631)
SPKR — 3.2 Ω speaker
B₁ — 9 V battery or power supply

Figure 15-35. Schematic and parts list for push-pull amplifier.

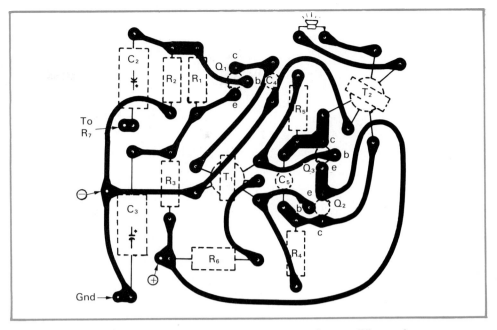

Figure 15-36. A suggested circuit board parts layout for amplifier project.

SUMMARY

- A power amplifier is designed to deliver power to a load. These amplifiers are normally used as the final stage to drive a power device such as a loudspeaker.
- A preamplifier amplifies weak signals so they are strong enough to drive the power amplifier.
- Methods of coupling transistor amplifier stages include RC coupling, transformer coupling, and direct coupling.
- The large-signal amplifier is usually mounted on a heat sink.
- Collector efficiency is the ratio of ac output power to dc input power.
- For an RC-coupled, class-A amplifier, maximum collector efficiency is 25%. Maximum efficiency of a transformer-coupled, class-A amplifier is 50%.
- The class-B amplifier obtains greater efficiency by moving the bias point to the cutoff point.
- In push-pull operation, two transistors are connected in parallel to allow an output current for both positive and negative halves of a signal. The output waveform is a reproduction of the original input wave, with some distortion.
- A standard push-pull amplifier consists of two transistors of the same type and a center-tapped transformer.
- Crossover distortion occurs in push-pull circuits where the EB junction is zero biased.
- A phase splitter supplies two outputs in antiphase to drive a push-pull amplifier.
- A circuit in which an npn and a pnp transistor with matched characteristics are connected in parallel is a complementary-symmetry amplifier.
- A speaker converts electrical energy into sound energy.
- Three types of speakers include permanent magnet, electromagnet, and electrostatic.

KEY TERMS

Each of the following terms has been used in this chapter. Do you know their meanings?

audio frequency
blocking capacitor
cascade
complementary-symmetry
  amplifier
condensation
coupling capacitor
crossover distortion
crossover network
diaphragm
direct coupling
electrodynamic speaker
electrostatic speaker

excursion
input characteristic curve
large-signal amplifier
midrange
multistage amplifier
permanent-magnet speaker
phase inverter
phase splitter
power amplifier
preamplifier
push-pull amplifier
push-pull operation

rarefaction
RC coupling
single-ended amplifier
sound wave
speaker
speaker voice coil
stage
transducer
transformer coupling
tweeter
universal output transformer
woofer

TEST YOUR KNOWLEDGE

Please do not write in this text. Place your answers on a separate sheet of paper.

1. A device that is used to block dc between stages while permitting ac to pass is termed a _____ capacitor.
2. An RC coupling circuit has a value of R of 500 Ω. What value of C should be used at a frequency of 400 Hz?
3. What method usually is used to overcome loss of gain due to mismatch with RC coupling?
   a. Add another transistor stage.
   b. Remove a transistor stage.
   c. Bypass the capacitor.
4. What are some advantages of transformer coupling?
5. A _____ turns ratio is required for a transformer to match 1600 Ω to 4 Ω.
6. What is the average power of a signal with V_{p-p} = 10 V and I_{p-p} = 0.4 A?
7. What is the collector efficiency of a power transistor when its ac output power is 100 mW and its dc input power is 300 mW?
8. A small forward bias is applied to transistors in a push-pull amplifier to reduce crossover distortion. True or False?
9. What is one advantage of the complementary-symmetry amplifier?
10. Why is a phase splitter required for the standard push-pull amplifier?
11. What is the difference between a PM speaker and an electrodynamic speaker?
12. A tweeter is a _____ frequency speaker.
13. A woofer is a _____ frequency speaker.
14. Why are crossover networks used with speakers?
15. What is the purpose of a heat sink?
    a. To store heat.
    b. To dissipate heat.
    c. To preheat transistor.

Section IV

SEMICONDUCTORS, POWER SUPPLIES, AND AMPLIFIERS

SUMMARY

Important Points

☐ Semiconductors are made of materials that are somewhere between conductors and insulators in terms of resistance.

☐ The invention of the transistor was announced in 1948.

☐ Diodes are rectifiers, which are able to pass current easily in one direction and block it in the other direction.

☐ Diodes have high resistance in the reverse-biased direction and low resistance in the forward-biased direction.

☐ Transistors are semiconductor amplifying devices. Bipolar junction transistors are either npn or pnp.

☐ Vacuum tubes were the forerunners of transistors. They have been largely replaced by transistors.

☐ A power supply is an electronic circuit that provides ac and dc voltages for equipment operation.

☐ Power supplies can:
  • Step up or step down ac line voltage to the required voltage using transformer action.
  • Change ac voltage to a pulsating dc voltage by either half-wave or full-wave rectification.
  • Filter pulsating dc voltage to produce a more pure dc for equipment use.
  • Provide voltage division for equipment use.
  • With the proper components, regulate output in proportion to the applied load.

☐ Most power supplies have a rectifier and filter. Transformers and voltage regulators are also used in some power supplies.

☐ Amplifiers control large output signals using small input signals.

☐ Forward bias results in low resistance to current. Reverse bias results in high resistance to current.

☐ Class-A amplifiers are biased so that output current flows for 360° of the input signal cycle.

☐ Class-B amplifiers are biased so that the output current flows for 180° of the input signal cycle.

☐ Class-C amplifiers are biased so that the output current flows less than 180° of the input signal cycle.

☐ Amplifiers can be connected in a number of different ways to increase gain.

☐ Power amplifiers are used to amplify large signals. They are usually the final output stage.

1. The short lead of an LED is the _____.
2. A rectifier changes _____ to _____.
3. A filter changes _____ dc to _____ dc.
4. The transistor replaced the _____ _____.
5. Draw the symbol for a diode and show the direction of forward-biased electron flow.
6. An amplifier has a change in ac output voltage of 50 mV and a change in ac input voltage of 10 μV. What is the voltage gain?
7. A device that controls a large output signal with a smaller input signal is an _____.
8. What is the voltage regulation in a power supply with a no-load voltage of 25 V and a full-load voltage of 24 V?
9. What is the percent ripple of an ac signal with a ripple value of 0.44 Vrms and total voltage of 72 Vpeak.
10. _____-_____ amplifiers are biased at cutoff.
11. Another name for amplification is _____.
12. Two transistors connected in parallel in an output stage operate as a _____-_____ amplifier.
13. What are the PIV and I_O ratings of a 1N4001 diode?

Section V COMMUNICATION

Communication is a vital part of our lives. Communication is the process of sending and receiving information. The information moves from a source, through a transmitter, to a receiver, and, finally, to a destination.

There are many ways to communicate. Methods that use electronics include human-to-machine, machine-to-human, and machine-to-machine. In electronic communication, the transmitted signal is often fed through a channel using waves or cable. Computers are an excellent example of an electronic communication medium.

Chapter 16 covers the generation of signals or waveforms in electronics. A number of types of oscillators are discussed.

Chapter 17 discusses radio-wave transmission and reception. Amplitude modulation (AM) and frequency modulation (FM) are explained. Also discussed is the transmission of electromagnetic waves. Radio-wave receivers are presented including: AM and FM receivers; superheterodyne receivers; citizens-band receivers. There are some excellent projects in Chapter 17.

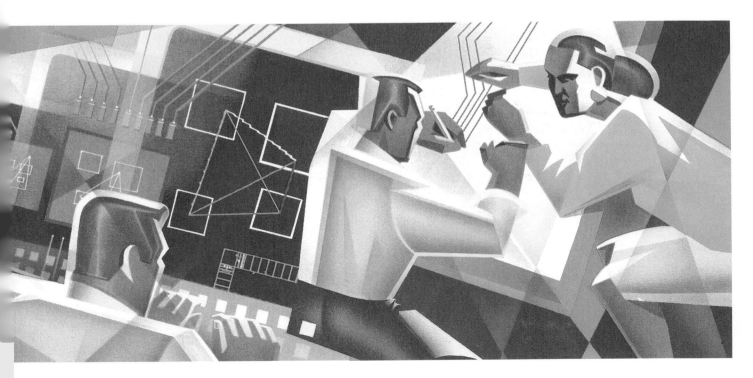

440

Applications of Electronics Technology:
This 18″ satellite dish enables consumers to receive 150 TV channels from the nation's first high-powered direct broadcast satellite distribution services. (RCA)

Chapter 16

OSCILLATORS

The pendulum on a grandfather's clock swings back and forth, marking the seconds. This is the way the clock keeps time. The swinging pendulum can be thought of as an *oscillator*. The main spring, which is wound with a key, is used to move the pendulum. The spring feeds energy to the pendulum to keep it oscillating.

After studying this chapter, you will be able to:
□ *Identify the conditions that must exist in a circuit to have oscillation.*
□ *Explain the use of a transistor as a switching device.*
□ *Describe some of the more common oscillator circuits.*
□ *Discuss the differences between crystal, RC, and tunnel-diode oscillators.*
□ *Discuss special multivibrator circuits used in computers.*

INTRODUCTION

In electronics, an **oscillator** is a circuit that produces a repetitive output waveform with only a dc input voltage. Like amplifiers, they convert dc *power* into ac power. However, the oscillator makes this conversion without the aid of an ac input.

Oscillators can be mechanical or electronic. An alternator is a good example of a mechanical oscillator. This device is limited mechanically to lower frequencies. More than a few thousand hertz and it would self-destruct. Electronic oscillators can generate frequencies in the megahertz range and higher.

An oscillator depends on feedback to sustain oscillation. Energy from the output of the system is fed back to the input and reamplified. The feedback must reinforce the input signal, or be *regenerative*. We call this **regenerative feedback, positive feedback,** or **regeneration.** It is the opposite of **negative feedback,** or **degeneration;** wherein, the signal feedback opposes the input signal. If feedback is degenerative, oscillation will not be sustained.

The classic example of feedback under exaggerated conditions, or feedback gone wild, is illustrated by the sound amplification system found at many public events. If the microphone gets too close to the speaker system, it will pick up music, speech, or just noise from the speaker. These sounds will be amplified and reproduced by the speaker, only to be picked up again and reamplified; around and around it goes, Figure 16-1. A disagreeable *howl* is produced.

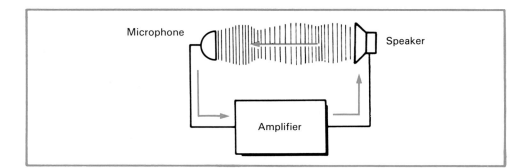

Figure 16-1. This amplifier produces a howl due to regenerative feedback.

In the case of the speaker system, regenerative feedback is undesirable. However, instead of the entire output signal, we may tap off a small portion of the signal and feed it back to the input. In this way, we can reinforce the input signal. With the right amount of feedback, an output signal of a desired amplitude and frequency can be sustained.

There are many types of oscillators. Before discussing them, though, a quick review of the principles of a parallel tuned circuit, or tank circuit, is in order.

PRINCIPLES OF OPERATION

In the circuit of Figure 16-2, capacitor C is charged when the switch is closed. When it is opened, C starts its discharge cycle, Figure 16-3, and a magnetic field is produced around L. When C is fully discharged, there is no more discharge current and the magnetic field collapses. This collapse induces a current, which recharges C. The current is in such a direction that C is charged to the polarity opposite to its previous charge, Figure 16-4.

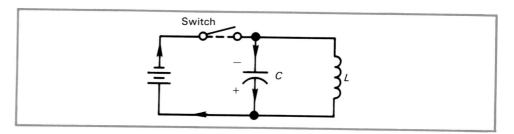

Figure 16-2. When the switch is closed, capacitor C charges.

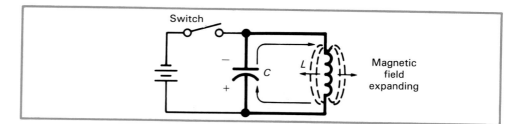

Figure 16-3. C discharges through L and produces a magnetic field.

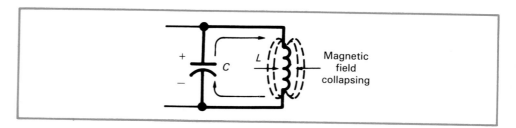

Figure 16-4. The collapsing magnetic field will charge C to the opposite polarity.

Now C attempts to discharge in the opposite direction, as in Figure 16-5. Again, a magnetic field is produced by the discharge current. Again, the magnetic field collapses after C has discharged, inducing a current that charges C to its original polarity. See Figure 16-5.

This current will continue to circulate until the energy originally supplied

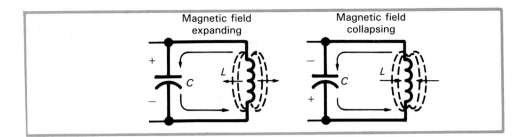

Figure 16-5. C discharges through L. As the field around L collapses, it charges C to the opposite polarity.

to charge C is dissipated, or used up, by the resistance of the circuit. The action of the current is described as **flywheel action.** The tuned circuit, as you know, is a tank circuit.

The circulating current and the voltage across L or C can be plotted as shown in Figure 16-6. Each successive wave is less in amplitude due to the loss of energy. This oscillation is called a **damped wave.**

The frequency of the wave output will depend upon the values of L and C. You will recall the resonant frequency formula as:

$$f_r = \frac{1}{2\pi\sqrt{LC}} = \frac{0.159}{\sqrt{LC}}$$

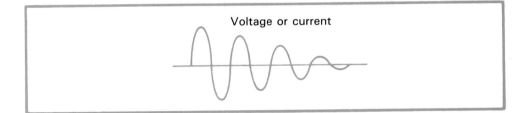

Figure 16-6. A damped oscillation successively decreases in amplitude until it dies out.

You might conclude that if a little bit of energy were added to the tuned oscillating circuit at the proper time, it would no longer decay to zero, but would continue to oscillate. This is quite correct. Then, the output of the circuit would appear as a **continuous wave (CW),** as pictured in Figure 16-7.

THE TRANSISTOR AS A SWITCH

So far, we have shown the oscillator circuit activated by manually operating a switch. Opening and closing a switch by hand is impractical for an actual

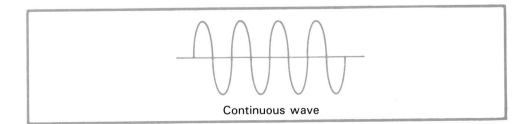

Figure 16-7. A continuous wave of constant amplitude and frequency.

working oscillator. For this purpose, a transistor is used as an extremely fast electronic switch.

Switching is not a mysterious term. The typical example is the ON/OFF light switch found most anywhere. Switching, as you know, may also be expressed as *opening* or *closing* a circuit. With a switch open, a circuit offers infinite resistance with no current. A closed switch offers near zero resistance and maximum current.

A mechanical light switch requires manual effort to make it "change its state" from on to off or vice versa. In the case of an electromagnetic switch, such as a relay, the change of state can be accomplished by applying a current through its operation coil. Refer to Chapter 5. This could be a steady current or a pulse of current.

Nevertheless, in either of these applications, the switching rate is relatively slow due to the inertia of the mechanical parts. It would be difficult to turn a light on or off manually more than two or three times per second. Using a transistor, though, circuits may be switched at remarkable speeds.

An examination of the curves of a typical switching transistor, Figure 16-8, discloses two states at which a transistor could operate as a switch:
- *On state*—with the transistor at saturation current with minimum resistance. I_B somewhere over 200 μA (in this case).
- *Off state*—with the transistor cut off and minimum current. Any current would be leakage current.

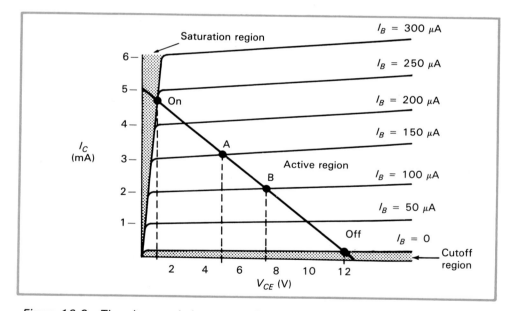

Figure 16-8. The characteristic curves of a switching transistor showing saturation region, active region, and cutoff.

In earlier discussions of transistor action, you discovered that these conditions may be obtained by changing the bias of the transistor. Actually, in the cutoff state, the emitter-base junction is in reverse bias state. In the saturation state, both emitter-base and collector-base junctions are forward biased. Operated in this manner, the circuit would be called a *saturated switching circuit*. This method has a disadvantage, since there is a slight delay to change a transistor from conducting in saturation to its cutoff state.

For extremely rapid switching applications, it is better not to drive the transistor to saturation. Rather, it should approach some point in the active region of the transistor. See points A and B in Figure 16-8. Depending on the particular requirements of a given application or device, one of these bias points might be enough to produce a voltage change (ΔV_{CE}) or a current change (ΔI_C) large enough for a useful output.

For example, consider point A. Assume a voltage pulse of $\Delta V_C \cong 7$ V will provide our required level of output. To achieve this, a pulse of base current of 150 μA is needed. As we can see from our load line, a 150-μA base current would produce a 3-mA collector current and a voltage change of about 7 V. If, instead, the transistor was driven to point B by a base current of 100 μA, we would not get the required level of output voltage. The roughly 5-V output pulse might be suitable for another application, however.

Conditions at saturation and cutoff

A change in the dc level of voltage applied as a bias between the emitter-base of a transistor could change its state from ON to OFF or vice versa. In Figure 16-9, a basic common-emitter switching circuit is drawn. Input pulses are indicated with the corresponding output pulses (values unrelated to Figure 16-8). Follow the action:

- When the input is at 0 V, the emitter-base voltage is also zero, and the transistor is cut off. With no collector current, there is no voltage drop across R_L and the collector voltage approaches the value of $-V_{CC}$.
- A pulse of -6 V applied to the input makes the base -6 V with respect to the emitter. This is severe forward bias, and the transistor is driven to saturation. The collector current causes a voltage drop across R_L close to the value of $-V_{CC}$, and the collector voltage approaches zero. Moving from $-V_{CC}$ to 0 V is in a positive direction, and the output is inverted when compared to the input pulse.

The resistor in the base circuit of Figure 16-9 is a current-limiting resistor. To compute the value of this resistor, assume a value of $-V_{CC}$ as -6 V. Ideally, at saturation, V_{R_L} equals $-V_{CC}$, and V_{CE} equals 0 V. I_C at saturation can be found, since we know that R_L equals 1000 Ω.

$$I_C = \frac{V_{CC}}{R_L} = \frac{6 \text{ V}}{1000 \text{ } \Omega} = 6 \text{ mA}$$

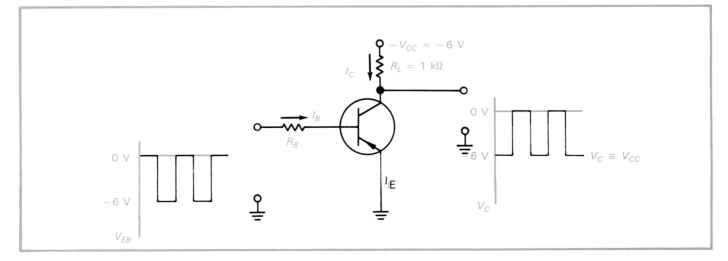

Figure 16-9. The common-emitter circuit is used to demonstrate switching action. R_B is a base, current-limiting resistor.

What base current is required to produce a collector current of 6 mA? A review of Chapter 13 will show that the dc current gain (β_{dc}, or h_{FE}) of the CE transistor is:

$$\beta_{dc} = h_{FE} = \frac{I_C}{I_B} \bigg| \; V_{CE} \text{ constant}$$

Transposing the formula, we find that:

$$I_B = \frac{I_C}{h_{FE}}$$

Assuming the current gain of this selected transistor is 50 (this value of h_{FE} will be found in the specifications of the transistor), then the base current will be:

$$I_B = \frac{6 \text{ mA}}{50} = 120 \text{ } \mu A$$

The voltage drop across the emitter-base junction is very small and can be disregarded for practical purposes. The value of the current limiting resistor will become:

$$R_B = \frac{V_{RB}}{I_B} = \frac{6 \text{ V}}{120 \text{ } \mu A} = 50 \text{ k}\Omega$$

The transistor switch and oscillator feedback

Follow the action in Figure 16-10. Assuming proper bias voltages for the transistor: When the circuit is turned on, collector current starts to flow. The increasing I_C through L_1 produces a magnetic field that is coupled to L_2 by mutual induction. The increasing I_C produces a current in L_2 of such a polarity that I_B is increased, which causes I_C to increase further.

This positive feedback action continues until transistor Q is conducting at saturation and I_C can increase no further. Such being the case, there is no further *change* in I_C. As a result, the secondary voltage L_2 drops to zero, and

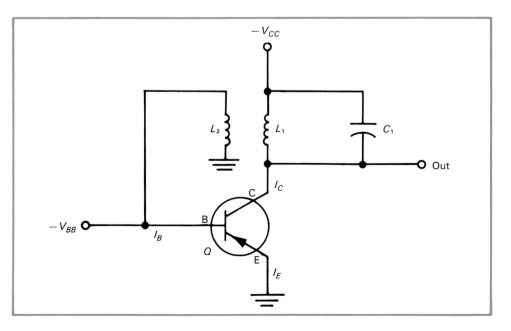

Figure 16-10. A typical oscillator circuit. Bias resistors are omitted for simplification.

the transistor returns to the dc bias level. Collector current I_C decreases as it returns from saturation to the Q-point. The changing I_C causes magnetic coupling again, but of such a polarity in L_2 that I_B is decreased. This causes another change in I_C, which causes another decrease in I_B until cutoff is reached. At cutoff, I_C again ceases to change, and there is no magnetic coupling. The transistor returns back up the load line to the Q-point. In the process, the cycle is repeated.

The frequency of the output depends upon the values of L and C in the tank circuit. Notice that it is the tank circuit that controls the points at which energy is added to the circuit. In the earlier example, energy was added to C at times determined by an outside source. In this circuit, energy is added at times determined by a transistor that is controlled by magnetic feedback. Coil L_2 is sometimes called a **tickler coil.** The output waveform of this circuit is a stable oscillation, Figure 16-11.

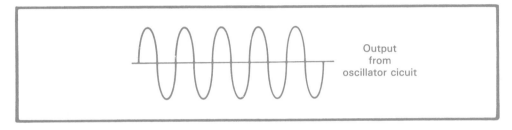

Figure 16-11. The output from an oscillator system with transistor switching is a continuous wave.

A basic block diagram for an oscillator is shown in Figure 16-12. Every oscillator circuit has a wave producing circuit, an amplifier, and a feedback circuit. The feedback circuit sends a small electrical signal back to the wave producing circuit to sustain the oscillations.

BASIC OSCILLATOR DESIGNS

There are a number of different oscillators, and each oscillator has numerous variations. The *Armstrong, Clapp, Colpitts, Hartley,* and *Pierce* name just a few of the variations. To gain a better understanding of oscillators, you will examine the operation of three simple designs—the Armstrong, the Hartley, and the Colpitts oscillators.

The Armstrong oscillator

An **Armstrong oscillator** circuit is shown in Figure 16-13. From this circuit, the basic theory of oscillators can be shown. Notice the tuned tank circuit,

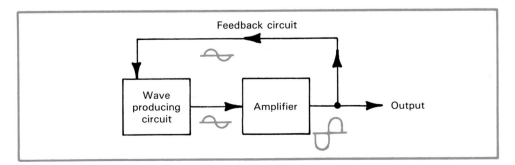

Figure 16-12. Block diagram for a basic oscillator.

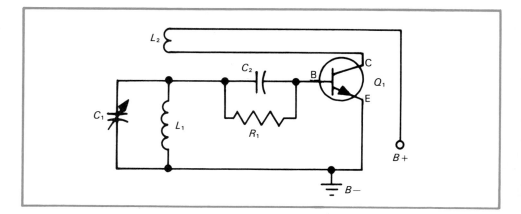

Figure 16-13. Schematic of an Armstrong oscillator.

$L_1 C_1$, which determines the frequency of the oscillator. Follow the action for this circuit.

When the voltage is applied to the circuit, current flows from the negative terminal of the battery $(B-)$, through the transistor and L_2, to the positive terminal of the battery $(B+)$, Figure 16-14. L_2 is closely coupled to L_1. The expanding magnetic field of L_2 makes the collector end of L_1 positive. C_1 charges to the polarity shown. The base of Q_1 also collects electrons. It charges C_2 in the polarity shown.

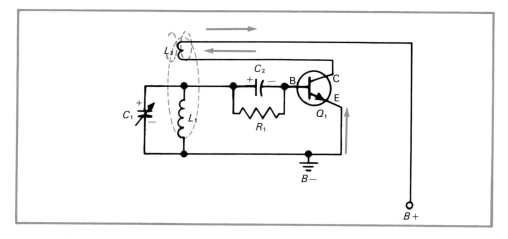

Figure 16-14. Armstrong oscillator operation. Current flows from $B-$ to $B+$. Energy is added to tank circuit through tickler coil.

When Q_1 reaches its saturation point, there is no longer a change of current in L_2. This is shown in Figure 16-15. Magnetic coupling to L_1 drops to zero. The negative charge on the base side of C_2 is no longer opposed by the induced voltage of L_1. The negative charge drives the transistor to cutoff. This rapid decrease in current through the transistor and L_2 causes the base end of L_1 to become negative. This increases the reverse bias on Q_1. C_1 discharges through L_1. C_2 bleeds off its charge through R_1.

Q_1 is held at cutoff until the charge on C_2 is bled off to above cutoff. At that time the transistor starts conduction and the cycle is repeated.

There are a few points to remember in Armstrong oscillator operation:
• The voltage developed across L_1 first opposes and then adds to the bias

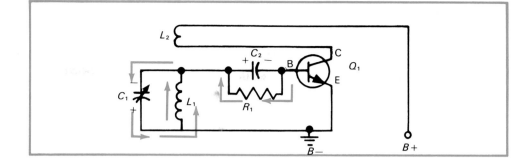

Figure 16-15. Armstrong oscillator operation—the first half-cycle of oscillation.

developed by the R_1C_2 combination.

- The energy added to the tuned tank circuit, L_1C_1, by the tickler coil, L_2, is great enough to offset the energy lost in the circuit due to resistance. The coupling between L_1 and L_2 can be adjusted.
- The combination R_1C_2 has a fairly long time constant. It sets the operating bias for the transistor. Q_1 is operated class C.

Study the voltage waveform on the base of Q_1 in Figure 16-16. The shaded portion is transistor conduction. At point A the bias is negative (reverse). This results from the charge on C_2 plus the induced voltage across L_1. Interval A to B denotes the time that passes while C_2 discharges through R to the cutoff point and conduction begins for the next cycle.

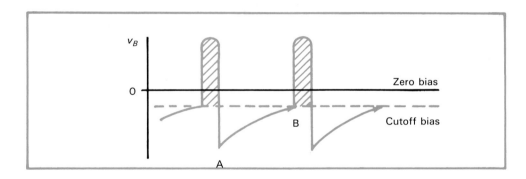

Figure 16-16. Voltage waveform on base of oscillator transistor. Compare it to the description of circuit action.

The Hartley oscillator

The **Hartley oscillator** circuit, Figure 16-17, depends upon magnetic coupling for its feedback. Its operation and theory is the same as the tickler coil and transformer. The Hartley oscillator is identified by a tapped tank coil or autotransformer (L_1 and L_2 in Figure 16-17). Follow the action for this circuit.

As the circuit is turned on, current through L_1 induces a voltage at the top of L_2. This makes the base of transistor Q more negative. This drives the transistor to saturation. At saturation, there is no longer a change in current, and the coupling between L_1 and L_2 falls to zero. The less negative voltage at the base of Q causes the transistor to decrease in conduction. This decrease, in turn, induces a positive voltage at the top end of L_2. This is reverse bias for the transistor, so the transistor is quickly driven to cutoff. Then, the cycle is repeated. The tank circuit is energized by pulses of current. The transistor alternates between saturation and cutoff at the same frequency.

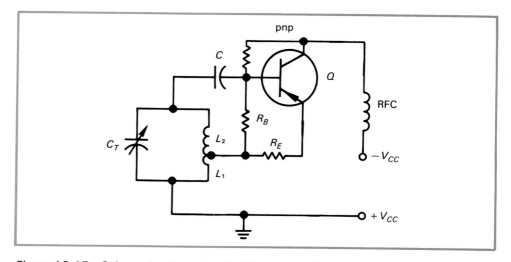

Figure 16-17. Schematic of a series-fed Hartley oscillator.

The at-rest, or quiescent, bias condition of the transistor is established by resistors R_B and R_E. The radio frequency choke (RFC) blocks the RF signal (coming from the oscillator) from the power source of the circuit. In this circuit, note that the coil L_1 is in series with the collector circuit of the transistor. It is called a *series-fed* Hartley oscillator.

In Figure 16-18, a *shunt-fed* Hartley oscillator is shown. The operation is the same. Note that the dc path for the emitter-collector current is not through the coil L_1. The ac signal path, however, is through C and L_1. At point A, the two current components are separated and required to take parallel paths. This oscillator, like the series-fed Hartley, receives its feedback energy by means of magnetic coupling.

The Colpitts oscillator

Feedback to sustain oscillation may also be accomplished by means of an electrostatic field, as developed in a capacitor. If the tapped coil from the Hartley oscillator is replaced with a split stator capacitor, oscillation can also be made to occur. This circuit is called the **Colpitts oscillator,** Figure 16-19.

The Colpitts oscillator is similar to the Hartley oscillator in theory of operation. However, in the Colpitts setup, the signal is coupled back to C_1 of the

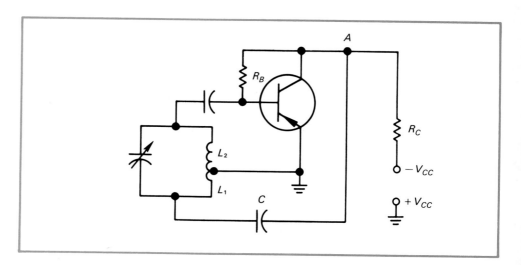

Figure 16-18.   Schematic of a shunt-fed Hartley oscillator.

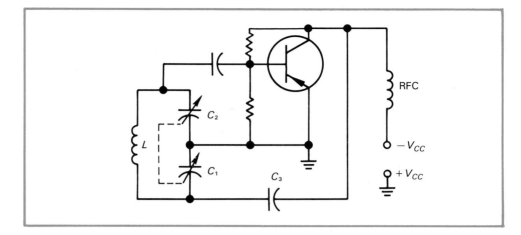

Figure 16-19. Schematic diagram of a Colpitts oscillator.

tank circuit through coupling capacitor C_3. A changing voltage at the collector appears as a voltage across the tank circuit LC_1C_2 in the proper phase to be a regenerative signal. The amount of feedback depends upon the ratio of C_1 to C_2. Often, this ratio is fixed, and both capacitors, C_1 and C_2, are adjusted by a single shaft. Note that adjustable capacitors that work this way are called **ganged capacitors**.

The tuned tank consists of C_1 and C_2 in series and L. Note that the circuit is shunt fed. Series feed is impossible due to the blocking of dc by the capacitors.

VARIATIONS OF BASIC OSCILLATOR DESIGNS

There are a number of different electronic components that may be used to set a system into oscillation. *Piezoelectric crystals*, *RC combinations*, and *tunnel diodes* are three items that use their different characteristics to induce circuits to oscillate. Further examination of these setups is needed.

Crystal oscillators

The piezoelectric effect of crystals was studied in Chapter 2. When a crystal is energized or excited, it vibrates at a frequency that is set by its physical dimensions. The crystals are cut from quartz, and their characteristics will depend on how they are cut. By inserting a crystal into an oscillator setup, a **crystal oscillator** can be created.

Before examining a crystal oscillator, it is well to look at the electrical equivalent of the crystal, as shown in Figure 16-20. A crystal is placed between

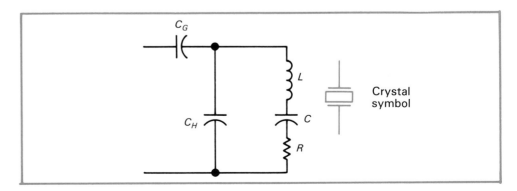

Figure 16-20. The electrical equivalent circuit of a crystal. At right is its schematic symbol.

two metallic holders. This forms a capacitor C_H with the crystal itself as the dielectric. C_G represents the series capacitance between the metal holding plates and the air gap between them as a dielectric. L, C, and R represent the characteristics of the crystal. Of special importance is the similarity of the equivalent crystal circuit to a tuned circuit. They both have a resonant frequency.

A typical mounted crystal is displayed in Figure 16-21. A crystal ground to vibrate at a frequency of 1 MHz would be approximately 1 inch² and 0.1125 inch thick. The frequency would increase if the crystal were ground thinner.

Figure 16-21. Typical crystal used to fix the transmitting frequency of a radio station. (Texas Crystals)

A transistor crystal oscillator circuit is drawn in Figure 16-22. Compare this circuit with Figure 16-17. They are the same except a crystal has been added to the feedback circuit of the new oscillator. The crystal acts as a series resonance circuit and determines the frequency of the feedback currents. The tank circuit must be tuned so that the feedback reinforces the crystal's oscillation.

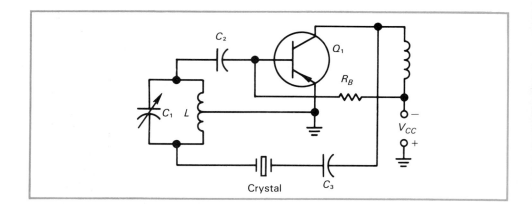

Figure 16-22. A crystal-controlled Hartley oscillator circuit.

In the circuit of Figure 16-23, the crystal is used in place of the tuned circuit in a Colpitts oscillator. This version has been named the **Pierce oscillator.** Compare the circuit to Figure 16-19.

The amount of feedback needed to keep the crystal oscillating, again, depends upon the ratio of C_1 to C_2. These capacitors form a voltage divider across the base-emitter junction of the transistor. The Pierce oscillator circuit is a very stable circuit under varying circuit conditions and changes. Another useful feature of the Pierce oscillator is the ease in changing oscillation frequencies. A new crystal may be inserted for a new frequency.

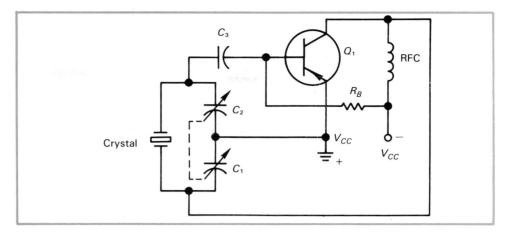

Figure 16-23. A crystal-controlled Pierce oscillator circuit.

RC oscillators

If part of the output of an amplifier is fed back to the input such that it is in phase or regenerative, another type of oscillator, an **RC oscillator,** may be created. Refer to Figure 16-24, in which two stages of CE amplifiers are coupled together.

First, consider the circuit without C_F, the coupling feedback capacitor. A positive-going signal into Q_1 produces a negative going signal at the collector of Q_1. The signal is inverted. A negative-going signal at Q_2 produces a positive-going signal at the output of Q_2. The signal now is in phase with the input to Q_1. A positive-going signal at the collector will produce a negative-going signal at the emitter of Q_2. This voltage is coupled to the emitter of Q_1, which decreases the forward bias of Q_1. This has exactly the same effect as a positive input to Q_1. Thus, it is regenerative feedback. The circuit oscillates with a frequency that depends on the values of R and C. However, this design is not stable. Usually, the circuit can operate over a range of frequencies. More practical RC circuits can be designed.

In Figure 16-25, a single transistor RC oscillator is diagrammed. This setup is called a **phase-shift oscillator.** This oscillator takes advantage of the fact that an RC circuit produces a phase shift. That is, in a capacitive circuit, the

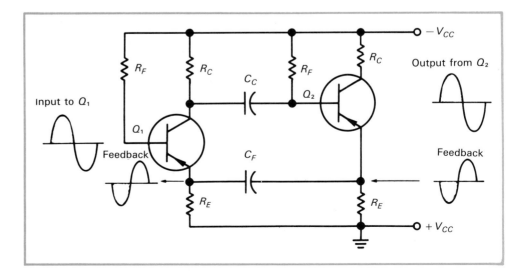

Figure 16-24. A two-stage RC oscillator.

phase of the current leads the applied voltage by an angle of 90° or less. In this circuit, R_1C_1 shifts the phase by 60°, R_2C_2 shifts it another 60°, and R_3C_3 shifts it a final 60° for a 180° phase shift. As a result, a positive-going signal is inverted by the phase-shift network. This produces a regenerative signal for the transistor input. The conditions for oscillation are satisfied.

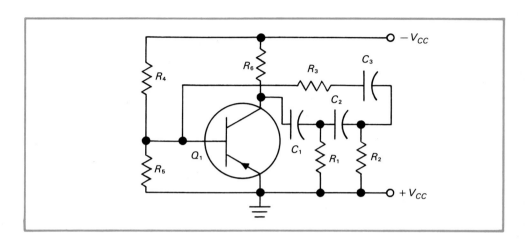

Figure 16-25. A phase-shift oscillator circuit.

Tunnel-diode oscillators

Another circuit can create oscillations with the use of a tunnel diode. Before exploring the workings of a *tunnel-diode oscillator*, a review of the concept of negative resistance is necessary.

Negative resistance. Ordinarily, we describe resistance in terms of Ohm's law. An increase in resistance would produce a decrease in current; an increase in voltage would produce an increase in current. There are some effects that seemingly contradict this law. An example of this is the characteristic of negative resistance. It is seen when current decreases with an increase in voltage. A typical characteristic curve of a negative-resistance device is drawn in Figure 16-26.

A tuned circuit will not continuously oscillate unless energy is added to overcome the resistance of the tank circuit. Why not add negative resistance to cancel the positive resistance? Then, the circuit would oscillate continuously.

Tunnel diode. In Chapter 12, we discovered that a pn junction conducts when biased in a forward direction. Conduction is low until the bias voltage

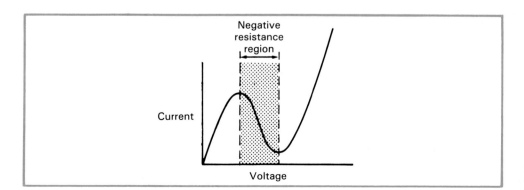

Figure 16-26. In the negative-resistance region, the current decreases as voltage is increased.

reaches a level sufficient to overcome the barrier potential. To express it another way, the carriers must gain sufficient energy to cross the depletion region. The characteristic curve for a *typical* diode is drawn in Figure 16-27.

The typical diode does not exhibit the peculiar characteristic of negative resistance. However, among devices that do is the **tunnel diode**. (It is also known as the *Esaki tunnel diode*, named after its inventor). The principle underlying this device is that current carriers may tunnel through the depletion region even though they have insufficient energy to cross over it. The requirement made on the device is that the depletion region be made very narrow. A tunnel diode is made of two very heavily doped crystals. The p crystal is rich in holes, and the n crystal is rich with free electrons. Consequently, the depletion region is very narrow, and the carriers can tunnel through the region and cross the junction.

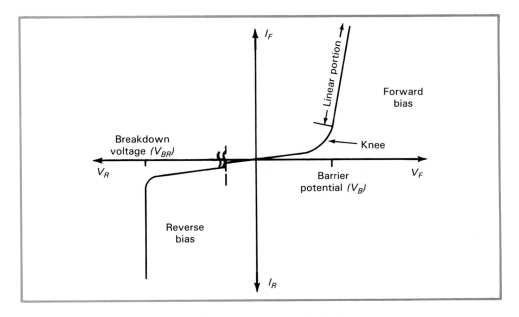

Figure 16-27. The characteristic curve of a typical diode.

The voltage-current characteristics of a typical tunnel diode are shown in Figure 16-28. Note that when a small voltage is applied, a small current starts to flow. From zero to *peak*, the current is a result of carriers tunneling through the depletion region. A further increase in voltage causes the current to decrease. At the *valley* voltage, the diode starts to act with a positive resistance and current rises again as would be expected for a conventional diode. The current after the valley is the result of carriers crossing and joining at the junction. It is the area of negative resistance which interests us at the moment.

The negative resistance may be used in parallel with a tuned circuit or in series with a tuned circuit. In either case, its purpose is to cancel the positive resistance of the oscillatory circuit. The tunnel diode will be biased near the center of the negative resistance portion of its curve.

Tunnel-diode oscillator operation. Study the circuit in Figure 16-29, and follow its action. R is selected to provide a voltage across diode D (V_D) near the center of its negative-resistance curve. When the switch is closed, current rises to a value determined by the resistance of R plus the resistance of D. The voltage divides across D and R according to the ratio of their resistances.

However, as V_D passes point P, it enters the negative area, and resistance

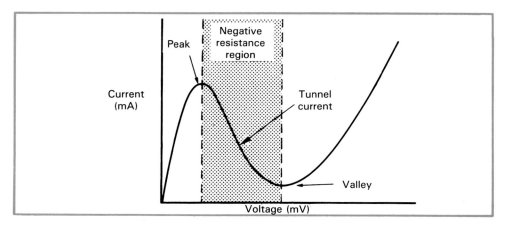

Figure 16-28. The characteristic curve of a tunnel diode.

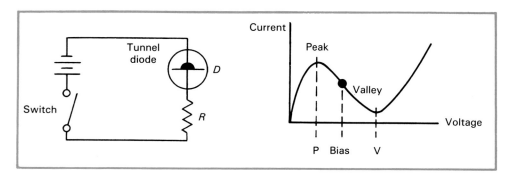

Figure 16-29. Simple diagram for explaining tunnel-diode oscillation.

starts to increase. Therefore, a larger voltage appears across D, and its resistance continues to increase until V_D reaches point V on the curve. At this point, a further increase in voltage drives the diode into its positive resistance region. The resulting increase in current increases the voltage across R (V_R), and V_D will again decrease into its negative resistance region. This decrease in V_D decreases its resistance, and circuit current continues to rise until point P is again reached.

This describes one cycle of operation. The circuit will continue to oscillate back and forth through the negative resistance region. The output across R is similar to a sine wave.

A practical circuit using the tunnel diode is shown in Figure 16-30. In this case, negative resistance is used in series with a tuned circuit consisting of capacitor C and the speaker voice coil.

First, the switch is closed, and the voltage is adjusted across R_1 and R_3 until the proper bias point is reached. This point can be determined using an oscilloscope connected across the speaker. An audible tone should be heard from the speaker. The oscillations of the tunnel diode are now energizing the tuned circuit, and a continuous wave is produced at its output. In this circuit, R_3 sets the proper bias level for the diode; R_1, in parallel with the tank, sets the proper current level for the diode.

MULTIVIBRATORS

Multivibrators are oscillators that couple the output of each of two electronic devices, commonly transistors, to the input of the other. Multivibrators produce a square wave signal. They are designed to have zero (*astable*), one (*monostable*), or two (*bistable*) stable output conditions. All three designs are

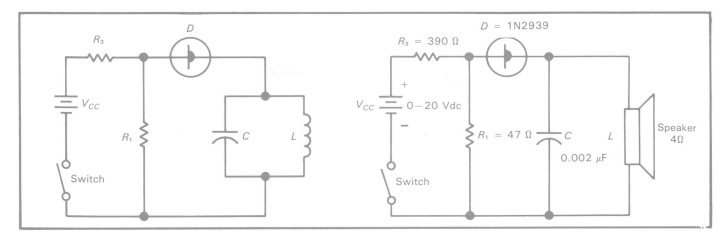

Figure 16-30. *A schematic of a practical tunnel-diode oscillator circuit is shown on the left, and an audio application of the circuit is shown on the right.*

worth studying. Note that all three are commonly available as ICs. First, a little background on the pulse signal.

Pulse terminology

Look at the ideal pulse wave in Figure 16-31. Then, familiarize yourself with the following terms.

- **Leading edge:** Moving from left to right, this is the edge where the pulse rises from zero to a maximum.
- **Trailing edge:** This is the edge where the pulse drops from a maximum to zero.
- **Peak value:** The maximum value of a pulse.
- **Average value:** Value of current or voltage found by dividing the area under one alternation by the distance along the *x*-axis. This level is indicated by the dashed line in Figure 16-31.
- **Pulse duration (t_d):** The time that a pulse remains above a certain percent of its maximum value (usually 90%).
- **Resting period:** The time interval between pulses.
- **Cycle:** One complete current alternation, from the beginning of one pulse to the beginning of the next pulse.
- **Pulse repetition rate (prr):** The number of cycles per unit time.
- **Pulse recurrence time (prt):** The time interval for one cycle. The reciprocal of pulse repetition rate.
- **Duty cycle:** This is the ratio between the average value of the pulse and the peak value of the pulse.

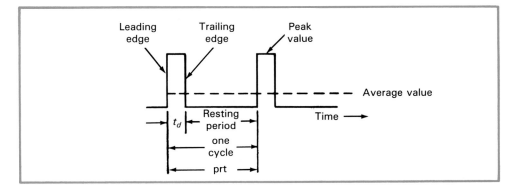

Figure 16-31. *Terminology of an ideal pulse wave.*

$$\text{duty cycle} = \frac{\text{average value}}{\text{peak value}}$$

Expressed another way, duty cycle refers to the ratio of *on* time to *total* time.

$$\text{duty cycle} = \frac{\text{pulse duration}}{\text{pulse recurrence time}} = \frac{t_d}{\text{prt}}$$

Since pulse repetition rate is the reciprocal of pulse recurrence time:

$$\text{duty cycle} = t_d \times \text{prr}$$

- **Rise time (t_r):** The time required for the buildup or rise of a wave. It is the time period between a voltage change from 10% to 90% of peak value. See Figure 16-32.
- **Fall time (t_f):** The time required for a pulse to drop or decay from 90% of its peak value to 10%.

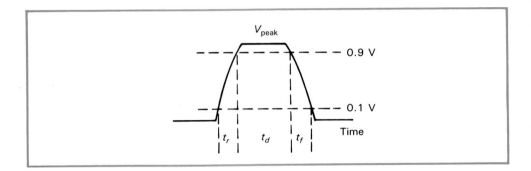

Figure 16-32. *Terminology of the rise and decay of a pulse.*

Astable multivibrators

A free-running multivibrator circuit is called an **astable multivibrator.** This circuit could be called an oscillator with a square wave output. A circuit for the oscillator is drawn schematically in Figure 16-33. It uses common-emitter resistance for feedback. When this circuit is energized, emitter currents of both transistors flow through R_E. Current causes a voltage drop across collector load resistors R_{C_1} and R_{C_2}. The collector voltages become less negative by the amounts of the voltage drops. Which transistor will conduct first? Follow the action.

When the circuit is first turned on, the capacitor instantaneously attempts to charge to the value of $-V_{CC}$. This charging current will make the base end of R_B more negative than the emitter of Q_2. This quickly drives Q_2 to saturation. The emitter end of R_E is driven quite negative due to the current of Q_2, and Q_1 is cut off.

Now, Q_2 conducts fully while C is charging. However, as C approaches full charge at an exponential rate, the current decays at the same rate. When the current drops to zero, the base end of R_B approaches zero, and Q_2 conducts less. A decrease in current causes the emitter end of R_E to become less negative. This starts Q_1 conducting. The collector of Q_1 becomes more positive, which is coupled back to the base of Q_2.

Q_2, then, is driven to cutoff, and C will discharge to the new value of the Q_1 collector voltage. This discharge current makes the base of Q_2 even more positive. Q_1 remains conducting until C has discharged. Then, the reverse bias is removed from Q_2, and conduction is switched to Q_2. This oscillation con-

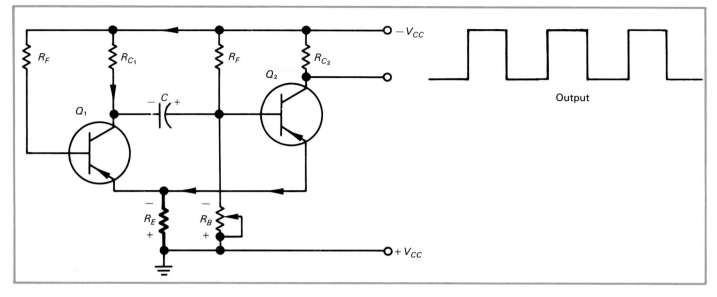

Figure 16-33. An emitter-coupled astable multivibrator circuit.

tinues. Its frequency is dependent upon the circuit time constants.

Astable multivibrator project. A second method of designing an astable multivibrator is shown in Figure 16-34. Construct this circuit for yourself. Connect oscilloscope input probes across transistor's collector and emitter terminals and examine the operation.

The multivibrator begins oscillating because of a slight difference between

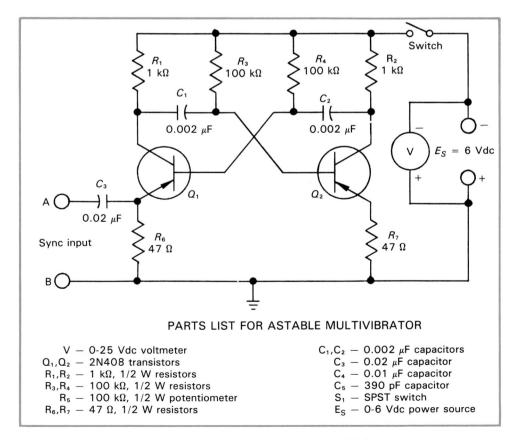

PARTS LIST FOR ASTABLE MULTIVIBRATOR

V — 0-25 Vdc voltmeter	C_1, C_2 — 0.002 μF capacitors
Q_1, Q_2 — 2N408 transistors	C_3 — 0.02 μF capacitor
R_1, R_2 — 1 kΩ, 1/2 W resistors	C_4 — 0.01 μF capacitor
R_3, R_4 — 100 kΩ, 1/2 W resistors	C_5 — 390 pF capacitor
R_5 — 100 kΩ, 1/2 W potentiometer	S_1 — SPST switch
R_6, R_7 — 47 Ω, 1/2 W resistors	E_S — 0-6 Vdc power source

Figure 16-34. Schematic and parts list for an astable multivibrator.

the conduction of Q_1 and Q_2. Assume that Q_1 starts to conduct slightly more than Q_2. The conduction of Q_1 causes a larger voltage drop across R_1 and, consequently, a less negative voltage at the collector of Q_1. This positive-going signal is coupled through C_1 to the base of Q_2. This decreases the forward bias of Q_2. Q_2 is quickly driven to cutoff. A decreasing current in Q_2 causes its collector to become more negative. This negative-going voltage is coupled to the base of Q_1, which is quickly driven into saturation.

At this time, C_2 charges to the collector voltage of Q_2, and C_1 discharges to the collector voltage of Q_1 as governed by the time constant of the circuit. After an interval, the base end of C_2 becomes positive, and the base end of C_1 becomes negative. This decreases the forward bias of Q_1 and decreases the conduction of Q_1. Also, the negative voltage at the base of Q_2 increases the forward bias, and Q_2 is driven into conduction. Action continues with Q_1 and Q_2 alternating conduction. A square wave is produced. The frequency is dependent upon the time constant of the circuit.

This circuit can be kept synchronized by means of an input pulse at a frequency a little higher than the circuit's free-running frequency. To understand this action. assume a positive pulse voltage is applied across R_6. This would drive Q_1 immediately into conduction by increasing the forward bias of its EB junction. When this occurs, even if Q_1 is not quite ready to conduct, it will be forced into conduction. The oscillator will lock-in on the pulse frequency.

Pulse synchronization is used in your TV set to produce a stable picture. The horizontal hold and vertical hold controls on the TV set make slight adjustments to the horizontal and vertical oscillator frequencies so that they can lock-in on the pulse sent by the TV station.

Monostable multivibrators

A **monostable multivibrator** circuit has one stable state. The circuit is frequently referred to as a **one-shot multivibrator.** A pulse from an external source causes the transistors to have a change in conduction. After an interval of time, depending on the circuit time constant, it will switch back to its original state. See Figure 16-35.

Assume that Q_1 is conducting at saturation, and Q_2 is cut off. This is the

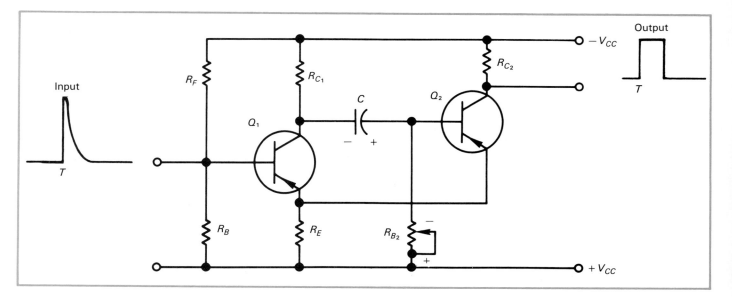

Figure 16-35. A monostable multivibrator circuit.

situation as voltage dividers $R_F R_B$ set up a forward bias for Q_1, while Q_2 has no forward bias. The collector voltage of Q_1 is less negative due to the large voltage drop across R_{C_1}.

When a positive pulse is applied to input across R_B, it places a reverse bias on Q_1 and cuts it off. The collector voltage of Q_1 approaches $-V_{CC}$ since there is no drop across R_{C_1}. A large voltage appears across R_{B_2} in the polarity shown in Figure 16-35 (due to charging current for C). Q_2 is driven to conduct and continues conducting until C is charged to the value of $-V_{CC}$. At that time, the voltage across R_{B_2} disappears, Q_2 is cut off, and the circuit reverts to its original state with Q_1 conducting.

The duration of time that Q_2 conducts depends upon the time constant of the RC circuit. For the multivibrator to go through its entire cycle using one input pulse, the time required for Q_2 to turn on and off must be greater than the time duration of the triggering pulse.

Bistable multivibrators (flip-flops)

The **bistable multivibrator,** or **flip-flop,** is widely used in computer circuitry. The circuit requires two pulses to return it to its original state. Remember, a transistor can be either *on* or *off* in its stable state. Therefore, the on and off can be related to the 1 and 0 of the *binary number system*.

Figure 16-36 diagrams the circuit. Biasing resistors for each transistor are matched. The circuit is symmetric.

Assume that Q_1 is on and Q_2 is off. A positive pulse to the base of Q_1 will reverse bias the EB junction, and Q_1 is cut off. The collector voltage of Q_1 approaches $-V_{CC}$. Now, R_1 and R_{B_2} form a bias voltage divider for Q_2. The base of Q_2 becomes more negative because of the increased divider current. Q_2 turns *on*. The collector voltage V_C of Q_2 becomes more positive, and voltage at the base of Q_1 becomes less negative, so transistor Q_1 is cut off.

These conditions can be tabulated as follows:

Q_1	Q_2	V_{C_1}	V_{C_2}	V_{B_1}	V_{B_2}
ON	OFF	$\cong 0$	$-V_{CC}$	Negative	Positive
OFF	ON	$-V_{CC}$	$\cong 0$	Positive	Negative

You will observe that a positive pulse switches conduction from Q_1 to Q_2. In order to switch the circuit back, a negative pulse may be applied to Q_1, or a positive pulse may be applied to Q_2.

By using several of the flip-flops in cascade, a counter circuit can be made with a binary number readout. The output of each flip-flop is wired to the input of the succeeding flip-flop, Figure 16-37.

In each flip-flop, a light is placed in the collector circuit of Q_1. When Q_1 is conducting, the light will glow. Assume that all transistors begin off. See the binary pulse table in Figure 16-38.

In this table, note that:
- FF1 switches with every pulse.
- FF2 switches with every second pulse.
- FF3 switches with every fourth pulse.
- FF4 switches with every eighth pulse.

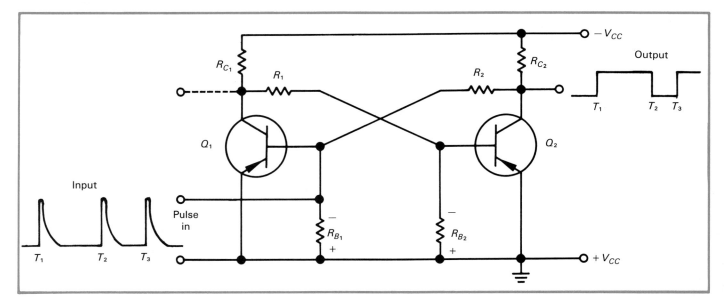

Figure 16-36. A bistable multivibrator, or flip-flop, circuit.

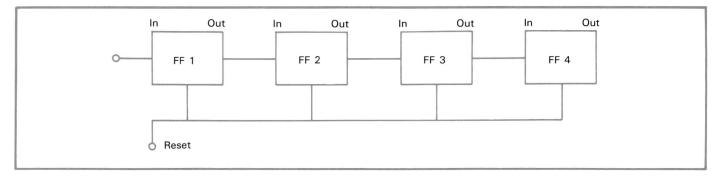

Figure 16-37. Block diagram of a counter circuit made of cascaded flip-flops.

DIGITAL	PULSES IN	LIGHTS ON	BINARY
0		0000	0000
1		0000●	0001
2		000●0	0010
3		000●●	0011
4		00●00	0100
5		00●0●	0101
6		00●●0	0110
7		00●●●	0111
8		0●000	1000

CAN CONTINUE TO 15 PULSES

Figure 16-38. Binary pulse table for a four flip-flop setup.

At 16 pulses, this counter returns to zero. All lights are out. Additional stages can be added to count to any number desired.

In Chapters 19 and 20, binary number systems will be presented in greater detail. A knowledge of number systems other than base 10, binary in particular, is important to electronic engineers and technicians.

Other types of waveforms

An ideal square wave is a difficult form to create. The actual wave may resemble a wave similar to Figure 16-39 (right). To understand this, you must realize that the leading edge and trailing edge of a wave cannot be perfectly vertical. A minute fraction of time *must* elapse as the wave builds up or decays. When a rise time is extremely rapid, the leading edge approaches a straight vertical line.

There are many other types of waveforms that find wide application in electronic circuitry. We will not explore deeply into the generation of these waves, but rather give you a speaking knowledge of them. Some have already been discussed in earlier chapters in this text. Study Figures 16-39 to 16-45.

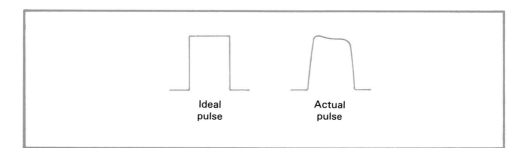

Figure 16-39. Comparison of an ideal pulse to an actual pulse.

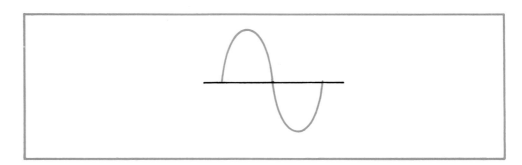

Figure 16-40. **Sine wave.** *Curve is a function of the angle θ. See Chapter 6.*

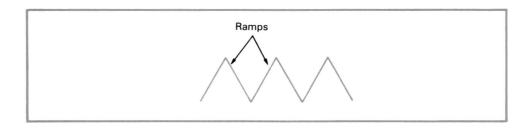

Figure 16-41. **Triangular wave.** *The positive and negative ramps have the same slant and time duration.*

Figure 16-42. **Sawtooth wave.** *The wave is similar to the triangular wave except the positive ramps and negative ramps have unequal slopes. The rise time is greater than the fall time.*

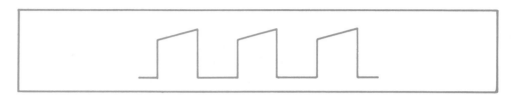

Figure 16-43. **Trapezoidal wave.** *This wave consists of a positive step and a positive ramp followed by a negative step.*

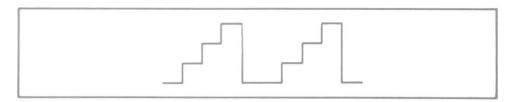

Figure 16-44. **Staircase wave.** *This wave is a series of positive steps followed by a negative step.*

Figure 16-45. **Exponential wave.** *The rise and fall of this wave is dictated by a mathematical equation involving a variable exponent.*

MORSE-CODE OSCILLATOR PROJECT

To experiment with a basic oscillator, like those you have been studying, you may wish to build the Morse-code oscillator pictured in Figure 16-46. The schematic and parts list is given in Figure 16-47. By plugging in the key, Figure 16-46, the circuit is activated. Morse-code signals can be produced by repeated pressing of the key. You can control the audio signal through R_1.

SUMMARY

- An oscillator depends on regenerative feedback to sustain oscillation.
- Transistors are commonly used as rapid switching devices in oscillators.
- The Hartley oscillator uses magnetic coupling to receive its feedback.

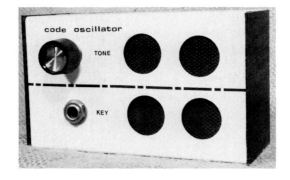

Figure 16-46. Morse-code oscillator.

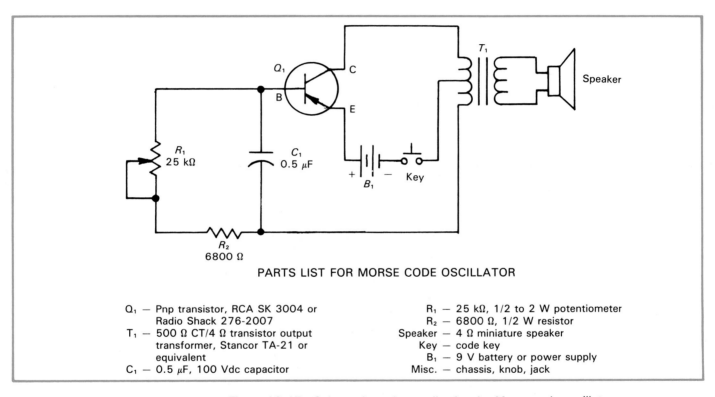

PARTS LIST FOR MORSE CODE OSCILLATOR

Q_1 — Pnp transistor, RCA SK 3004 or
    Radio Shack 276-2007
T_1 — 500 Ω CT/4 Ω transistor output
    transformer, Stancor TA-21 or
    equivalent
C_1 — 0.5 μF, 100 Vdc capacitor

R_1 — 25 kΩ, 1/2 to 2 W potentiometer
R_2 — 6800 Ω, 1/2 W resistor
Speaker — 4 Ω miniature speaker
    Key — code key
    B_1 — 9 V battery or power supply
    Misc. — chassis, knob, jack

Figure 16-47. Schematic and parts list for the Morse-code oscillator.

- The Colpitts oscillator obtains feedback through an electrostatic field developed in capacitors.
- The piezoelectric effect of quartz crystals is used to create oscillations in many different types of oscillators. The physical dimensions of the crystal determine its oscillating characteristics.
- A phase shift oscillator takes advantage of the phase shift characteristics of an RC circuit. The output of the circuit is shifted 180° and then fed back into the input.
- A tunnel diode allows some current carriers to *tunnel* through the barrier at the pn junction. This gives the diode its negative-resistance characteristics.
- The negative resistance of a tunnel diode can be used to create oscillations in a circuit.
- Multivibrators produce a square wave signal with zero, one, or two stable states.

- An astable multivibrator circuit produces an oscillating square wave.
- A bistable multivibrator, or flip-flop, has two stable states that can represent the 1 and 0 states in the binary number system.

KEY TERMS

Each of the following terms has been used in this chapter. Do you know their meanings?

Armstrong oscillator
astable multivibrator
average value
bistable multivibrator
Colpitts oscillator
continuous wave (CW)
crystal oscillator
cycle
damped wave
degeneration
duty cycle
exponential wave
fall time (t_f)
flip-flop
flywheel action

ganged capacitor
Hartley oscillator
leading edge
monostable multivibrator
multivibrator
negative feedback
one-shot multivibrator
oscillator
peak value
phase-shift oscillator
Pierce oscillator
positive feedback
pulse duration (t_d)
pulse repetition rate (prr)
pulse recurrence time (prt)

RC oscillator
regeneration
regenerative feedback
resting period
rise time (t_r)
sawtooth wave
sine wave
staircase wave
tickler coil
trailing edge
trapezoidal wave
triangular wave
tunnel diode
tunnel-diode oscillator

TEST YOUR KNOWLEDGE

Please do not write in this text. Place you answers on a separate sheet of paper.

1. How does an electronic oscillator differ from an amplifier?
2. Energy from the output of a sound amplification system is fed back to the input and reamplified. This is called:
   a. Feedback.
   b. Phase shift.
   c. Flip-flop.
3. Switching may be expressed as _____ or _____ a circuit.
4. There are two states at which a switching transistor will operate. Name them.
5. What method of feedback is used in the Hartley oscillator?
6. What method of feedback is used in the Colpitts oscillator?
7. Explain the concept of negative resistance.
8. What causes a wave in a tank circuit to dampen out?
9. A Hartley oscillator has a tank circuit consisting of a 100-μH coil and a capacitor of 159 pF. What is its frequency?
   a. 1.26 kHz.
   b. 12.6 kHz.
   c. 126 kHz.
10. In pulse terminology, _____ is one complete current alternation, from the beginning of one pulse to the beginning of the next pulse.
11. _____ is the time interval between pulses.
    a. Pulse pause.
    b. Resting time.
    c. Fall time.

12. _____ is the time required for buildup of the wave.
    a. Rise time.
    b. Gain.
    c. Amplitude.
13. A free-running multivibrator is called _____.
14. A monostable multivibrator circuit has one stable state. A _____ multivibrator circuit has two stable states.

Applications of Electronics Technology:
This full-color video phone allows users to view each other while they talk on the phone. It works over existing telephone lines. (AT&T)

Chapter 17

ELECTRONIC COMMUNICATION: TRANSMISSION AND RECEPTION

Humans have directly communicated with each other for thousands of years. In more recent years, communication has been enhanced through electronics. It has enabled people to communicate easily over long distances. Further, electronics has made it possible for people to communicate with machines. In the field of electronic communication, the study of reception and transmission of waves is vital.

After studying this chapter, you will be able to:
☐ *Discuss the frequency spectrum and frequency location of various services.*
☐ *Cite the principles of amplitude modulation.*
☐ *Cite the principles of demodulation, or detection.*
☐ *Draw block diagrams of AM and FM radio receivers.*
☐ *Describe the assembly of selected units of a radio receiver.*
☐ *Cite the principles of frequency modulation.*

INTRODUCTION

Technology has allowed humans to change their lives and change the world in which they live. **Communication** means sharing information. It is the purposeful transmission and reception of messages from one source to another. Usually, we think of communication from person to person; however, in this age, it is also possible for a person to communicate with a machine or for one machine to communicate with another machine.

Technology has greatly broadened the ways that we communicate. Early means of communication over distances included drums and signal fires. A later advance in distance communication was the French invention of the semaphore telegraph, in 1794, by Claude Chappe. Basically, this telegraph consisted of 120 wood towers spaced 3 – 6 miles apart. Each tower was equipped with an operator, a telescope for sighting between towers, and a *semaphore—* an apparatus with movable arms. In transmitting a message, a signal would be relayed between towers. The position of the movable arms constituted a signal. The arms of the apparatus would be moved to some other position, and the next signal would be relayed. The combination of signals held an encoded message. This system proved to be about 90 times faster than sending messages by horseback; 50 signals could be initiated (and received) in 1 hour.

An electrical telegraph was soon in the works. The first commercially practical telegraph instrument was designed by Samuel F. B. Morse, in 1832. In 1844, a 40-mile telegraph line was built from Baltimore to Washington. The system quickly won widespread acceptance; however, it used coded signals— dots and dashes. Both the sender and the receiver had to know the code to understand the message.

Electronic communication was vastly improved after the invention of the telephone, in 1875, by Alexander Graham Bell. Some historians include this device among the most important inventions ever. Of course, the telephone did not require special training to understand messages sent by way of it. Shortly after its invention, it was reported that there was one telephone for every 1000 people in the U.S. This soon grew to one for every 10 people. Today, we average over one telephone per person.

Another instrument of electronic communication came with the invention of the wireless telegraph in 1895. The wireless telegraph is often accredited to Guglielmo Marconi. It led the way for modern radio. This important invention used *electromagnetic radiation* to send messages.

ELECTROMAGNETIC RADIATION

In general, **electromagnetic radiation** is the radiation of energy by oscillation of an electric charge. It is a phenomenon of electric and magnetic fields. It is, perhaps, most frequently thought of in terms of transmission of radio waves or television waves. Such alternating current waves **(electromagnetic waves)** are transmitted from an antenna into space. Currently, certain frequency ranges are designated for transmitting, and these are controlled by the Federal Communications Commission (FCC).

The **frequency spectrum** is the entire range of frequencies of electromagnetic radiation. It is convenient to group certain bands of frequencies and label them for reference purposes. Common identifications for these bands can be seen in Figure 17-1.

Imagine what would happen if broadcasting companies were allowed to use as much of the frequency spectrum as desired. Radio and television signals would be badly distorted as the many broadcasts interfered with each other.

FREQUENCY AND WAVELENGTH

As you know, any particular waveform can be identified by its frequency *(f)*, in hertz, or cycles per second. So too, as presented in Chapter 6, it can be identified by its wavelength, in meters. Remember that radio waves travel through space at the speed of light: 3×10^8 m/s, or 186,000 mi./sec. Thus, it is possible to compute wavelength from the distance a wave travels during one time period of its cycle:

$$\lambda \text{ (in meters)} = \frac{3 \times 10^8}{f \text{ (in Hz)}} = \frac{300}{f \text{ (in MHz)}}$$

Further:

$$f \text{ (in Hz)} = \frac{3 \times 10^8}{\lambda \text{ (in meters)}}$$

Problem:

What is the wavelength of a 4-MHz signal?

Solution:

$$\lambda = \frac{300}{4} = 75 \text{ m}$$

Problem:

What is the frequency of a 10-meter radio wave?

Solution:

$$f = \frac{3 \times 10^8}{10} = 3 \times 10^7 \text{ Hz} = 30 \text{ MHz}$$

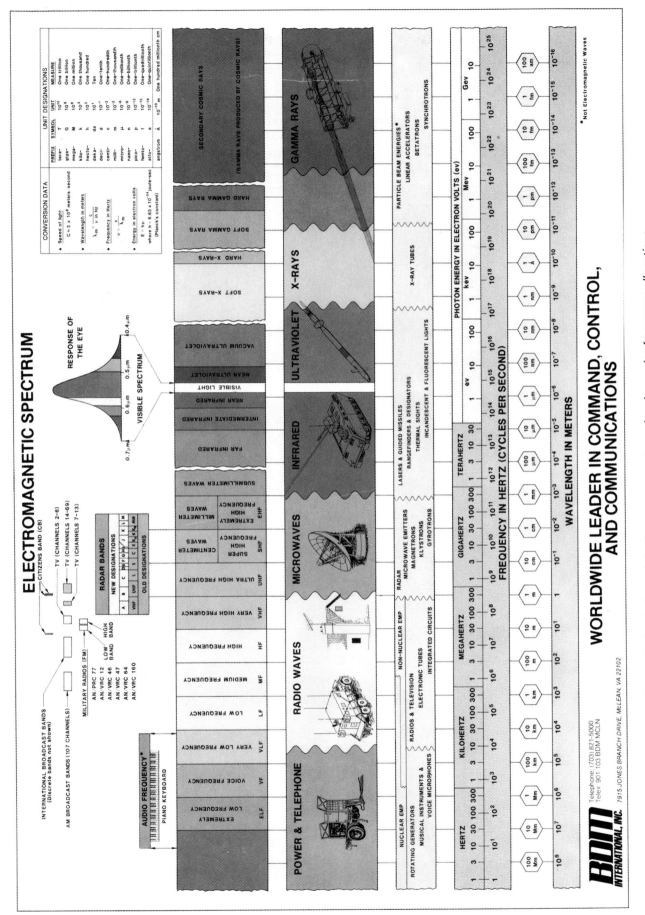

Figure 17-1. A chart of the frequency spectrum showing major frequency allocations.

CONTINUOUS AND INTERRUPTED CONTINUOUS WAVES

Oscillators, studied in Chapter 16, were designed to produce a **continuous wave (CW)**. When these oscillators are connected to an antenna system, they will radiate the waves they produce into space. If the oscillator is keyed (turned on and off), it is possible to send out, or radiate, spurts of energy. This is how **Morse code** is produced. Morse code is an early method of sending messages, but it is still in use. Figure 17-2 shows the continuous radio wave, then an **interrupted continuous wave (ICW)**. Morse code is illustrated in Figure 17-3.

One of the most exciting adventures in electronics is found in **amateur radio**, or **ham radio**. *Ham operators* can send radio messages by voice or Morse code to other operators throughout the world. To operate a ham radio in the U.S., you must be licensed by the FCC. A license is obtained by passing a written examination and by demonstrating ability to send and receive Morse code. To find out more about the process of obtaining a ham license, write to:

American Radio Relay League
225 Main Street
Newington, Connecticut 06111

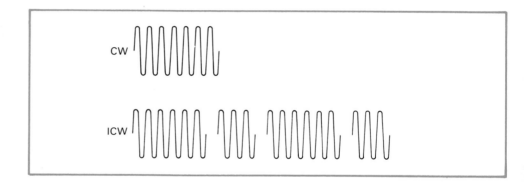

Figure 17-2. An ICW can conform to the Morse code for communication. Shown here is Morse code for the letter C.

A	●—	K	—●—	U	●●—
B	—●●●	L	●—●●	V	●●●—
C	—●—●	M	——	W	●——
D	—●●	N	—●	X	—●●—
E	●	O	———	Y	—●——
F	●●—●	P	●——●	Z	——●●
G	——●	Q	——●—		
H	●●●●	R	●—●	PERIOD	●—●—●—
I	●●	S	●●●	COMMA	——●●——
J	●———	T	—	QUESTION	●●——●●

1	●————	5	●●●●●	8	———●●
2	●●———	6	—●●●●	9	————●
3	●●●——	7	——●●●	0	—————
4	●●●●—				

Figure 17-3. The character set for Morse code.

HETERODYNING

In a radio receiver, two signals are combined together in a process known as **heterodyning**, or **mixing**. This signal conversion process is performed by a circuit called a **mixer**. The two signals consist of an incoming signal known as a **radio frequency (RF) signal** and one from a *local oscillator* within the receiver. When the two signals are mixed, four signals will appear at the output. These include the original two signals, a *sum signal*, and a *difference signal*. For example, assume a 1000-kHz incoming signal is mixed with a 999-kHz local signal. The output contains:

    1000 kHz — original signal
     999 kHz — original signal
    1999 kHz — sum of two signals
       1 kHz — difference of two signals

In this example, the incoming 1000-kHz signal is being mixed with a 999-kHz signal coming from the **beat frequency oscillator (BFO),** an oscillator built into the receiver. A BFO oscillates at a frequency such that when it is mixed with an incoming wave, the difference frequency is in the audible range.

When a CW transmitter is sending spurts of energy in Morse code, a receiver needs to change the signal into a sound you can hear. A 1000-kHz signal is well beyond our audio range. The oscillator signal is mixed with the incoming signal, and an audible difference signal can be heard through earphones or speakers. If a 400-Hz tone was desired, the BFO in the previous example would be tuned to produce a 999.6-Hz signal. At this point, it is important to notice that the l-kHz signal is in the audio range.

The BFO may have a variable panel control to produce an audible tone that is more pleasant to listen to. Also, the control could help distinguish a signal from other tones and interference. A basic block diagram of a transmitter and receiver are illustrated in Figure 17-4. Notice the positions of the mixer and the BFO.

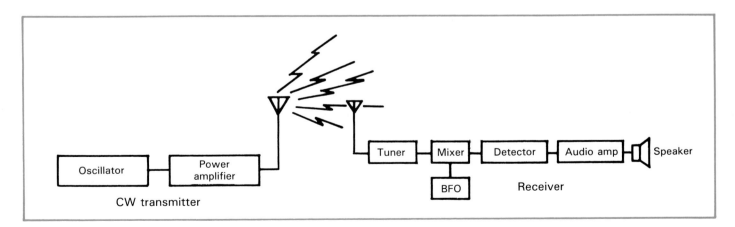

Figure 17-4. A basic block diagram of a CW transmitter and receiver.

AMPLITUDE MODULATION

Most of us enjoy listening to the radio. The sound you hear is information that has been superimposed upon a radio frequency for transmission through a process called **modulation**. Assume a radio transmitter is operating on a frequency of 1000 kHz. This 1000-kHz frequency is the frequency to which you would tune your radio. It is called the **carrier frequency**. A musical tone of 1000 Hz is to be used for modulation. The CW and the audio signal are illustrated in Figure 17-5. For **amplitude modulation (AM)**, using a device called a **modulator**, the amplitude of the carrier wave is made to vary at the rate of the audio signal.

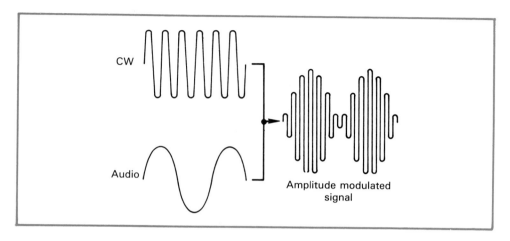

Figure 17-5. A CW and an audio wave are mixed to form an amplitude-modulated wave.

To help understand, look at the modulation process another way. The mixing of a 1000-Hz wave with a 1000-kHz wave produces a sum wave and a difference wave that are also in the radio frequency range. These two waves are at 1001 kHz and 999 kHz. These are known as sideband frequencies. The 1001-kHz and the 999-kHz signals are the **upper sideband (USB)** and the **lower sideband (LSB)**, respectively. In Figure 17-6, this modulation is shown.

The algebraic sum of the carrier wave and its sidebands results in the amplitude-modulated wave. Note that the same audio tone intelligence is present in both sidebands, since either sideband is the result of modulating

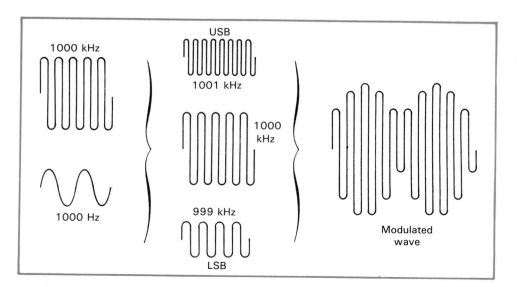

Figure 17-6. When the two waves are mixed, upper and lower sidebands are created. The algebraic sum of the waves forms the modulation envelope.

a 1000-kHz carrier wave with a 1000-Hz tone.

In Figure 17-7, the waves are shown on a frequency scale. If a 2000-Hz tone was used for modulation, then sidebands would appear at 998 kHz and 1002 kHz. In order to transmit a 5000-Hz tone of a piccolo or a violin using AM, sidebands at 995 kHz and 1005 kHz would be required. This would represent a necessary frequency bandwidth of 10 kHz to transmit a 5000-Hz musical tone.

There is not enough space in the spectrum to permit all broadcasters to transmit. The broadcast band for AM radio extends between 535 kHz and 1605 kHz and is divided into 106 channels 10 kHz wide. A station must be licensed to operate at a frequency in one of these channels. The broadcast channels are carefully allotted to stations that are sufficient distances from each other in order to prevent interference.

In order to improve the fidelity and quality of music under the limitations of a 10-kHz band occupancy, a **vestigial sideband filter** is used to remove a large portion of one sideband. Remember, both sidebands contain the same information. This allows one sideband to be cut without loss of valuable information. Using this method, some of the frequency band is opened up allowing room for frequencies higher than 5 kHz to be used in modulation. This improves the fidelity of radio broadcasts.

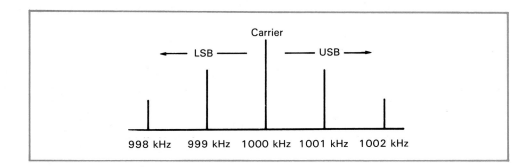

Figure 17-7. Carrier and sideband locations for a modulation tone of 1 kHz and 2 kHz.

MODULATION PATTERNS

With *100% modulation*, wave variation is from zero to two times the peak value of the carrier wave. With **undermodulation**, the peak value of the wave will be less than two times the carrier frequency. **Overmodulation** has the wave *exceeding* 100% modulation. By law, a radio transmitter is not permitted to exceed 100% modulation. This means that the modulation signal *cannot* cause the carrier signal to vary by over 100% of its unmodulated value. Examine the patterns in Figure 17-8. Note the amplitudes of the modulated waves.

When overmodulation occurs, modulation increases the carrier wave to over two times its peak value. At negative peaks the waves cancel each other, leaving a straight line of zero. Overmodulation will cause distortion and interference called **splatter.**

In AM, the **percent of modulation** may be computed using the formula:

$$\% \text{ Modulation} = \frac{e_{max} - e_{min}}{2e_{car}} \times 100\%$$

where:

e_{max} = the maximum amplitude of the modulated wave
e_{min} = the minimum amplitude of the modulated wave
e_{car} = the amplitude of the unmodulated wave

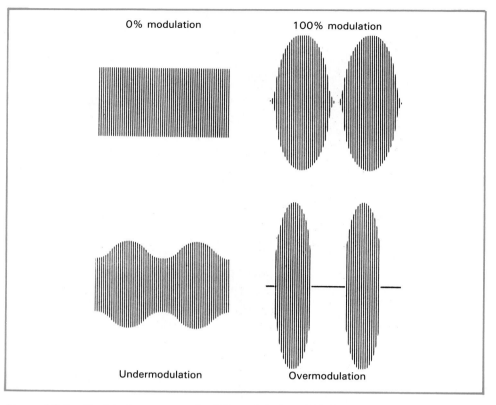

Figure 17-8. Modulation patterns for 0% modulation, 100% modulation, 50% modulation (undermodulation), and overmodulation.

Problem:

A carrier wave has a peak value of 500 V. A modulating signal causes amplitude variation from 200 V to 800 V. What is the percent of modulation?

Solution:

$$\% \text{ Modulation} = \frac{800 \text{ V} - 200 \text{ V}}{2 \times 500 \text{ V}} \times 100\% = \frac{600 \text{ V}}{1000 \text{ V}} \times 100\% = 60\%$$

INPUT POWER

To effectively use a transmitter, sufficient power must be supplied to the modulator. To compute the power required by a modulator (P_{audio}) use this formula:

$$P_{\text{audio}} = \frac{m^2 P_{\text{dc}}}{2}$$

where:

m = the percent of modulation, expressed as decimal
P_{dc} = the input power to the final amplifier

Problem:

What power is required to modulate a transmitter with a dc power input of 500 W to 100%?

Solution:

$$P_{\text{audio}} = \frac{(1)^2 \ 500 \text{ W}}{2} = 250 \text{ W}$$

This represents a total input power of 750 W ($P_{\text{audio}} + P_{\text{dc}}$). Note what happens under conditions of 50% modulation:

$$P_{\text{audio}} = \frac{(0.5)^2 \ 500 \text{ W}}{2} = 62.5 \text{ W}$$

Now, the total input power is only 562.5 W.

When the percent of modulation is reduced to 50%, the power is reduced to 25%. This is a severe drop in power that considerably decreases the broadcasting range of the transmitter. It is good operation to maintain a transmitter as close to, without exceeding, 100% modulation.

In discussing the power, the term *input power* is used. This is because any final amplifier is far from 100% efficient. In Chapter 15, we learned that efficiency, the amount of power drawn from the supply that is actually delivered to a load, is:

$$\% \ \eta_C = \frac{P_{\text{out}}}{P_{\text{dc}}} \times 100\%$$

If a power amplifier has a 60% efficiency and a P_{dc} of 500 W, its output power would be:

$$P_{\text{out}} = \eta_C \ (P_{\text{dc}}) = 0.6 \times 500 \text{ W} = 300 \text{ W}$$

Ham radio stations, for example, are limited by law to 1000 W input power. Their output power, then, is considerably less.

SIDEBAND POWER

Next, consider a transmitter with 100% modulation with a power of 750 W. Our previous calculations show that 500 W of this power are in the carrier wave and 250 W are in the sidebands. Therefore, there are 125 W, or one-sixth of the total power, in each sideband.

Note that the sidebands contain all of the information. In addition, each sideband contains the same information. So, why waste power sending the entire signal? With this in mind, single sideband transmission was created. With **single sideband transmission (SSB),** the carrier and one sideband are suppressed, and only one sideband is radiated. This saves power. At the receiver end, the carrier is reinserted and the difference signal (the audio signal) is detected and reproduced.

This text does not cover the methods of sideband transmission and reception, but you may wish to look into this very popular communication system.

AM DETECTION

Detection, or **demodulation,** is the process of removing the audio signal from a modulated radio wave. An AM receiver produces half-wave pulses from the transmitted signal. It is a form of rectification.

Study the waveforms in Figure 17-9. The detected AM waves are half-wave pulses of radio waves that vary in amplitude. Between the rectified pulses, the signal falls to zero. Detection is accomplished by applying the same principles used in filtering the output of a power supply. A demodulated wave can be filtered. This removes the pulses and raises the average value of the wave. This improvement is illustrated in Figure 17-10.

There are two different types of detectors used to receive AM signals, *passive detectors* and *active detectors*. A **passive detector** uses a *passive device* to detect the signal; an **active detector** uses an *active device*. **Passive devices** are devices that do not require an operating voltage to function. Resistors, capacitors, and diodes are all passive devices. **Active devices** require an operating voltage to function. Transistors and ICs are active devices.

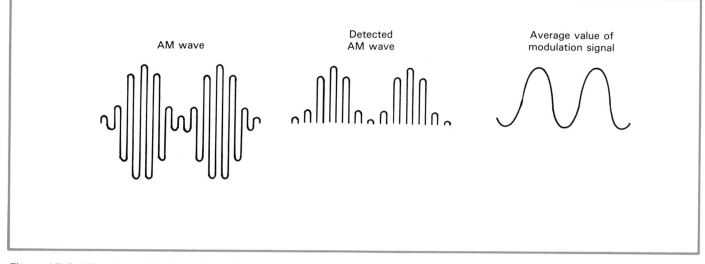

AM wave

Detected
AM wave

Average value of
modulation signal

Figure 17-9. The demodulation of an AM wave.

The diode detector

The semiconductor diode is used in a simple and common passive detector application. This detector is called a **diode detector**, or **crystal detector**. A point-contact diode is used to avoid shunting a radio frequency signal around a diode by its junction capacitance. (The reactance of this capacitance would be at a low value in the radio frequency range.) See Chapter 12 to review information on the point-contact diode.

In Figure 17-11, a circuit of a simple radio receiver is illustrated. Radio waves of many frequencies, from as many broadcasting stations, cut across and induce a very small voltage in the antenna. This causes currents to move up and down from antenna to ground. The antenna lead-in wire is connected to L_1, which is part of an antenna coil in the receiver. The other end of L_1 is connected to ground.

The oscillating currents of the many frequencies produce varying magnetic fields around L_1. L_1 is closely linked to L_2. Therefore, these voltages are induced across L_2. Now, the tank circuit L_2C_1 can be tuned to the desired frequency of a station chosen. At this resonant frequency, the tank circuit

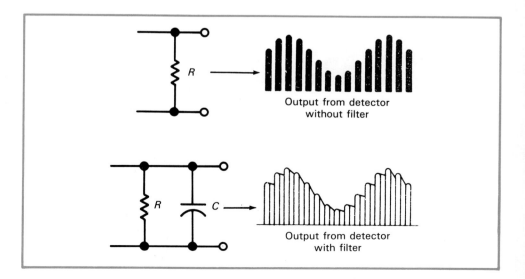

R

Output from detector
without filter

R   *C*

Output from detector
with filter

Figure 17-10. Compare the output from AM detectors before and after filtering.

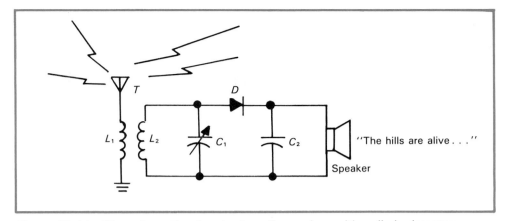

Figure 17-11. The schematic of a simple radio receiver with a diode detector.

will have maximum response, and its developed voltage is magnified many times. The amount of magnification depends on the Q of the circuit. (Review Chapter 11.) This signal is now detected by diode D. The audio output is filtered by C_2 and reproduces the "Sound of Music" in the speaker.

Usually, the coils of L_1 and L_2 are wound on a single core. In a parts catalog, it is called an **antenna transformer** or an **antenna coil,** Figure 17-12. You must remember that the selection of L_2 and C_1 depends upon the band of frequencies you wish to hear.

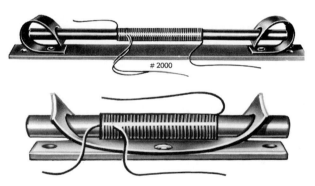

Figure 17-12. Antenna coils from radio receivers.

The transistor detector

A transistor is used in a simple active detector. The transistor has the added advantage of signal amplification. A transistor used in this application must be biased at cutoff or operate as class B. You will recall that class-B amplification is the same as half-wave rectification. (See Chapter 14.)

The circuit of a **transistor detector,** Figure 17-13, is a simple amplifier circuit; currently, a variety of these active detector circuits can be found in IC form. The emitter is at ground potential, and the voltage divider R_FR_B will establish the proper bias. This bias holds the transistor just slightly above cutoff.

An incoming AM wave will cause the emitter-base voltage and current to vary. The transistor will conduct only when its base is made more negative. The positive half-cycles of the AM wave will be cut off. Capacitor C serves as a bypass for radio frequencies.

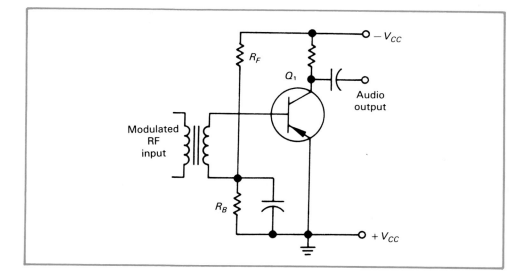

Figure 17-13. The circuit of a transistor detector.

THE BASIC TRANSISTOR RADIO

In Figure 17-14, several of your previous lessons have been applied. This small radio receiver employs diode detection, two stages of transistor amplification, and transformer coupling. You will recognize the tuning circuit L_1C_1. The tuned signal is detected by diode D_1. It is necessary to have C_2 in the circuit to block the dc bias of Q_1 from grounding out through antenna coil L_1.

The remainder of the circuit shown in Figure 17-14 is straightforward. Its components have been discussed in Chapter 14, except for potentiometer R_2. This is a gain control, a variable resistance shunted across the primary of T_1. Potentiometer R_2 determines the amount of signal voltage transferred to the next stage by transformer T_1.

AM SUPERHETERODYNE RECEIVER

Many years ago, it was found that the gain and selectivity of a radio could be easily designed with acceptable performance for one tuned frequency. When a station of another frequency was desired, less than maximum performance resulted unless separate tuning was performed on each of the receiver's stages. This proved difficult and inefficient. This *tuned radio frequency (TRF)* machine

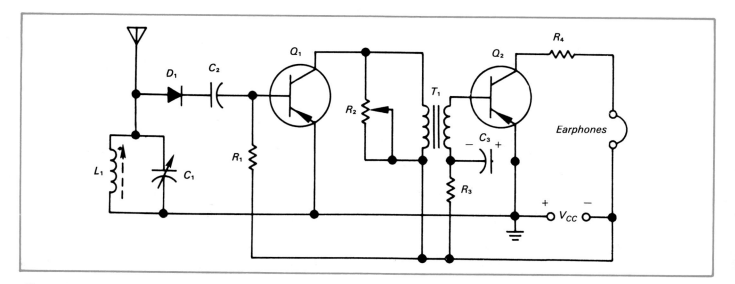

Figure 17-14. The schematic of a transistor radio with two stages of amplification.

is now a museum piece or a collector's item.

It was discovered that if all tuned incoming signals were converted to a single frequency, stages and coupling circuits could be used to produce the best results for many signals of many frequencies. This single frequency is called the **intermediate frequency (IF)**. The added convenience of using one dial for tuning made it possible for anyone to tune a radio, regardless of any technical knowledge.

This type of radio receiver employs the **superheterodyne** circuit. In this setup, the station is tuned using the methods previously discussed. Then, the signal is mixed with a signal from a **local oscillator**, which produces a constant intermediate frequency. The rest of the receiver is designed to work around this IF.

When a station is tuned, the local oscillator is varied in frequency such that the difference signal, the IF, is always the same. This is done by tuning the stages simultaneously, or **gang tuning**. The difference frequency is 455 kHz for home radios.

Assume that a radio is tuned to a 1000-kHz station. The local oscillator will be tuned to 1455 kHz. The difference frequency IF is 455 kHz. If you tune to a station at 600 kHz, the oscillator will tune to 1055 kHz. The IF is still 455 kHz. If you tune a station at 1400 kHz, the oscillator will be 1855 kHz, and the IF, again, is 455 kHz.

Beyond the convenience that the superheterodyne receiver introduced, there are other advantages to the system. When the IF amplifiers are designed for maximum gain at 455 kHz, all incoming signals tuned in by the radio will have maximum response. See the block diagram with signals, Figure 17-15. Notice that many superheterodyne receivers will have two IF amplifier stages for additional gain.

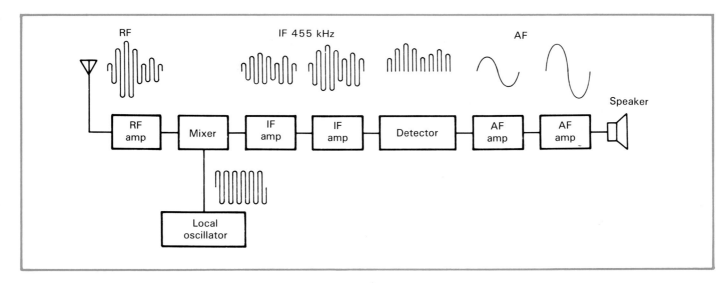

Figure 17-15. A block diagram of an AM superheterodyne receiver.

AUTOMATIC GAIN CONTROL

Unless the broadcasting stations are quite close to your radio, the radio signals will fluctuate in strength. This variation is due to fading and atmospheric conditions. Without correction, frequent adjustment of the manual volume control would be necessary to assure pleasant listening. Consequently, the volume of a receiver is automatically adjusted by electronic circuitry. Any level of volume will be maintained without further manual control. This is called **automatic gain control (AGC)**.

In transistor circuits, AGC is easy to set up since the gain of a transistor amplifier depends upon the input current. However, a junction transistor is a current-controlled device, and power from some source must be used.

In most cases, the AGC voltage is secured from the detector stage of a transistor circuit. The average value of the detected signal varies according to the strength of the incoming signal. The current through the detector diode produces a voltage across the detector load resistor that may be fed back to the previous IF amplifiers to regulate their gain. A stronger signal will reduce the gain. A weaker signal will increase the gain. A partial diagram, Figure 17-16, shows the connection to an IF amplifier to produce the results.

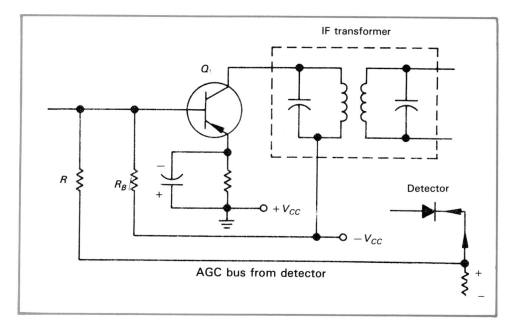

Figure 17-16. This schematic shows the connection of an AGC line to an IF amplifier.

A strong positive voltage on the AGC line will make the base of Q_1 more positive. This reduces forward bias, reduces emitter and collector current, and reduces amplifier gain. Resistor R is a limiting resistance to effectively control the amount of AGC feedback voltage. Similar circuits are used in television to keep the video signal at a constant level.

In this discussion on gain variation as the result of emitter-collector current, changes in input and output impedances have not been considered. More advanced texts will discuss these impedances that also affect amplifier gain.

FREQUENCY MODULATION

Frequency modulation (FM) is another popular method of radio communication. FM employs an entirely different method of superimposing intelligence, such as voice or music, on a radio frequency wave. FM carries its information in frequency deviations. The capabilities of FM allow relatively high audio sound to be transmitted, while still remaining within the legal spectrum of space assigned to a broadcast station. FM's advantages in noise reduction and its quick adaption to stereo sound have made it the choice medium for radio stations with music formats.

FM begins with a constant amplitude continuous wave signal. The frequency of the signal is then made to vary at an audio rate. This is shown graphically in Figure 17-17.

Each FM broadcasting station, like AM, is assigned a center frequency, or carrier frequency. To help understand FM, study the waveforms in Figure 17-18. The amount of frequency variation on each side of the carrier frequency is called the **frequency deviation**. This deviation is determined by the amplitude of the audio modulating wave. In Figure 17-18, a small audio signal causes the frequency of the carrier wave to vary between 100.01 MHz and 99.99 MHz. The deviation is ±10 kHz.

In the second example, Figure 17-18, a stronger audio signal causes a frequency swing between 100.05 MHz and 99.95 MHz, a deviation of ±50 kHz. The stronger audio modulation signal produces a greater departure from the carrier frequency, and, consequently, occupies a larger portion of the frequency band.

The rate at which the frequency varies from its highest to its lowest frequency

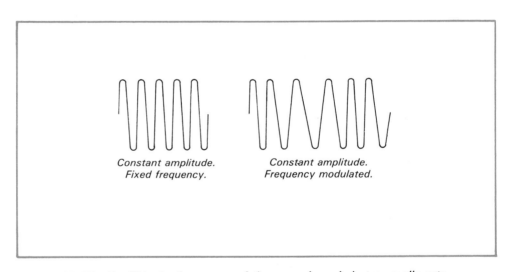

Constant amplitude.
Fixed frequency.

Constant amplitude.
Frequency modulated.

Figure 17-17. For FM, the frequency of the wave is varied at an audio rate.

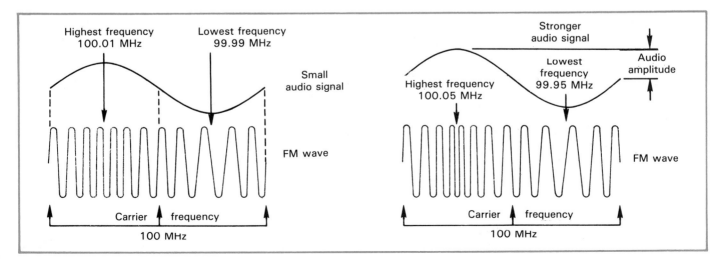

Figure 17-18. The amplitude of the modulating signal determines the frequency swing from the carrier.

depends upon the frequency of the audio modulating signal. See the examples in Figure 17-19.

If the audio signal is 1000 Hz, the carrier wave goes through its maximum deviation 1000 times per second. If the audio signal is 100 Hz then the frequency changes at a rate of 100 times per second. Notice that the modulating frequency does *not* change the amplitude of the carrier wave.

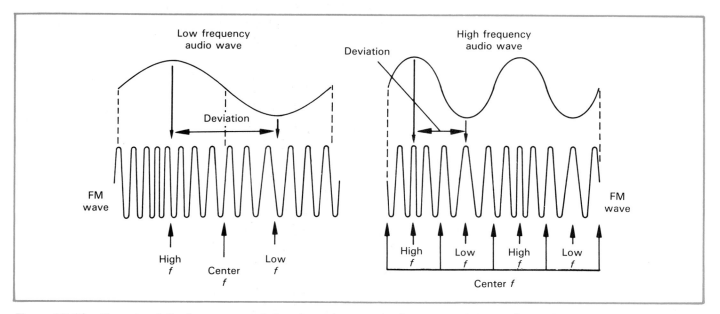

Figure 17-19. The rate of the frequency variation depends upon the frequency of the audio modulating frequency.

As with amplitude modulation, when a signal is frequency modulated, there is the formation of sidebands. However, with frequency modulation the number of sidebands produced depends upon the frequency and amplitude of the modulating signal. Each sideband on either side of the carrier frequency is separated by an amount equal to the frequency of the modulating signal. This is illustrated in Figure 17-20.

Note that the power of the carrier frequency is considerably reduced by the formation of sidebands, as the sidebands take power from the carrier. The amount of power taken depends upon the maximum deviation of the bands and the modulating frequency.

The formation of sidebands is the determining factor of the bandwidth required for transmission. In FM, bandwidth is specified by the frequency range between the upper and lower significant sidebands. A **significant sideband** is one that has an amplitude of at least 1% of the unmodulated carrier.

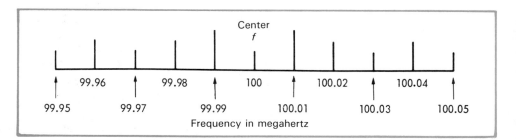

Figure 17-20. Shown are the sidebands generated by a 10-kHz modulating signal on a 100-MHz carrier wave.

MODULATION INDEX AND PERCENTAGE

The **modulation index** is the relationship between the maximum frequency deviation of the carrier and the modulating frequency (f_{mod}). It is expressed as:

$$\text{modulation index} = \frac{\text{maximum frequency deviation}}{\text{modulating frequency}}$$

Through the use of this index, the number of significant sidebands and the bandwidth of the FM signal may be calculated. A complete index may be found in reference texts. Examples of the use of the modulation index are given in the following table:

Modulation Index	Number of Sidebands	Bandwidth
0.5	2	$4 \times f_{mod}$
1	3	$6 \times f_{mod}$
5	8	$16 \times f_{mod}$
10	14	$28 \times f_{mod}$

In FM, *percent of modulation* is the ratio of the actual frequency deviation to the frequency deviation defined as 100% modulation. The deviation for commercial FM radio of ± 75 kHz is defined as 100% modulation; for the FM sound transmission in television, ± 25 kHz is.

Problem:

The amplitude of a modulating signal causes a maximum deviation of 10 kHz. The frequency of the modulating signal is 1000 Hz. What is the bandwidth of the FM signal?

Solution:

Step 1. Determine the modulation index.

$$\text{modulation index} = \frac{10,000 \text{ kHz}}{1000 \text{ Hz}} = 10$$

Step 2. Knowing the modulation index, we can consult the table to determine bandwidth.

$$\text{BW} = 28 f_{mod} = 28(1000 \text{ kHz}) = 28 \text{ kHz}$$

The FM signal has 14 significant sidebands and occupies a bandwidth of 28 kHz.

FM DETECTION

With AM radio, a detector has to be sensitive to the amplitude variations of a wave. An FM detector must be sensitive to frequency variations. The detector must remove the intelligence from the frequency variations in an FM wave. In other words, the FM detector must produce a varying amplitude and frequency audio signal from an FM wave.

Consider the diagram in Figure 17-21. Assume a circuit has a maximum response at its resonant frequency. All frequencies other than resonance will have a lesser response. So, if the carrier frequency of an FM wave is on the *slope* of the resonant response curve, a higher frequency will produce a *higher response in voltage,* and a lower frequency will produce a *lower voltage response.* Examining Figure 17-21, you will see that the amplitude of the output

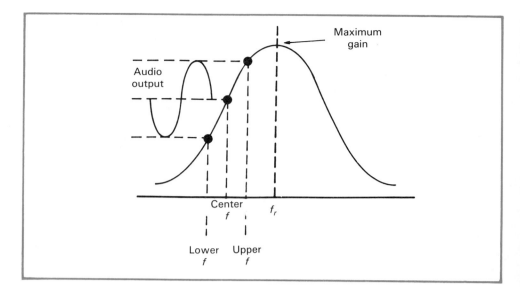

Figure 17-21. This diagram demonstrates slope detection.

wave is the result of the maximum deviation of the FM signal. The frequency of the audio output depends on the rate of change of the frequency of the FM signal.

The frequency discriminator

The **frequency discriminator** was the first FM detector to gain popularity. For ease in understanding, the discriminator in Figure 17-22 uses three tuned circuits. In this circuit, L_1C_1 is tuned to the carrier frequency. L_2C_2 is tuned to above, and L_3C_3 is tuned to below, the carrier frequency by an equal amount.

At the carrier frequency, equal voltages are developed across the tuned circuits, and D_1 and D_2 conduct equally. The voltages across R_1 and R_2 are equal and opposite in polarity. The circuit output is zero.

If the input frequency increases above center, L_2C_2 will develop a higher voltage. Then, D_1 will conduct more than D_2 and unequal voltages will develop across R_1 and R_2. The difference between these voltage drops will be the audio signal.

Thus, the output is a voltage wave that varies at the rate of the frequency change in the input. Its amplitude depends upon the maximum deviation. The capacitors across the output of the discriminator are to filter out any remaining radio frequencies.

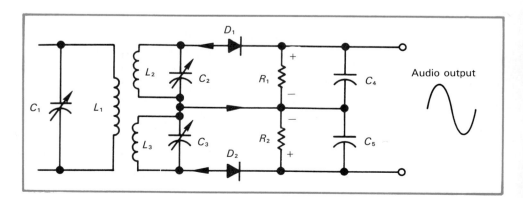

Figure 17-22. A schematic diagram of a frequency discriminator circuit.

The discriminator in Figure 17-23 is a typical circuit. L_1 and C_1 are tuned to the carrier frequency. At frequencies above resonance, the tuned circuit becomes more inductive. At frequencies below resonance, the circuit becomes more capacitive. The out-of-phase conditions produce resultant voltages that determine how each diode conducts. The output is an audio wave, as in the earlier discussion.

One note of interest is that each diode in the discriminator must have equal conduction capabilities. This means that the diodes used must be in matched pairs.

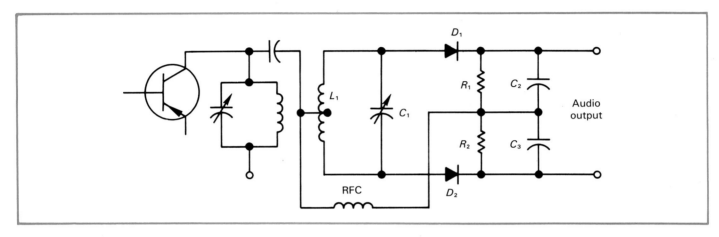

Figure 17-23. A typical Foster-Seeley discriminator using a special transformer designed for this purpose.

The ratio detector

The FM detector most commonly used in radios is illustrated in Figure 17-24. It is called the **ratio detector.** In this circuit, the diodes are connected in series with the tuned circuit. At center frequencies, both diodes conduct during half-cycles. The voltage across R_1 and R_2 charges C_1 to the output voltage. C_1 remains charged because the time constant of C_1R_1 is much longer than the period of the incoming waves. C_2 and C_3 also charge to the voltage of C_1.

When both D_1 and D_2, in Figure 17-24, are conducting equally, the charge on C_2 equals C_3, and they form a voltage divider. At the center point between C_2 and C_3, the voltage is effectively zero.

A frequency shift either below or above the carrier frequency will cause one diode to conduct more than the other. As a result, the voltages of C_2 and C_3 become unequal. Yet, they will always total the voltage of C_1. This change of voltage at the junction of C_2 and C_3 is the result of the *ratio* of the unequal division of charges between C_2 and C_3. This charge will vary at an audio rate equal to the rate of change of the FM signal.

Again in Figure 17-24, note the charge on C_1. It is the result of the amplitude of the carrier wave. It is charged by half-wave rectification of the FM signal. Therefore, it is a fine point to pick off an automatic gain control voltage to regulate the gain of previous stages.

NOISE LIMITING

Since FM radio receivers detect frequency variations as opposed to amplitude variations, FM is unaffected by most interference. Most noise and interference to radio reception is in the form of **noise spikes.** These are amplitude variations, and they have little effect on the FM detectors. Thus, FM reception is relatively free of noise and disturbance.

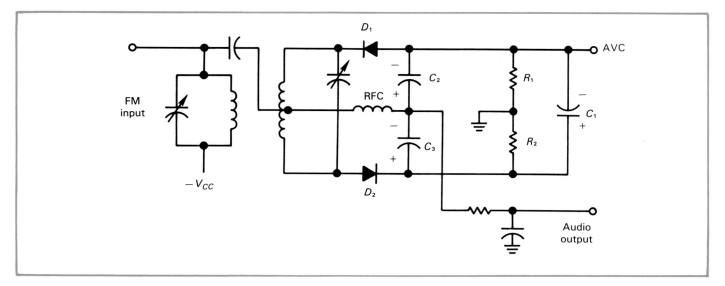

Figure 17-24. A schematic diagram of a typical ratio detector circuit.

To keep the FM signal at a constant amplitude before detection in a discriminator circuit, a **limiter** is used in the previous stage. A schematic of one type of limiter stage is shown in Figure 17-25.

This limiter is nothing more than an overdriven amplifier stage. If the incoming signal reaches a certain amplitude in voltage, it drives the transistor to cutoff, or to saturation when the voltage is opposite in polarity. At either of these points, gain cannot increase. Consequently, the output is confined within these limits. Any noise spikes in the form of amplitude modulation are clipped off.

FM SUPERHETERODYNE RECEIVER

A block diagram of a complete FM receiver is illustrated in Figure 17-26. Each block is labeled to designate its function in the system. Each of these blocks has been described in previous sections in this text. The same heterodyne principles involved in the AM systems apply to the FM systems. The intermediate frequency used in FM radio is 10.7 MHz.

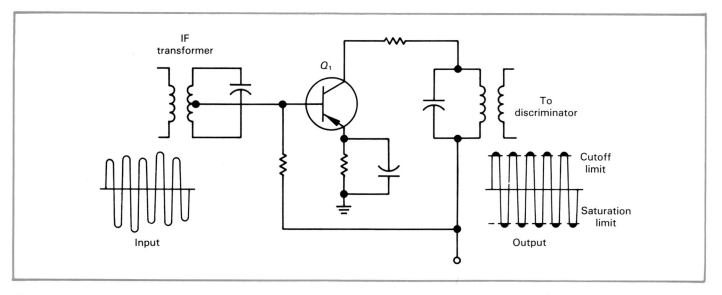

Figure 17-25. A circuit from a limiter stage before a discriminator.

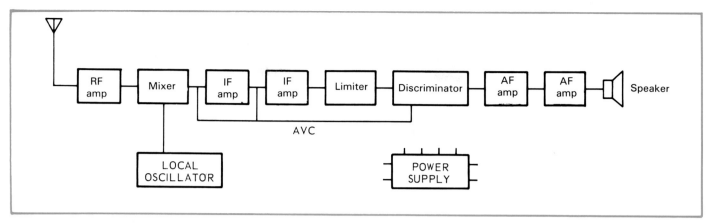

Figure 17-26. The block diagram of a typical FM superheterodyne receiver.

PULSE CODE MODULATION

Another important method of coding information on a waveform is called **pulse code modulation (PCM)**. Before the invention of PCM, all modern high-speed information, such as AM and FM, was transmitted in analog form. With analog technology, the electrical signals are continuously varying, as opposed to the discrete pulses in digital technology.

PCM introduced digital technology to electronic communications. There are three processes of PCM that are characteristic of its operation: sampling, quantizing, and coding. Digital technology is studied in more detail in Chapter 20.

CITIZENS BAND RADIO

In 1947, the FCC established the Citizens Radio Service to permit personal short range radio communications, signaling, and remote control by radio signals. **Citizens band (CB)** provided a convenience to practically anyone for business or personal activities. A list of CB 10 signals appears in Figure 17-27.

The original class D band of CB had 23 channels. During the early years of citizens band, there was limited usage. In the early 1970s, during the energy crisis, citizens band became very popular. In July, 1976, the FCC increased the number of CB channels from 23 to 40, Figure 17-28. The output power of a 40-channel CB is 4 watts.

10-1	Receiving Poorly	10-21	Call By Telephone	10-41	Please Tune To Channel...
10-2	Receiving Well	10-22	Report In Person To...	10-42	Traffic Accident At...
10-3	Stop Transmitting	10-23	Stand By	10-43	Traffic Tieup At...
10-4	OK, Message Received	10-24	Completed Last Assignment	10-44	I Have A Message For You (or...
10-5	Relay Message	10-25	Can You Contact...	10-46	Assist Motorists
10-6	Busy, Stand By	10-26	Disregard Last Information	10-50	Break Channel...
10-7	Out Of Service, Leaving Air	10-27	I Am Moving To Channel...	10-70	Fire At...
10-8	In Service, Subject To Call	10-28	Identify Your Station	10-73	Speed Trap At...
10-9	Repeat Messsage	10-29	Time Is Up For Contact	10-75	You Are Causing Interference
10-10	Transmission Completed, Standing By	10-30	Does Not Conform To FCC Rules	10-77	Negative Contact
10-11	Talking Too Rapidly	10-33	EMERGENCY TRAFFIC AT THIS STATION	10-82	Reserve Room For...
10-12	Visitors Present			10-84	My Telephone Number Is...
10-13	Advise Weather/Road Conditions	10-34	Trouble At This Station, Help Needed	10-85	My Address Is...
10-16	Make Pickup At...			10-89	Radio Repairman Needed At...
10-17	Urgent Business	10-35	Confidential Information	10-92	Your Transmistter Is Out Of Adjustment
10-18	Anything For Us?	10-36	Correct Time Is...		
10-19	Nothing For You, Return To Base	10-37	Wrecker Needed At...	10-93	Check My Frequency On This Channel
10-10	My Location Is...	10-38	Ambulance Needed At...		
		10-39	Your Message Delivered	10-200	Police Needed At...

Figure 17-27. List of CB 10 signals.   (KRAKO)

40 CHANNEL CB OPERATION (TRANSMIT)

CHANNEL NO.	VCO OUTPUT MHz	TX OSCILLATOR MHz	PLL MIXER OUTPUT MHz	CHANNEL NO.	VCO OUTPUT MHz	TX OSCILLATOR MHz	PLL MIXER OUTPUT MHz
1	26.965	29.515	2.55	21	27.215	29.515	2.30
2	26.975	29.515	2.54	22	27.225	29.515	2.29
3	26.985	29.515	2.53	23	27.255	29.515	2.26
4	27.005	29.515	2.51	24	27.235	29.515	2.28
5	27.015	29.515	2.50	25	27.245	29.515	2.27
6	27.025	29.515	2.49	26	27.265	29.515	2.25
7	27.035	29.515	2.48	27	27.275	29.515	2.24
8	27.055	29.515	2.46	28	27.285	29.515	2.23
9	27.065	29.515	2.45	29	27.295	29.515	2.22
10	27.075	29.515	2.44	30	27.305	29.515	2.21
11	27.085	29.515	2.43	31	27.315	29.515	2.20
12	27.105	29.515	2.41	32	27.325	29.515	2.19
13	27.115	29.515	2.40	33	27.335	29.515	2.18
14	27.125	29.515	2.39	34	27.345	29.515	2.17
15	27.135	29.515	2.38	35	27.355	29.515	2.16
16	27.155	29.515	2.36	36	27.365	29.515	2.15
17	27.165	29.515	2.35	37	27.375	29.515	2.14
18	27.175	29.515	2.34	38	27.385	29.515	2.13
19	27.185	29.515	2.33	39	27.395	29.515	2.12
20	27.205	29.515	2.31	40	27.405	29.515	2.11

Figure 17-28. Table of the current 40 CB channels.

Basically, citizens band radios come in three styles:

- Automobile type (12 Vdc). See Figure 17-29.
- Base station type or the type used at home (117 Vac). See Figure 17-30.
- Pack unit or walkie-talkie type. See Figure 17-31.

Modern CB receivers are complex instruments. Manufacturers choose advanced circuitry for superior performance. Single sideband models have the capability of converting 40 channels to 120 channels.

There are many different types of antennas designed for CB use. Selection should be based on the type of installation or car mount desired and the antenna specifications. A vertical whip normally has a 360° radiation pattern. It can be mounted on the rear bumper, rear fender, rear deck lid. It is also important to use the correct type of transmission line for the antenna chosen. It should be a coaxial line and have an impedance equal to the antenna impedance. See Figure 17-32 for a typical automobile CB antenna. Generally, the better the antenna, the better the distance communications.

A full 1/4 wavelength antenna is usually more efficient than the shorter versions equipped with a loading coil to electrically make up for their shorter length. However, the shorter antennas can provide adequate service and are less prone to damage from contact with external obstructions. Some short antennas can be more centrally located on the car. The car body acts as a ground plane and tends to shift the radiation pattern to favor a diagonal line. The line runs from right to left for an antenna mounted on the right front or left rear portion of the car. For an antenna mounted on the left front or right rear of the car, the pattern will follow the diagonal line from left to right. For a more circular pattern, the antenna would have to be centrally mounted on the car.

Figure 17-29. A 40-channel citizens band radio. (Cobra Div. of Dynascan Corp.)

Figure 17-30. A base station citizens band radio. (Cobra Div. of Dynascan Corp.)

Figure 17-31. A pack-unit type of citizens band radio, also called a walkie-talkie. (Cobra Div. of Dynascan Corp.)

Figure 17-32. A mobile CB antenna.   (Panasonic)

AM TRANSMITTER PROJECT

You can build a simple AM transmitter. This is a design that requires a minimum of electronic parts. The transmitting distance is approximately 50 ft. See Figures 17-33 and 17-34. A printed circuit layout for the transmitter appears in Figure 17-35. Parts placement for the printed circuit is shown in Figure 17-36.

Once you have completed the circuit, connect a crystal microphone to the input. Connect a battery or power supply to the transmitter. Then attach a short length of wire ($\cong$ 3 ft.) or a small telescoping antenna to it.

To operate the transmitter, speak into the microphone and tune your AM radio across the band. When you find the spot on the dial where you receive the signal, adjust C_5 to a point where the signal is coming in the clearest.

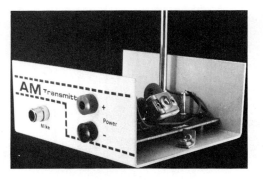

Figure 17-33. AM transmitter project.

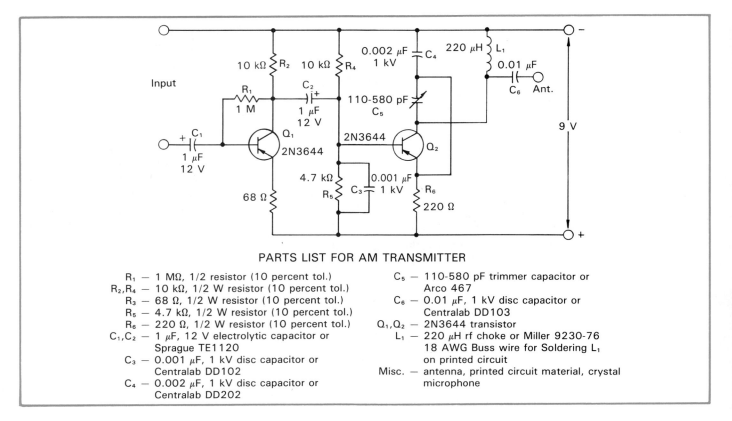

PARTS LIST FOR AM TRANSMITTER

R₁ — 1 MΩ, 1/2 resistor (10 percent tol.)
R₂,R₄ — 10 kΩ, 1/2 W resistor (10 percent tol.)
R₃ — 68 Ω, 1/2 W resistor (10 percent tol.)
R₅ — 4.7 kΩ, 1/2 W resistor (10 percent tol.)
R₆ — 220 Ω, 1/2 W resistor (10 percent tol.)
C₁,C₂ — 1 μF, 12 V electrolytic capacitor or
Sprague TE1120
C₃ — 0.001 μF, 1 kV disc capacitor or
Centralab DD102
C₄ — 0.002 μF, 1 kV disc capacitor or
Centralab DD202

C₅ — 110-580 pF trimmer capacitor or
Arco 467
C₆ — 0.01 μF, 1 kV disc capacitor or
Centralab DD103
Q₁,Q₂ — 2N3644 transistor
L₁ — 220 μH rf choke or Miller 9230-76
18 AWG Buss wire for Soldering L₁
on printed circuit
Misc. — antenna, printed circuit material, crystal
microphone

Figure 17-34. Schematic and parts list for AM transmitter.

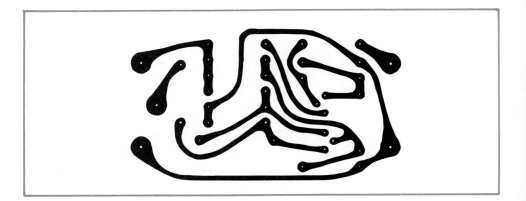

Figure 17-35. One possible circuit board layout for the AM transmitter.

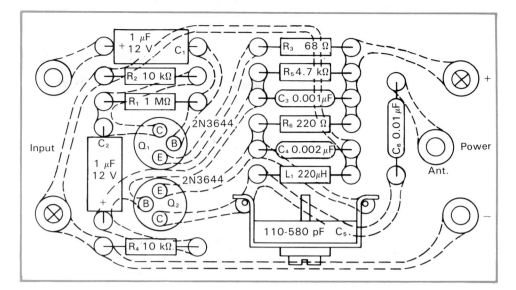

Figure 17-36. Parts placement diagram for the AM transmitter.

AM RECEIVER PROJECT

Here is an AM receiver to go with the AM transmitter you just built. This project includes an AM tuner, detector, and a one-stage amplifier. You may use it with an earphone. Stations will be much louder when the tuner is fed directly into an amplifier, such as the basic transistor amplifier in Chapter 14.

A good antenna will improve the performance of this radio, but the radio can work effectively using 4−5 ft. of wire as an antenna. The parts list and diagram are given in Figure 17-37. This project may be made on an etched-circuit board, Figure 17-38.

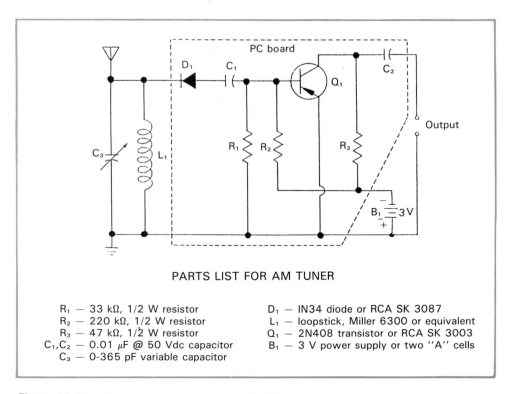

PARTS LIST FOR AM TUNER

R_1 — 33 kΩ, 1/2 W resistor
R_2 — 220 kΩ, 1/2 W resistor
R_3 — 47 kΩ, 1/2 W resistor
C_1, C_2 — 0.01 µF @ 50 Vdc capacitor
C_3 — 0-365 pF variable capacitor

D_1 — IN34 diode or RCA SK 3087
L_1 — loopstick, Miller 6300 or equivalent
Q_1 — 2N408 transistor or RCA SK 3003
B_1 — 3 V power supply or two "A" cells

Figure 17-37. Schematic and parts list for AM tuner.

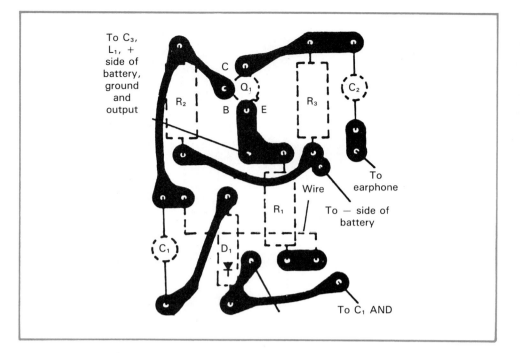

Figure 17-38. One possible circuit board layout for the AM tuner.

FM TRANSMITTER PROJECT

An FM transmitter is an excellent project for the experimenter. This project uses many of the same component parts that were in the AM transmitter. The schematic for the FM transmitter and the parts list are given in Figure 17-39. The component parts placement is shown in Figure 17-40.

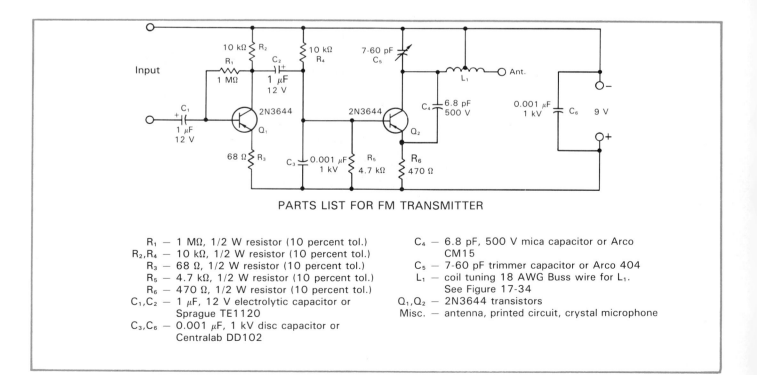

PARTS LIST FOR FM TRANSMITTER

R₁ — 1 MΩ, 1/2 W resistor (10 percent tol.)
R₂,R₄ — 10 kΩ, 1/2 W resistor (10 percent tol.)
R₃ — 68 Ω, 1/2 W resistor (10 percent tol.)
R₅ — 4.7 kΩ, 1/2 W resistor (10 percent tol.)
R₆ — 470 Ω, 1/2 W resistor (10 percent tol.)
C₁,C₂ — 1 μF, 12 V electrolytic capacitor or
        Sprague TE1120
C₃,C₆ — 0.001 μF, 1 kV disc capacitor or
        Centralab DD102

C₄ — 6.8 pF, 500 V mica capacitor or Arco
     CM15
C₅ — 7-60 pF trimmer capacitor or Arco 404
L₁ — coil tuning 18 AWG Buss wire for L₁.
     See Figure 17-34
Q₁,Q₂ — 2N3644 transistors
Misc. — antenna, printed circuit, crystal microphone

Figure 17-39. Schematic and parts list for an FM transmitter.

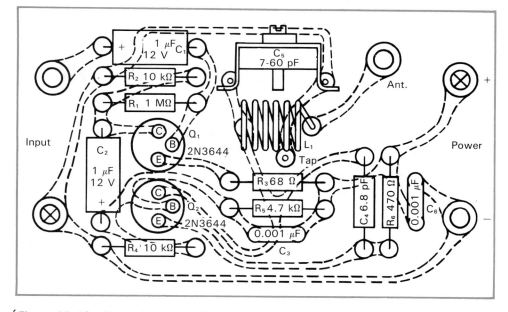

Figure 17-40. Parts placement diagram for the FM transmitter.

A printed circuit layout is shown in Figure 17-41. See Figure 17-42 for details of how to make coil L_1. See Figure 17-43 for a top view of the circuit board with the components laid out.

The operation of this project is similar to the AM transmitter. Place an FM receiver nearby and turn it on. Adjust the tuner to a spot in the middle of the FM broadcast range. Connect a power supply or battery to the power terminals and a 24-36 in. antenna to the unit. Hook up a crystal microphone to the input, then adjust the C_5 unit. The signal will be heard in the FM receiver.

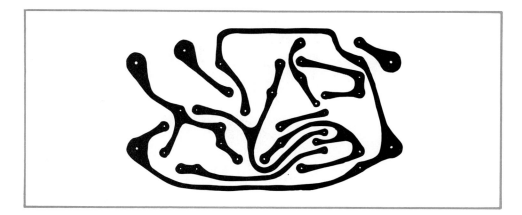

Figure 17-41. One possible printed circuit board layout for the FM transmitter. *(Hickok)*

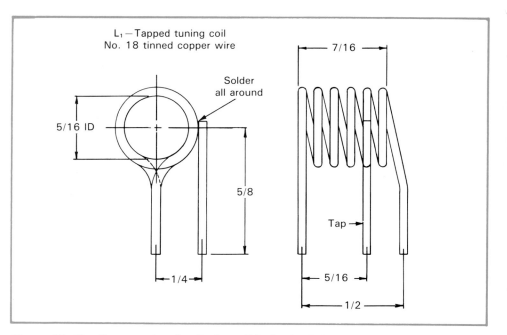

Figure 17-42. *Construction details for coil L₁ in the FM transmitter.*   (Hickok)

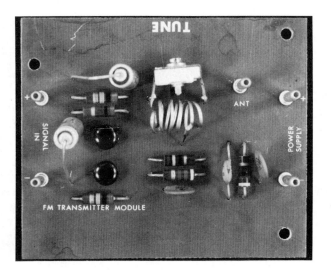

Figure 17-43. *Top view of the setup for the FM transmitter.*   (Hickok)

SUMMARY

- The frequency spectrum covers the entire range of electromagnetic radiation.
- Heterodyning, or mixing, is the process of combining two signals together. The output consists of the two input signals, a sum signal, and a difference signal.
- In amplitude modulation, the information is coded by having the amplitude of the carrier wave vary at the rate of the audio signal.
- In amplitude modulation, each sideband contains the same information.
- Diode detectors (passive) and transistor detectors (active) use a form of rectification to detect AM signals.
- Superheterodyne receivers convert all incoming signals to some intermediate frequency. Using this method, a receiver can produce a maximum response for many incoming frequencies.

- The volume output from a radio is kept constant, despite changing atmospheric conditions, through feedback from an automatic gain control circuit.
- In frequency modulation, the information is coded by having the frequency of the signal vary at an audio rate.
- The modulation index is the relationship between the maximum frequency deviation and the modulating frequency. It may be used to determine the number of significant sidebands of a signal.
- FM detectors are biased so that the signals received fall on the slope of their maximum response curve.
- Citizens band radio offers a range of frequencies for anyone to send personal short range signals.

KEY TERMS

Each of the following terms has been used in this chapter. Do you know their meanings?

active detector	frequency deviation	Morse code
active device	frequency discriminator	noise spike
amateur radio	frequency modulation (FM)	overmodulation
amplitude modulation	frequency spectrum	passive detector
antenna coil	gang tuning	passive device
antenna transformer	ham radio	percent of modulation
automatic gain control (AGC)	heterodyning	pulse code modulation (PCM)
beat frequency oscillator (BFO)	intermediate frequency (IF)	radio frequency (RF) signal
carrier frequency	interrupted continuous wave (ICW)	ratio detector
citizens band (CB)	limiter	significant sideband
communication	local oscillator	single sideband transmission (SSB)
continuous wave (CW)	lower sideband (LSB)	splatter
crystal detector	mixer	superheterodyne
demodulation	mixing	transistor detector
detection	modulation	undermodulation
diode detector	modulation index	upper sideband (USB)
electromagnetic radiation	modulator	vestigial sideband filter
electromagnetic wave		

TEST YOUR KNOWLEDGE

Please do not write in this text. Place your answers on a separate sheet of paper.

1. What is the wavelength of a 21-MHz radio wave?
   a. 1.428 meters.
   b. 14.28 meters.
   c. 142.8 meters.
2. The frequency of an 11-meter wave is _____ MHz.
3. A radio-frequency wave of 455 kHz must be mixed with what frequency to produce a 500-Hz audio tone signal?
4. Modulation power of _____ watts is required for 100% modulation of a 1000-watt amplifier.
5. Modulation power of _____ watts is required for 75% modulation of a 1000-watt amplifier.
6. Modulation power of _____ watts is required for 50% modulation of a 1000-watt amplifier.
7. Draw waveforms showing 100% modulation, overmodulation, and undermodulation.

8. Why is demodulation similar to half-wave rectification?
9. What is the main principle of the superheterodyne receiver?
10. What are some of the advantages of FM over AM?
11. The discriminator is a typical circuit you will encounter in _____ receivers.
12. What IF is used in FM radio?
    a. 10.7 MHz.
    b. 17.7 MHz.
    c. 21.7 MHz.
13. _____ is the process of removing the audio signal from a modulated radio wave.
14. Name the three styles of citizens band radios.
15. When did citizens band radio become popular?

Applications of Electronics Technology:
For baseball fans — this pocket-sized electronic encyclopedia answers literally billions of baseball questions. (Franklin)

Section V COMMUNICATION

SUMMARY

Important Points

- [ ] An oscillator is an electronic circuit that generates an ac signal at a desired frequency.
- [ ] Oscillators are commonly classified based on feedback method.
- [ ] The basic parts of an oscillator are the wave-producing circuit, amplifier, and feedback circuit.
- [ ] A transmitter is an electronic device that sends information toward its destination.
- [ ] A receiver is an electronic device that picks up transmitted waves.
- [ ] Modulation is the process of adding an audio wave to a carrier wave.
- [ ] In amplitude modulation (AM), the amplitude of the carrier wave is varied at the audio rate.
- [ ] In frequency modulation (FM), the frequency of the carrier wave is varied at the audio rate.
- [ ] Demodulation, or detection, is the process of removing the audio portion of a signal from the carrier wave.
- [ ] A tuned radio frequency (TRF) receiver picks up a transmitted RF wave, amplifies it, detects or demodulates it, and amplifies the audio wave.
- [ ] A superheterodyne receiver mixes the incoming modulated signal with an unmodulated local oscillator signal. The result is an intermediate frequency signal that carries the message. This signal is then amplified and demodulated to retrieve the signal.

1. Name four basic types of oscillators.
2. A fundamental wave-producing circuit made up of a capacitor and inductor connected in parallel is the _____ circuit.
3. The frequencies assigned by the FCC to the AM broadcast band are _____ kHz to _____ kHz. For FM, the broadcast band is from _____ MHz to _____ MHz.
4. The process for mixing two signals together is _____.
5. The process of removing an audio signal from a modulated RF signal is _____.
6. What major advantage do FM receivers have over AM receivers?
7. A wave has a frequency of 54 MHz. It has a wavelength of _____ meters.

Section VI INTEGRATED CIRCUITS AND COMPUTERS

We are living in the information age, which has been made possible mostly because of semiconductor technology. The integrated circuit was developed in the late 1950s. It is regarded as one of the most important inventions of the 20th century. The integrated circuit is usually referred to as an IC, or a *chip*. It might be considered the heart of the computer. Crude computers existed in the day of the vacuum tube. These computers filled large rooms and consumed large amounts of power. They were also slow as compared to today's standards. The IC has been largely responsible for revolutionary improvements in medicine, transportation, manufacturing, communications, and education, to name a few.

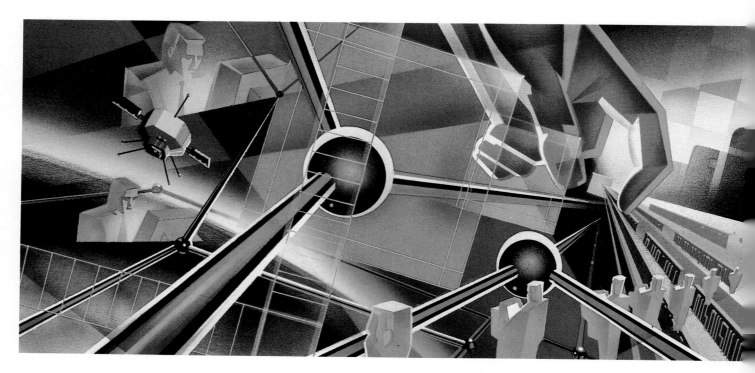

Chapter 18 discusses the development and evolution of the IC. An excellent description of how chips are made and types of ICs are also presented.

In Chapter 19, such integrated circuits as operational amplifiers, voltage regulators, and analog-to-digital and digital-to-analog converters are presented.

Digital circuits, presented in Chapter 20, are most important to the operation of today's computers. Digital numbering systems used in computers as well as digital logic gates are discussed.

Chapter 21 presents a fascinating overview of the modern computer. Computer inputs, arithmetic and control processes, memory, and outputs are explained.

Applications of Electronics Technology:
This very popular handheld video game system uses a color LCD screen. (Sega)

Chapter 18   INTEGRATED CIRCUITS

Probably one of the most amazing achievements in the field of electronics is the miniaturization of circuits. These circuits are so small, in fact, that their actual construction must be done by technicians using microscopes. These tiny circuits, which have revolutionized the electronics field, are called **integrated circuits (ICs)**.

After studying this chapter, you will be able to:
☐ *Compare the advantages and disadvantages of ICs to discrete components.*
☐ *Discuss how integrated circuits are made.*
☐ *Explain the difference between two types of ICs.*
☐ *Give examples of where ICs are used.*
☐ *Discuss special handling and mounting precautions required by integrated circuits.*

INTRODUCTION

Prior to development of ICs in the late 1950s, electronic equipment circuits consisted solely of *discrete components* interconnected by means of wiring or printed circuit board. (**Discrete components** are *individual* electrical devices such as resistors, diodes, or transistors that are made prior to circuit construction.) Now, with the development of ICs, most required electronic components and interconnections can be fabricated on a single substrate.

The substrate is made of a **wafer,** or thin slice, of semiconductor material, most often silicon, which is an excellent semiconductor element. The very small **microcircuits** making up the IC are made up of transistors, resistors, diodes, capacitors, and similar components. All of these components are fabricated on the wafer of silicon, or substrate. The substrate and microcircuit make up what is called a **chip**. The chip, however, is not ready for use until it is packaged and provided with terminal connections. See Figure 18-1.

Figure 18-1. ICs, or chips, made on a round silicon wafer, shown here in the background of the photo. Each chip is mounted in an IC package, which allows for connection to an external circuit. The whole package is shown in the foreground. (Intel Corp.)

There are advantages and a few disadvantages to the IC. ICs will perform the same function as *discrete circuits*. At the same time, they are more reliable. This means that ICs usually will work for longer periods of time without giving trouble. An IC logic gate is about 100,000 times more reliable than a vacuum tube logic gate. It is about 100 times more reliable than a transistor logic gate.

ICs are many, many times smaller than like circuits of discrete components, Figure 18-2. Too, manufacturers are able to pack much more circuitry into an IC. It is possible to place over a million components on a single chip. This cuts down on the size and weight of the product. See Figure 18-3.

Cost is another advantage of integrated circuits over discrete components. By mass production, an IC is now comparable in cost to a single transistor.

Integrated circuits do have some limitations. Inductors are not suited for integrated circuits. ICs are limited to lower operating voltages. Also, they are limited in the amount of current that they can handle. ICs, like diodes and transistors, are quite delicate. They cannot withstand rough handling or excessive heat. A heat sink must be used when soldering ICs to a circuit board. (Preferably, connection to a circuit board is made though a socket.) In summary, the advantages of the IC far outweigh the disadvantages.

Figure 18-2. A typical IC compared to the size of a daisy. (Intel Corp.)

*Figure 18-3. A worker inspects a **recticle**, which is like an IC photographic negative. It, and others like it, are used to pack more than 70,000 active microelectronic elements onto a 1/4-inch-square microprocessor chip.   (NCR Corp.)*

TYPES OF INTEGRATED CIRCUITS

ICs may be classified in a couple of ways. One way is in terms of their function — as to whether they are *linear IC devices* or *digital IC devices*. **Linear ICs**, also called **analog ICs**, have variable outputs, which are controlled by variable inputs. They are not on/off devices; they do not operate as switches. Some typical examples are audio amplifiers, IF amplifiers, and operational amplifiers. Linear amplifiers are explained in detail in Chapter 19.

Digital ICs perform repetitive operations. These ICs are on/off devices. They operate as switches and are sometimes called switching circuits. A digital signal varies in *discrete* steps (i.e., pulses or ON/OFF operation) whereas an analog signal has *continuous* values. See Chapter 20 for more information on digital ICs.

Digital ICs, then, may be further classified in terms of their **component density**. This is the number of components contained within a given package or chip. The ability of electronic manufacturers to put more and more components into a small space is astounding. Since they were invented, the number of components that could fit into a standard size integrated circuit has dramatically increased. A method of classifying by component density is as follows:

- **Small-scale integration (SSI):** 0-11 gates per IC
- **Medium-scale integration (MSI):** 12-99 gates per IC
- **Large-scale integration (LSI):** 100-999 gates per IC
- **Very-large-scale integration (VLSI):** 1000+ gates per IC

(Note that these groupings will vary from one manufacturer or source to another.) The chronological development of these semiconductor technologies in the electronics field is shown in Figure 18-4.

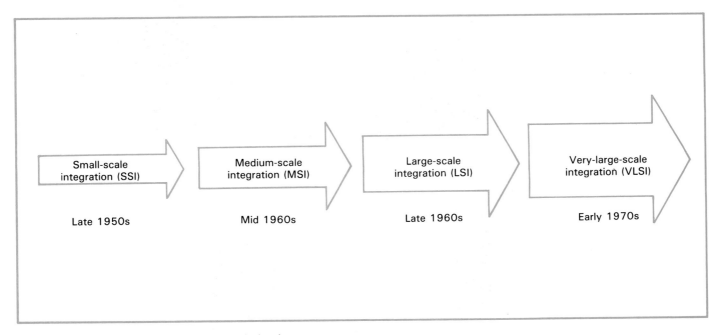

Figure 18-4. Development of integrated circuits.

IC CONSTRUCTION

Making an integrated circuit is a detailed process. Many steps are involved in creating this tiny device. The process is explained in Figures 18-5 through 18-28.

HOW INTEGRATED CIRCUITS ARE MADE _____

The production of semiconductor components is one of the most sophisticated processes known in modern industry. The processing of silicon into solid-state discrete or integrated circuit devices requires a great depth of knowledge and skill in such disciplines as physics, chemistry, metallurgy, optics, and photolithography.

While thought of primarily as an assembly industry, the semiconductor business actually involves more chemical and photographic processing than anything else. Read on to find out how integrated circuits are made.

(Photographs and narrative courtesy of Motorola, Inc.)

Figure 18-5.   This microprocessor (computer chip) is encased in a 64-pin ceramic dual in-line package. It is uncapped, revealing the highly complex chip.

Figure 18-6.   Advances in semiconductor technology have provided the capability to make a microprocessor of a single silicon chip. They are at least an order of magnitude higher in performance and circuit complexity than had been available before. The Motorola MC68000 microprocessor shown combines advanced circuit design techniques with computer sciences to achieve a 16-bit microprocessor.

Figure 18-7.   The processing of semiconductors starts with a chemical that when reacted with hydrogen in a high temperature reactor forms a shiny material called polycrystalline silicon. From this raw material, which is shown in the photo, single crystal silicon wafers are made.

Figure 18-8.   The silicon is melted down in large crystal-growing furnaces, which operate at 1440°C. A seed of single crystal silicon, mounted on a rod, is inserted into the molten mass and slowly rotated and withdrawn.

Figure 18-9.   After several hours, a nearly 2-foot-long ingot of single crystal silicon has been formed. Depending on the equipment used, the diameter of the ingot can range from 2-6 inches. This ingot is 6 inches in diameter.

Figure 18-10.   After the ingots are tested for their electrical properties, they are sawed into wafers about 25 mils thick. Diamond-toothed, inside-diameter cutting saws are used in this operation.

*Figure 18-11.   The wafers are then lapped and polished to a mirror finish. Now, the wafers are ready for **photolithographic processing**. In this process, the pattern of identical circuits is reproduced many times over on the face of the wafer.*

*Figure 18-12.   These are epitaxial furnaces, which deposit epitaxial layers on the silicon substrates. (An **epitaxial layer** is one with the same crystal orientation as the parent material.) With these furnaces, layers can be grown with many combinations of dopants, resistivity, and thickness requirements.*

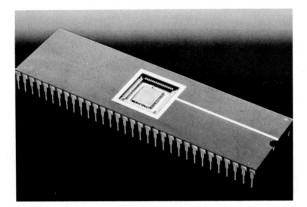

Figure 18-5.

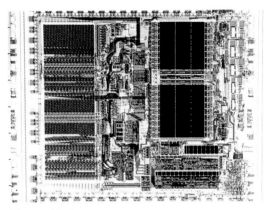

Figure 18-6.

Figure 18-7.

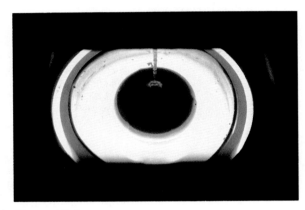

Figure 18-8.

Figure 18-9.

Figure 18-10.

Figure 18-11.

Figure 18-12.

Figure 18-13. The photolithographic process starts with the preparation of highly complex artwork for the circuit. This includes detailed artwork of each specific device used. Much of the detailed drawing is completed by automated drafting equipment, such as a pattern generator shown here. The machine pictured takes a magnetic tape and transforms its information into an image on a glass plate.

Figure 18-14. The 10X reticle (left) shows a ten times enlarged image of the repeated patterns shown on the wafer (right).

*Figure 18-15. This step-and-repeat machine takes the image on the 10X reticle and reduces it down to actual size. It repeatedly prints it in rows and columns over an entire glass plate. This plate is called a **mask**. This mask and other matching masks in a set are then given to the wafer processing area.*

Figure 18-16. The mask is used at the wafer processing area to project its pattern onto a silicon wafer. After a layer of silicon dioxide is grown on the surface, the wafer is coated with a photosensitive material. The circuit pattern is then printed in this photosensitive layer on the wafer by exposing light through the mask onto the coated wafer.

Figure 18-17. The portions of the silicon dioxide no longer covered by the photosensitive material are now etched away. The photosensitive material is then removed, having served its purpose. The clean wafer now has areas free of silicon dioxide. These areas are often referred to as windows.

Figure 18-18. An operator (left) in a clean room is positioned at an etch station. Here, the unwanted portion of the wafer is etched away by means of acid. The required pattern is left behind. Another operator (right) is measuring the very thin film of oxide on a wafer.

Figure 18-19. Diffusion furnace operating at 1200°C. These furnaces are typical of those used to grow a masking silicon dioxide layer on a wafer prior to doing the photolithography and mask pattern alignment. Similar furnaces are used to drive dopant atoms into silicon wafers.

Figure 18-20. Ion implantation equipment. It is used to implant the right amount of specific elements into regions of the silicon. This establishes the electrical characteristics of the silicon.

Figure 18-13.

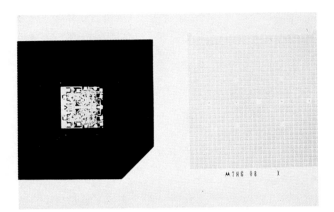

Figure 18-14.

Figure 18-15.

Figure 18-16.

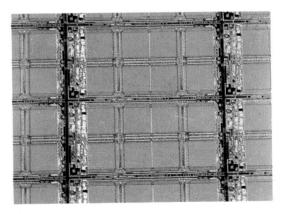

Figure 18-17.

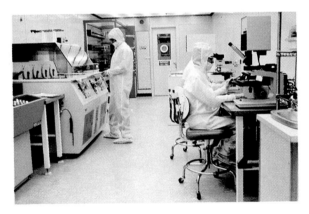

Figure 18-18.

Figure 18-19.

Figure 18-20.

Figure 18-21. The operator is making a final quality assurance check before the wire bonding operation.

Figure 18-22. This is a close-up of a portion of a wafer after it has been separated into hundreds of transistor chips. The separation has been completed by a semiautomatic diamond saw, much in the manner of cutting glass.

Figure 18-23. This assembly step shows a bonding operator placing the chip in position on the IC-device package frame. A vacuum needle is used by the operator to pick up each chip and place it onto the frame where it is fixed into position.

*Figure 18-24. The next step is to connect the chip to the exterior posts of the package. This will eventually permit the device to be wired into a circuit board. This critical step is called **wire bonding**. In this case, two silicon aluminum wires are attached between the chip and the posts through the use of a semiautomatic bonding machine.*

Figure 18-25. This is a close-up of the package frame. It shows the chip fixed into position and connecting wires attached. The wire bonding is done ultrasonically. The silicon aluminum wire is about 1 mil (0.001 inch). It is less than one-third the diameter of a human hair!

Figure 18-26. This is called an RF final test machine. This unit automatically tests the semiconductor parts for electrical specifications. Each of the bins at the base of the machine accepts devices having various electrical properties. One bin takes rejects.

Figure 18-27. The semiconductor devices are now ready for encapsulation in either plastic, metal, or ceramic. This final covering seals the device and protects it from damage, or from the effects of heat, water, and air. Shown here is a wide variety of discrete packages.

Figure 18-28. Shown here is a great variety of integrated circuit packages in plastic, metallic, and ceramic materials. After undergoing a number of electrical, mechanical, and environmental tests, the devices are ready for packing and shipping.

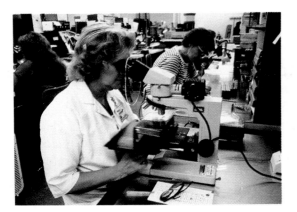

Figure 18-21.

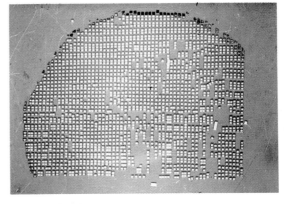

Figure 18-22.

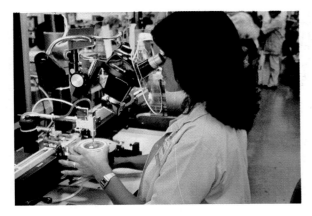

Figure 18-23.

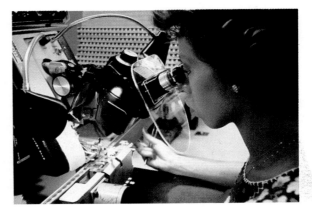

Figure 18-24.

Figure 18-25.

Figure 18-26.

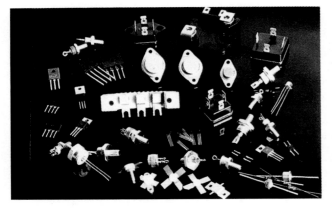

Figure 18-27.

Figure 18-28.

IC PACKAGES

Integrated circuits come in various sizes and shapes. Package designs for ICs are usually:

- Round (*transistor-outline,* or *TO, can*).
- Square (*flat pack*).
- Rectangular (*dual in-line package,* or *DIP*).

Figure 18-29 shows these typical IC package designs.

All manufacturers use a standardized pin numbering system for their devices. Typical pin connections for IC units are shown in Figure 18-30. Note the starting dot, notch, or tab on each device. These serve as an index, or reference point. The pins are numbered counterclockwise as you view the IC from the top, beginning at the reference point. ICs typically have 8, 10, or 16 pins. A special 88-pin IC package is shown in Figure 18-31.

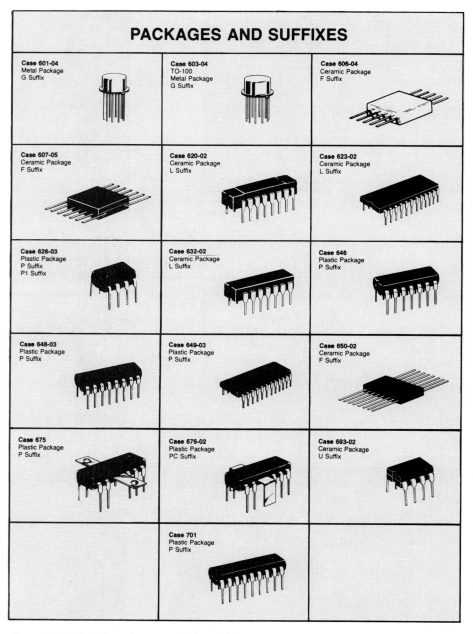

Figure 18-29. IC packages.   (Motorola)

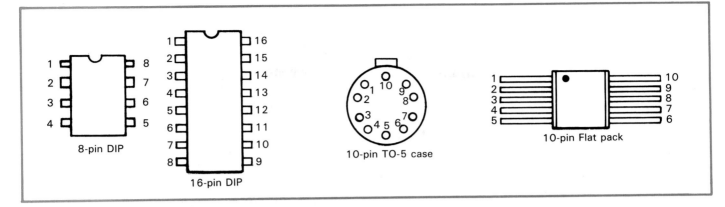

Figure 18-30. Pin connections for IC packages.

TYPICAL APPLICATIONS

Microelectronics, the body of electronics connected with extremely small electronic parts, has soared in the past decade. Smaller and smaller electronic components perform more and more complex electronic functions. We can see many benefits in our lives. There are many electronic consumer products on the market today that did not exist in the recent past. The applications are far-reaching.

A common use of microelectronics is seen in the digital watch, Figure 18-32. These watches are controlled by ICs. Digital readouts may be either LED or, more often, LCD 7-segment display formats. See Figure 18-33.

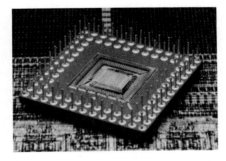

Figure 18-31. An 88-pin bed-of-nails grid array package frames an integrated circuit. (Monolithic Memories, Inc.)

Figure 18-32. This digital watch also has a temperature data/memory function, which allows the user to keep track of temperature changes. (Casio, Inc.)

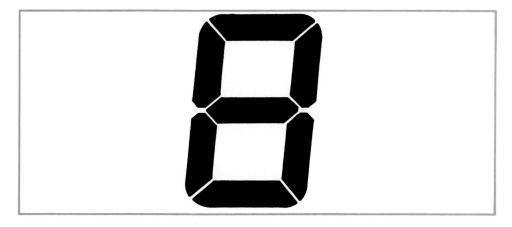

Figure 18-33. The 7-segment display comes as a single unit. It will display digits 0-9, depending upon which segments are on.

Another popular application of microelectronics is the calculator. These range in complexity from models that perform basic arithmetic to the programmable, scientific pocket calculator. See Figure 18-34.

Figure 18-34. This scientific calculator has the storage and display capabilities of a personal computer and provides engineers with the ability to develop, solve and analyze problems. It can be connected to a computer or used alone.   (Hewlett-Packard)

Modern electronic test instruments use electronic circuits. Digital meters, frequency counters, and logic probes are some examples of test equipment that use integrated circuits. Figure 18-35 shows a miniature digital multimeter being used for equipment testing.

Integrated circuits are the heart of computers today. Due to their large capacity, speed, accuracy, and reliability, personal computers have become very common. Figure 18-36 shows a laptop computer that uses a liquid crystal display. A personal communicator is shown in Figure 18-37.

AM/FM Radio

The AM/FM radio integrated circuit was developed by General Electric. It is an excellent example of how microcircuits can be used to reduce size and cost and provide improved performance over a transistor radio. The AM/FM IC contains most of the radio circuits active components. One of the interesting features of this integrated circuit is its ability to operate over a wide supply

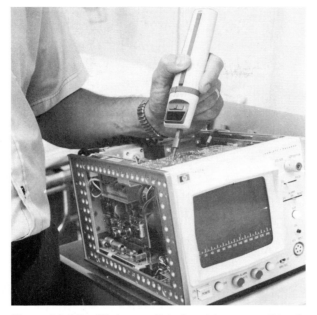

Figure 18-35. Miniature digital multimeter. (Hewlett-Packard)

Figure 18-36. This laptop computer runs all day long on batteries and weighs just 2.9 pounds. Notice the graphics on the LCD display screen. (Hewlett-Packard)

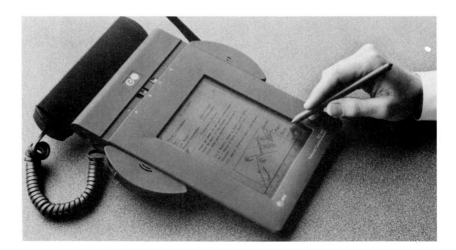

Figure 18-37. Shown here is a fully integrated communications device that combines the power of cellular phones, fax machines, and personal computers. The pen allows the user to "write" on the screen. (AT&T)

voltage range. Typically, it can operate from voltages up to 13 Vdc and down to as little as 2 Vdc. Figure 18-38 shows a photo of the IC and a pin identification diagram. Figure 18-39 shows the IC internal circuitry, and Figure 18-40 shows the AM/FM IC and external circuitry.

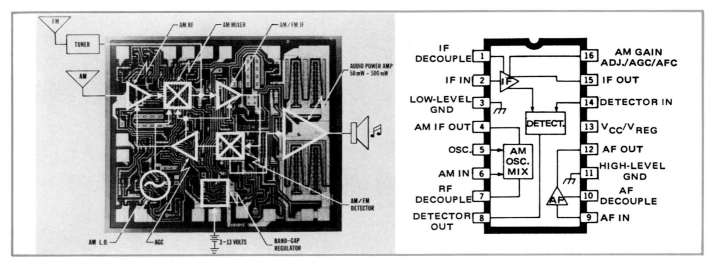

Figure 18-38. Photo and pin diagram of the AM/FM integrated circuit. (Sprague Electric Co.)

WORKING WITH IC DEVICES

When building test circuits that have ICs, be very careful when you are soldering them into a printed circuit. It is always a good idea to use a small soldering pencil of 30 W or less to solder ICs, transistors, diodes, etc. If possible, use a heat sink to avoid damage due to excessive heat. Even better, use a plug-in type IC (such as a DIP) and socket arrangement. See Figure 18-41.

There are other things to watch out for. Some ICs can be damaged by static electricity. Read the manufacturer's specification sheet and observe any special handling suggestions. Be sure to observe correct polarity of the battery or power supply when making connections to an integrated circuit. Use the index mark or notch on the package for correct terminations. Also, take care to insert plug-in type ICs into the circuit properly. To avoid pin damage, assure that pins are aligned before applying pressure to the IC.

SUMMARY

- Integrated circuits are microcircuits made up of transistors, resistors, diodes, capacitors, and similar components. All of these components are fabricated on a wafer of silicon.
- ICs are also called chips.
- Among advantages of the IC are: reliability; low cost; high component density; low power consumption; small size. Among disadvantages are: only certain parts can be built into an IC; voltage and current limitations; fragile.
- Making an integrated circuit is a detailed process. Many steps are involved in creating this tiny device.
- ICs may be classified as either linear or digital devices. Linear ICs have variable outputs. Digital ICs have discrete outputs.
- ICs come in various sizes and shapes. Package designs for ICs are usually TO can, flat pack, and DIP.
- All manufacturers use a standardized pin numbering system for their devices.

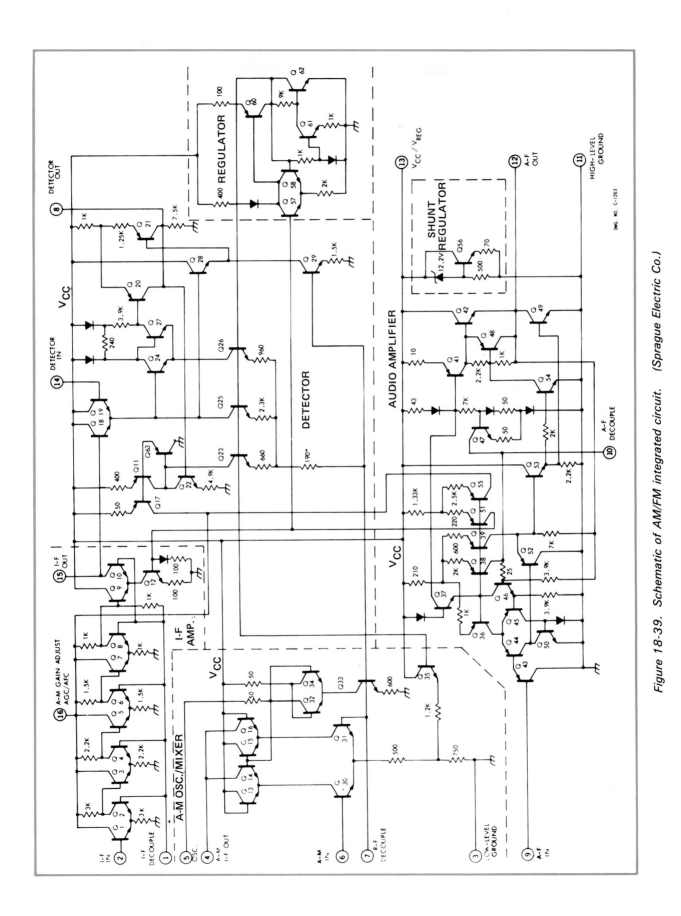

Figure 18-39. Schematic of AM/FM integrated circuit. (Sprague Electric Co.)

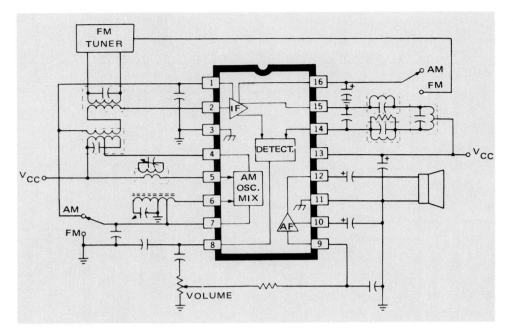

Figure 18-40. AM/FM radio system.   (Sprague Electric Co.)

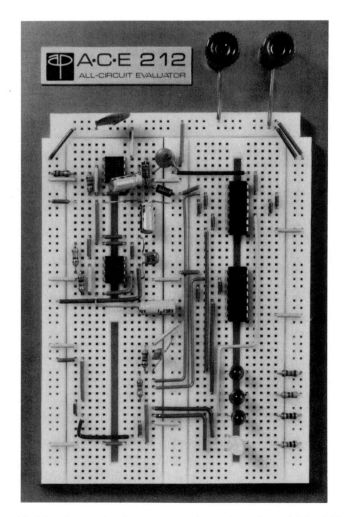

Figure 18-41. A test circuit constructed on a breadboard. The DIP-type ICs shown are connected to the rest of the circuit simply by plugging them in. Soldering is not required. (AP Products)

KEY TERMS

Each of the following terms has been used in this chapter. Do you know their meanings?

analog IC
chip
component density
digital IC
discrete component
epitaxial layer
integrated circuit (IC)

large-scale integration (LSI)
linear IC
mask
medium-scale integration (MSI)
microcircuit
microelectronics

photolithographic processing
recticle
small-scale integration (SSI)
very-large-scale integration (VLSI)
wafer
wire bonding

TEST YOUR KNOWLEDGE

Please do not write in this text. Place your answers on a separate sheet of paper.

1. It is possible to place over a million components on a single chip. True or False?
2. Inductors are well suited for ICs. True or False?
3. What are four advantages of integrated circuits over discrete circuits?
4. Name two basic types of integrated circuits and explain the function of each type.
5. Which of the following is used as integrated circuit substrates?
   a. Carbon.
   b. Silicon.
   c. Boron.
6. What is the primary difference between the various types of IC integration (SSI, MSI, LSI, or VLSI)?
7. Draw a symbol for an 8-pin integrated circuit in a rectangular package showing the locator index.
8. Name some IC applications that are discussed in this chapter. Also name some applications that are not listed in this chapter.
9. A soldering pencil of less than _____ should be used to solder integrated circuits on a printed circuit.
10. Are precautions necessary when handling ICs? Explain.
11. In the process of _____ _____, a pattern of identical circuits is reproduced many times over the face of a wafer.
12. Describe, in a paragraph or two, the process of IC construction.

Applications of Electronics Technology:
The CAD system shown here is being used for golf course design. Notice the background trees are the result of digitizing a photographic image. The water and green are computer generated. (Integraph)

Chapter 19

LINEAR INTEGRATED CIRCUITS

Linear ICs had a big part in the enormous growth of technology in the latter half of the 20th century. Presently, these devices find application in all market segments including motor controls, instrumentation, aerospace, automotive, telecommunication, medical, and consumer products. See Figure 19-1.

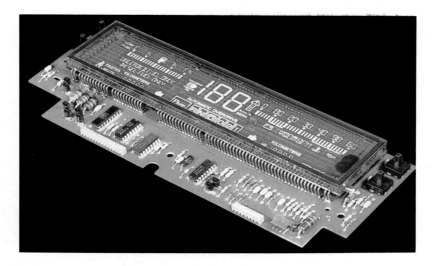

Figure 19-1. Linear ICs are used in this automotive instrument cluster to drive display and indicator lights.   (General Motors Corp.)

After studying this chapter, you will be able to:
☐ *Explain the function of an operational amplifier.*
☐ *Identify three basic op amp circuits.*
☐ *Discuss IC voltage regulators.*
☐ *Name two devices used to interface analog and digital signals.*

INTRODUCTION

As already stated, linear integrated circuits are one of the two basic types of IC. These ICs process **analog signals** — signals that are **continuous**, which means they can vary between any range of values. Analog signals usually represent physical variables, such as voltage, sound, temperature, etc.

The purpose of a linear IC is quite different from that of its digital counterpart. Most commonly, linear ICs are used as signal amplifiers, Figure 19-2. Other examples of use include oscillators, mixers, modulators, limiters, detectors, and voltage regulators. They are frequently found in automotive and industrial control circuits and audio, radio, and TV circuits.

In this chapter, we will introduce two linear IC devices. One is the *operational amplifier*. The other is the IC voltage regulator. We will also touch upon analog-digital conversion. ICs that perform this function employ operational amplifiers to bridge the gap between analog and digital worlds.

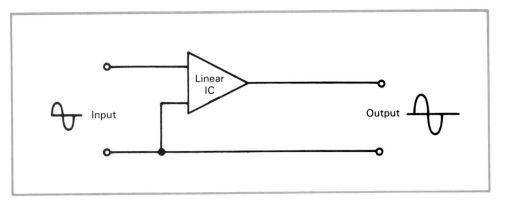

Figure 19-2. This linear IC is a signal amplifier. Note the use of the triangle for the IC. This symbol exclusively means amplifier.

OPERATIONAL AMPLIFIERS

The **operational amplifier**, or **op amp**, as it is usually called, is probably the most widely used linear IC today. The device uses external feedback to control response characteristics. It can achieve very high gain, with 500,000 to 1,000,000 being common. Ideally, this device has infinite gain. Op amps are used as basic amplifier circuits. They are used in two-way intercoms, function generators, cable TV video line drivers, and servo amplifiers, Figure 19-3. They are also used in analog computers and for scale changing in some electronic instruments.

Figure 19-3. These servo amplifiers, *which use op amps, are designed to control* servo-valves. *These valves maintain a fluid flow that is proportional to the signal.* (Rexroth Corp.)

Op amps were developed in the 1940s to be used with analog computers. This, in fact, is where the word *operational* comes from. The device was developed to perform certain mathematical operations. The original op amps were vacuum-tube operated and had all of the problems associated with vacuum tubes. The first IC op amp was developed by Fairchild Semiconductors in the late 1960s. It had all of the advantages of solid-state technology and proved to be much more reliable.

The symbol for an op amp is shown in Figure 19-4. Notice that the device has two input terminals, one plus ($+$), one minus ($-$). The ($+$) lead is the noninverting input. A signal applied to this terminal will appear both in phase and amplified at the output. The ($-$) lead is the inverting output. A signal applied to this terminal will appear amplified but inverted at the output.

The power supply leads are indicated $+V$ and $-V$. Sometimes, you may see schematics where these leads are not marked. In such cases, these power connections have been omitted to simplify the drawing. Unless otherwise stated, you can assume that the device still has these leads, and that they still require power. When connecting the power supply leads, never reverse the polarity. Doing so, even for a moment, will cause a destructive current through the device.

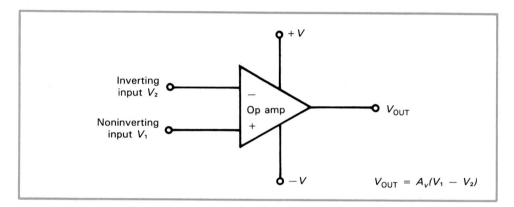

Figure 19-4. Op amp symbol.

The schematic of a general purpose op amp is shown in Figure 19-5. Also shown are its pin configuration and packaging information. This IC has only one op amp. It is also possible to have ICs that house two or four op amps in one package. See Figure 19-6.

These popular ICs offer many features that make their application almost foolproof. This includes overload protection on both the input and output and a compensation feature, which controls unwanted oscillation. They also have a no-*latch-up* feature for when the common mode range is exceeded. This way, circuits will always revert to their intended mode.

BASIC OP AMP CIRCUITS

There are a number of op amp circuits. Let us now discuss three of them. These are the *inverting amplifier*, the *noninverting amplifier*, and the *unity gain amplifier*.

Inverting amplifier

Figure 19-7 shows an op amp used as a basic **inverting amplifier**. A signal is fed to the device through R_1 and the inverting ($-$) lead. The ($+$) lead is usually grounded or tied to the input and output circuit common. R_3 is the bias current balancing resistor. Its purpose is to minimize error in the output voltage.

Part of the output signal is fed back through resistor R_2 to the input. The purpose of this is to control or reduce the signal amplification as well as to stabilize the circuit. The output signal is antiphase (180° out of phase) to the input signal. As a result, the feedback signal is negative, and it subtracts from the input signal. The reduced input, in turn, reduces the output signal.

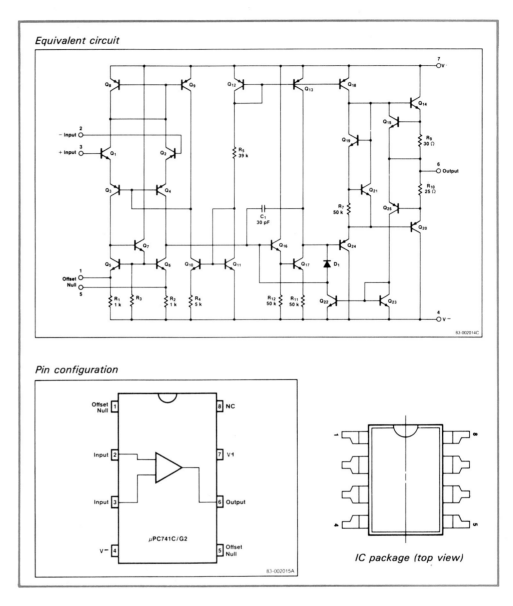

Figure 19-5. μPC741 op amp circuit. Depending on the manufacturer, these may also be referred to as LM741s, or simply, 741s.   (NEC Electronics, Inc.)

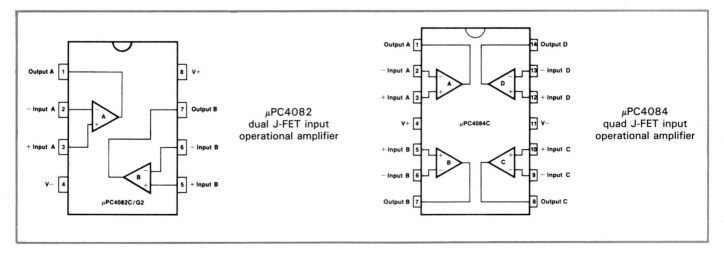

Figure 19-6. Dual and quad op amps.   (NEC Electronics, Inc.)

By choosing the correct values for R_1 and R_2, the op amp voltage gain can be accurately controlled to whatever level is needed. The formula for voltage gain of the inverting amplifier in Figure 19-7 is:

$$A_v = \frac{R_2}{R_1} = \frac{\text{feedback resistance}}{\text{input bias resistance}}$$

Thus, using the circuit values in Figure 19-7, the gain is:

$$A_v = \frac{R_2}{R_1} = \frac{82 \text{ k}\Omega}{10 \text{ k}\Omega} = 8.2$$

Since gain is a ratio, it does not have a unit.

By choosing the correct value of R_3, the voltages appearing at ($+$) and ($-$) may be made equal, resulting in a zero output voltage error. For them to be equal, the drop across R_3 must equal the drop across the parallel combination of R_1 and R_2. Thus, the value of R_3 for this particular circuit is found to be:

$$R_3 = \frac{R_1 R_2}{R_1 + R_2} = \frac{(10,000 \ \Omega)(82,000 \ \Omega)}{10,000 \ \Omega + 82,000 \ \Omega} = 8.9 \text{ k}\Omega$$

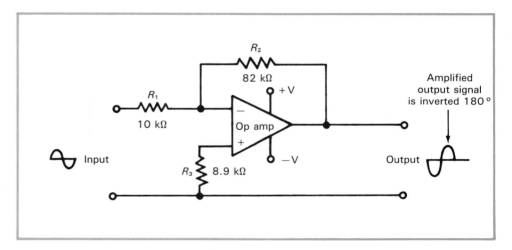

Figure 19-7. Inverting op amp circuit.

Noninverting amplifier

The op amp can also be used as a **noninverting amplifier**, Figure 19-8. This is done by feeding the signal into the ($+$) input lead. The ($-$) input lead, then, is tied to the input and output common through R_1. The output signal is the difference between the ($+$) input signal and the ($-$) feedback signal.

The formula for voltage gain of the noninverting amplifier in Figure 19-8 is:

$$A_v = 1 + \frac{R_2}{R_1} = 1 + \frac{\text{feedback resistance}}{\text{input bias resistance}}$$

Thus, using the circuit values in Figure 19-8, the gain is:

$$A_v = 1 + \frac{500 \text{ k}\Omega}{100 \text{ k}\Omega} = 6$$

Unity gain amplifier

The op amp can also be used as a **unity gain amplifier**. This special circuit has neither input resistance nor feedback resistance. It is used as a buffer or separator of one circuit from another. In unity gain amplifiers, the feedback circuit is connected through a path of no resistance (shorted) back to one of

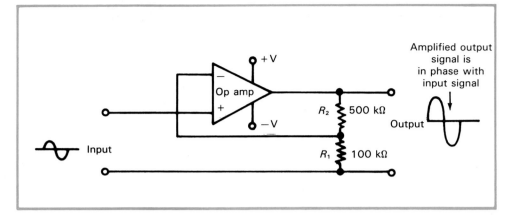

Figure 19-8. Noninverting op amp circuit.

the input leads. Figure 19-9 shows a typical noninverting unity gain circuit. The output is unamplified; there is no gain. In addition to being a buffer, this circuit may be used to match impedance from one circuit to another. This amplifier is also called a **voltage follower**.

IC VOLTAGE REGULATORS

Several types of IC voltage regulators are available. In Chapter 12, **fixed-voltage, three-terminal regulators** were briefly discussed. Another type of IC regulator is the **precision voltage regulator**, which has more than three terminal leads. The purpose of these and other regulators is to provide fixed output voltages. These devices are considered to be part of the family of linear ICs.

The trend in voltage regulation has been toward smaller, low-cost, low-current, fixed-voltage IC regulators. These devices, which use high-gain amplifiers and negative feedback, achieve much better regulation than simple zener-diode voltage regulators. Usually, these devices require very little or no heat sinking for operation. They also have few or no external components. Voltage regulators are available in a variety of packages. See Figures 19-10 and 19-11.

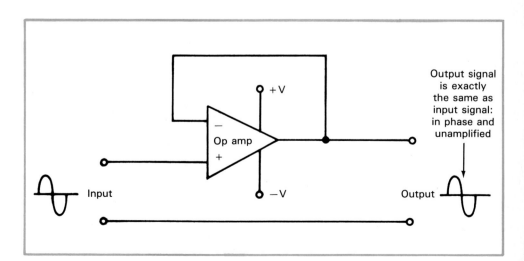

Figure 19-9. Unity gain op amp circuit. Note that the feedback circuit is "short" from output to input.

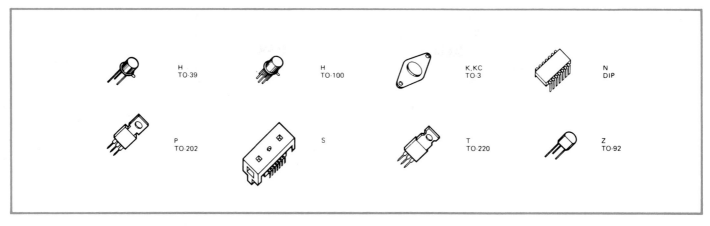

Figure 19-10. An assortment of regulator packages. (National Semiconductor Corp.)

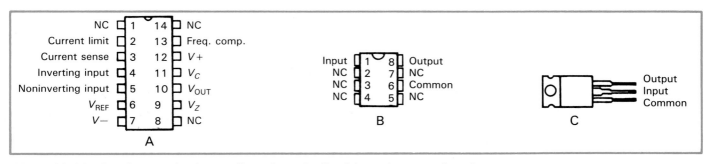

Figure 19-11. Regulator packaging configurations. A—Precision voltage regulator in a dual in-line package. B—Three-terminal regulator in a dual in-line package. C—Three-terminal regulator with TO-style package. (Note that NC means not connected.)

BASIC CONFIGURATION IN POWER-SUPPLY CIRCUIT

Figure 19-12 shows an IC voltage regulator as a black box. Note that one lead of the regulator is the input lead from the unregulated dc power-supply filter circuit. The other lead is the output lead, which provides either the positive or negative regulated output voltage. The third lead is a ground connection, which is common to both the input and output circuits.

The capacitors in Figure 19-12 are external components. The input capacitor provides additional filtering for the input signal, which comes from the power-supply filter circuit. The output capacitor is not necessary but improves the *transient response* of the circuit. In other words, it helps to stabilize the circuit when abrupt line or load changes occur.

DESCRIPTION OF OPERATION

A basic circuit function diagram of the voltage regulator is shown in Figure 19-13. *Series-pass transistor* Q_1 acts as a variable voltage dropping element in series with R_L, the load resistance $(V_{CE} = V_{IN} - V_{R_L})$. The V_{REF} block is a temperature-stabilized *reference voltage*. It is held at a nearly fixed value. A zener diode can be used to achieve this. Likewise, a special *band gap circuit*, the operation of which will not be explored here, is very often used to achieve the reference voltage.

The **voltage sampling network**, R_1-R_2, takes a fraction of the load voltage V_{R_L} and feeds it back to the inverting terminal of the **error amplifier**. It is a **differential amplifier**, as it compares two inputs. It compares the *feedback voltage* V_f, from the voltage sampling network, to the reference voltage V_{REF}. The difference $(V_{REF} - V_f)$ is amplified and used to control the base current to Q_1.

As long as the difference between the two inputs is constant, say for example, 2 V, the error-amplifier output and, therefore, V_{CE} remain constant. If V_{OUT} tries to drop, either from a drop in V_{IN} or a rise in I_{R_L}, V_f will decrease, and the difference, $V_{REF} - V_f$, will increase. The output of the error amplifier, then, will increase significantly. In turn, the base current to Q_1 will increase, increasing emitter current. The attempted decrease in output voltage is offset. If V_{OUT} attempts to rise, the reverse action happens. Thus, output voltage is kept nearly constant.

All the regulator protection circuits such as the current limit, the safe area, and the thermal shutdown circuit, when activated, limit or turn off the base drive in the series-pass transistor. The output current is either limited or the

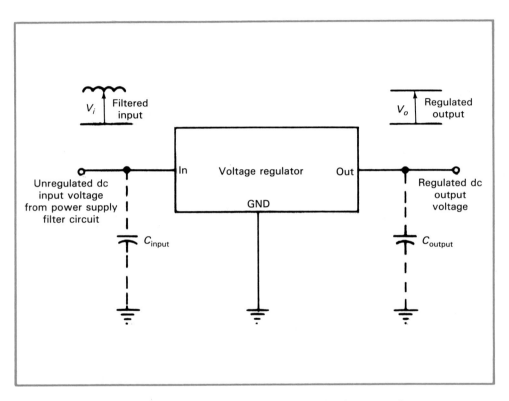

Figure 19-12. Block representation of a three-terminal voltage regulator.

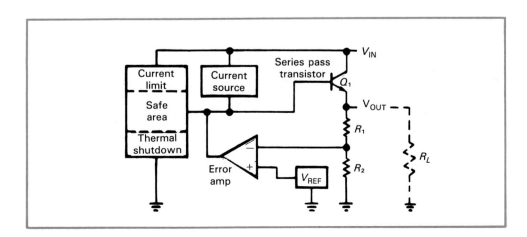

Figure 19-13. Block diagram depicting the internal structure of an IC voltage regulator.

series-pass transistor is completely turned off if there are protection problems. A complete schematic of a popular voltage regulator is shown in Figure 19-14.

ANALOG-DIGITAL CONVERSION

These ICs are neither true analog nor true digital devices. They are used to bridge analog and digital signals. An **analog/digital (A/D) converter** converts analog signals to digital form. Conversely, a **digital/analog (D/A) converter** converts digital signals to analog form.

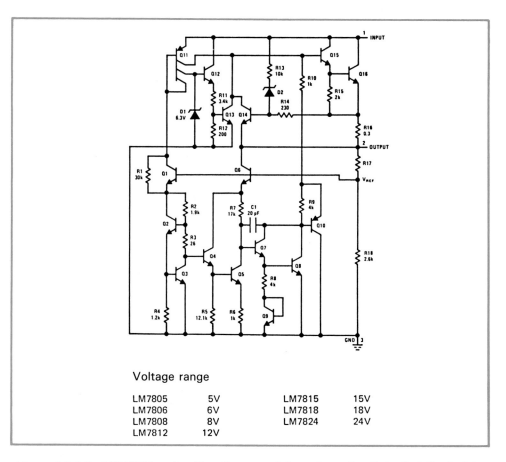

Voltage range

LM7805	5V	LM7815	15V
LM7806	6V	LM7818	18V
LM7808	8V	LM7824	24V
LM7812	12V		

Figure 19-14. LM78XX series IC voltage regulator schematic. Note that the last two digits of the part number give the regulator output voltage. (National Semiconductor Corp.)

A/D CONVERSION

The basic process of converting an analog to a **digital signal** — that is, a signal represented by discrete values — involves sampling the analog signal at regular intervals. The amplitude value of the signal is **quantized**, which means it is converted into a digital value. The approximate amplitude of the signal at each sample interval represents the digital value.

Figure 19-15 shows how an analog signal looks in digital form. The original analog signal is *mapped* into a series of pulses. The scale on the right of the plot is in the **binary number system**, which is a *base 2* system. The system of numbers that you are familiar with is the **decimal number system**, which is a *base 10* system. Whereas the decimal system operates with 10 digits, the binary system uses just two — 0 and 1. Binary numbers are much more efficient for electronic machines. The two numbers can be represented by two voltage states: OFF and ON. This was seen in Chapter 16 in the discussion on flip-flops. Chapter 20 will give more information on the binary numbering system so that you can better understand it.

Figure 19-15 shows that by increasing the sampling rate and increasing the number of **quantization intervals** (increments of amplitude), we get a closer approximation of the original signal. However, we can never totally encode a continuous analog function. The analog waveform has an infinite number

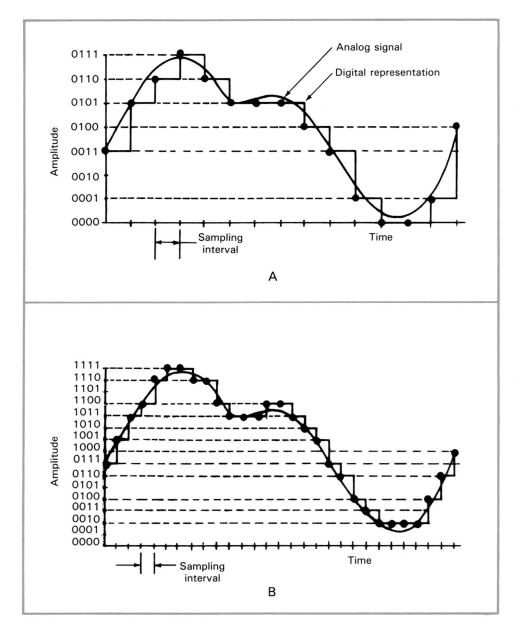

Figure 19-15. An analog audio signal and its digital counterpart. A—Digital signal characterized by low sampling rate and few quantization intervals. B—Increasing sampling rate and number of quantization intervals produces closer approximation of analog waveform.

of values. In converting to digital, we can only choose from a finite number of increments.

Figure 19-16 is a simple *black-box* diagram of an A/D converter. It shows

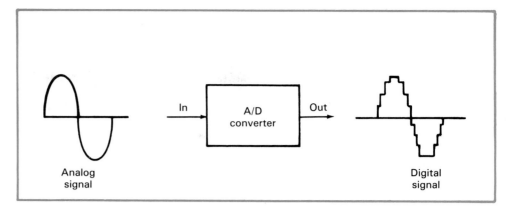

Figure 19-16. Converting an analog signal to a digital signal.

an analog signal being fed into the device. The output is the signal converted into digital form.

D/A CONVERSION

The D/A converter converts a digital input into a voltage whose magnitude is proportional to the value of the digital signal. For example, if the digital input of 0001 causes a 1-V output, then a digital input of 0010 would cause a 2-V analog output. An input of 0011 would cause a 3-V output. (Note that 0001 is equal to 1 in the decimal system, 0010 is equal to 2, and 0011 is equal to 3.)

The simplest D/A converter contains a series of resistors and switches, Figure 19-17. There is a switch for each input *bit*. (For example, the binary number

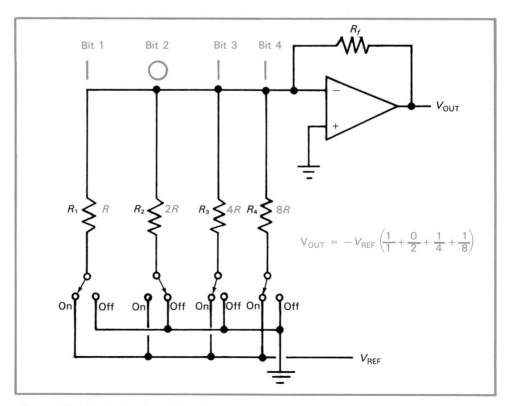

$$V_{OUT} = -V_{REF} \left(\frac{1}{1} + \frac{0}{2} + \frac{1}{4} + \frac{1}{8} \right)$$

Figure 19-17. A simple weighted-resistor D/A converter showing position of switches for the binary number 1011. Resistors increase in size from left to right, allowing less current for each digit moving right—thus, giving more weight to each digit moving left. The more current reaching the op amp, the higher the output voltage.

1011 is a 4-*bit* number.) If our digital values are all 4-bit numbers (0000—1111), then four switches are required. A corresponding resistor represents the value associated with each bit. Each resistor is twice the size of its neighbor. A reference voltage is used to generate current in the resistors. Only a digital 1 will close a switch and cause current. No current flows for a digital 0. An op amp sums the currents and converts them to a voltage.

AUDIO AMPLIFIER PROJECT

An audio amplifier can be built using an op amp. In this project, you will construct a simple, yet effective, circuit for an audio amplifier. It is capable of amplifying frequencies from 20 Hz—20,000 Hz. The circuit schematic and parts list are shown in Figure 19-18. The parts placement as well as the printed circuit board (PCB) layout are shown in Figure 19-19. A photo of the completed project is shown in Figure 19-20. When complete, the circuit may be encased in a metal or plastic chassis if desired.

When construction of your audio amplifier is complete, place the PCB on a nonconductive surface and connect it to the 9-Vdc battery or power supply. Connect the output of the audio frequency (AF) generator to the input of the audio amplifier. Adjust the frequency of the AF generator to 1 kHz and the amplitude to a very low setting. Place switch S_1 in the ON position.

You should now hear the 1-kHz tone and be able to adjust its volume with R_5. If your audio amplifier does not operate in this manner, immediately remove the power input leads from the power supply. In this event, check all solder connections. Poorly soldered connections must be reheated to form

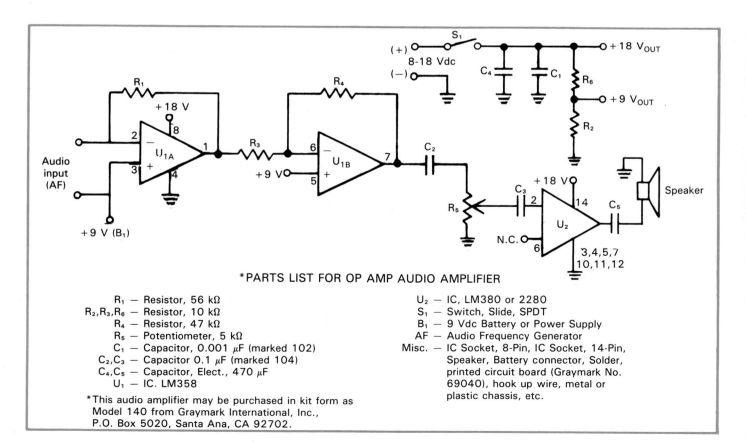

*PARTS LIST FOR OP AMP AUDIO AMPLIFIER

R_1 — Resistor, 56 kΩ	U_2 — IC, LM380 or 2280
R_2, R_3, R_6 — Resistor, 10 kΩ	S_1 — Switch, Slide, SPDT
R_4 — Resistor, 47 kΩ	B_1 — 9 Vdc Battery or Power Supply
R_5 — Potentiometer, 5 kΩ	AF — Audio Frequency Generator
C_1 — Capacitor, 0.001 μF (marked 102)	Misc. — IC Socket, 8-Pin, IC Socket, 14-Pin,
C_2, C_3 — Capacitor 0.1 μF (marked 104)	Speaker, Battery connector, Solder,
C_4, C_5 — Capacitor, Elect., 470 μF	printed circuit board (Graymark No.
U_1 — IC. LM358	69040), hook up wire, metal or
	plastic chassis, etc.

*This audio amplifier may be purchased in kit form as Model 140 from Graymark International, Inc., P.O. Box 5020, Santa Ana, CA 92702.

Figure 19-18. Schematic and parts list for op amp audio amplifier. (Graymark Electronics)

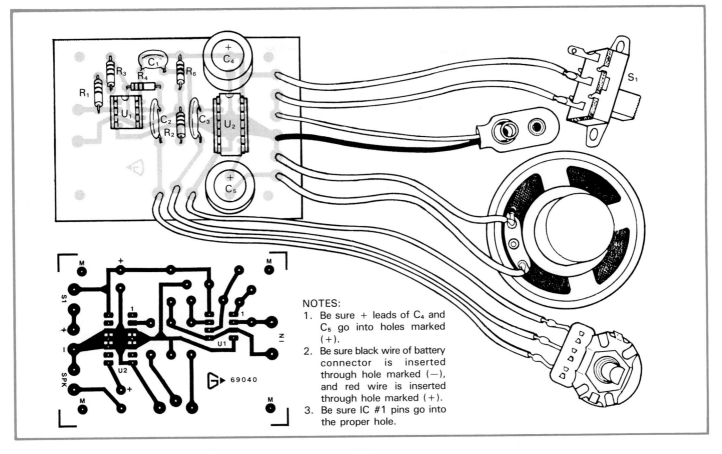

NOTES:
1. Be sure + leads of C_4 and C_5 go into holes marked (+).
2. Be sure black wire of battery connector is inserted through hole marked (−), and red wire is inserted through hole marked (+).
3. Be sure IC #1 pins go into the proper hole.

Figure 19-19. Op amp audio amplifier parts placement and PCB layout. (Graymark Electronics)

Figure 19-20. Op amp audio amplifier. (Graymark Electronics)

good joints. Check the placement of all component leads and wire leads as shown in the diagram. Be sure that solder has not accidentally bridged between two copper islands. Be sure that the copper trace on your PCB has not been etched away or broken.

There are many uses for an audio amplifier and the following are only a few suggestions:

- To use it as a P.A. amplifier, connect either a crystal or dynamic microphone to the input leads.
- To use it as a tape playback amplifier, connect the magnetic tape head output to the input leads.
- To use it as a telephone amplifier, connect a telephone *pickup coil* to the input leads. Place the pickup coil next to the handset near the earpiece. (The pickup coil may be purchased at most any electronic parts store.)
- To use it as a phono amplifier, connect the phono cartridge output to the input leads.
- To use it as a signal tracer, connect a 100-pF capacitor and a diode in series with one of the input leads.

LED BATTERY-CHECKER PROJECT

Another op amp circuit, the LED battery checker, provides an indication of a battery's condition, while placing the battery under load at the same time. The circuit schematic and parts list are shown in Figure 19-21. The parts placement as well as the printed circuit board (PCB) layout are shown in Figure

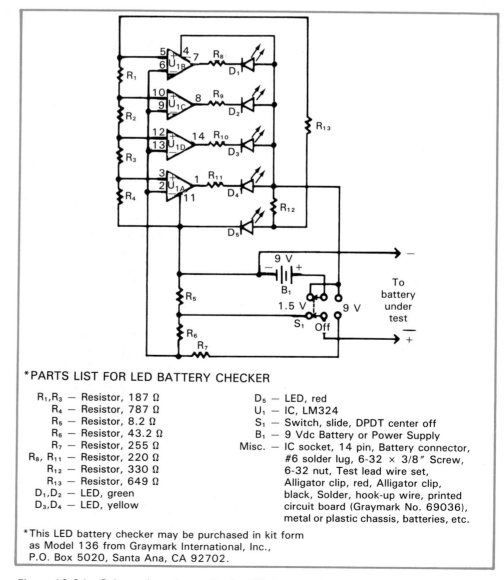

*PARTS LIST FOR LED BATTERY CHECKER

R_1,R_3 — Resistor, 187 Ω	D_5 — LED, red
R_4 — Resistor, 787 Ω	U_1 — IC, LM324
R_5 — Resistor, 8.2 Ω	S_1 — Switch, slide, DPDT center off
R_6 — Resistor, 43.2 Ω	B_1 — 9 Vdc Battery or Power Supply
R_7 — Resistor, 255 Ω	Misc. — IC socket, 14 pin, Battery connector,
R_8,R_{11} — Resistor, 220 Ω	#6 solder lug, 6-32 × 3/8″ Screw,
R_{12} — Resistor, 330 Ω	6-32 nut, Test lead wire set,
R_{13} — Resistor, 649 Ω	Alligator clip, red, Alligator clip,
D_1,D_2 — LED, green	black, Solder, hook-up wire, printed
D_3,D_4 — LED, yellow	circuit board (Graymark No. 69036),
	metal or plastic chassis, batteries, etc.

*This LED battery checker may be purchased in kit form
as Model 136 from Graymark International, Inc.,
P.O. Box 5020, Santa Ana, CA 92702.

Figure 19-21. Schematic and parts list for LED battery checker.   (Graymark Electronics)

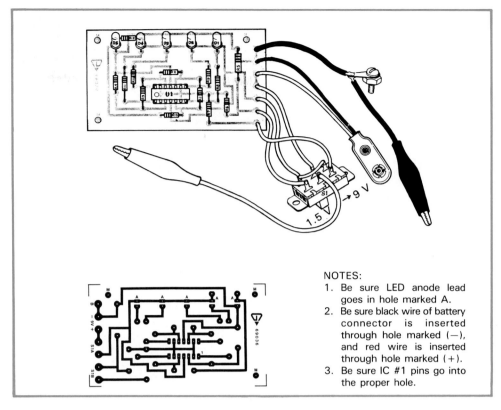

NOTES:
1. Be sure LED anode lead goes in hole marked A.
2. Be sure black wire of battery connector is inserted through hole marked (−), and red wire is inserted through hole marked (+).
3. Be sure IC #1 pins go into the proper hole.

Figure 19-22. LED battery checker parts placement and PCB layout. (Graymark Electronics)

19-22. A photo of the completed project is shown in Figure 19-23. When complete, the circuit may be encased in a metal or plastic chassis if desired.

When construction of your LED battery checker is complete, place the PCB on a nonconductive surface and connect it to the 9-Vdc battery or power supply. Place switch S_1 in the 9-V position. The red LED should light. Connect the red test lead to the positive lead of your 9-V source. Connect the black test lead to the negative lead of your 9-V source.

All LEDs should be lit. If your LED battery checker does not operate in this manner, immediately remove the power input leads from the power supply. In this event, troubleshoot as recommended for the audio amplifier: Check all solder connections. Check the placement of all component leads and wire leads as shown in the diagram. Be sure that solder has not accidentally bridged between two copper islands. Be sure that the copper trace on your PCB has not been etched away or broken.

Batteries are used to supply electrical energy for many different devices. The design of each device determines the minimum battery voltage required, as well as the amount of current drawn from the battery. It is, therefore, impossible to design a battery for all possible uses.

Note that an LED battery tester provides only an indication of a battery's relative condition as compared to its specified rating. See Figure 19-24. It will not, however, tell you whether or not it can power any given device. It is possible that a battery that reads "Very Weak" will work alright in some devices. Likewise, a battery that tests "Very Good" may not be good enough for some devices. It all depends on the particular requirement of the device.

Note, also, that the red "Replace" LED serves two purposes. First, if it is the only LED lit, it indicates that the battery under test should be replaced.

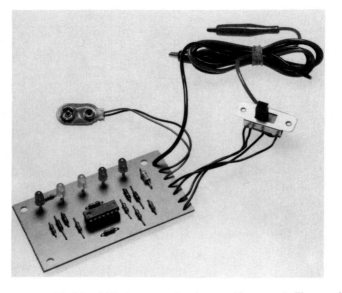

Figure 19-23. LED battery checker.   (Graymark Electronics)

NUMBER LED'S LIT	BATTERY CONDITION
5	Very Good
4	Good
3	Weak
2	Very Weak
1	Replace

Figure 19-24. Table to determine battery condition.   (Graymark Electronics)

Second, if it fails to light when the switch is in either the 1.5-V or 9-V position, it indicates that the 9-V battery that supplies power to the battery tester should be replaced.

To test a 1.5-V battery, be sure a 9-V battery is connected to the battery tester. Place the switch in the 1.5-V position. Connect a 1.5-V battery to the two test leads. Be sure the red test lead goes to the (+) side of the battery and the black test lead goes to the (−) side. Observe the number of LEDs that are lit, and refer to Figure 19-24 to determine the condition of the battery. (Do not connect a 9-V battery to the test leads when S_1 is in the 1.5-V position, or resistor R_5 will overheat.)

To test a 9-V battery, be sure a 9-V battery is connected to the battery tester. Place the switch in the 9-V position. Connect a 9-V battery to the two test leads. Observe the number of LEDs that are lit. Refer to Figure 19-24 to determine the condition of the battery.

SUMMARY

- Linear ICs process analog signals.
- Operational amplifiers and voltage regulators are two types of linear integrated circuit.
- ICs that bridge the gap between analog and digital worlds are analog/digital

converters and digital/analog converters.

- Op amps, which are used as basic amplifier circuits, ideally, have infinite gain. The device uses external feedback to control response characteristics.
- Op amps have two signal input terminals. One is an inverting input. One is a noninverting input.
- IC voltage regulators achieve much better regulation than simple zener-diode voltage regulators by using high-gain amplifiers and negative feedback.
- An IC voltage regulator uses a transistor as variable voltage dropping element V_{CE} in series with the load resistance. An error amplifier detects changes in regulator output voltage. Its output controls the transistor base current. In this way, it varies V_{CE} to hold the output voltage nearly constant.
- Converting an analog to a digital signal involves sampling the analog signal at regular intervals. The signal amplitude is converted into a digital value. The approximate amplitude of the signal at each interval represents the digital value.
- Increasing sampling rate and number of quantization intervals gives a closer approximation of an analog signal. However, we can never totally encode a continuous analog function.
- A D/A converter converts a digital input into a voltage whose magnitude is proportional to the value of the digital signal.
- A weighted-resistor D/A converter has resistors that progressively increase in size, giving more weight to digits of higher place value. This allows more output current and, therefore, higher output voltage.

KEY TERMS

Each of the following terms has been used in this chapter. Do you know their meanings?

analog/digital (A/D) converter	digital signal	operational amplifier
analog signal	error amplifier	precision voltage regulator
binary number system	fixed-voltage, three-terminal	quantization intervals
continuous signal	regulator	quantize
decimal number system	inverting amplifier	unity gain amplifier
differential amplifier	noninverting amplifier	voltage follower
digital/analog (D/A) converter	op amp	voltage sampling network

TEST YOUR KNOWLEDGE

Please do not write in this text. Place your answers on a separate sheet of paper.

1. Linear ICs process _____ signals.
2. Name two types of linear IC.
3. An analog signal has discrete values. True or False?
4. _____ _____ are used as basic amplifier circuits.
5. Draw the symbol for an op amp and name the basic leads.
6. The output signal of an _____ amplifier is antiphase to the input signal.
7. A noninverting op amp has an input resistance of 5 kΩ and a feedback resistance of 500 kΩ. For an input voltage of 0.1 V, calculate the output voltage.
8. What does an IC voltage regulator do?
9. An A/D converter changes analog values to the _____ number system.
10. A D/A converter has the following switch positions: OFF, OFF, ON, ON. What number does this equate to in the decimal system?

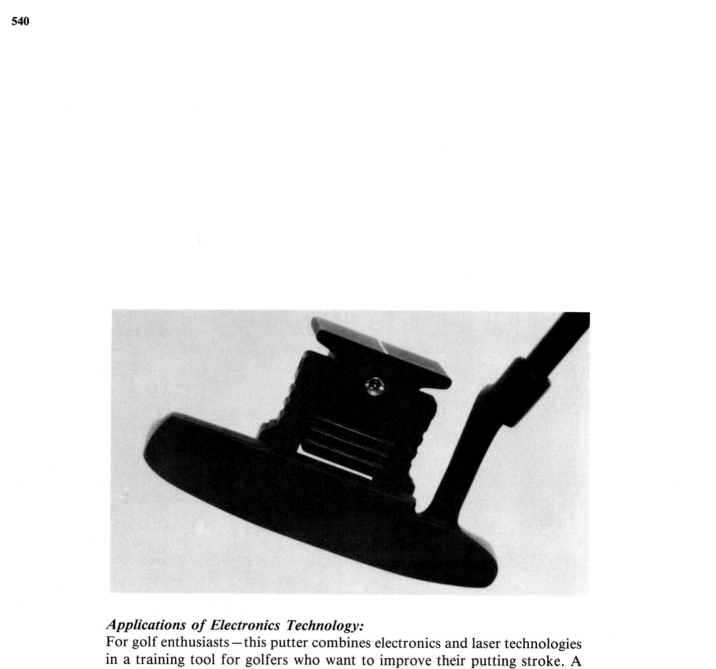

Applications of Electronics Technology:
For golf enthusiasts—this putter combines electronics and laser technologies in a training tool for golfers who want to improve their putting stroke. A brilliant red laser dot is projected on a regulation flagstick up to 35′ away so the golfer can tell if the putter face is on line to the hole. (Lyte Electronics)

Chapter 20

DIGITAL INTEGRATED CIRCUITS

Digital ICs are the basis of the modern computer as well as many other electronic goods today. Digital-based electronics relies on two circuit states or values—either ON or OFF. These values can be described in the binary number system as either 1 (ON) or 0 (OFF).

After studying this chapter, you will be able to:
□ *Define terminology used in digital electronics.*
□ *Explain basic number systems used in digital circuits.*
□ *Identify the basic gate circuits used in digital electronics.*
□ *Demonstrate how digital circuits can be used in combination to perform certain logic functions.*

INTRODUCTION

The most basic digital circuit could be illustrated by a simple circuit consisting of a single-pole, single-throw switch, a battery, and a lamp. See Figure 20-1. When the switch is OFF, the lamp is off, which represents a 0 condition. A 1 condition is represented when the switch is ON, allowing current, which results in the lamp burning. In reality, it is not practical to build digital electronic circuits using manual switches. However, they do provide a good basis for understanding.

Digital integrated circuits handle digital information using switching circuits. The most common electronic component that can be used as a switch is a transistor. Simple circuits made up of diodes, transistors, and resistors can perform the basic *Boolean logic functions*. We will discuss this concept later in this chapter.

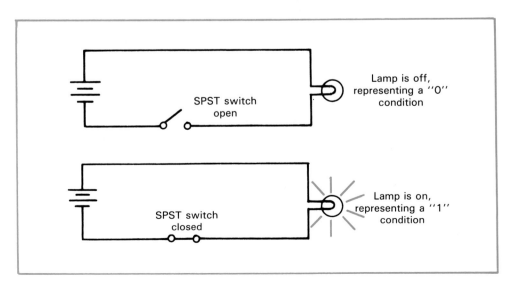

Figure 20-1. Simple digital circuit.

NUMBER SYSTEMS

A basic understanding of number systems is key to the understanding of digital electronics. The early Romans and Egyptians had their own number systems. We refer to ours as the *Arabic number system* because it uses the Arabic symbols 0 through 9. Perhaps more commonly, our number system is referred to as the **decimal number system.**

In the computer world, the *binary number system* is used. All data is stored and manipulated inside the computer in binary. Other number systems include *octal, hexadecimal,* and *binary-coded decimal.* Data may be entered into the computer using these other number systems for efficiency; however, these must be translated by an *assembler software program* into *machine language* (binary) before they can be understood by the computer. We should take a moment to explore these systems in a bit more detail.

DECIMAL NUMBER SYSTEM

This system is the most common of all number systems in use. It is, as mentioned previously, based on ten digits — 0 through 9. For a quick review, the single-digit numbers 0 through 9 make up the set of numbers called *units*, or *ones*. A count greater than 9 units is *carried over* to the left. The same digits, 0 through 9, are used for this second position. Such digits occupy the *tens* position. The digit indicates the number of 10s. A count greater than 9 tens and 9 units (99) is carried over, again, to the left to occupy the third position. Again, the same digits are used, but this time the *hundreds* position is occupied. The digit indicates the number of 100s. On it goes to *thousands, ten thousands,* etc.

Using exponents, the units position is 10^0, the tens position is 10^1, the hundreds position is 10^2, etc. It is important to note that each position, or *place value*, to the left increases by a factor of 10. Thus, the decimal number system is a base 10 system.

We can analyze the sample decimal number 9,876,543.210 of Figure 20-2 as follows:

$$
\begin{array}{llll}
9 \text{ millions} & = 9 \times 10^6 = & 9{,}000{,}000. \\
+ \ 8 \text{ hundred thousands} & = 8 \times 10^5 = & 800{,}000. \\
+ \ 7 \text{ ten thousands} & = 7 \times 10^4 = & 70{,}000. \\
+ \ 6 \text{ thousands} & = 6 \times 10^3 = & 6{,}000. \\
+ \ 5 \text{ hundreds} & = 5 \times 10^2 = & 500. \\
+ \ 4 \text{ tens} & = 4 \times 10^1 = & 40. \\
+ \ 3 \text{ units} & = 3 \times 10^0 = & 3. \\
+ \ 2 \text{ tenths} & = 2 \times 10^{-1} = & 0.2 \\
+ \ 1 \text{ hundredths} & = 1 \times 10^{-2} = & 0.01 \\
+ \ 0 \text{ thousandths} & = 0 \times 10^{-3} = & 0.000 \\
\hline
& & 9{,}876{,}543.210 \\
& & \text{(in basic units)}
\end{array}
$$

In addition, this sample number can also be expressed using scientific or engineering notation. Figure 20-2 indicates each position of the decimal point to the right of the corresponding place value. For example, we can write our sample number in a variety of ways including (but not limited to):

$$
\begin{aligned}
9{,}876{,}543.210 &= 9.876543210 \times 10^6 \\
&= 9.876543210 \text{ mega} \\
&= 9876.543210 \times 10^3 \\
&= 9876.543210 \text{ kilo}
\end{aligned}
$$

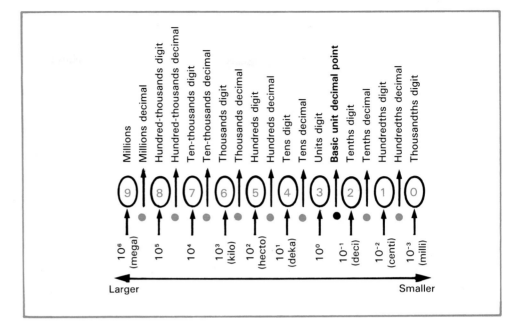

Figure 20-2. The decimal number system.

BINARY NUMBER SYSTEM

In previous chapters, you were introduced to the binary number system. It is based on two digits—0 and 1. The binary system is much simpler than the decimal system. It is very compatible with digital electronic circuits that are either ON or OFF. The two **binary digits** of 0 and 1 are sometimes called **bits**, which evolves from the two words *bi*nary dig*its*.

Binary-decimal conversion

A table showing the relationships between the first 27 numbers (including zero) of the binary and decimal systems is shown in Figure 20-3. Note that the place-value positions in the binary system are: *ones* (2^0), *twos* (2^1), *fours* (2^2), *eights* (2^3), and so on.

Converting decimal to binary. The binary number for a larger number, like decimal 143, is shown in Figure 20-4. The binary equivalent for this number was obtained by a method called **successive division by two**. The procedure is as follows: Divide the decimal number by two. Use the remainder from this division to fill in the ones place value (rightmost position). Continue division by two on each resulting quotient. Each remainder fills the next higher place value. Procedure is over when no further divisions are possible (division of zero).

The binary equivalent for 143 is determined as follows:

$$143 \div 2 = 71 \text{ remainder } 1 \implies \quad \text{1s place}$$
$$71 \div 2 = 35 \text{ remainder } 1 \implies \quad \text{2s place}$$
$$35 \div 2 = 17 \text{ remainder } 1 \implies \quad \text{4s place}$$
$$17 \div 2 = 8 \text{ remainder } 1 \implies \quad \text{8s place}$$
$$8 \div 2 = 4 \text{ remainder } 0 \implies \quad \text{16s place}$$
$$4 \div 2 = 2 \text{ remainder } 0 \implies \quad \text{32s place}$$
$$2 \div 2 = 1 \text{ remainder } 0 \implies \quad \text{64s place}$$
$$1 \div 2 = 0 \text{ remainder } 1 \implies \text{128s place}$$
$$0 \div 2$$

DECIMAL		BINARY NUMBER				
10^1	10^0	2^4	2^3	2^2	2^1	2^0
10s	1s	16s	8s	4s	2s	1s
	0					0
	1					1
	2				1	0
	3				1	1
	4			1	0	0
	5			1	0	1
	6			1	1	0
	7			1	1	1
	8		1	0	0	0
	9		1	0	0	1
1	0		1	0	1	0
1	1		1	0	1	1
1	2		1	1	0	0
1	3		1	1	0	1
1	4		1	1	1	0
1	5		1	1	1	1
1	6	1	0	0	0	0
1	7	1	0	0	0	1
1	8	1	0	0	1	0
1	9	1	0	0	1	1
2	0	1	0	1	0	0
2	1	1	0	1	0	1
2	2	1	0	1	1	0
2	3	1	0	1	1	1
2	4	1	1	0	0	0
2	5	1	1	0	0	1
2	6	1	1	0	1	0

Figure 20-3. Decimal-binary conversion table.

2^7	2^6	2^5	2^4	2^3	2^2	2^1	2^0	
128s	64s	32s	16s	8s	4s	2s	1s	
1	0	0	0	1	1	1	1	← 143

Figure 20-4. Binary number for 143.

Note that the binary number 10001111 is the result.

Converting binary to decimal. Had we been converting binary 10001111 to decimal, the conversion could have been performed as follows:

$$
\begin{aligned}
1 \times 2^7 &= 128 \\
+\, 0 \times 2^6 &= 0 \\
+\, 0 \times 2^5 &= 0 \\
+\, 0 \times 2^4 &= 0 \\
+\, 1 \times 2^3 &= 8 \\
+\, 1 \times 2^2 &= 4 \\
+\, 1 \times 2^1 &= 2 \\
+\, 1 \times 2^0 &= \underline{1} \\
&\quad\ \ 143 \text{ (in decimal system)}
\end{aligned}
$$

Arithmetic in binary

Arithmetic operations can be performed on binary numbers just as on decimal numbers. The rules for simple binary addition include:

$$\begin{array}{ccccc} 0 & & 0 & & 1 \\ +0 & \text{or} & +1 & \text{or} & +0 \\ \hline 0 & & 1 & & 1 \end{array}$$

When 1 is added to 1 in binary, a 1 is carried to the next column (2^1); the original, or units, column (2^0) reverts back to 0:

$$\begin{array}{cc} 2^1 & 2^0 \\ \hline & 1 \\ + & 1 \\ \hline 1 & 0 \end{array}$$

A comparison of addition using decimal numbers and equivalent binary numbers is shown in Figure 20-5. We can then convert the decimal sum to binary or the binary sum to decimal, as shown. Either way, we see that their sums are equivalent.

For binary multiplication, all possible products are:

$$0 \times 0 = 0 \qquad 0 \times 1 = 0 \qquad 1 \times 0 = 0 \qquad 1 \times 1 = 1$$

This is the same as in decimal multiplication. When more digits are involved, perform as for decimal multiplication—multiply partial products and add them.

This discussion on binary arithmetic is by no means complete. A more detailed study would include subtraction, division, *two's complement*, signed numbers, etc. You may wish to explore these at a later time.

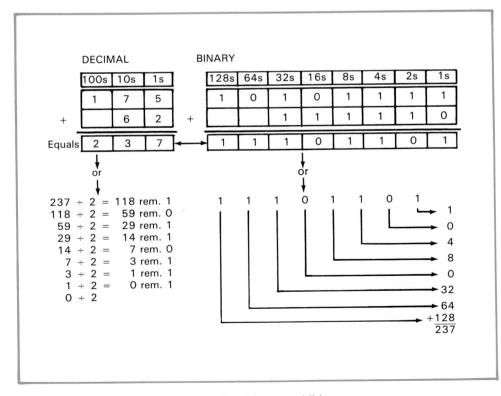

Figure 20-5. Comparison of decimal and binary addition.

OCTAL NUMBER SYSTEM

Normally, data is fed to a computer in some system other than binary. This is because entering data in binary is time-consuming and prone to error. Also, binary numbers are difficult to remember and express in words. The decimal system is one alternative. The **octal number system**, which is based on eight digits, is another. A table showing the relationship between the different number systems is shown in Figure 20-6. Note that after the first eight digits (0-7) in the octal system, the next number is 10, which is an 8 in the decimal system.

Octal is more closely related to binary than is decimal. The relationship permits easy conversion between the systems. Binary numbers can be converted to octal by inspection; calculations are not necessary. To convert an octal number to a binary number, simply convert each octal digit to its three-digit binary equivalent and record it from left to right. For example:

325 (octal) = 011 010 101 (binary) => 11010101 (binary)

100 (octal) = 001 000 000 (binary) => 1000000 (binary)

To convert a binary number to octal, group the binary number in threes from right to left. Then, convert each group to octal. For example:

10001 (binary) => 010 001 (binary) = 21 (octal)

1110111 (binary) => 001 110 111 (binary) = 167 (octal)

Decimal	Binary	Octal	Hex	BCD
0	0	0	0	0
1	1	1	1	1
2	10	2	2	10
3	11	3	3	11
4	100	4	4	100
5	101	5	5	101
6	110	6	6	110
7	111	7	7	111
8	1000	10	8	1000
9	1001	11	9	1001
10	1010	12	A	1 0000
11	1011	13	B	1 0001
12	1100	14	C	1 0010
13	1101	15	D	1 0011
14	1110	16	E	1 0100
15	1111	17	F	1 0101
16	10000	20	10	1 0110
17	10001	21	11	1 0111
18	10010	22	12	1 1000
19	10011	23	13	1 1001
20	10100	24	14	10 0000
21	10101	25	15	10 0001
22	10110	26	16	10 0010
23	10111	27	17	10 0011
24	11000	30	18	10 0100
25	11001	31	19	10 0101
26	11010	32	1A	10 0110
27	11011	33	1B	10 0111
28	11100	34	1C	10 1000
29	11101	35	1D	10 1001
30	11110	36	1E	11 0000
31	11111	37	1F	11 0001
32	100000	40	20	11 0010

Figure 20-6. Comparison of decimal, octal, binary, hexadecimal, and binary-coded decimal number systems.

HEXADECIMAL NUMBER SYSTEM

Another number system, which is used in computers, is the **hexadecimal number system**. This system is based on 16 digits. It is sometimes called *hex*, for short. The prefix *hexa* stands for six. *Decimal* implies ten. Thus, *hexa* + *decimal* (6 + 10) implies a base *16* number system. The first ten digits in hex are the same as those in the decimal system (0−9). However, for the decimal numbers 10−15, the hexadecimal number system uses *letters* A−F. Refer again to Figure 20-6. Since many personal computers today use a 16-bit *assembly language*, the hexadecimal number system has become popular.

Hex is also more closely related to binary than decimal. Again, the relationship permits easy conversion between the systems. Binary numbers can be converted to hexadecimal by inspection. To convert hex to a binary number, simply convert each hex digit to its *four*-digit binary equivalent and record it from left to right. For example:

D5 (hexadecimal) = 1101 0101 (binary) =〉 11010101 (binary)

40 (hexadecimal) = 0100 0000 (binary) =〉 1000000 (binary)

To convert a binary number to hex, group the binary number in fours from right to left. Then, convert each group to hexadecimal. For example:

10001 (binary) =〉 0001 0001 (binary) = 11 (hexadecimal)

1111110 (binary) =〉 0111 1110 (binary) = 7E (hexadecimal)

BINARY-CODED DECIMAL NUMBER SYSTEM

As we have seen, direct conversion between binary and decimal is not possible. To overcome this limitation, the **binary-coded decimal (BCD) number system** was developed. This system is based upon a code. While the code permits direct conversion between decimal and binary (of sorts), it is not a true binary system.

Refer again to Figure 20-6. The table shows the method of counting in BCD. Note that from 0−9 (decimal), BCD follows the rules of true binary. Thereafter, it does not. In making the conversion, each decimal digit is changed to its *four*-digit binary equivalent and recorded from left to right. A space is left between every four bits to eliminate confusion between BCD and straight binary. For example:

31 (decimal) = 0011 0001 (BCD)

597 (decimal) = 0101 1001 0111 (BCD)

To convert a BCD to decimal, group the BCD number in fours from right to left. Then, convert each group to decimal. For example:

0001 0001 (BCD) = 11 (decimal)

0011 0010 0111 (BCD) = 327 (decimal)

Note that the BCD number system provides a handy interface between computer and human operator. Within a computer, however, it is usually an inconvenient way to represent numbers. Binary is usually better. BCD requires special arithmetic. It also requires more space for memory, or storage.

BITS, BYTES, NIBBLES, AND WORDS

As stated earlier, a bit is a binary digit. It has a value of 0 or 1. A bit is the smallest piece of information handled by a computer. Eight bits make up a **byte**. Either 0s or 1s can make up a byte. One byte can express 256, or 2^8, different values. A **nibble** is 4 bits, which is also 1/2 byte. A **word** is a collection of one or more bytes that have some significance.

Computer storage is given in terms of bytes. One byte is a single unit of memory in a computer. Computer memory is given in terms of so many *k* (short for *kilo*bytes) or so many *meg* (short for *mega*bytes). Most computers can now hold millions of bytes, or megabytes, of storage. To give this some meaning, two average novels can be stored on 256k of memory, which is a standard unit of memory.

Octal and hexadecimal systems are well suited to a computer's structure, since computers tend to be byte oriented. For example, word size for a computer is usually a multiple of eight bits (16, 32, 64, etc.). BCD, on the other hand, is nibble oriented. The storing of a single 4-bit BCD digit in an 8- or 16-bit memory location results in inefficient use of memory. (This is another disadvantage of BCD.) To overcome this problem, digits from more than one BCD number are *packed* into a single memory location. For example, four BCD digits can be packed into 16 bits. This is not without cost, however. Programming steps are needed for packing and unpacking.

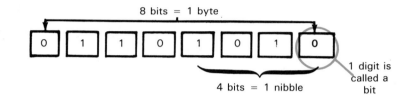

Figure 20-7. There are 8 bits in 1 byte, 4 bits in 1 nibble.

BOOLEAN ALGEBRA AND LOGIC GATES

In the 19th century, a self-taught British mathematician named George Boole developed an alternate form of algebra. **Boolean algebra**, as it came to be known, was based on propositions of logic that could be answered with either of two responses — true, or false. For this, the binary number system is a natural fit, with 1 standing for *true* and 0 standing for *false*. About a century later, an American mathematician named Claude Shannon was able to put Boole's algebra to use as the basis for carrying out complex calculations in digital computers. Through Boolean algebra and the development of solid-state technology in the 1940s – 1960s, the modern computer came to be.

Boolean algebra is based on commonsense logic, and it has three basic functions — *AND, OR,* and *NOT*. In addition, these functions used in some combination yield *NAND, NOR, XOR,* and *XNOR* logic functions. These **Boolean logic functions** may be achieved either through *discrete* switching circuits or through *IC* **logic gates**. These are IC switching circuits that decide whether inputs will pass to output or be stopped. Logic gates can be used to add, subtract, multiply, divide, and perform logic functions, which modern computers do with great ease.

When switches are connected together, they create digital circuits that can perform both logical and mathematical operations. Boolean algebra is used to design and explain digital circuits. Binary arithmetic is used to count and do mathematical operations. Although they both use 1s and 0s, keep in mind that binary arithmetic and Boolean algebra are different concepts.

Using Boolean algebra, the equal (=) sign used in equations means *equal to* or *results in*. A multiplication (× or •) sign is used to signify **AND**. The sign does *not* mean multiply; it should not be read as "times." Often, the sign is omitted. An AND gate produces an output only when all of its inputs are energized. A circuit application might be to start a fan when pressure switch (input *A*) *and* flow switch (input *B*) are closed. The Boolean statement for this would be:

pressure switch *(A) AND* flow switch *(B)* energizes fan motor (Output)

We could express this:

$A \times B$ = Output or $A \cdot B$ = Output or AB = Output

What this means is that there is an output when (and only when) *both* switches are closed. An output, in this case, means the fan starts.

A plus (+) sign is used to signify **OR**. The sign does *not* mean add; it should not be read as "plus." An OR gate produces an output whenever any one of its inputs is energized. A circuit application might be to turn on a heating element when push-button switch (input *A*) *or* thermostat switch (input *B*) is closed. The Boolean statement for this would be:

push-button switch *(A) OR* thermostat switch *(B)* energizes heater (Output)

We could express this:

$A + B$ = Output

What this means is that there is an output when *either* or *both* switches are closed. An output, in this case, means we get heat.

NOT is the other logic function of Boolean algebra. This function inverts the input; the output is always opposite of the input. NOT is expressed as some symbol with a bar (‾) over it. For example, if we are to invert an input *A*, then our output would be $\overline{A}$, which would be read *NOT A*. Thus:

$A = 1$	$\overline{A} = 0$	Output = 0
A = true	$\overline{A}$ = false	Output = false
A = on	$\overline{A}$ = off	Output = off

Conversely:

$A = 0$	$\overline{A} = 1$	Output = 1
A = false	$\overline{A}$ = true	Output = true
A = off	$\overline{A}$ = on	Output = on

A normally closed switch can be used for NOT logic. Consider a NC flow switch. With no flow, the switch is not activated; contacts are closed—there is an output. Input is 0; output is 1. With flow, the switch is activated; contacts are open—there is no output. Input is 1; output is 0. The NOT gate is commonly called an **inverter**.

Boolean algebra goes beyond the simple Boolean logic statements just presented. In addition, there are a number of Boolean identities and properties. A chief purpose of Boolean algebra is simplifying logic circuits. Through it, we can design less expensive circuits with fewer gates. Figure 20-8 is given merely for example. Explanation of the Boolean identities and properties is considered beyond the scope of this text. We will, however, explore logic gates in greater detail in upcoming paragraphs.

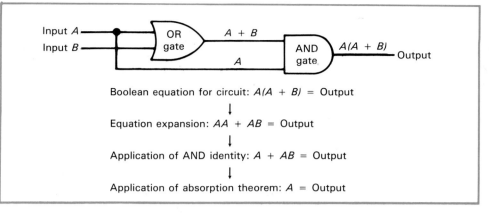

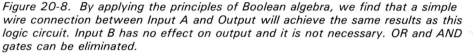

Boolean equation for circuit: $A(A + B) = $ Output

$\downarrow$

Equation expansion: $AA + AB = $ Output

$\downarrow$

Application of AND identity: $A + AB = $ Output

$\downarrow$

Application of absorption theorem: $A = $ Output

Figure 20-8. By applying the principles of Boolean algebra, we find that a simple wire connection between Input A and Output will achieve the same results as this logic circuit. Input B has no effect on output and it is not necessary. OR and AND gates can be eliminated.

THE AND GATE

The AND gate may be explained with the switching circuit of Figure 20-9. In this circuit, both switch *A* and switch *B* must be closed to light the lamp. A *truth table* can be constructed to describe the circuit action, Figure 20-10. A **truth table** is a tabular way of describing circuit output for all possible combinations of inputs. The *A* and *B* columns of this table indicate the switch positions. The *X* column is the output, or the lamp.

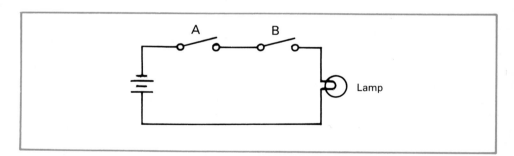

Figure 20-9. This switching circuit describes an AND gate.

Figure 20-10 shows different ways of representing the same input and output conditions. All four tables describe the exact same circuit action. Figure 20-10A states the physical condition of the circuit. Figure 20-10B is the binary representation of the circuit. Most truth tables that you see in digital electronics are in this form. ON is also defined as a *true* (T) condition, and OFF, as a *false* (F) condition. Thus, we can use *T*s and *F*s in a truth table, as seen in Figure 20-10C. Finally, in Figure 20-10D, *yes* is used for a 1 state; *no* is used for a 0 state.

In addition, a truth table could give voltage conditions. In **positive logic**, which is used in this text, this would mean *high* (ON) for a 1 (yes) state, and *low* (OFF) for a 0 (no) state. (Conversely, in **negative logic**, high [ON] is a 0 [no] state; low [OFF] is a 1 [yes] state.) These high/low voltage conditions do not apply to our *switch-based* circuit of Figure 20-9, since the switches control *current*. However, in *IC* logic, *voltage*, not current, is used for signal input. Commonly, switches either place 0 V (low) or +5 V (high) between the input terminal and ground, depending on their position.

Inputs		Output
A	*B*	*X*
Off	Off	Off
Off	On	Off
On	Off	Off
On	On	On
A		

Inputs		Output
A	*B*	*X*
0	0	0
0	1	0
1	0	0
1	1	1
B		

Inputs		Output
A	*B*	*X*
F	F	F
F	T	F
T	F	F
T	T	T
C		

Inputs		Output
A	*B*	*X*
No	No	No
No	Yes	No
Yes	No	No
Yes	Yes	Yes
D		

Figure 20-10. Truth tables for two-input AND gate. A—Physical representation. B—Binary representation. C—True/False representation. D—Yes/No representation.

The Boolean equation for the *two*-input AND gate is $AB = X$. This means that *A and B* equal output *X*. Rather than draw the circuitry for the gate, a symbol is used. Figure 20-11 shows two-input and *three*-input AND gates. The truth table for the three-input AND gate is drawn in Figure 20-12. The Boolean equation for the three-input AND gate is $ABC = X$.

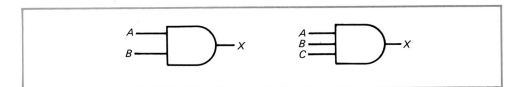

Figure 20-11. Logic symbols for AND gate.

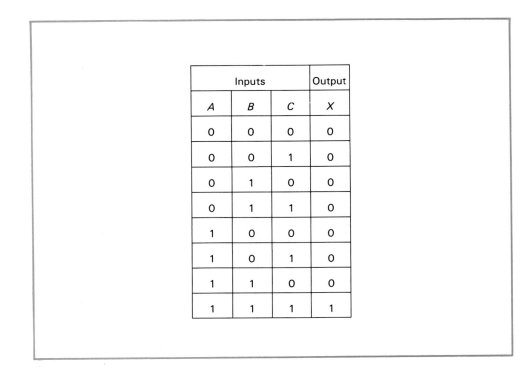

Inputs			Output
A	*B*	*C*	*X*
0	0	0	0
0	0	1	0
0	1	0	0
0	1	1	0
1	0	0	0
1	0	1	0
1	1	0	0
1	1	1	1

Figure 20-12. Truth table for three-input AND gate.

Figure 20-13 presents a circuit that will perform AND logic by using diodes. Follow the action:
* In its present state, both diodes D_1 and D_2 are conducting. The voltage across R is nearly 5 V. Thus, the voltage at output X is nearly 0 V with no input at A and B.
* A 5-V pulse applied to A will cut off D_1, but D_2 will still conduct. Voltage at X will still be near 0 V.
* A 5-V pulse applied to B will cut off D_2, but D_1 will still conduct. Voltage at X will still be near 0 V.

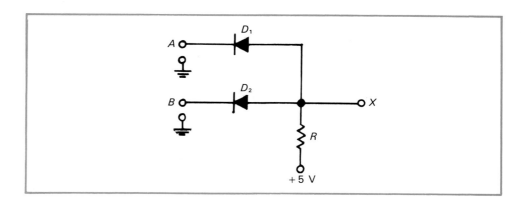

Figure 20-13. An AND gate circuit using diodes.

* When 5-V pulses are applied to both A and B, then D_1 and D_2 are cut off. No current flows. There is no voltage drop across R, and the voltage at X is 5 V. Thus, an output is seen at X only with inputs at both A and B.

Figure 20-14 shows the circuit of the AND gate using transistors. The transistor has the advantage of signal amplification with each gate. Follow the action:
* In its present state, both npn transistors are cut off. They appear as open circuits. No voltage will appear at X.

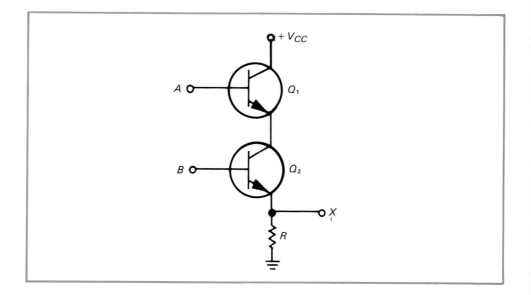

Figure 20-14. An AND gate circuit using transistors.

- A positive pulse applied to A would forward bias the EB junction of Q_1 and permit it to conduct. However, Q_2 is cut off, so no current will flow in the series circuit. No voltage will appear at X.
- A positive pulse applied to B would forward bias the EB junction of Q_2 and permit it to conduct. However, Q_1 is cut off; again, no current will flow in the circuit; no voltage will appear at X.
- When positive pulses are applied to both A and B, both transistors conduct, and a voltage appears at X equal to the drop across R.

THE OR GATE

The OR gate may be explained with the switching circuit of Figure 20-15. In this circuit, the lamp lights when either switch A or switch B is closed. The truth table describing the circuit action is shown in Figure 20-16.

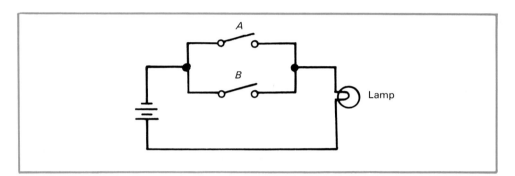

Figure 20-15. This switching circuit describes an OR gate.

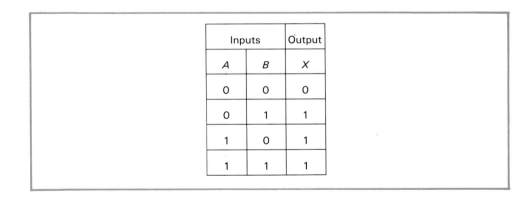

Inputs		Output
A	B	X
0	0	0
0	1	1
1	0	1
1	1	1

Figure 20-16. Truth table for two-input OR gate.

The Boolean equation for the *two*-input OR gate is $A + B = X$. This means that either A *or* B equals output X. Figure 20-17 shows the symbols for two- and three-input OR gates. The truth table for the three-input OR gate is drawn in Figure 20-18. The Boolean equation for the three-input OR gate is $A + B + C = X$.

Figure 20-19 presents a circuit that will perform OR logic by using diodes. Follow the action:

- In its present state, neither diode D_1 nor D_2 are conducting. Without an input, there is no current, and the voltage at output X is 0 V. Thus, no input yields no output.
- Diode D_1 will conduct with a positive pulse (say $+5$ V) applied to A. A voltage will appear at X equal to the drop across R, $I_T \times R$. Since most

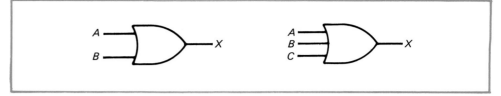

Figure 20-17. Logic symbols for OR gate.

	Inputs		Output
A	B	C	X
0	0	0	0
0	0	1	1
0	1	0	1
0	1	1	1
1	0	0	1
1	0	1	1
1	1	0	1
1	1	1	1

Figure 20-18. Truth table for three-input OR gate.

of the drop is across R, the voltage will be close to $+5$ V.

• Diode D_2 will conduct with a positive pulse (say $+5$ V) applied to B. A voltage will appear at X equal to the drop across R, $I_T \times R$, which will be close to $+5$ V.

• When positive pulses (say $+5$ V) are applied to both A and B, both diodes conduct. A voltage will appear at X equal to the drop across R, $I_T \times R$, which will be close to $+5$ V. Thus, an output is seen at X with inputs at A, at B, or at both A and B.

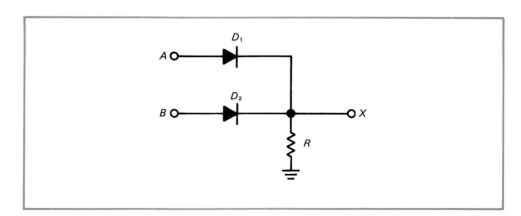

Figure 20-19. An OR gate circuit using diodes.

Figure 20-20 shows the circuit of the OR gate using transistors. Follow the action:

* In its present state, both npn transistors are cut off. They appear as open circuits. No voltage will appear at X.
* A positive pulse applied to A will forward bias the EB junction of Q_1 and permit it to conduct. An output will appear at X equal to the drop across R, $I_T \times R$.
* A positive pulse applied to B will forward bias the EB junction of Q_2 and permit it to conduct. An output will appear at X equal to the drop across R, $I_T \times R$.
* When positive pulses are applied to both A and B, both transistors conduct. Again, a voltage appears at X. It is equal to the drop across R, $I_T \times R$. Thus, equal outputs are seen at X with inputs at A, at B, or at both A and B.

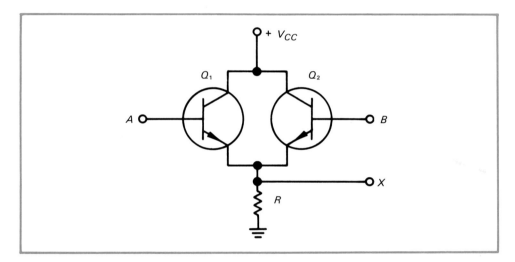

Figure 20-20. An OR gate circuit using transistors.

THE NOT GATE

Many times in computer logic circuits it is necessary to invert a signal. For this, the inverter, or NOT, gate is used. The Boolean equation that describes this function is $\overline{A} = X$. The truth table describing the circuit action is shown in Figure 20-21. Figure 20-22 shows the logic symbol for the NOT gate.

The circuit to perform this simple function has already been studied. We know that a common-emitter transistor circuit will invert its input signal. Consider Figure 20-23.

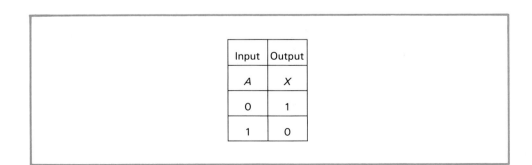

Input	Output
A	*X*
0	1
1	0

Figure 20-21. Truth table for NOT gate.

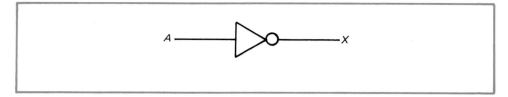

Figure 20-22. Logic symbol for NOT gate.

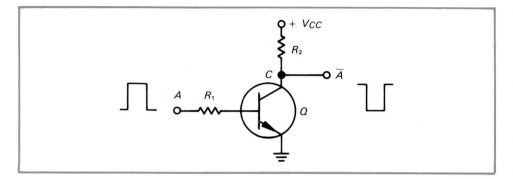

Figure 20-23. A common-emitter inverter circuit.

With no signal applied to the base of Q, the transistor is cut off, and the voltage at C is equal to V_{CC}. A positive pulse to the base of Q will turn on the npn transistor. The voltage at C will drop and move in the direction of zero. Thus, the input pulse gives rise to a corresponding inverted pulse at the circuit output.

THE NAND GATE

A variation of the AND gate is the *NOT AND*, or **NAND,** gate. This circuit simply inverts the output of an AND gate. See Figure 20-24. From the truth table, notice the output is opposite to an AND output. The Boolean equation that describes the NAND function is $\overline{AB} = X$. Figure 20-25 shows the logic symbol for the NAND gate.

Figure 20-26 shows the circuit of the NAND gate using transistors. This circuit combines AND and inverter circuit configurations. With an AND gate by itself, only a signal at *A and B* will produce an output of 1. With the inverter added, a 1 output is inverted to a 0 and vice versa. A signal at *A* and *B* permits current. Most of the voltage is dropped across *R,* so voltage at *C*

Inputs		Output
A	*B*	*X*
0	0	1
0	1	1
1	0	1
1	1	0

Figure 20-24. Truth table for NAND gate.

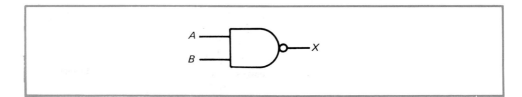

Figure 20-25. Logic symbol for NAND gate.

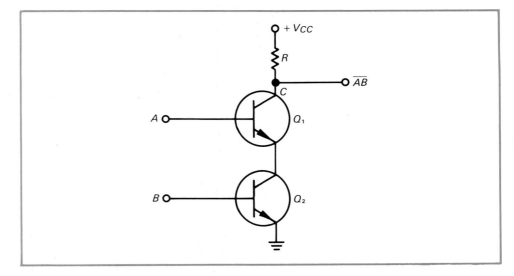

Figure 20-26. A NAND gate circuit using transistors.

is close to 0 V. On the other hand, *no* signal at *A* or *B* or both presents an open circuit condition. Voltage at *C* is equal to V_{CC}, producing a logic 1 in these instances.

THE NOR GATE

The *NOT OR* gate, or **NOR**, gate is a variation of the OR gate. This circuit inverts the output of an OR gate. See Figure 20-27. From the truth table, notice the output is opposite to an OR output. The Boolean equation that describes the NOR function is $\overline{A + B} = X$. Figure 20-28 shows the logic symbol for the NOR gate.

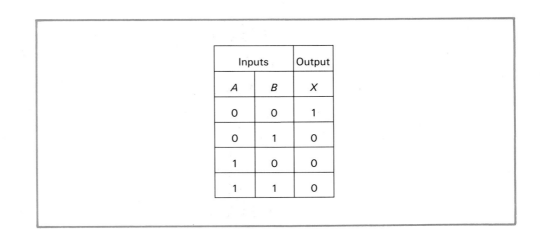

Inputs		Output
A	*B*	*X*
0	0	1
0	1	0
1	0	0
1	1	0

Figure 20-27. Truth table for NOR gate.

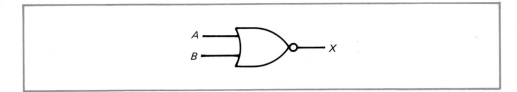

Figure 20-28. Logic symbol for NOR gate.

Figure 20-29 shows the circuit of the NOR gate using transistors. This circuit combines OR and inverter circuit configurations. With an OR gate by itself, a signal at *A* or *B* would produce an output of 1. With the inverter added, a signal at *A* or *B* produces *no* output because of the voltage drop across *R*. On the other hand, with *no* signals at *A* and *B*, an open circuit condition exists. Output voltage is equal to V_{CC}, producing a logic 1.

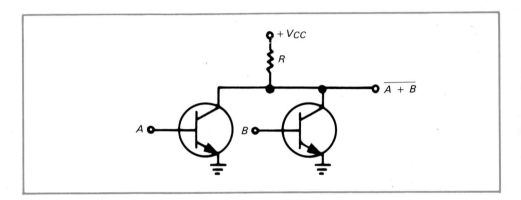

Figure 20-29. A NOR gate circuit using transistors.

THE XOR GATE

The *Exclusive OR*, or **XOR**, gate operates somewhat like an OR gate. The primary difference is that when inputs at both *A* and *B* are 1, the output is 0. See Figure 20-30. With the XOR gate, a value of 1 in one input or the other, but not both, will result in an output of 1. The XOR function is expressed by the equation $A \oplus B = X$. The logic symbol for the XOR gate is shown in Figure 20-31.

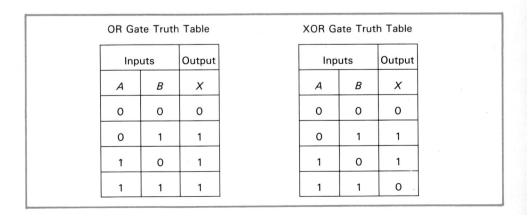

OR Gate Truth Table				XOR Gate Truth Table		
Inputs		Output		Inputs		Output
A	B	X		A	B	X
0	0	0		0	0	0
0	1	1		0	1	1
1	0	1		1	0	1
1	1	1		1	1	0

Figure 20-30. Comparison of OR and XOR gate truth tables.

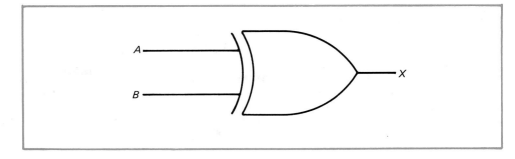

Figure 20-31. Logic symbol for XOR gate.

The switching circuit of Figure 20-32 shows how two SPDT switches can be connected to function much like an XOR gate. Note that the lamp lights if the switches are either in the $A\overline{B}$ position, or if they are in the $\overline{A}B$ position. The lamp does *not* burn if the switches are in the AB position or the $\overline{A}\overline{B}$ position. An everyday example of this is one light being controlled by switches at two different locations of a room, hallway, or stairwell.

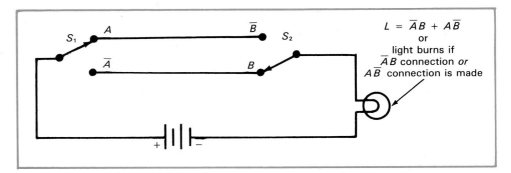

Figure 20-32. This switching circuit describes an XOR gate.

Converting truth table

Boolean equations may be derived from truth tables. The easiest method of doing so is called the *sum-of-products* method. The Boolean equation for the XOR gate obtained by this method is $\overline{A}B + A\overline{B} = X$. Note that in the **sum-of-products** method, the elements of each input set (one row of truth table) that generate a 1 output are ANDed. Each of these sets are then ORed. An explanation follows:

In sum-of-products, only input sets that result in an output of 1 are considered. Every 1 that exists in the output column is determined by a unique set of input conditions. We cannot have two sets of conditions at the same time. Thus, we will have one set of input conditions producing an output of 1 *OR* another set *OR* another, etc. Further, it is the *combination* of input conditions that determines this output—input A *AND* input B, for example. For this reason, inputs are ANDed. Each input set would be fed to an AND gate; the output of each AND gate would be fed to an OR gate.

As an example, let us see how the Boolean equation for the XOR gate is derived from its truth table. Two input sets result in an output of 1. In row 2, the first input set for our purposes, the input conditions are $A = 0$ *AND* $B = 1$. If we are to obtain a 1 from an AND gate, all inputs must be 1. In this case then, input A must first go through an inverter. From this, it follows that the inputs to the AND gate for this combination of inputs would be $\overline{A}$ and B. The gate output would be $\overline{A}B$.

In row 3, the other input set, inputs are $A = 1\ AND\ B = 0$. Now, input B must be routed through an inverter to achieve a 1 output. The inputs to the AND gate for this input set would A and $\overline{B}$. The gate output would be $A\overline{B}$.

The two outputs, $\overline{A}B$ and $A\overline{B}$, become the inputs to an OR gate. Putting all of this together gives us the sum-of-products Boolean equation $\overline{A}B + A\overline{B}$ = X and the logic diagram of Figure 20-33. (Note that there are Boolean rules that allow for manipulation of the equation in this form. In the $A \oplus B$ form, there are not.)

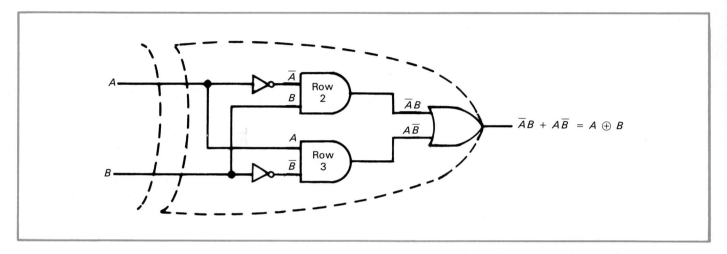

Figure 20-33. Logic diagram and Boolean expression converted from an XOR truth table. The diagram shows an XOR gate can be made from two AND gates, two NOTs, and one OR gate.

THE XNOR GATE

A gate similar to the XOR gate is the *Exclusive NOR*, or **XNOR**, gate. It is the XOR gate with the output inverted. There is an output only if both inputs are ON or both inputs are OFF. The truth table is shown in Figure 20-34. The symbol for the XNOR gate is shown in Figure 20-35. The XNOR function may be expressed by the equation $\overline{A \oplus B} = X$.

Inputs		Output
A	B	X
0	0	1
0	1	0
1	0	0
1	1	1

Figure 20-34. Truth table for XNOR gate.

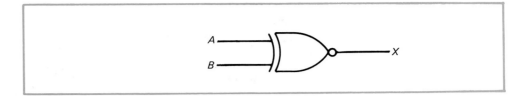

Figure 20-35. Logic symbol for XNOR gate.

COMBINATIONAL LOGIC

By combining gates together, binary numbers can be added. Of course, adding is very important in computers. Adding (very quickly) a number over and over again becomes multiplication. The reverse of the addition process is subtraction, and the reverse of the multiplication process is division. Thus, by using different combinations of logic gates, the basic mathematical processes can be accomplished with great speed and accuracy.

The combining of logic gates to perform a given function is referred to as **combinational logic**. With combinational logic, much more can be accomplished in a computer. By connecting the various gates together in different combinations, new tasks may be accomplished. Recall how we made an XOR gate from AND, OR, and NOT gates.

For more examples of combinational logic, review logic diagrams and corresponding, expanded truth tables in Figures 20-36 through 20-43. An expanded

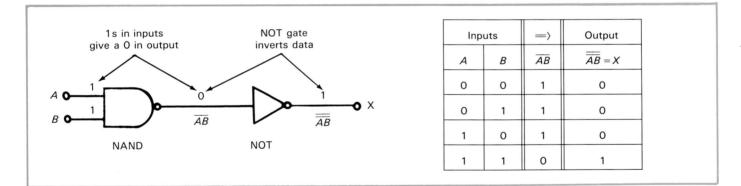

Inputs		$\Rightarrow$	Output
A	B	$\overline{AB}$	$\overline{\overline{AB}} = X$
0	0	1	0
0	1	1	0
1	0	1	0
1	1	0	1

Figure 20-36. Creating an AND gate from a NAND and a NOT gate.

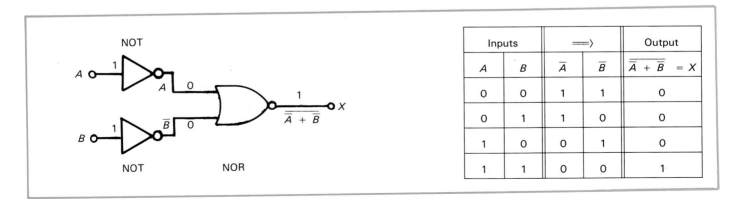

Inputs		$\Rightarrow$		Output
A	B	$\overline{A}$	$\overline{B}$	$\overline{A} + \overline{B} = X$
0	0	1	1	0
0	1	1	0	0
1	0	0	1	0
1	1	0	0	1

Figure 20-37. Making an AND gate from two NOT gates and a NOR gate.

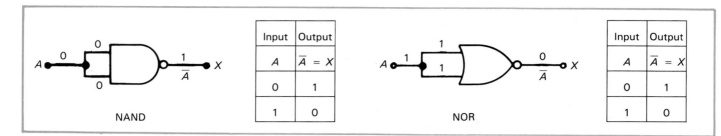

Figure 20-38. Making a NOT gate from a NAND gate or a NOR gate.

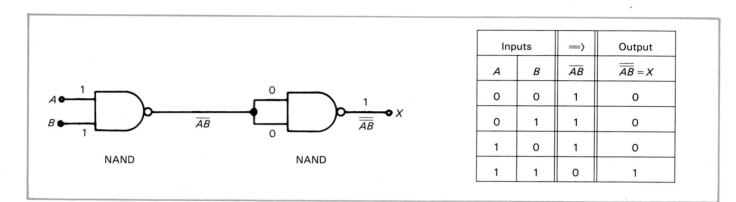

Figure 20-39. Creating an AND gate from two NAND gates.

Inputs		==>	Output
A	B	$\overline{AB}$	$\overline{\overline{AB}} = X$
0	0	1	0
0	1	1	0
1	0	1	0
1	1	0	1

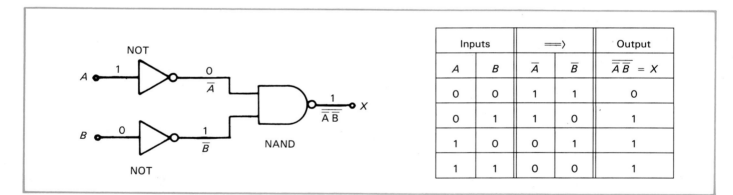

Figure 20-40. Creating an OR gate from two NOT gates and a NAND gate.

Inputs		==>		Output
A	B	$\overline{A}$	$\overline{B}$	$\overline{\overline{A}\,\overline{B}} = X$
0	0	1	1	0
0	1	1	0	1
1	0	0	1	1
1	1	0	0	1

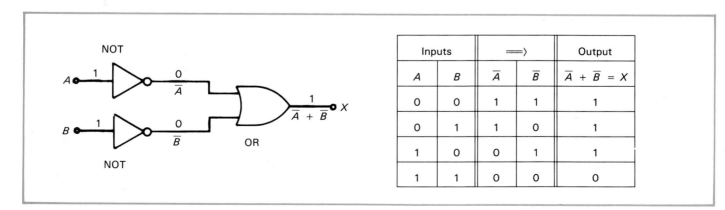

Figure 20-41. Creating a NAND gate from two NOT gates and an OR gate.

Inputs		==>		Output
A	B	$\overline{A}$	$\overline{B}$	$\overline{A} + \overline{B} = X$
0	0	1	1	1
0	1	1	0	1
1	0	0	1	1
1	1	0	0	0

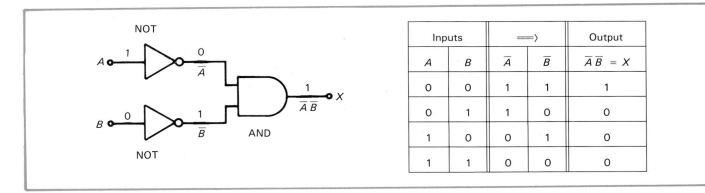

Figure 20-42. A NOR gate made from two NOT gates and an AND gate.

Inputs		==>		Output
A	B	$\overline{A}$	$\overline{B}$	$\overline{A}\,\overline{B}$ = X
0	0	1	1	1
0	1	1	0	0
1	0	0	1	0
1	1	0	0	0

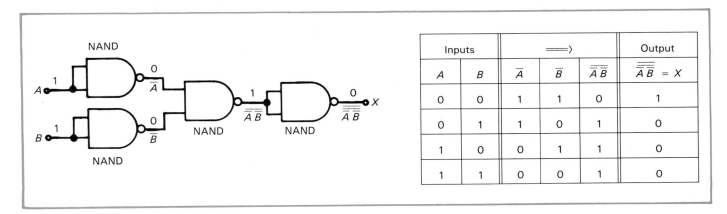

Figure 20-43. Four NAND gates used to create a NOR gate.

Inputs		==>			Output
A	B	$\overline{A}$	$\overline{B}$	$\overline{A}\,\overline{B}$	$\overline{\overline{A}\,\overline{B}}$ = X
0	0	1	1	0	1
0	1	1	0	1	0
1	0	0	1	1	0
1	1	0	0	1	0

truth table will show, in tabular form, circuit output and also circuit action each step of the way. Otherwise, to determine circuit output, you could individually evaluate each input set, jotting down 0s and 1s, as appropriate, as you follow through the circuit logic. As an example, one input set has been chosen for each circuit.

Combinational logic is very important in the operation of modern computers. Chapter 21 will discuss the overall parts and operation of basic digital computers. In addition, further study might include the detailed operation of: *sequential logic circuits*, including flip-flops; *shift registers*; counters; *half adders* and *full adders; decoding* and *encoding; multiplexers*; D/A and A/D converters; memories (*ROM, RAM, PROM, EPROM,* etc.) and storage retrieval systems.

TIMING DIAGRAMS

We have covered three different ways of describing a digital circuit. Included here are logic diagrams, truth tables, and Boolean expressions. You will note that all of these representations describe static conditions. They do not take *time* into account. In digital circuits, dynamic conditions are often important. For this, we use *timing diagrams*.

Timing diagrams express logic values as a function of time. Circuit inputs and resulting output are plotted for any given set of conditions. In most timing diagrams, the output graph is drawn below the input graphs. Time is plotted horizontally, as usual, and signal level is plotted vertically. Figure 20-44 gives a number of examples.

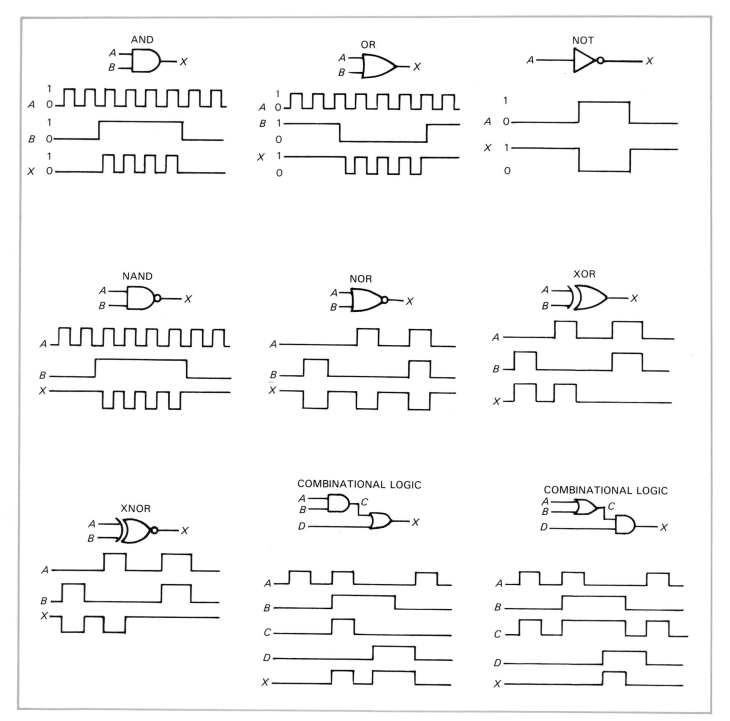

Figure 20-44. Typical timing diagrams.

LOGIC FAMILIES

Modern computers are built around the integrated circuit. ICs make up the heart of the machine. They are building blocks of logic, connected together to perform mathematical and logic operations.

Manufacturing techniques have a major impact on the arrangement of digital circuits into groups or families. It is crucial that the traits of one logic family match the traits of another family when a number of digital ICs are used in a piece of equipment. Two of the more common *families* of IC are the *transistor-transistor logic* family and the *metal oxide semiconductor* family.

THE TTL FAMILY

A widely used family of ICs used for logic functions is the family of **transistor-transistor logic (TTL) ICs.** These ICs are based on bipolar technology, much like npn and pnp transistors. TTLs operate on +5 V. Logic high (1) for these ICs is about 2.5 V to 5 V. Logic low (0) is about 0 V to 0.8 V.

TTL ICs work very quickly; however, they tend to be big power consumers. Component density, then, is limited by the amount of heat that may be properly dissipated. This means, to accommodate, a chip must be made larger or contain less circuitry. One TTL subfamily, the *low-power Schottky,* uses less power and is currently very popular. The TTL IC family is usually designated with a "74" prefix, followed by two more digits. Examples of popular TTL ICs are shown in Figure 20-45.

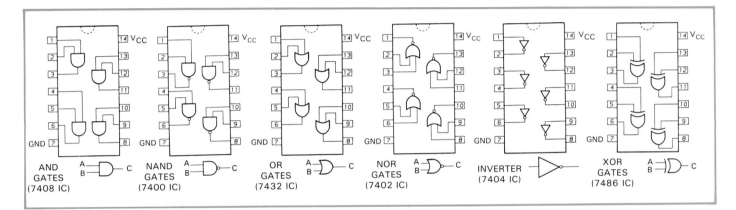

Figure 20-45. Four-gate (quad) and six-gate (hex), TTL, 14-pin IC packages. Note that supply voltage (V$_{CC}$) to these devices is 5 Vdc.

THE MOS FAMILY

Another popular family used in logic circuits is the family of **metal oxide semiconductor (MOS) ICs.** These ICs are based on MOS technology, much like that used in producing the MOSFET. These ICs can operate on a wide range of supply voltages—from 3 V to 18 V. For a 10-V supply, logic high would be about 7 V to 10 V. Logic low would be about 0 V to 3 V.

MOS ICs have low power consumption. As a result, they have more components, such as diodes, transistors, and resistors, packed into a smaller area. Also, they draw less current. A popular type of MOS IC is the **complementary metal oxide semiconductor (CMOS) IC.** CMOS ICs are identified by a 4000 series number, such as 4001, 4011, and 4013.

CMOS ICs can be easily damaged by static electricity. Care must be used in their storage and handling. They should be stored in a conductive foam pad until ready for installation. IC circuit boards should be wrapped in a material such as aluminum foil, which will protect them from static electricity. Also, people and equipment coming in contact with them must be properly grounded. This is to eliminate the possibility of static discharge and consequent damage to the device. Special bracelets connected to ground may be worn for this purpose.

SUMMARY

- Digital-based electronics relies on two circuit states — either ON or OFF. These values can be described in the binary number system as either 1 (ON) or 0 (OFF).
- All data is stored and manipulated inside the computer in binary. Data may be entered into the computer using decimal, octal, or hexadecimal systems; however, these must be translated into binary before being understood by a computer.
- The decimal system is based on 10 digits, binary on 2, octal on 8, and hexadecimal on 16 digits.
- Binary-coded decimal permits direct conversion between decimal and binary. It provides a handy interface between operator and computer; however, it is not a true binary system.
- A bit is a binary digit. Eight bits make up a byte. One byte is a single unit of memory in a computer.
- The three basic logic functions are AND, OR, and NOT. Other logic functions are NAND, NOR, XOR, and XNOR.
- An AND gate produces an output only when all of its inputs are energized. The Boolean equation for the AND gate is $AB = X$.
- An OR gate produces an output whenever any one of its inputs is energized. The Boolean equation for the OR gate is $A + B = X$.
- NOT is the other logic function of Boolean algebra. This function inverts the input; the output is always opposite of the input. The Boolean equation for the NOT gate is $\overline{A} = X$.
- Four ways of describing a digital circuit include truth tables, logic diagrams, Boolean expressions, and timing diagrams.
- The NAND gate is an inverted AND gate. The NOR gate is an inverted OR gate.
- An XOR gate produces an output when one input or the other is energized, but not both. The XNOR gate is the XOR gate with the output inverted.
- Boolean equations may be derived from truth tables using the sum-of-products method. In this method, elements of input sets are ANDed. Each set is ORed.
- Combinational logic is the combining of logic gates to perform a given function.
- TTL ICs work very quickly but they have high power dissipation. This means they produce a lot of heat, which limits the number of components a TTL device may hold.
- MOS ICs have a low power consumption, which allows them a high component density. These ICs are easily damaged by static electricity, and they require special handling precautions.

KEY TERMS

Each of the following terms has been used in this chapter. Do you know their meanings?

AND
binary digit
binary-coded decimal (BCD)
  number system
bit
Boolean algebra
Boolean logic function
byte
combinational logic
complementary metal oxide
  semiconductor (CMOS) IC
decimal number system

hexadecimal number system
inverter
logic gates
metal oxide semiconductor
  (MOS) IC
NAND
negative logic
nibble
NOR
NOT
octal number system

OR
positive logic
successive division by two
sum-of-products
timing diagram
transistor-transistor
  logic (TTL) IC
truth table
word
XNOR
XOR

TEST YOUR KNOWLEDGE

Please do not write in this text. Place your answers on a separate sheet of paper.

1. Give the number of digits in each of the following number systems:
   a. Decimal.
   b. Binary.
   c. Octal.
   d. Hexadecimal.
   e. BCD.
2. The decimal number for the binary number 0101001 is _____.
3. The octal number for the binary number 0101001 is _____.
4. Bit stands for _____ _____.
5. One nibble is _____ bits, while one byte is _____ bits.
6. The system of mathematics that serves as the basis for digital computers is known as _____ _____.
7. Name the three basic gates.
8. A common-emitter transistor circuit can function as a(n) _____ gate.
9. The ladder diagram shown depicts _____ logic.

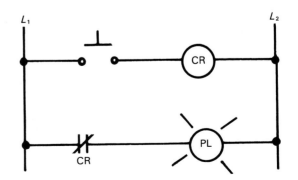

10. Write a Boolean equation to express the circuit logic given by the ladder diagram.
11. Draw the symbols for:
   a. A NAND gate.
   b. A NOR gate.
   c. An XOR gate.
   d. A NOT gate.
12. What is the difference between an OR gate and an Exclusive OR gate?
13. In an AND gate, input *A* must be _____ and input *B* must be _____ for the output to be 1.
14. Complete the timing diagram pictured.

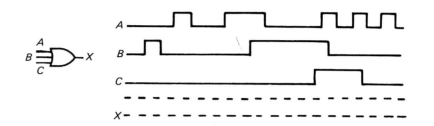

15. What safeguards are required for CMOS ICs and why?
16. Draw the logic diagram for this Boolean equation:

$$(A + B)(C + D) = X$$

17. Draw the logic diagram for this Boolean equation:
$$AB + CD = X$$

18. Draw the logic diagram for this Boolean equation:

$$AB + C + D = X$$

19. Write the Boolean equation for this logic diagram:

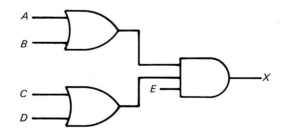

20. Draw a logic diagram for an XNOR gate using the basic logic gates. Express the output with a Boolean statement. (Avoid using the circled + [$\oplus$] notation.)

Chapter 21

COMPUTERS

Since the beginning of time, humans have always been fascinated with making complicated tasks easier. In the past few decades, the computer has helped humans extend their knowledge to a great degree. It is not an exaggeration to say that most of the technological achievements in recent years have depended on the computer. It was the computer that ushered us out of the *Industrial Age* and into the *Information Age.*

After studying this chapter, you will be able to:
☐ *Summarize the history and development of the computer.*
☐ *Give examples of computers in application.*
☐ *Name and describe the different types of computers available.*
☐ *Draw a diagram of computer architecture.*
☐ *Give examples of computer hardware.*
☐ *List and discuss four general types of programs for computer use.*

INTRODUCTION

Early attempts to perform arithmetic by use of an instrument, or counting device, date back to ancient times. One such device, the *abacus*, Figure 21-1, is still around today. An abacus consists of a frame with movable beads, which are strung in columns on parallel rods. With the beads, decimal numbers can be represented. Place value is designated by column position. Digit value is given by the bead count. Beads above the crossbar have a value of 5; beads below, a value of 1. Beads against the outside frame are in the zero position. By moving the beads, numbers can be added, subtracted, multiplied, divided, squared, and square rooted.

In 1617, John Napier, inventor of logarithms, invented a pocket-sized device for multiplication. It was called *Napier's bones.* Numbers were printed on square rods, which could be moved. Answers were found by adding numbers in horizontal, side-by-side sections.

The first calculating machine was invented by French mathematician Blaise Pascal. In 1642, he invented a machine that used *cogged* wheels to add and subtract. Pascal's calculating machine used the decimal number system. Each wheel had ten cogs, or teeth, representing digits 0-9.

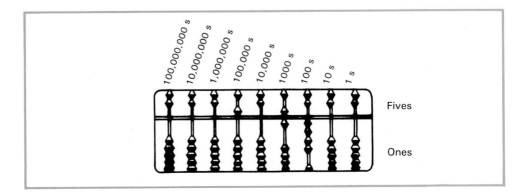

Figure 21-1. In earlier times, the abacus was routinely used for performing calculations. Note that the decimal number 506,400 is represented here.

In 1671, Gottfried Leibniz, a German mathematician, devised a calculating machine that could also multiply and divide. In addition, Leibniz developed the binary number system. He attempted to work out symbolic logic. However, it was not perfected until nearly two centuries later. This is when George Boole came out with what became known as Boolean logic. The binary number system and Boolean logic are at the core of computer operation.

In 1822, an English inventor named Charles Babbage began work on the design of a mechanical computing machine. It was called the *difference engine*. It was designed to compute mathematical tables of logarithms and such. However, it never worked due to mechanical problems in machining accuracy.

In 1833, he designed another device called the *analytical engine*. It was designed to add numbers and store the results at each stage of the calculations. The machine was to be directed by means of *punched cards*, which could store partial answers for operations to be performed later. The idea of punched cards came from an earlier inventor, J. M. Jacquard. In 1801, Jacquard had designed a system of binary-coded cards—hole, or no hole—for controlling textile machinery.

Babbage never completed his machine. Just recently, however, using parts of the required precision, it was finished for the sake of seeing if it would work. It did! Babbage's design contained all of the basic parts of a modern digital computer, and, although his efforts were somewhat fruitless, his ideas and research became the basis for modern computers.

A skilled mathematician and close friend of Charles Babbage was Ada Augusta, better known as Lady Lovelace, the daughter of Lord Byron. Lovelace developed the essential ideas of **programming**—the process of writing a detailed set of instructions under which a machine operates. She is credited with being the first programmer.

The first successful commercial adding machine was produced in 1885 by William S. Burroughs, a bookkeeper and founder of the Burroughs Corporation. In 1890, Herman Hollerith, an engineer working for the U.S. Census Bureau, designed an improved process for tabulating numbers that employed punched cards. The cards held census figures, which were in coded form through holes punched in the cards. Nails were used in the tabulating machine to electrically *read* the data. If a nail passed through a hole, it would complete a circuit. The pulses of current that were generated triggered counters, which would then be indexed to the next higher number. The punched cards enabled data to be processed much more quickly. Hollerith formed a company that later became part of the IBM Corporation. *Punch cards* and *punch tape*, closely related, later became the media for conveying information to computers.

In 1938, Claude Shannon of the Massachusetts Institute of Technology (M.I.T.) stated that relays could be used in switching circuits to evaluate logic statements. ON and OFF positions of the relays could be used to represent corresponding logic values of 1 and 0. In 1944, *Mark I*, the first *general-purpose computer*, was built. It was conceived by Howard Aiken of Harvard University and built as a joint venture between Harvard and IBM Corporation. It used electromechanical relays to compute by binary arithmetic. This room-sized computer used punched cards and was a modern version of Babbage's work. The Mark I had the usual problems that mechanical relays have—relatively slow operating speed and short life.

Between 1946 and 1952, a number of electronic calculators and computers emerged in rapid order. The world's first large electronic computer was developed in 1946 at the University of Pennsylvania by J. Presper Eckert and

John Mauchly. It was called the *Electronic Numerical Integrator And Calculator*, better known as the *ENIAC*. The ENIAC did not use relays, it used vacuum tubes. The vacuum tubes established it as a **first-generation computer**.

The ENIAC weighed 30 tons. It contained 18,000 vacuum tubes. The vacuum tubes acted somewhat like relays in that they could be biased to conduct (ON, or logic 1) or not conduct (OFF, or logic 0). The ENIAC was swift as compared to the Mark I. It could do 5000 additions per second; the Mark I could do 10. It was not able to store a *program*. The specific sequence of calculations was *hard-wired* into the machine. If a different program was required, the ENIAC had to be rewired.

MODERN COMPUTERS

The birth of the *modern computer* really began back in 1945. This is when John von Neumann came up with his design for a total computer system. His design pretty much defines the computer as we know it. That is, a **computer** is a device having four basic units. These are a *central processing unit (CPU)*, an *input device*, an *output device*, and *storage*. See Figure 21-2. This is sometimes called **von Neumann architecture**. He said that the operating instructions (the program), as well as data, could be stored in the computer *memory*. The computer, then, could be made to modify these instructions under the control of the program. Von Neumann architecture was not a part of the ENIAC, since it could not store a program.

*Figure 21-2. In this age, a single chip can hold CPU, memory, and input/output interface circuits as does this 8-bit **single-chip microcomputer**, or computer-on-a-chip. Note that this photograph is magnified many times. (Motorola, Inc.)*

In 1951, the *UNIVAC I* (*UNIV*ersal *A*utomatic *C*omputer), designed by Eckert and Mauchly, emerged as the first commercially available computer. This machine used vacuum tubes. It was used for single job operations. The UNIVAC I had all the elements of von Neumann architecture. Not long after it became operational, automatic programming was developed.

Vacuum tubes had certain drawbacks. They were big and bulky. They required large amounts of power and gave off great amounts of heat. They were fragile and unreliable. Transistors, of the **second-generation computer**, solved these problems. This device was 1/200th the size of the vacuum tube. It was also faster, more rugged, and more reliable. During operation, it created much less heat than a vacuum tube. Second-generation computers also had faster and cheaper storage and memory devices.

In 1958, the integrated circuit was developed, but it took a few years before the device was used in a computer. This did not happen until the mid 1960s, which brought forth the **third-generation computer**. The IC made computers faster, more reliable, and cheaper to operate than earlier computers. Obviously, they also allowed them to be smaller, and, in 1967, the first handheld calculator was developed at Texas Instruments.

The 1970s was a decade of innovation and invention for the computer. Large-scale integration (LSI) was newly developed. Out of this came the first *microprocessor*, invented by Ted Hoff at Intel in the early 1970s. A **microprocessor** is a complete processing unit contained on a single chip. This single invention revolutionized the way computers were designed and applied. From it, the **fourth-generation computer** was born.

The first microprocessor measured just 1/8 inch by 1/6 inch. It was a *4-bit device* (meaning it processed strings of 4 bits at a time) made up of about 2300 MOS transistors. It was equal in computing power to the 5-ton ENIAC with its 18,000 vacuum tubes. The *Intel 4004*, as it was called, could execute 60,000 operations per second, which is primitive by today's standards.

In 1972, Intel came out with an 8-bit microprocessor, the *Intel 8008*. This device was costly. It had only a 16k memory. About a year later, it came out with a much better microprocessor, the *Intel 8080*. This IC had a 64k memory and could execute about 290,000 operations per second. The 8080 used the innovative n-channel MOS process, whereas the 4004 utilized p-channel MOS technology. This brought it vast gains in speed, power, capacity, and density.

The first serious commercial **personal computer (PC)**, the *Altair*, was introduced in 1975. PCs are the desktop or laptop computers that you no doubt are familiar with. The Altair was introduced by *Micro Instrumentation and Telemetry Systems (MITS)* through an article in *Popular Electronics*. It was a mail-order item that sold in kit form. It used an Intel 8080 IC. Thus, began the PC market, and out of this, a new breed of computer hobbyist evolved.

In 1979, the *Motorola 68000* VLSI microprocessor was developed. It was a 16-bit chip and had about 70,000 components. The 68000 was very popular for PC use. It could multiply two 16-bit numbers in 3.2×10^{-6} second. This chip provided the technology that led to the birth of a number of other personal computers.

Commodore began in 1977 with the introduction of its *Personal Electronic Transactor (PET)* computer. The Tandy-Radio Shack *TRS-80 Model I* and *Apple II* computer were also introduced in 1977. The *IBM Personal Computer*, or *IBM PC*, was introduced later in 1981. See Figure 21-3. In the decade of the 1970s, the computer industry evolved from a handful of hobbyists to a vital industry in the world marketplace. The microelectronics revolution had

Figure 21-3. Personal computers, like the IBM PC shown here, gave computer access to businesses and individuals who could not afford mainframes or did not need their immense computing power.   (IBM Corp.)

brought us personal computers, with far more computing capability than the first computer, the enormous ENIAC.

In the 1980s, the computer industry greatly expanded. Microprocessors grew in speed, memory, and quality. See Figure 21-4. Personal computers became available with, among other things, *color monitors, mouse input systems, expanded memories*, and *laser printers*. A host of *software programs* for the PC also became available. Offices and factories could be tied together, or *networked*, by computers worldwide. See Figure 21-5.

USES OF COMPUTERS

There are so many uses of computers today. They are invaluable in helping in the learning process. At home or in school, personal computers are used to assist learners of all ages, Figure 21-6. They may assist you in your education.

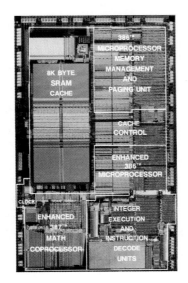

Figure 21-4. This Intel486™ microprocessor contains close to 1.2 million transistors. The different sections of the circuit are denoted here in this magnified view.   (Intel Corp.)

Figure 21-5. Computers networked in an office.   (Apple Computer, Inc.)

Figure 21-6. Examples of computers used at home and in the classroom.
(Sofsource, Inc., IBM Corp.)

Computers are so deeply interwoven into the fabric of science and engineering, Figure 21-7, that many universities now require each student to have one.

Computers can assist with planning of family meals or special diets. Household members may enjoy playing computer games. You may use computers to help keep a checkbook or savings account accurate. Banks use them to process checks and accurately keep account balances, Figure 21-8.

Figure 21-7. Computers are vital in science and engineering. (IBM Corp., Sharp)

Figure 21-8. Banks use computers for many purposes. (IBM Corp.)

Computers are revolutionizing the field of medicine. They are found in pacemakers and prostheses, for example. Also, computers have become valuable tools in medical examinations and diagnostics. For example, they are used in the laboratory for blood chemistry analyses, for *computerized axial tomography (CAT scans)*, and for *magnetic resonance imaging (MRIs)*.

Computers are assisting in the manufacturing arena. Included among new manufacturing techniques is *computer-integrated manufacturing (CIM)*. Also included is *computer-aided manufacturing (CAM)*, which is a branch of CIM. The computer is also a valuable tool in helping designers better design products through *computer-aided design (CAD)*. See Figure 21-9.

Energy-management systems use computers to control interior environments—temperature, for example. Your automobile may have a computer-controlled engine that will adjust the ignition timing and air/fuel mixture to the driving conditions. Many types of telephones are computer

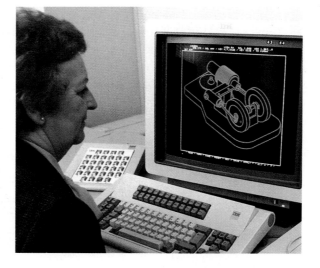

Figure 21-9. Computer-aided design.   (IBM Corp.)

controlled. Cellular phones, for example, have a small computer. *Voice-command* phones are available to place calls upon command.

The evolution of computer technology over the last few years has been very rapid. Since the advent of transistors and integrated circuits, we have seen some fascinating developments. The future of the field of electronics would appear no less exciting.

TYPES OF COMPUTERS

As you can see, computers vary widely in their function. There are countless applications for computers. In addition, they vary widely in size and speed. We can classify computers by application as either *general purpose* or *special purpose*. **General-purpose computers** are versatile. They can be used in many different applications. **Special-purpose computers** are dedicated to a specific task. Although the distinctions are fading as the technology evolves, we can also classify computers by the system size and usage. This would be according to the following categories:

- *Mainframe computers.*
- *Minicomputers.*
- *Microcomputers.*
- *Handheld calculators.*

Mainframe computers

A **mainframe computer** is a large computer with the capability of serving hundreds of users at one time. Most mainframe users have ready access to a number of software programs. These include programs for word processing, memorandum systems, data bases, etc. Mainframes process data very rapidly, and they have extensive storage. Among mainframe computers are *super-computers*. These computers are extremely fast and have tremendous storage capability. Mainframes are frequently seen in large commercial and industrial firms. They are usually supported and maintained by personnel employed solely for that purpose.

The mainframe computer will, typically, be housed in several large cabinets. Computer *terminals*, consisting of *keyboards* and *CRT monitors*, are usually connected to mainframe computers by coaxial or *fiber-optic* cable, either directly or through a *modem*. Associated with mainframes are *high-speed printers* and *laser printers*. A typical mainframe computer is shown in Figure 21-10.

Figure 21-10. A mainframe computer.   (IBM Corp.)

Minicomputers

A **minicomputer** is a surprisingly powerful, midsize computer. In some cases, minicomputers are as powerful as mainframes. These machines will have several terminals networked to a central computer unit. They are most often used in smaller businesses and in schools, where networking capability is desired. Figure 21-11 shows a typical minicomputer environment.

Microcomputers

Microcomputers are smaller, *stand-alone* (as opposed to networked) computers. The "guts" of a microcomputer include a CPU, memory, and input/output interface circuits. The CPU of the microcomputer is a microprocessor. All of these can be contained on one or two printed-circuit boards or a single integrated circuit, as in the single-chip microcomputer seen in Figure 21-2. Also, like other computers, microcomputers have a source for input and a target for output of information.

Figure 21-11. Minicomputer systems such as this are sometimes used in schools and businesses.   (IBM Corp.)

Most often, microcomputers are thought of as small machines that perform single tasks for one user at a time. However, *multitasking* and *multiuser* systems have emerged. Microcomputers may be classified as personal computers, Figure 21-12, or as **embedded computers**. These special-purpose computers control the operation of machines. Automobiles, video cassette recorders (VCRs), and microwave ovens all have embedded computers.

Handheld calculators

Handheld calculators are considered computers. They have all the hardware elements of the digital computer. Of course, they typically do not have all of the ancillary equipment, or *peripherals*. Calculators may or may not be programmable. These devices are an absolute must for many people.

Figure 21-12. A personal computer, like the one shown here, is a microcomputer. It can be difficult to tell the difference between a microcomputer and a minicomputer. (Apple Computer, Inc.)

COMPUTER ARCHITECTURE AND HARDWARE

Modern computers are designed with von Neumann architecture. The diagram of Figure 21-13 shows the basic organization of a computer. It shows the computer divided into two parts. These include a **central computer unit** and **peripheral equipment**, or **I/O (input/output) units**. The central computer unit is divided into a *central processing unit* and *primary storage*, or *memory*. The *central processing unit* is made up of a *control unit* and an *arithmetic/logic unit*. The I/O units include *input, output*, and *auxiliary storage devices*. Finally, these units are connected through *modems* and *networks*.

The physical equipment just described makes up what is called computer **hardware**. See Figure 21-14. Two other computer-related terms are *firmware* and *software*. **Firmware** is instructional data built into a system, which is placed in the chip by the manufacturer. It tells the computer how to perform certain operations. It cannot be changed. **Software** is the instructional data used to tell the computer what to do during a given operation. It is the computer *program*. Generally, it *can* be changed.

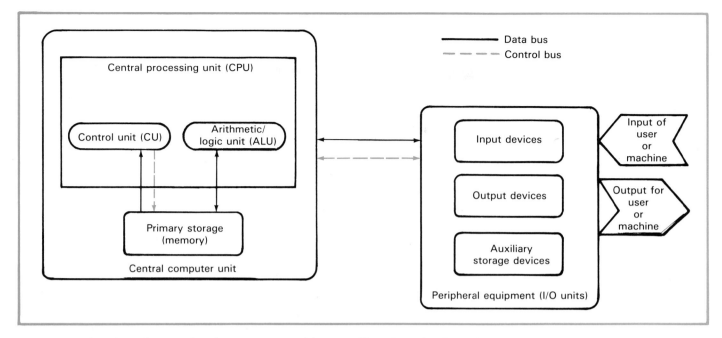

Figure 21-13. Block diagram showing computer architecture. Note that a **bus** *is an electrical channel in a computer along which information is transmitted.*

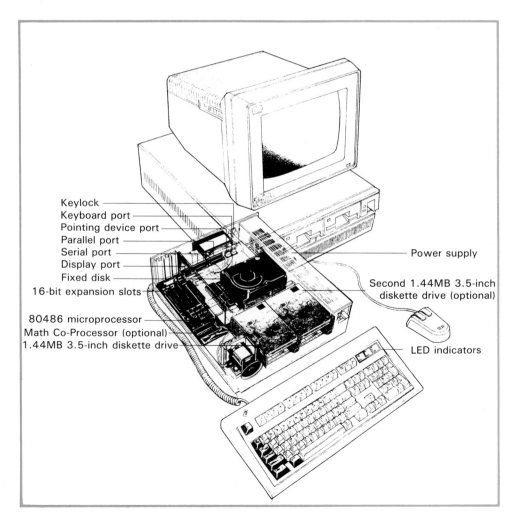

Figure 21-14. Typical hardware components of a personal computer. (IBM Corp.)

We will say more about software later. Now, let us discuss computer hardware in greater detail, including the following:

- Central processing unit.
- Memory.
- Input devices.
- Output devices.
- Auxiliary storage devices.
- Modems and networks.

CENTRAL PROCESSING UNIT

The **central processing unit (CPU)** is the heart of the computer. The CPU performs arithmetic and logic operations, controls instruction processing, and supervises the entire operation of the computer. In many computers, it is comprised of a single microprocessor. Recall that the CPU is made up of a control unit and an arithmetic/logic unit.

The **control unit (CU)** in the computer takes the instructions from the input and moves them into memory. It generates the necessary signals to make the computer accomplish a given task. It takes these instructions and organizes them into an exact sequence of synchronized steps or operations. It is the "traffic cop" in the computer that instructs or controls where the data goes and what will happen to it.

The **arithmetic/logic unit (ALU)** executes what the control unit sends to it. It performs basic arithmetic operations of addition and subtraction. Other operations—multiplication, division, square root, and integration, for instance—are obtained from programs that use these two basic arithmetic operations to achieve the answer. Sometimes these programs must be furnished by the programmer. Other times, they are programmed directly into the microprocessor. The ALU also performs logic functions. Comparing numbers is an example.

MEMORY

The **main memory** (hereinafter **memory**) of a computer is an integral part of the central computer unit. Comprised of a very high-speed integrated circuit, it is where numbers or instructions are stored, usually in 16-bit (or more) words. One word occupies one storage location in a computer. This location is designated by a unique **address**. Memory is also called **primary**, or **internal**, **storage**.

There are two types of memory—*volatile* and *nonvolatile*. **Volatile memory** receives information from an input device, auxiliary storage, or the CPU. The contents of this type of memory are lost when power is removed. **Nonvolatile memory** contains information stored by the manufacturer. This information helps the computer operate. Nonvolatile memories do not lose their contents when power is turned off.

Two terms associated with memory are *read* and *write*. **Reading** is the extracting of information from storage (memory, auxiliary storage, or temporary storage *buffer*) by the CPU. **Writing** is essentially the opposite. It is the sending of information to storage by the CPU. Memories may be classified by whether they are alterable or permanent, according to the following groups:

- *Random-access memory.*
- *Read-only memory.*
- *Programmable read-only memory.*
- *Erasable programmable read-only memory.*

Random-access memory

Random-access memory (RAM) is sometimes called **read/write**, or **R/W, memory**. This is because with RAM, data may not only be read, but it may also be *over*written. In other words, the RAM IC may be readily altered. New data entered into RAM erases old data. This memory is temporary. Data can be lost if power is interrupted. This makes RAM volatile. Some computers have built-in batteries to provide backup voltage if the power is lost.

Information in RAM is extracted in any order, or *at random*. This is opposed to the **serial-access memory**, where information is available for reading only in a certain order. Since a sequence is not followed for RAM, this type of memory is much faster for extracting information. A computer will copy current programs from auxiliary storage to RAM, and operate from RAM since data can be accessed much more readily.

RAMs may be classified as *dynamic* or *static*. **Dynamic RAMs (DRAMs)** have memory cells, each of which consists of one capacitor and one transistor. Bits are stored as voltage (or no voltage) on these very small capacitors. DRAMs are very compact and fast. **Static RAMs (SRAMs)** are made up of flip-flops, wherein the data bits are stored. They are at least ten times faster than DRAMs but provide less storage.

Read-only memory

Read-only memory (ROM) is unalterable memory. Contents of ROM cannot be changed, only read. ROM contains firmware, the information programmed by the manufacturer. Bits are stored in a memory circuit of transistors that are permanently locked either on or off. ROM, like RAM, is also accessed in a random fashion. Inexpensive pocket calculators have ROMs. Computers have them, too. ROM is nonvolatile memory. Contents of ROM are not lost if the power supply is turned off, or if the equipment is unplugged.

Programmable read-only memory

Programmable ROM (PROM) is programmed by the user after the IC is made. To program the PROM, sufficient current is applied to *burn* the instructions into the chip. The current melts a *fusible link*, making this an irreversible process. Once the chip has been made, the information is there to stay.

Erasable programmable read-only memory

An **erasable PROM (EPROM)** uses MOS cells, which are programmed by applying a high voltage and trapping electrons, or stored charge. A window is provided on the IC package of the EPROM. To erase the EPROM, the chip is exposed to ultraviolet light through the window. This excites the electrons to an energy level to where they can escape. The EPROM may be reprogrammed after it has been erased.

More recently, the **electrically erasable PROM (EEPROM)** has been developed. In the EEPROM, there is an electric path for charge removal. By changing the voltage within that memory section of the computer, the EEPROM can be erased. This memory is easier to work with, since errors in programming may be quickly corrected.

INPUT DEVICES

Input devices are used to feed data, images, sound, commands, programs, and other information into the central computer unit. The information that is entered goes to a temporary storage location within the central computer unit called an **input/output buffer**. From here, it is moved into memory by the CPU. There are a number of input devices available.

A **keyboard** is a standard input device. Almost everyone is familiar with this device used to input data and text. Another common input device is a **mouse**. This device is moved across a flat surface to move a *cursor* across a display screen (output device). It is used to locate relative coordinates for a drawing. The device can also be used to select commands from an on-screen *menu*. Figure 21-15 shows a computer with a keyboard and a mouse.

Figure 21-15. In foreground, keyboard and mouse input devices. (Apple Computer, Inc.)

Other input devices include *digitizing tablets, touch screens, light pens*, and different types of *scanners*. Digitizing tablets use devices that look and function somewhat like a mouse. Touch screens provide input in response to touch. Light pens input data when they are held up to the computer screen. Laser scanners are typically used in grocery stores to read bar codes and are tied-in to a computer. Other scanners capture text and pictures and translate them into digital form for computer use. Figure 21-16 is an example of a kind of scanner used with a personal computer.

Automatic control systems, with embedded computers, may have inputs coming from other devices. This might be analog data from a *sensor* about physical conditions such as temperature, pressure, and humidity converted

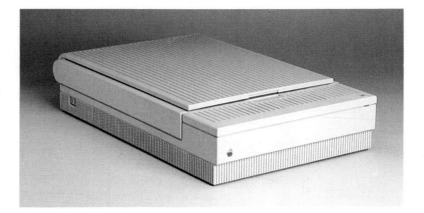

Figure 21-16. Scanner used to input text and pictures into a computer. (Apple Computer, Inc.)

to digital form in an A/D converter. It might be a signal from contacts of, for instance, a control relay, a limit switch, or a flow switch. Such a signal could be used to indicate *status* of some condition: pump *on* or *off*, damper *open* or *closed*, *flow* or *no flow*, etc.

OUTPUT DEVICES

Output devices are targets for information held in the central computer unit. The equipment presents the binary signals of the computer in a translated, usable form that can be understood by the user. With output, information is moved by the CPU from memory into the buffer. From here, it gets written to the output device. Output may take many forms. Visual and audio outputs communicate information. Other outputs activate or actuate mechanical equipment. There are a number of output devices available.

A standard output device is a **video monitor**. This device provides a temporary visual output. It is also referred to as a **monitor**, a **CRT (cathode-ray tube)**, or a **VDT (video display terminal)**. *Flat-screen displays* also provide a temporary visual output. These are commonly LCD screens, like those of laptop computers. See Figure 21-17. Monitors may display text or graphics. There are *color monitors*, and there are *monochrome monitors*, which are one color — generally, black and white. A monitor is usually accompanied by a keyboard. The two devices together make up a **computer terminal**.

Figure 21-17. This laptop computer is optimized for travel. It includes a mouse that is built in and performs like a standard desktop mouse.   (Hewlett-Packard)

Another standard output device is a **printer**. This device enables a user to obtain a printout, or *hardcopy*, of the computer output. A **hardcopy** is a permanent record on paper. For PCs, there are a few basic types of printers — *dot-matrix, ink-jet,* and *laser*.

The **dot-matrix printer**, Figure 21-18, uses electromechanical *printheads* with 9 or 24 pins. This type of printer is fast, reliable, and relatively inexpensive. The **ink-jet printer** actually squirts ink directly onto the paper in the shape of letters. This type provides much better quality printing than the dot-matrix. The **laser printer**, Figure 21-19, uses a laser beam, controlled by the computer, to fuse ink onto the surface of the paper. The process is similar to that of the office copy machine. Its advantage is primarily in print quality. However, laser printers continue to be more expensive.

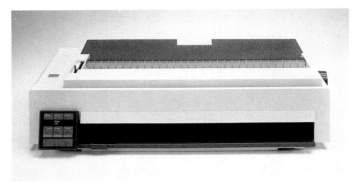

Figure 21-18. Dot-matrix printer. (IBM Corp.)

Figure 21-19. Laser printer. (Apple Computer, Inc.)

Plotters are hardcopy devices. They are used with computers to transfer computer graphics to paper. They are an indispensable part of a CAD system. Very large drawings can be made with plotters. Drawings may be one color or multicolor. Figure 21-20 shows a desktop color plotter, and Figure 21-21 shows a larger floor-model plotter.

Where microcomputers are used for automatic control systems, an output device would be whatever is being controlled. In temperature control, for example, this is often affected by a control relay. Output may be sent to control relays to which, perhaps, fans, pumps, or two-position dampers are connected. The output energizes the relays to start and stop the fans or pumps or open

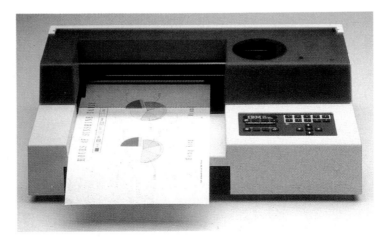

Figure 21-20. Tabletop color plotter. (IBM Corp.)

Figure 21-21. Floor-model color plotter. (IBM Corp.)

and close the dampers. Where an analog output is required, for example, to control a variable speed motor, the digital output of the computer must first be converted to analog. This is accomplished by a D/A converter.

AUXILIARY STORAGE DEVICES

Auxiliary storage, also called **external**, or **mass**, **storage**, is used for storing data and programs. The central computer unit can extract information from external storage to perform its operation. It can send results back to storage to update the file. In this way, it reads and writes to the storage medium. The most common storage media are *magnetic disks, magnetic tapes*, and *optical disks*. Most commercial programs are purchased on these formats. Others types included the old *punch cards* and *punch tapes*. Common to all of these are the *drive units*, the hardware used to access the storage.

Magnetic disks

A **magnetic disk** is a rotating circular plate with a magnetizable surface. It is used by a computer for storing information. Disks are *random-access* devices. Files may be accessed without reading through the other files stored on them. The two types of magnetic disks are *floppy* and *hard disks*.

Floppy disks. Also called **diskettes**, these are pliable, or "floppy," disks, as the name implies. Common disk sizes are 3.5″ and 5.25″. See Figure 21-22. They are made of a polyester film that is only 75 *microns* (0.075 mm) thick. The disk has a coating of fine, uniformly applied, magnetic particles, which enable information to be recorded on the surface. The information is stored within *sectors* on concentric *tracks* lying on the surface.

The floppy disk is enclosed in a square case, or *jacket*. Obviously, the jacket helps to protect the disk from damage due to dirt and scratches. Even fingerprints and smoke particles can contaminate a disk. To protect the disk from such contaminants, a friction-free liner is placed inside the jacket. The liner cushions the ride and cleans the magnetic disk. It will not protect the disk from havoc caused by an external magnetic field, however, which can wipe out everything contained on the disk.

A floppy disk is basically made of the same material as the tape inside audio and video cassettes. Of course, the material is a flat disk rather than a wound tape. Magnetic recording media such as cassette tapes, videotapes, and floppy

A   B

Figure 21-22. Floppy disks come in different sizes. A—5.25″ disk (in jacket). B—3.5″ disk (in jacket).   (Maxell Corp.)

disks use the principle of electromagnetism to magnetize the ultra-fine magnetic particles spread over the surface of the film. In this process, sounds, images, and information can be accurately recorded and played back at any time.

Hard disks. These come as a unit containing from one to a number of aluminum plates that spin on a common shaft. Hard disks are also called **fixed disks** because they are built into the drive unit. The disks also have a magnetic coating, and data storage is essentially accomplished via the same method used for floppy disks. Hard disks have a much larger capacity for storing information than floppy disks. This is due to greater surface area and more dense packing of information. Typical units provide from 20 to hundreds of megabytes of storage. In addition, they are much faster in accessing information. This advantage is attributed to the precision of the drive unit mechanisms, which hard disks afford.

Hard disks do have certain advantages, but so do floppies. While hard disks have larger capacity and are faster, floppy disks are convenient for portability and organization. Further, floppy disks can act as a "safety net." It is possible to lose the entire contents of a fixed disk due to a computer malfunction. With a backup on floppy disk, the contents can be restored.

Magnetic tapes

This type of auxiliary storage is used in the mainframe computer as well as some personal computers. Styles include reel-to-reel and various types of cassettes, or cartridges. **Magnetic tape** is made of a magnetically coated polyester film—again, like the material used inside audio and video cassettes. The same principles are involved in recording. Magnetic tape is useful in its ability to store large amounts of information at low cost. However, they are slower than magnetic disks. This is due in part because they are serial-access devices. The tapes must be wound to the location where the information resides.

Optical disks

Optical disk technology is a dramatic advancement in data storage. **Optical disks** have information embedded in their surface in digital form. The encoded information, represented by submicroscopic *pits*, can be read by a laser beam. A single **compact disc (CD)**, a type of optical disk, may hold billions of pits on its surface. Optical disks are useful in their ability to store large amounts of data in a small package at low cost.

Two types of optical disk for computer use are *CD ROM* and *WORM.* **CD ROM** is becoming a common auxiliary storage medium for computers. It functions like a read-only memory. A CD ROM is programmed by the manufacturer. Currently, entire encyclopedias can be purchased on these disks.

WORMs can be *w*ritten *o*nce by the user and *r*ead *m*any times. Once the information or data is recorded, it can be read off the disk repeatedly. However, it cannot be changed or erased.

In addition, other optical systems combine the technologies of electromagnetic and optical storage to perform read/write functions. An example of this is the *magneto-optical* disk and drive shown in Figure 21-23.

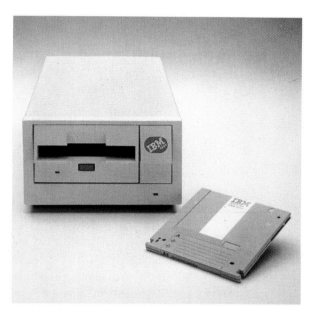

Figure 21-23. Optical disk cartridge and drive.   (IBM Corp.)

Drive units

As briefly discussed, information can be stored by magnetism. An elemental explanation can be offered by two ferromagnetic rings, each strung on separate wires as in Figure 21-24. Passing an electric current through the wires magnetizes the rings. If the currents are of opposite direction, the magnetic rings will have opposite magnetic polarities. Their magnetic fields could be used to represent bits — logic 0 for one polarity, logic 1 for the opposite. This essentially describes the workings of a *core memory*, a type of computer memory that is no longer in use. It is also the underlying principle involved with magnetic disks and tapes; wherein, 0s and 1s are represented by microscopic magnets pointing in opposite directions.

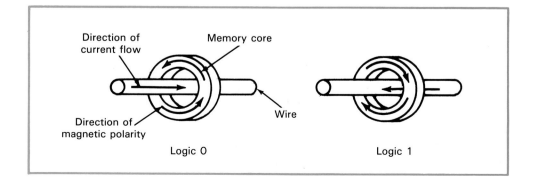

Figure 21-24. Magnetized particles of opposite polarity represent 0s and 1s.

Recording (writing) onto a magnetic medium and playback (reading) are carried out using binary information, as stated. All data transferred between the computer and disk or tape is in the form of 0s and 1s. We will explain this process later in more detail. Reading and writing are carried out by **drive units**. In magnetic storage, these include **disk drives** and **tape drives**. The function of the drive corresponds broadly to that of a cassette deck in an audio system. Drive units are actually a combination input and output device. A typical PC floppy disk drive is shown in Figure 21-25.

Figure 21-25. Personal computer disk drive.   (IBM Corp.)

An important part of the disk or tape drive is the **magnetic head**. On a floppy disk, there is a window in the disk jacket through which this *read/write head* gains access to the disk. (In addition, there is a *write-protect* feature on the jacket. This feature is used, when desired, to prevent writing of information to the disk.) Drive units for hard disks **(hard drives)** may have multiple read/write heads. The magnetic head of the drive is placed on or near the magnetic recording surface. By so doing, the transfer of information can occur.

When writing to a disk or tape, information gets imprinted on the media surface as designated by instructions from the computer. Once the magnetic pattern has been recorded, the surface retains the magnetism in the same way that a permanent magnet would. This is the principle of recording. See Figure 21-26. Even if the recording surface has been magnetized by the recording process, it can be newly magnetized again. In this regard, it is like an audio or video cassette tape.

Reading from disk or tape involves the opposite process. The magnetic head senses the magnetic pattern on the recorded surface. It transmits this magnetic binary code back to the computer in the form of electrical impulses. Depending on the circumstance, the information read might then be used to perform calculations. It might be translated into alphanumeric characters and displayed on a CRT, or translated into a graphic and plotted on hardcopy, or whatever.

There is one big difference between recording with magnetic computer media and recording with the typical audio or video cassette tape. The latter records analog information — a continuous waveform describing the music or image. Computer disks and tapes, on the other hand, store discrete information — 0s and 1s. These two recording systems function on completely different principles.

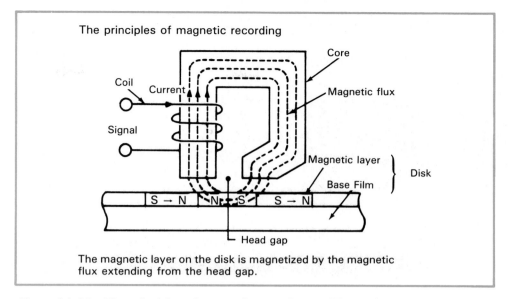

The principles of magnetic recording

The magnetic layer on the disk is magnetized by the magnetic flux extending from the head gap.

Figure 21-26. The principles of magnetic recording. (Maxell Corp.)

Let us use audio cassette tape as an example with which to explain analog recording. A music signal is made up of a continuous waveform of varying amplitude (level). In recording, this waveform is converted to an electrical signal, which is sent to the magnetic recording head of the cassette deck. The magnetic output of the head varies with the music signal, and the cassette tape is magnetized accordingly. The music signal continuously varies in amplitude producing a continuously varying magnetic pattern on the recorded tape.

On the other hand, information that is to be recorded digitally is represented as a continuous stream of 0s and 1s. This is convenient since the individual bits can be clearly distinguished at playback. A digital recording system is only required to record, or store, two distinct states rather than continuous waveforms. As such, it is in principle much simpler than an analog recording system.

Unfortunately, if an error is made in recording a 0 or 1, or if one digit is dropped, the meaning of the data may be completely changed. Consequently, it is important that the performance of the magnetic head of the disk or tape drive be optimized. Also, floppy disk drives must be designed to minimize wear where disk and magnetic head are in contact. Here, wear on the disk due to friction and warped surfaces are a concern.

To some extent, errors can be found and corrected. Error-correction codes exist to identify exact error locations. When the error is a problem of an inverted bit, the code inverts the erroneous bit back to its intended state. Correction of multiple errors is possible, but the method is more complex.

As mentioned, wear of magnetic storage media is a problem that can lead to errors in the recording of digital information. In optical disk technology, there is no physical contact between the disk and read mechanism. This prevents disk wear. In this system, data is read by a laser beam. The beam hits the disk and is reflected through a prism to a light-sensitive diode (photodiode). Current through the photodiode varies with light intensity. With this setup, light signals are transmitted as one of two discrete electrical signals. This is because there are two types of surfaces on the laser disk — abruptly changing and flat. The laser detects the abrupt *changes* caused by the submicroscopic pits on the disk surface. These changes are transmitted as logic 1. Flat stretches, which include the disk surface and bottoms of pits, are transmitted as logic 0.

MODEMS AND NETWORKS

Modems and *networks* fall under the broader category of *data communication*. This involves sending and receiving of computer signals. A telephone line is a typical channel for data communication to remote computers. However, telephone lines are intended for analog signals and computers use digital signals. For this reason, **modems** are used at each end to interface the two systems. These input/output devices convert digital to analog (*mo*dulate) and analog back to digital (*dem*odulate). From these two functions, the term *modem* is derived.

With older technology, modems were acoustically coupled through the handset of a telephone. With current technology, modems are connected directly to the telephone line. The speed of transmission is very fast. Speed is given as a **baud rate**, wherein 1 *baud* equals 1 bit per second. Typical rates for PCs are 1200 baud and 2400 baud. Larger computers use modems that transmit at even higher baud rates.

The use of communication channels to join geographically separated computers and peripheral devices is called **networking**. Modems, therefore, are used for networking. *Networks* are convenient for sharing computer software and hardware. A **wide area network (WAN)** connects computers over a large geographic area. This would be used, for example, by the transportation industry for scheduling. A **local area network (LAN)** connects computer workstations within a single site, such as different buildings of a college campus. A diagram of a LAN is shown in Figure 21-27.

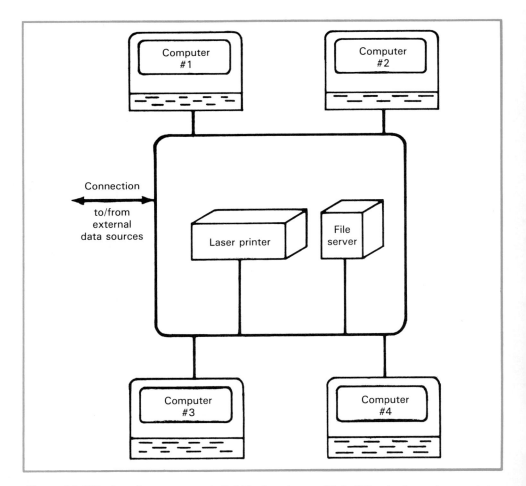

*Figure 21-27. Local area network (LAN) showing multiple PCs sharing a laser printer and **file server**. This is a large storage device, usually 50 megabytes or more, that holds programs and data to be shared by workstations connected to the LAN.*

COMPUTER SOFTWARE

Modern computers can be described in terms of hardware and software. As discussed, hardware is the physical equipment associated with the computer. Software refers to the stored **program**, or step-by-step sequence of instructions, given to the computer. The program contains the words that tell the computer what to do. Programs may be used for processing, documenting, and operating.

Software is kept in auxiliary storage and *loaded* into the computer memory prior to use. Software programs may be purchased for word processing, accounting, data-base management, graphics, etc. These are called **applications software packages**. Software packages are designed to be convenient for the user, or *user friendly*. A person can use these programs without having to know a particular computer *language*. However, behind every applications software package is a detailed program written by a computer programmer in some computer language.

Some common computer programming languages are *BASIC, FORTRAN, COBOL, Pascal,* and *C*. These are called **high-level languages**. These are *procedure-oriented*. That is, instructions are written as procedures that are more meaningful to people. For example, READ, WRITE, GOTO, and END are some common procedural commands. Programs written in high-level languages must be translated into machine language for the computer. This is done by programs called **compilers** and **assemblers**.

Finally, to coordinate all other software programs and hardware devices, an *interactive* **operating-system (OS) program** is required. The OS program guides the input and output devices and assists in the operation of the computer in general. *DOS, UNIX,* and *CMS* are common operating systems.

SUMMARY

- Early computers date back to the abacus.
- Some early inventors of calculating machines were Pascal, Napier, Leibniz, and Babbage.
- Charles Babbage's analytical engine was designed to add numbers and store the results at each stage of the calculations.
- Herman Hollerith successfully devised a counting machine that used punched cards to tabulate census figures.
- John von Neumann developed modern computer architecture.
- In 1944, the first general-purpose computer (Mark I) was built. It used electromechanical relays. The first electronic computer was the ENIAC, developed in 1946. It used 18,000 vacuum tubes. In 1951, the UNIVAC I, which used vacuum tubes, emerged as the first commercially available computer.
- Transistors replaced vacuum tubes in computers, beginning the second generation of computers. In the early 1960s, ICs replaced transistors. This began the third generation. The fourth generation of computers began in the 1970s with the arrival of the microprocessor.
- A microprocessor is a complete CPU contained on one computer chip.
- The first serious PC, the Altair, was introduced in 1975.
- Microcomputers are computers built around a microprocessor.
- Computers may be classified into four groups: mainframe computers, minicomputers, microcomputers, and handheld calculators.
- Computer hardware includes CPU, memory, and input, output, and auxiliary storage devices. Modems and networks are also types of computer hardware.

- The CPU performs arithmetic and logical operations, controls instruction processing, and supervises the entire operation of the computer. It is made up of control and arithmetic/logic units.
- Two main types of memory are RAM and ROM. RAM is working storage. It is volatile memory. ROM stores operating instructions. It is nonvolatile memory.
- Input devices feed information into a computer. Examples are keyboards and digitizing tablets.
- Output devices are used to produce information held in the central computer unit. Examples are CRTs and printers.
- Auxiliary storage devices are used for storing data and programs. Examples are magnetic disks and tapes and their drives.
- Modems allow computers to communicate with each other over telephone lines. Networks allow multiple computers to share hardware and software.
- Firmware is instructional data built into a system. Software tells a computer what to do by way of a computer program.
- High-level languages are used to write applications software. These are translated into machine language for the computer in assemblers and compilers. Operating system software controls the entire computer operation.

KEY TERMS

Each of the following terms has been used in this chapter. Do you know their meanings?

address	fourth-generation computer	operating system (OS) program
applications software package	general-purpose computer	optical disk
arithmetic/logic unit (ALU)	hard disk	output device
assembler	hard drive	peripheral equipment
auxiliary storage	hardcopy	personal computer (PC)
baud rate	hardware	plotter
bus	high-level language	primary storage
CD ROM	ink-jet printer	printer
central computer unit	input device	program
central processing unit (CPU)	input/output buffer	programmable ROM (PROM)
compact disc (CD)	input/output (I/O) unit	programming
compiler	internal storage	random-access memory (RAM)
computer	keyboard	reading
computer terminal	laser printer	read-only memory (ROM)
control unit (CU)	local area network (LAN)	read/write (R/W) memory
cathode-ray tube (CRT)	magnetic disk	second-generation computer
disk drive	magnetic head	serial-access memory
diskette	magnetic tape	single-chip microcomputer
dot-matrix printer	main memory	software
drive unit	mainframe computer	special-purpose computer
dynamic RAM (DRAM)	mass storage	static RAM (SRAM)
electrically erasable PROM (EEPROM)	memory	tape drive
embedded computer	microcomputer	third-generation computer
erasable PROM (EPROM)	microprocessor	video display terminal (VDT)
external storage	minicomputer	video monitor
file server	modem	volatile memory
firmware	monitor	von Neumann architecture
first-generation computer	mouse	wide area network (WAN)
fixed disk	networking	WORM
floppy disk	nonvolatile memory	writing

TEST YOUR KNOWLEDGE

Please do not write in this text. Place your answers on a separate sheet of paper.

1. Draw a basic block diagram for a computer.
2. Computers can be divided into four groups. What are these?
3. What is the difference between a minicomputer and a microcomputer?
4. The physical equipment that makes up a computer is known as _____.
5. Instructional data built into a system is known as _____.
6. A floppy disk is an example of main memory. True or False?
7. In a computer, a _____ is a group of 16 or more bits occupying one storage location in a computer.
8. What is the purpose of a control unit?
9. Information held in memory may be *readily* altered in:
   a. ROM.
   b. PROM.
   c. RAM.
   d. EPROM.
10. The exact location of a word in memory is given by its _____.
11. _____ is the extracting of information from storage or input device by the CPU.
12. _____ is the sending of information to storage or output device by the CPU.
13. ROMs lose their data whenever power is turned off in the circuit. True or False?
14. List three input devices for computers.
15. List three output devices for computers.
16. What is a computer network?
17. High-level languages are human oriented; therefore, they must be translated into machine language to be understood by a computer. True or False?
18. A program used to control general computer operation is an:
   a. Applications software package.
   b. Assembler or compiler.
   c. Operating system.

Section VI

INTEGRATED CIRCUITS AND COMPUTERS

SUMMARY

Important Points

□ Integrated circuits are complete electronic circuits contained in one package. Usually included in an IC are many transistors, diodes, resistors, and capacitors connected together. ICs are also called chips.

□ ICs may be either linear or digital.

□ Linear ICs have variable outputs.

□ Digital ICs have only two outputs — ON or OFF.

□ The binary number system has only two states — 0 or 1. For example, the binary number for 22 (base 10) is 10110.

□ A bit is a single binary digit (0 or 1). A byte is 8 bits.

□ An inverter changes the value of an incoming signal: 0s are changed to 1s; 1s are changed to 0s.

□ Categories of computers include mainframe, minicomputer, microcomputer, and handheld calculators.

□ The CPU of a microcomputer is a microprocessor.

□ A minicomputer is a midsize computer wherein several terminals are networked to a central computer unit.

□ A microprocessor is a complete processing unit contained on a single chip.

□ Basic computer hardware consists of a central processing unit memory, and peripheral equipment consisting of input, output, and auxiliary storage devices.

□ Software is a set of instructions given to a computer to tell it what to do. Generally, software can be changed.

□ Boolean algebra is used as the basis for computer logic gates.

1. Outline the many steps involved in the making of an IC.
2. Name common uses of linear ICs.
3. What is the binary number for 256 (decimal)?
4. Draw the truth tables for:
   a. A NAND gate.
   b. An AND gate.
   c. An XOR gate.
   d. An OR gate.
   e. A NOT gate.
5. Develop a time line showing key developments in the history of the computer.
6. What major computer functions are contained in the CPU?
7. Summarize the different types of computer memory.
8. Name four computer input devices.
9. Name three computer output devices.
10. Expand upon the subject of computer *languages*.

Applications of Electronics Technology:
An entry from a nine-million-word encyclopedia held on CD ROM is shown
on this computer display screen. (Compton's NewMedia)

Chapter 22

CAREER OPPORTUNITIES IN ELECTRONICS

Since 1960, the worldwide electronics market has grown from $25 billion to over $500 billion. Projections indicate that in the next century it will become the world's largest industry. Career opportunities in the electronics industry are almost unlimited. The field of electronics offers many advantages to a person who is interested in an excellent career.

PERSONAL QUALITIES

A person aiming at a career in the electronics field, especially in the areas of research and development, should possess a number of personal qualities:
- An intense curiosity about the nature of things and how they work.
- An analytical mind and a desire to solve problems.
- A capacity for hard work and self-discipline.
- A positive attitude and an ability to take failure in stride and learn from it.
- An interest in math and science and an aptitude suited for electronics.
- An avid interest in the subject — perhaps even, electronics as a hobby.

AREAS OF ELECTRONICS

If we were to subdivide the entire electronics industry, we might break it down into four areas. The first three could be the market areas of government, industrial or commercial, and consumer products and services, according to who is using a product. The fourth area could be the components that are manufactured for the other three areas.

The government purchases all kinds of different products. Examples include missile and space guidance systems, communications systems, and electronic goods used in medicine, education, crime detection, and traffic control. Some examples of electronic research and products used by the government are shown in Figures 22-1 through 22-3.

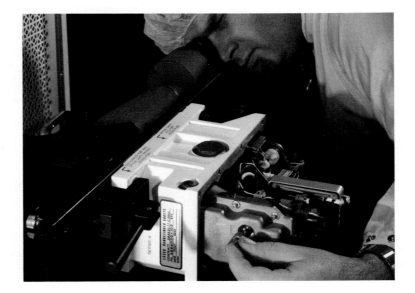

Figure 22-1. Energy output is measured on the laser rangefinder for a military battle tank. The light and compact system is designed to provide accurate and almost instantaneous range information to the tank's fire-control computer. (Hughes Aircraft Co.)

Figure 22-2. Electrical installation in progress of components in research and development phase of superconducting supercollider (SSC). The SSC was a basic research tool to study the fundamental nature of matter and energy. The project was cancelled before completion due to budget restrictions. (U.S. Department of Energy)

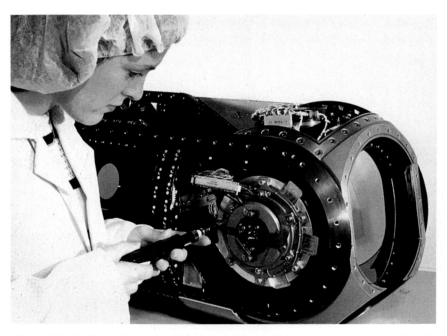

Figure 22-3. Technician working on a military night-vision system. (Martin Marietta)

Electronic products have also become an important part of daily business operations. Computers and fax machines, Figure 22-4, are examples. Radio and television broadcasting equipment and production control equipment would also fall under the category of commercial products. Service robots, another example, are foreseen to command a multibillion dollar market. See Figure 22-5.

Consumer products probably are the most familiar types of electronic

Figure 22-4. A fax (facsimile) machine will send letters and photos over telephone lines.   (Ricoh Corp.)

Figure 22-5.   Security robots, such as this one, are used as industrial security guards. (Denning Mobile Robotics Inc.)

products. Every day, thousands of people buy televisions, radios, microwave ovens, stereos, and calculators, to name a few. Personal computers are very popular. See Figures 22-6 through 22-8.

Components are needed to manufacture and repair electronic products. Some of the well-known components are capacitors, switches, transistors, relays, amplifiers, and integrated circuits. See Figures 22-9 through 22-12.

*Figure 22-6. Personal computers being used by students as an educational tool.
(IBM Corp.)*

Figure 22-7. A musician using a PC to compose music.   (IBM Corp.)

Figure 22-8. A college student using a PC for word processing.   (IBM Corp.)

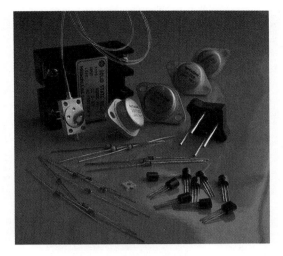

Figure 22-9. Solid-state components. (Hitachi Ltd.)

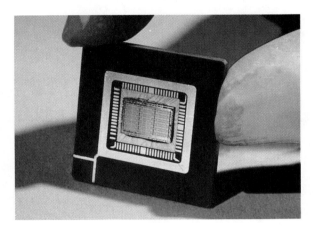

Figure 22-10. This microelectronic chip has circuitry equal to 72 high-speed processors. (Note the use of the protective glove.) (Martin Marietta)

Figure 22-11. A design engineer examines the diagram of a very large-scale integrated circuit. Eventually, it will be reduced to a single chip layer with circuit lines one-fiftieth the thickness of a human hair. (Martin Marietta)

Figure 22-12. The protective coating on an electronic circuit is touched up under ultraviolet light during quality-control checks. The light will reveal the slightest flaws. (Martin Marietta)

PLANNING A CAREER IN ELECTRONICS

Planning a career should begin as early as possible. The career you desire has a direct impact on what you need to do in preparation for that career. See Figure 22-13. If electronics is your choice, make plans carefully. Talk about your plans with those who know you well. Talk with people who work in the electronics field. Talk to your instructors and read as much as you can about the subject. School and public libraries contain many helpful references.

An excellent reference on careers in many industries is the *Occupational Outlook Handbook*. It is published by the U.S. Department of Labor, Bureau of Statistics. Most libraries have copies of this book. The book can also be ordered by writing: Superintendent of Documents, U.S. Government Printing Office, Washington, DC 20402.

There are numerous electricity and electronics associations. Contact them for information. One such organization is the Electronics Industries Association, or EIA. They publish a career guidance brochure. It can be ordered by writing: Electronics Industries Association, 2001 Pennsylvania Avenue, N.W., Washington, DC 20006.

Career counselors and job placement centers can be another very good source of information. Some schools hold career seminars. Colleges and universities frequently host job fairs on campus. Also, each year, they usually have a part in "career nights" at the local high school. Talk to their representatives. If you are interested, you should plan a campus visit to learn more about what they offer. Take advantage of seminars, career days, or whatever is available that may help you to better plan a career in electronics.

There are many effective things you can do to prepare yourself for this career. You can develop or further develop an inquiring, logical mind, good study habits, and the ability to express yourself effectively. You could enroll in electricity and electronics courses. Join a related club. Get interested in electronic kit-building. (A number of low-cost project kits are available.) Consider becoming an amateur radio operator. Perhaps there is no more rewarding hobby or better way to prepare for a career in electronics.

IMPORTANCE OF
EDUCATION

Education is very important to anyone entering a technical field such as electronics. Today, a high school education has become a minimum standard. Ideally, as early as junior high or middle school, a student should take such courses as technology education, computer science, and physics. These will help in an electronics career. The high school graduate is in a better position in the job market, and one with a strong technical background is even more so.

Although training beyond high school has been the standard for some time for many professional occupations, other areas of work also require more than a high school diploma. As new automated equipment is introduced on

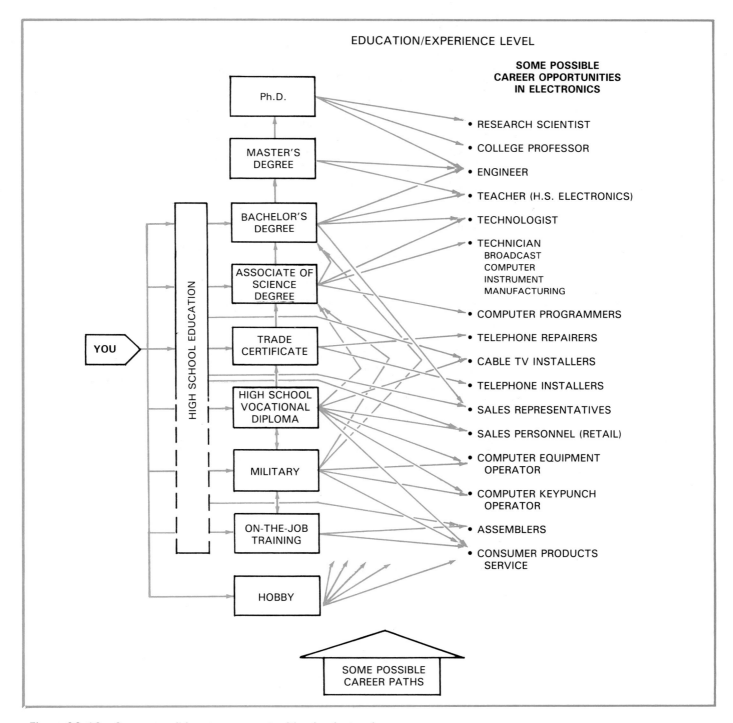

Figure 22-13. Some possible career opportunities in electronics.

a wider scale, skill requirements are rising for clerical and other jobs. Employers increasingly are demanding better trained workers to operate complicated machinery.

Developments in machine design, use of new materials, and the complexity of equipment are making greater technical knowledge a requirement in many areas of sales work. Many occupations are becoming increasingly complex and technical. As a result, occupational training such as that obtained through apprenticeship and post-secondary education is becoming more and more important.

Those people entering the job market who are not well-prepared or well-schooled will find the going more difficult in the years ahead. Employers will be more likely to hire workers who have at least a high school diploma. Furthermore, present experience shows that the less education and training workers have, the less chance they have for obtaining steady jobs.

In addition to its importance in competing for jobs, education makes a difference in lifetime income. According to the most recently available data, those who have college degrees can expect to earn:

* Three or four times the amount earned by workers with less than eight years of schooling.
* Nearly twice the amount earned by workers with one to three years of high school.
* More than 1 1/2 times as much as a high school graduate.

Clearly, the completion of college pays a dividend; even the completion of high school does. A high school graduate could expect to earn:

* Nearly twice the amount earned by workers who have completed elementary school only.
* Nearly 1 1/3 times the amount earned by workers with one to three years of high school.

In summary, people who have acquired skills and a good basic education will have a better chance for interesting work, good wages, and steady employment. Therefore, getting as much education and training as your abilities permit should be a top priority.

CAREERS AND EDUCATION

In both manufacturing and service industries, four classes of workers exist. These classes are *semiskilled, skilled, technical,* and *professional.* Career opportunities exist in each area. The training needed to perform jobs in each of these areas varies. Likewise, employment outlooks and salaries vary.

Semiskilled workers

Semiskilled workers perform jobs that do not require a high level of training. These workers are most often found working on assembly lines. Most semiskilled jobs are limited to a certain type and number of tasks. These tasks are often simple and repetitive. In general, semiskilled work is routine. Advance study and training are required to move up from a semiskilled position. There are very few electronics jobs done by semiskilled workers.

Skilled workers

Skilled workers have thorough knowledge of and skill in a particular area. This knowledge and skill are gained through additional study. Additional study is obtained through apprenticeship or community college programs. Some skilled workers may obtain training from the military services. Others receive specialized education in training programs, which some large companies maintain.

An apprenticeship is a period of time spent learning a trade from an experienced, skilled worker. This training is done on the job. It is usually combined with special self-study courses. Four years is the usual length of time for an apprenticeship program.

Many community colleges offer courses and programs in the electronics field. Courses can be taken to learn more about a particular subject, or a program may be followed to gain a particular skill. Further, the armed forces offer many specialized areas of study in the electronics field. The opportunities for learning a trade are very good.

These sources can provide a solid foundation of education. Many workers do not stop at this point, however. They continue to study and read to keep abreast of the many changes and new technologies forever developing in the industry.

Many jobs in the electronics field are skilled positions. Some of these include maintenance and construction electricians, assemblers, and quality control inspectors. Assemblers wire and solder various parts for televisions, stereos, printed circuit boards, computers, etc. See Figure 22-14. Quality control inspectors check the finished work of the assemblers.

Figure 22-14. Assembling electrical control panels.   *(Miller Electric Mfg. Co.)*

Technicians

Technicians are specially trained workers capable of doing complex, technical jobs. Many receive their training in two year programs at community colleges. Technicians work with electronic equipment and assist engineers. They have the training needed to service and repair complex machines and components. Engineers rely on technicians to help them conduct research, test machines and components, and design new devices. Therefore, technicians must keep up to date on developments in the electronics industry. Typical career positions in this area include broadcast technicians, robotics technicians, and computer technicians.

Professionals

Nearly all professional workers have four years of college training. Many have more advanced degrees, such as masters' and doctorates. Professionals have excellent opportunities for advancement.

One of the best known professional positions in the electronics industry is the engineer. Engineers do many things. A chief function is designing and monitoring the building and/or installation of new equipment. Their goal is to design equipment that runs smoothly and does the job intended. Once this goal is met, technicians are assigned to maintain the equipment. Engineers must have a solid background in math and science. This background allows them to visualize designs before putting them down on paper. See Figure 22-15.

Teaching is another professional position in the electronics industry. Teachers of electronics have the opportunity to challenge students interested in electronics. They can share their knowledge and interest in electronics with their students. The rewards of teaching are many.

ENTREPRENEURS

Entrepreneurs own and operate their own businesses. They usually start with an idea for filling a hole in the marketplace where a new product or service is needed. Then, a business plan is made. This plan outlines goals for the business, along with a timetable for meeting those goals. This plan is vital if the business is to succeed.

In addition to a sound business plan, a successful entrepreneur is knowledgeable. This person has knowledge of a certain industry, service, or product, which allows the owner to make smart business decisions about what is being sold. For instance, an appliance service technician needs knowledge of the appliance being serviced. If this is not the case, the business will fail.

The successful entrepreneur also has sound management skills. These skills allow the owner to manage money, time, and employees. Entrepreneurial skills are also very important to be successful. These skills allow the business owner to control the business and move it in the right direction.

Entrepreneurship opportunities are vast in the electronics industry. With the growth in consumer electronics products, similar growth has occurred in servicing these products. Support of the office products industry also allows for many business opportunities, and servicing of home appliances continues to be a sound business in the electronics industry.

Consulting is yet another growing entrepreneurial business in the electronics industry. Consultants work for clients on different projects. The specific job they do often depends on what work is needed. The consultant is paid by the client. When the job is completed, the consultant is free to move on to a new job and client.

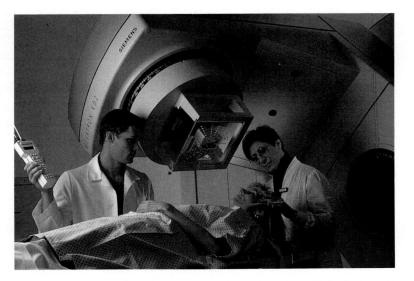

Figure 22-15. Medical diagnostic equipment such as this is designed in part by electrical engineers.

FUTURE CAREER OPPORTUNITIES

There are many opportunities for skilled workers, technicians, and professionals. Today, there is very little opportunity for the unskilled worker. Employment in electronics manufacturing is expected to increase faster than the average for all industries. In addition to the jobs resulting from this growth, large numbers of openings will arise as experienced workers retire or take jobs in other industries.

The majority of the nation's workers belong to the service sector. The production of goods requires only about a third of the country's work force. In general, job growth is expected to continue to be faster in the service sector than in the manufacturing sector.

Discoveries and developments in science and technology are made every day. Electronics and other fields of service play a major role in daily life. Think of common devices that have been improved with more complex machines. For example, the communication power of the telephone, television, and radio has been increased through the use of satellites. The safety of boats and airplanes has been improved through the use of sonar and radar devices.

Where else is the use of electronics apparent? Industry is electronically controlled. Robots and machines displace hundreds of production workers every day. In many businesses, computers can complete complex tasks in only a few minutes. Electronics has also entered the field of medicine. Using laser technology, surgeons can now perform surgery without cutting any skin. For each new device developed, many highly skilled technicians must be trained to perform maintenance and service, Figure 22-16. The need for people with electronics training will continue to grow into the next century.

Figure 22-16. Equipment troubleshooting underway.   (Miller Electric Mfg. Co.)

Knowing how to use a computer is vital in electronics careers. *(IBM Corp.)*

APPENDIX

Appendix A

SCIENTIFIC NOTATION, PREFIXES, AND CONVERSIONS

When working with formulas in electronics, manual computations can typically lead to many errors and the incorrect placement of decimal points. It is advisable to master the *power of ten*, or scientific notation. In its more simple form, a large number is converted to a number between 1 and 10 and multiplied by 10 at a certain power.

Example:

$$365 = 3.65 \times 10^2$$
$$3650 = 3.65 \times 10^3$$
$$36,500 = 3.65 \times 10^4$$

Example:

$$1 = 1 \times 10^0$$
$$10 = 1 \times 10^1$$
$$100 = 1 \times 10^2$$
$$1000 = 1 \times 10^3$$
$$10,000 = 1 \times 10^4$$
$$100,000 = 1 \times 10^5$$
$$1,000,000 = 1 \times 10^6$$

$$0.1 = 1 \times 10^{-1}$$
$$0.01 = 1 \times 10^{-2}$$
$$0.001 = 1 \times 10^{-3}$$
$$0.0001 = 1 \times 10^{-4}$$
$$0.00001 = 1 \times 10^{-5}$$
$$0.000001 = 1 \times 10^{-6}$$

When adding powers of 10, convert numbers so all have the same exponent. Then, add the decimal numbers. The powers of 10 remain the same.

Example:

$$
\begin{array}{r}
3.65 \times 10^3 \\
+ 9.37 \times 10^5 \\
\end{array}
\Rightarrow
\begin{array}{r}
3.65 \times 10^3 \\
937.00 \times 10^3 \\
\hline
940.65 \times 10^3 \\
\end{array}
$$

When subtracting powers of 10, convert numbers so all have the same exponent. Then, subtract the decimal numbers. The powers of 10 remain the same.

Example:

$$
\begin{array}{r}
5.97 \times 10^{11} \\
- 6.93 \times 10^{10} \\
\end{array}
\Rightarrow
\begin{array}{r}
59.70 \times 10^{10} \\
- 6.93 \times 10^{10} \\
\hline
52.77 \times 10^{10} \\
\end{array}
$$

When multiplying large numbers together, it can be helpful to first convert to scientific notation. Then, multiply the decimal numbers together and multiply the powers of 10 together by *adding* the exponents.

Examples:

$$4000 \times 600,000 \Rightarrow (4 \times 10^3)(6 \times 10^5) = 24 \times 10^8$$

$$0.081 \times 0.00062 \Rightarrow (8.1 \times 10^{-2})(6.2 \times 10^{-4}) = 50.22 \times 10^{-6}$$

$$3,200,000 \times 0.031 \Rightarrow (3.2 \times 10^6)(3.1 \times 10^{-2}) = 9.92 \times 10^4$$

$$0.0000013 \times 130 \Rightarrow (1.3 \times 10^{-6})(1.3 \times 10^2) = 1.69 \times 10^{-4}$$

When dividing numbers converted to scientific notation, divide the decimal numbers and divide the powers of 10 by *subtracting* the divisor (denominator) exponent from the dividend (numerator) exponent.

Examples:

$$\frac{24 \times 10^8}{6 \times 10^5} = 4 \times 10^3 \qquad\qquad \frac{50.22 \times 10^{-6}}{8.1 \times 10^{-2}} = 6.2 \times 10^{-4}$$

$$\frac{9.92 \times 10^4}{3.2 \times 10^6} = 3.1 \times 10^{-2} \qquad\qquad \frac{1.69 \times 10^{-4}}{1.3 \times 10^{-6}} = 1.3 \times 10^2$$

Further, when dividing numbers, the powers of 10 in the denominator can be moved to the numerator (or vice versa) by changing the sign of the exponent.

Examples:

$$\frac{4 \times 10^2}{2 \times 10^{-2}} = \frac{4 \times 10^2 \times 10^2}{2} = \frac{4 \times 10^4}{2} = 2 \times 10^4$$

$$\frac{(4 \times 10^2)(2 \times 10^{-6})}{(2 \times 10^{-3})(2 \times 10^3)} = \frac{4 \times 2}{2 \times 10^{-3} \times 2 \times 10^3 \times 10^{-2} \times 10^6} = \frac{8}{4 \times 10^4} = 2 \times 10^{-4}$$

Most formulas in electronics are written to use quantities in their basic units: current in amperes, voltage in volts, and resistance in ohms. If a quantity is given in other than its basic form, a conversion should be made before performing the calculation. For example, milliamperes should be changed to amperes, microfarads should be changed to farads, etc.

Examples:

$$E = 3 \text{ k}\Omega \times 5 \text{ }\mu\text{A} = (3 \times 10^3 \text{ }\Omega)(5 \times 10^{-6} \text{ A}) = 15 \times 10^{-3} \text{ V}$$

$$X_L = 2\pi(3 \text{ kHz})(2 \text{ mH}) = 2\pi(3 \times 10^3 \text{ Hz})(2 \times 10^{-3} \text{ H}) = 6 \text{ }\Omega$$

$$I = \frac{36 \text{ mV}}{12 \text{ M}\Omega} = \frac{36 \times 10^{-3} \text{ V}}{12 \times 10^6 \text{ }\Omega} = 3 \times 10^{-9} \text{ A}$$

Conversion to *basic* units is not necessary if adding or subtracting like units. However, prefixes should be the same.

Examples:

$$R_T = 10 \text{ k}\Omega + 3 \text{ k}\Omega = 13 \text{ k}\Omega$$

$$I_{R_1} = 9 \text{ mA} - 30 \text{ }\mu\text{A} = 9000 \text{ }\mu\text{A} - 30 \text{ }\mu\text{A} = 8970 \text{ }\mu\text{A}$$

Conversion to basic units is also not necessary when using the following common relationships:

$$\text{mA} \times \text{k}\Omega = \text{V} \qquad \text{V/k}\Omega = \text{mA} \qquad \mu\text{A} \times \text{M}\Omega = \text{V} \qquad \text{V/M}\Omega = \mu\text{A}$$

The common prefixes and their meanings are shown in Figure A-1 for your reference.

Prefix	Symbol	Power of 10	Multiply by
tera-	T	10^{12}	1,000,000,000,000
giga-	G	10^9	1,000,000,000
mega-	M	10^6	1,000,000
kilo-	k	10^3	1,000
(basic units)		10^0	1
milli-	m	10^{-3}	0.001
micro-	μ	10^{-6}	0.000001
nano-	n	10^{-9}	0.000000001
pico-	p	10^{-12}	0.000000000001

Figure A-1. Engineering notation.

Prefixes may be explained in another way by the prefix chart in Figure A-2.

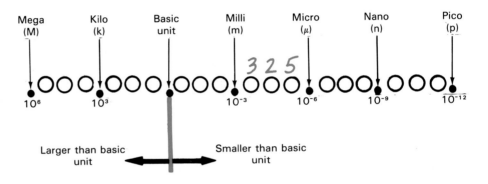

Figure A-2. Chart shows how prefixes of basic units are carried out by zeros in both directions.

Figure A-2 can be used to convert between scientific notation, engineering notation, and basic units. For example, consider a value of 0.325 mA. Using the conversion chart above, we can convert to other prefixes or scientific notation at a glance:

$$0.325 \text{ mA} = 0.325 \times 10^{-3} \text{ A}$$
$$= 0.000325 \text{ A}$$
$$= 325 \ \mu\text{A}$$
$$= 325 \times 10^{-6} \text{ A}$$
$$= 3.25 \times 10^{-4} \text{ A}$$

Appendix B

TRIGONOMETRY REVIEW

Trigonometry is used for finding the solutions of alternating current problems. It finds many uses in designing and understanding electronic circuits.

Basically, trigonometry is the relationship between the angles and sides of a triangle. These relationships are called trigonometric functions. They represent the numerical ratio between the two sides of the right triangle.

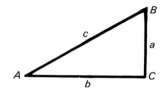

$$\sin A = \frac{\text{opposite side}}{\text{hypotenuse}} = \frac{a}{c}$$

$$\cos A = \frac{\text{adjacent side}}{\text{hypotenuse}} = \frac{b}{c}$$

$$\tan A = \frac{\text{opposite side}}{\text{adjacent side}} = \frac{a}{b}$$

There are other functions, but these three (sine, cosine, and tangent) are widely used in solving problems in electronics. Using these equations, if two values are known, the third may be found.

In addition, the Pythagorean theorem may come in handy. It is:

$$c = \sqrt{a^2 + b^2}$$

Example: In a triangle, side $a = 6$ and $b = 10$. Find angle A.

$$\tan A = \frac{a}{b} = \frac{6}{10} = .6$$

Look at the table of Natural Trigonometric functions, Figure B-1. Find the angle whose tangent is .6. It is 31°.

Example: Angle A is 45° and side a is 6. What is the value of side c? Use the sine equation.

$$\sin 45° = \frac{6}{c}$$

Look up the sine of 45°. It is .707. Then:

$$c = \frac{6}{.707} \cong 8.5$$

In electronics, the right triangle used for the previous examples can be labeled other ways. But the problems are worked out in the same manner.

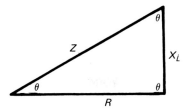

Example: A circuit contains 40 Ω of resistance and 50 Ω of inductive reactance. What is the circuit impedance?

Find angle θ:

$$\tan \theta = \frac{X_L}{R} = \frac{50}{40} = 1.25$$

Look up in the table:

$$\theta \cong 51°$$

Find Z:

$$\sin \theta = \frac{X_L}{Z}$$

Look up in the table:

$$.777 = \frac{50}{Z}$$

$$Z = \frac{50}{.777} = 64.3 \ \Omega$$

Alternately, using the Pythagorean theorem:

$$Z = \sqrt{(50)^2 + (40)^2} \cong 64 \ \Omega$$

Example: Peak ac voltage is 10 V. The instantaneous voltage at any point during the cycle is: $e = E_{peak} \sin \theta$. What is the voltage at 70°?

$$\sin 70° = .9397$$

$$e = .9397 \times 10 \text{ V} = 9.397 \text{ V}$$

These applications of trignometry must be thoroughly understood. The best way to gain this understanding is through practice.

			NATURAL TRIGONOMETRIC FUNCTIONS				
Angle	Sine	Cosine	Tangent	Angle	Sine	Cosine	Tangent
1°	.0175	.9998	.0175	46°	.7193	.6947	1.0355
2°	.0349	.9994	.0349	47°	.7314	.6820	1.0724
3°	.0523	.9986	.0524	48°	.7431	.6691	1.1106
4°	.0698	.9976	.0699	49°	.7547	.6561	1.1504
5°	.0872	.9962	.0875	50°	.7660	.6428	1.1918
6°	.1045	.9945	.1051	51°	.7771	.6293	1.2349
7°	.1219	.9925	.1228	52°	.7880	.6157	1.2799
8°	.1392	.9903	.1405	53°	.7986	.6018	1.3270
9°	.1564	.9877	.1584	54°	.8090	.5878	1.3764
10°	.1736	.9848	.1763	55°	.8192	.5736	1.4281
11°	.1908	.9816	.1944	56°	.8290	.5592	1.4826
12°	.2079	.9781	.2126	57°	.8387	.5446	1.5399
13°	.2250	.9744	.2309	58°	.8480	.5299	1.6003
14°	.2419	.9703	.2493	59°	.8572	.5150	1.6643
15°	.2588	.9659	.2679	60°	.8660	.5000	1.7321
16°	.2756	.9613	.2867	61°	.8746	.4848	1.8040
17°	.2924	.9563	.3057	62°	.8829	.4695	1.8807
18°	.3090	.9511	.3249	63°	.8910	.4540	1.9626
19°	.3256	.9455	.3443	64°	.8988	.4384	2.0503
20°	.3420	.9397	.3640	65°	.9063	.4226	2.1445
21°	.3584	.9336	.3839	66°	.9135	.4067	2.2460
22°	.3746	.9272	.4040	67°	.9205	.3907	2.3559
23°	.3907	.9205	.4245	68°	.9272	.3746	2.4751
24°	.4067	.9135	.4452	69°	.9336	.3584	2.6051
25°	.4226	.9063	.4663	70°	.9397	.3420	2.7475
26°	.4384	.8988	.4877	71°	.9455	.3256	2.9042
27°	.4540	.8910	.5095	72°	.9511	.3090	3.0777
28°	.4695	.8829	.5317	73°	.9563	.2924	3.2709
29°	.4848	.8746	.5543	74°	.9613	.2756	3.4874
30°	.5000	.8660	.5774	75°	.9659	.2588	3.7321
31°	.5150	.8572	.6009	76°	.9703	.2419	4.0108
32°	.5299	.8480	.6249	77°	.9744	.2250	4.3315
33°	.5446	.8387	.6494	78°	.9781	.2079	4.7046
34°	.5592	.8290	.6745	79°	.9816	.1908	5.1446
35°	.5736	.8192	.7002	80°	.9848	.1736	5.6713
36°	.5878	.8090	.7265	81°	.9877	.1564	6.3138
37°	.6018	.7986	.7536	82°	.9903	.1392	7.1154
38°	.6157	.7880	.7813	83°	.9925	.1219	8.1443
39°	.6293	.7771	.8098	84°	.9945	.1045	9.5144
40°	.6428	.7660	.8391	85°	.9962	.0872	11.4301
41°	.6561	.7547	.8693	86°	.9976	.0698	14.3006
42°	.6691	.7431	.9004	87°	.9986	.0523	19.0811
43°	.6820	.7314	.9325-	88°	.9994	.0349	28.6363
44°	.6947	.7193	.9657	89°	.9998	.0175	57.2900
45°	.7071	.7071	1.0000	90°	1.0000	.0000	

Figure B-1. The table of natural trigonometric functions.

Appendix C   SCHEMATIC SYMBOLS

The following are symbols found on electrical schematics. There has been an attempt to standardize these symbols, but there still exists much confusion among manufacturers and users of components. The following symbols are IEEE (Institute of Electrical and Electronic Engineers) approved.

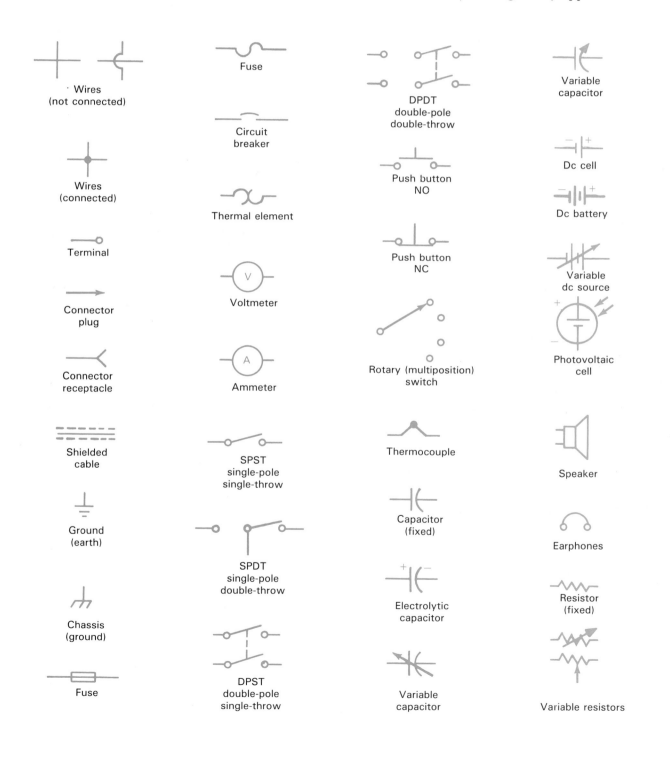

Variable resistor
(rheostat)

Inductor
(iron core)

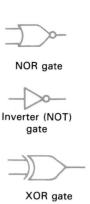

Field effect
transistor
(p-channel)

NOR gate

Tapped
resistor

Variable
inductor

Unijunction
transistor
(with n base)

Inverter (NOT)
gate

Piezoelectric
crystal

Incandescent
lamp

Unijunction
transistor
(with p base)

XOR gate

Neon
lamp

N channel MOS
enhancement type
field effect transistor

XNOR gate

Motor

R

N channel MOS
depletion type field
effect transistor

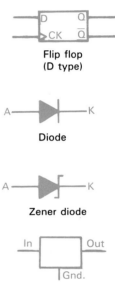

Flip flop
(D type)

Generator

LED indicators

A ——▷|—— K

Diode

Ac power
source

Photo diode

Antenna

A ——▷|—— K

Zener diode

Transformer
(air core)

Bipolar junction
transistor
(npn)

Single-
ended
amplifier

In [ ] Out
Gnd.

Voltage regulator

Transformer
(iron core)

Bipolar junction
transistor
(pnp)

Amp.
operational

Silicon
controlled
rectifier

Inductor

Field effect
Transistor
(n-channel)

AND gate

TRIAC

NAND gate

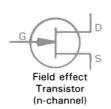

OR gate

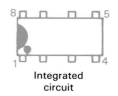

Integrated
circuit

Appendix D

THE GREEK ALPHABET

A	α	alpha
B	β	beta
Γ	γ	gamma
Δ	δ	delta
E	ϵ	epsilon
Z	ζ	zeta
H	η	eta
Θ	θ	theta
I	ι	iota
K	$\varkappa$	kappa
Λ	λ	lambda
M	μ	mu
N	ν	nu
Ξ	ξ	xi
O	o	omicron
Π	π	pi
P	ϱ	rho
Σ	σ	sigma
T	τ	tau
Υ	υ	upsilon
Φ	ϕ	phi
X	χ	chi
Ψ	ψ	psi
Ω	ω	omega

Appendix E

LETTER SYMBOLS AND STANDARD ABBREVIATIONS

A	ampere, unit of current
A_i	current amplification
A_p	power gain of amplifier
A·t	ampere-turn, SI unit of magnetomotive force
A_v	voltage amplification
α	temperature coefficient of resistance
α, h_{fb}	common-base small-signal current gain
α_{dc}, h_{FB}	common-base dc current gain
B	magnetic flux density
BW	bandwidth
β, h_{fe}	common-emitter small-signal current gain
β_{dc}, h_{FE}	common-emitter dc current gain
°C	degrees Celsius, unit of temperature
C	capacitance
C	coulombs, unit of charge
c	speed of light (3×10^8 m/s)
$\cos \theta$	power factor
dB	decibel, unit of gain
Δ	change in
E, V	voltage
e, v	instantaneous voltage
E_a	armature voltage
E_{avg}, V_{avg}	average voltage
E_{cemf}	cemf generated by a motor
E_{eff}, E_{rms}, V_{eff}, V_{rms}	effective, or root-mean-square, voltage
$e_{L_{avg}}$	average induced cemf in inductor
E_{max}, E_{peak}, E_0, V_{peak}	maximum instantaneous, or peak, voltage
E_p	primary voltage
$E_{p\text{-}p}$	peak-to-peak voltage
E_S	source voltage
E_s	secondary voltage
E_{sh}	shunt voltage
eV	electron volt (1.602×10^{-19} J)
ϵ	permittivity
ϵ_r	relative permittivity
ϵ_0	permittivity of free space (8.85×10^{-12} F/m)
η_C	collector efficiency
°F	degrees Fahrenheit, unit of temperature
F	farad, unit of capacitance
f	frequency
f_r	resonant frequency
G	conductance
G	gauss, cgs unit of magnetic flux density
Gb	gilbert, cgs unit of magnetomotive force
H	henry, unit of inductance

H	magnetic field intensity
h_i	ac resistance
h_{ie}	common-emitter ac input impedance
Hz	hertz, unit of frequency
I	current
i	instantaneous current
I_a	armature current
I_{avg}	average current
I_B	base dc current
i_b	base ac current
i_B	total base current
$I_{B_{max}}$	maximum dc base current
$I_{B_{min}}$	minimum dc base current
I_{BQ}	base dc current at quiescence
I_C	collector dc current
i_c	collector ac current
i_C	total collector current
I_{CBO}	collector cutoff current
$I_{C_{max}}$	maximum dc collector current
$I_{C_{min}}$	minimum dc collector current
I_{CO}	reverse saturation current
I_{CQ}	collector dc current at quiescence
I_E	emitter dc current
i_e	emitter ac current
i_E	total emitter current
I_{eff}, I_{rms}	effective, or root-mean-square, current
I_{FSM}	peak surge current
I_H	forward holding current
I_M	full-scale deflection current
I_{max}, I_{peak}	maximum instantaneous, or peak, current
I_O	average rectified forward current
I_p	primary current
$I_{p\text{-}p}$	peak-to-peak current
I_R	maximum reverse, or leakage, current
I_S	reverse saturation current
I_s	secondary current
I_{sh}	shunt current
I_{SL}	surface-leakage current
I_{ZK}	avalanche current, or zener knee current
J	joule, unit of work
L	inductance
λ	ac wavelength
M	mutual inductance
Mx	maxwell, cgs unit of magnetic flux
μ	permeability
μ_r	relative permeability
μ_0	permeability of free space ($4\pi \times 10^{-7}$ W/A·m)
N_p	number of turns in primary
N_s	number of turns in secondary
Oe	oersteds, cgs unit of magnetic field intensity
Ω	ohm, unit of resistance, reactance, or impedance
ω	angular velocity, or radian frequency
P	power, true power

p	instantaneous power
P_D	power dissipation
P_{dc}	dc input power
$P_{D_{max}}$	maximum power-dissipation rating
PIV	peak inverse voltage
P_{out}	maximum class-A ac output power
prr	pulse repetition rate
prt	pulse repetition time
Φ	magnetic flux
Q	electrical charge, quality factor, or figure of merit, reactive power
R	resistance
R_a	armature resistance
R_{eq}	equivalent resistance
R_{fs}	full-scale resistance
R_G	generator internal impedance
R_{hs}	half-scale resistance
R_i	internal resistance
R_L	load resistance
r_M	internal resistance of moving coil
R_{mult}	meter multiplier resistance
R_{sh}	shunt resistance
R_V	voltmeter resistance
$\mathcal{R}$	reluctance
ϱ	electrical resistivity
S	apparent power, stability factor
S	siemen, unit of conductance
S_S	synchronous motor speed
σ	conductivity
T	motor torque, period of ac waveform
T	tesla, SI unit of magnetic flux density
t	time, time constant
t_d	pulse duration
t_f	fall time
t_r	rise time
θ	phase angle
V	volt, unit of voltage
VA	volt-ampere, unit of apparent power
VAR	volt-ampere reactive, unit of reactive power
V_B, V_0	barrier, or contact, potential
V_B	base dc voltage to ground
v_b	base ac voltage to ground
v_B	base total voltage to ground
V_{BB}	base biasing voltage
V_{BR}, V_Z	breakdown, reverse-breakdown, or zener, voltage
$V_{(BR)F}$	forward breakover voltage
V_C	collector dc voltage to ground
v_c	collector ac voltage to ground
v_C	collector total voltage to ground
V_{CB}	collector-to-base dc voltage
v_{cb}	collector-to-base ac voltage
v_{CB}	collector-to-base total voltage
V_{CC}	collector biasing voltage
V_{CE}	collector-to-emitter dc voltage

v_{ce}	collector-to-emitter ac voltage
v_{CE}	collector-to-emitter total voltage
$V_{CE_{max}}$	maximum collector-to-emitter dc voltage
$V_{CE_{min}}$	minimum collector-to-emitter dc voltage
V_{CE_Q}	collector-to-emitter dc voltage at quiescence
V_{DD}	drain-supply dc voltage
V_{DS}	drain-source voltage
V_E	emitter dc voltage to ground
v_e	emitter ac voltage to ground
v_E	emitter total voltage to ground
V_{EB}	emitter-to-base dc voltage
v_{eb}	emitter-to-base ac voltage
v_{EB}	emitter-to-base total voltage
V_{EE}	emitter biasing voltage
V_F	forward voltage
V_f	feedback voltage
V_{GD}	gate-drain voltage
V_{GG}	gate-supply dc voltage
V_{GS}	gate-source voltage
V_p	pinch-off voltage
V_R	reverse bias voltage
V_{REF}	reference voltage
W	watt, unit of power, weber, SI unit of magnetic flux
X	reactance
X_C	capacitive reactance
X_L	inductive reactance
Z	impedance

Appendix F

DATA SHEETS

Motorola, Inc.
Used by Permission

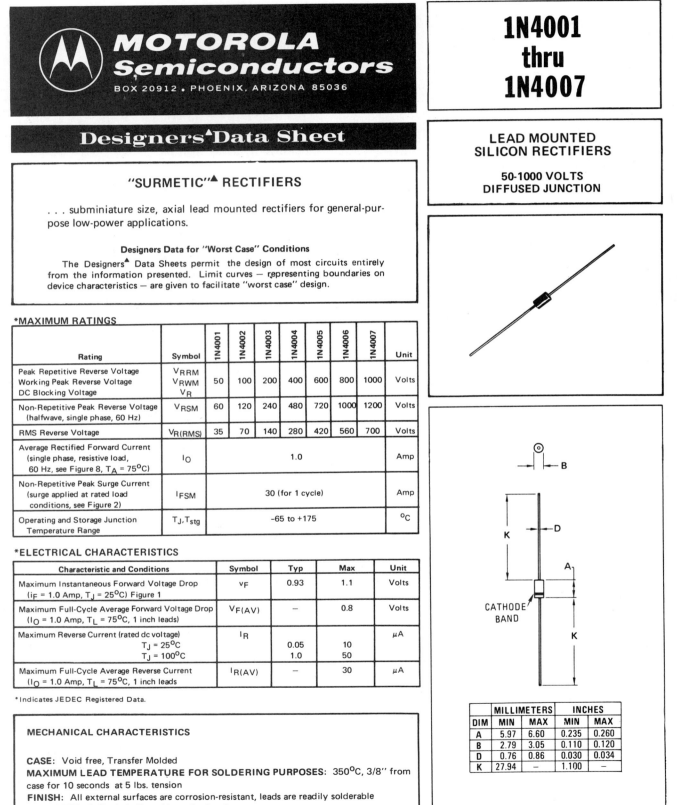

MOTOROLA Semiconductors
BOX 20912 • PHOENIX, ARIZONA 85036

Designers Data Sheet

"SURMETIC"▲ RECTIFIERS

. . . subminiature size, axial lead mounted rectifiers for general-purpose low-power applications.

Designers Data for "Worst Case" Conditions

The Designers▲ Data Sheets permit the design of most circuits entirely from the information presented. Limit curves — representing boundaries on device characteristics — are given to facilitate "worst case" design.

1N4001 thru 1N4007

LEAD MOUNTED SILICON RECTIFIERS

50-1000 VOLTS DIFFUSED JUNCTION

*MAXIMUM RATINGS

Rating	Symbol	1N4001	1N4002	1N4003	1N4004	1N4005	1N4006	1N4007	Unit
Peak Repetitive Reverse Voltage Working Peak Reverse Voltage DC Blocking Voltage	V_{RRM} V_{RWM} V_R	50	100	200	400	600	800	1000	Volts
Non-Repetitive Peak Reverse Voltage (halfwave, single phase, 60 Hz)	V_{RSM}	60	120	240	480	720	1000	1200	Volts
RMS Reverse Voltage	$V_{R(RMS)}$	35	70	140	280	420	560	700	Volts
Average Rectified Forward Current (single phase, resistive load, 60 Hz, see Figure 8, T_A = 75°C)	I_O	1.0							Amp
Non-Repetitive Peak Surge Current (surge applied at rated load conditions, see Figure 2)	I_{FSM}	30 (for 1 cycle)							Amp
Operating and Storage Junction Temperature Range	T_J, T_{stg}	–65 to +175							°C

*ELECTRICAL CHARACTERISTICS

Characteristic and Conditions	Symbol	Typ	Max	Unit
Maximum Instantaneous Forward Voltage Drop (i_F = 1.0 Amp, T_J = 25°C) Figure 1	v_F	0.93	1.1	Volts
Maximum Full-Cycle Average Forward Voltage Drop (I_O = 1.0 Amp, T_L = 75°C, 1 inch leads)	$V_{F(AV)}$	–	0.8	Volts
Maximum Reverse Current (rated dc voltage) T_J = 25°C T_J = 100°C	I_R	0.05 1.0	10 50	μA
Maximum Full-Cycle Average Reverse Current (I_O = 1.0 Amp, T_L = 75°C, 1 inch leads	$I_{R(AV)}$	–	30	μA

*Indicates JEDEC Registered Data.

MECHANICAL CHARACTERISTICS

CASE: Void free, Transfer Molded
MAXIMUM LEAD TEMPERATURE FOR SOLDERING PURPOSES: 350°C, 3/8" from case for 10 seconds at 5 lbs. tension
FINISH: All external surfaces are corrosion-resistant, leads are readily solderable
POLARITY: Cathode indicated by color band
WEIGHT: 0.40 Grams (approximately)

CATHODE BAND

DIM	MILLIMETERS		INCHES	
	MIN	MAX	MIN	MAX
A	5.97	6.60	0.235	0.260
B	2.79	3.05	0.110	0.120
D	0.76	0.86	0.030	0.034
K	27.94	–	1.100	–

CASE 59-04
Does Not Conform to DO-41 Outline.

▲Trademark of Motorola Inc.

© MOTOROLA INC., 1975   DS 6015 R3

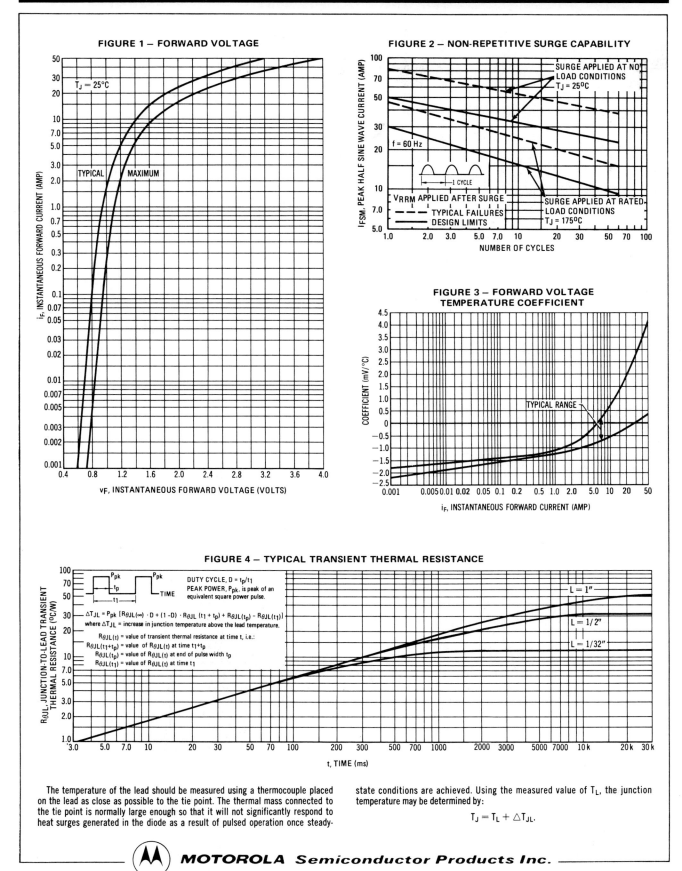

FIGURE 1 — FORWARD VOLTAGE

FIGURE 2 — NON-REPETITIVE SURGE CAPABILITY

FIGURE 3 — FORWARD VOLTAGE TEMPERATURE COEFFICIENT

FIGURE 4 — TYPICAL TRANSIENT THERMAL RESISTANCE

DUTY CYCLE, D = t_p/t_1
PEAK POWER, P_{pk}, is peak of an equivalent square power pulse.

$\Delta T_{JL} = P_{pk} [R_{\theta JL}(\infty) \cdot D + (1-D) \cdot R_{\theta JL}(t_1 + t_p) + R_{\theta JL}(t_p) - R_{\theta JL}(t_1)]$
where ΔT_{JL} = increase in junction temperature above the lead temperature.

$R_{\theta JL}(t)$ = value of transient thermal resistance at time t, i.e.:
$R_{\theta JL}(t_1+t_p)$ = value of $R_{\theta JL}(t)$ at time t_1+t_p
$R_{\theta JL}(t_p)$ = value of $R_{\theta JL}(t)$ at end of pulse width t_p
$R_{\theta JL}(t_1)$ = value of $R_{\theta JL}(t)$ at time t_1

The temperature of the lead should be measured using a thermocouple placed on the lead as close as possible to the tie point. The thermal mass connected to the tie point is normally large enough so that it will not significantly respond to heat surges generated in the diode as a result of pulsed operation once steady-state conditions are achieved. Using the measured value of T_L, the junction temperature may be determined by:

$$T_J = T_L + \triangle T_{JL}.$$

MOTOROLA *Semiconductor Products Inc.*

2N3903
2N3904

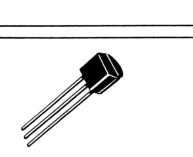

**NPN SILICON
SWITCHING & AMPLIFIER
TRANSISTORS**

NPN SILICON ANNULAR TRANSISTORS

... designed for general purpose switching and amplifier applications and for complementary circuitry with types 2N3905 and 2N3906.

- High Voltage Ratings — $V_{(BR)CEO}$ = 40 Volts (Min)
- Current Gain Specified from 100 μA to 100 mA
- Complete Switching and Amplifier Specifications
- Low Capacitance — C_{ob} = 4.0 pF (Max)

MAXIMUM RATINGS

Rating	Symbol	Value	Unit
*Collector-Emitter Voltage	V_{CEO}	40	Vdc
*Collector-Base Voltage	V_{CBO}	60	Vdc
*Emitter-Base Voltage	V_{EBO}	6.0	Vdc
*Collector Current — Continuous	I_C	200	mAdc
**Total Device Dissipation @ T_A = 25°C Derate above 25°C	P_D	625 5.0	mW mW/°C
Total Power Dissipation @ T_A = 60°C	P_D	450	mW
**Total Device Dissipation @ T_C = 25°C Derate above 25°C	P_D	1.5 12	Watts mW/°C
**Operating and Storage Junction Temperature Range	T_J, T_{stg}	− 55 to 150	°C

THERMAL CHARACTERISTICS

Characteristic	Symbol	Max	Unit
Thermal Resistance, Junction to Case	$R_{\theta JC}$	83.3	°C/W
Thermal Resistance, Junction to Ambient	$R_{\theta JA}$	200	°C/W

*Indicates JEDEC Registered Data.
**Motorola guarantees this data in addition to the JEDEC Registered Data.

EQUIVALENT SWITCHING TIME TEST CIRCUITS

FIGURE 1 — TURN-ON TIME

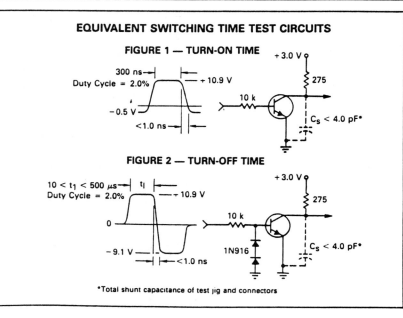

FIGURE 2 — TURN-OFF TIME

*Total shunt capacitance of test jig and connectors

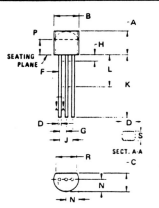

NOTES:
1. CONTOUR OF PACKAGE BEYOND ZONE "P" IS UNCONTROLLED.
2. DIM "F" APPLIES BETWEEN "H" AND "L". DIM "D" & "S" APPLIES BETWEEN "L" & 12.70 mm (0.5") FROM SEATING PLANE. LEAD DIM IS UNCONTROLLED IN "H" & BEYOND 12.70 mm (0.5") FROM SEATING PLANE.

DIM	MILLIMETERS MIN	MILLIMETERS MAX	INCHES MIN	INCHES MAX
A	4.32	5.33	0.170	0.210
B	4.44	5.21	0.175	0.205
C	3.18	4.19	0.125	0.165
D	0.41	0.56	0.016	0.022
F	0.41	0.48	0.016	0.019
G	1.14	1.40	0.045	0.055
H	—	2.54	—	0.100
J	2.41	2.67	0.095	0.105
K	12.70	—	0.500	—
L	6.35	—	0.250	—
N	2.03	2.67	0.080	0.105
P	2.92	—	0.115	—
R	3.43	—	0.135	—
S	0.36	0.41	0.014	0.016

All JEDEC dimensions and notes apply.

**CASE 29-02
(TO-226AA)**

ELECTRICAL CHARACTERISTICS (T_A = 25°C unless otherwise noted.)

Characteristic		Symbol	Min	Max	Unit
OFF CHARACTERISTICS					
Collector-Emitter Breakdown Voltage[1] (I_C = 1.0 mAdc, I_B = 0)		$V_{(BR)CEO}$	40	—	Vdc
Collector-Base Breakdown Voltage (I_C = 10 μAdc, I_E = 0)		$V_{(BR)CBO}$	60	—	Vdc
Emitter-Base Breakdown Voltage (I_E = 10 μAdc, I_C = 0)		$V_{(BR)EBO}$	6.0	—	Vdc
Collector Cutoff Current (V_{CE} = 30 Vdc, $V_{EB(off)}$ = 3.0 Vdc)		I_{CEX}	—	50	nAdc
Base Cutoff Current (V_{CE} = 30 Vdc, $V_{EB(off)}$ = 3.0 Vdc)		I_{BL}	—	50	nAdc
ON CHARACTERISTICS[1]					
DC Current Gain		h_{FE}			—
(I_C = 0.1 mAdc, V_{CE} = 1.0 Vdc)	2N3903		20	—	
	2N3904		40	—	
(I_C = 1.0 mAdc, V_{CE} = 1.0 Vdc)	2N3903		35	—	
	2N3904		70	—	
(I_C = 10 mAdc, V_{CE} = 1.0 Vdc)	2N3903		50	150	
	2N3904		100	300	
(I_C = 50 mAdc, V_{CE} = 1.0 Vdc)	2N3903		30	—	
	2N3904		60	—	
(I_C = 100 mAdc, V_{CE} = 1.0 Vdc)	2N3903		15	—	
	2N3904		30	—	
Collector-Emitter Saturation Voltage		$V_{CE(sat)}$			Vdc
(I_C = 10 mAdc, I_B = 1.0 mAdc)			—	0.2	
(I_C = 50 mAdc, I_B = 5.0 mAdc)			—	0.3	
Base-Emitter Saturation Voltage		$V_{BE(sat)}$			Vdc
(I_C = 10 mAdc, I_B = 1.0 mAdc)			0.65	0.85	
(I_C = 50 mAdc, I_B = 5.0 mAdc)			—	1.0	
SMALL-SIGNAL CHARACTERISTICS					
Current-Gain — Bandwidth Product		f_T			MHz
(I_C = 10 mAdc, V_{CE} = 20 Vdc, f = 100 MHz)	2N3903		150	—	
	2N3904		200	—	
Output Capacitance (V_{CB} = 5.0 Vdc, I_E = 0, f = 100 kHz)		C_{obo}	—	4.0	pF
Input Capacitance (V_{BE} = 0.5 Vdc, I_C = 0, f = 100 kHz)		C_{ibo}	—	8.0	pF
Input Impedance		h_{ie}			kΩ
(I_C = 1.0 mAdc, V_{CE} = 10 Vdc, f = 1.0 kHz)	2N3903		0.5	8.0	
	2N3904		1.0	10	
Voltage Feedback Ratio		h_{re}			X 10^{-4}
(I_C = 1.0 mAdc, V_{CE} = 10 Vdc, f = 1.0 kHz)	2N3903		0.1	5.0	
	2N3904		0.5	8.0	
Small-Signal Current Gain		h_{fe}			—
(I_C = 1.0 mAdc, V_{CE} = 10 Vdc, f = 1.0 kHz)	2N3903		50	200	
	2N3904		100	400	
Output Admittance (I_C = 1.0 mAdc, V_{CE} = 10 Vdc, f = 1.0 kHz)		h_{oe}	1.0	40	μmhos
Noise Figure (I_C = 100 μAdc, V_{CE} = 5.0 Vdc, R_S = 1.0 kΩ, f = 10 Hz to 15.7 kHz)	2N3903	NF	—	6.0	dB
	2N3904		—	5.0	
SWITCHING CHARACTERISTICS					
Delay Time	(V_{CC} = 3.0 Vdc, $V_{BE(off)}$ = 0.5 Vdc, I_C = 10 mAdc, I_{B1} = 1.0 mAdc)	t_d	—	35	ns
Rise Time		t_r	—	50	ns
Storage Time	2N3903	t_s	—	800	ns
	2N3904 (V_{CC} = 3.0 Vdc, I_C = 10 mAdc,		—	900	
Fall Time	I_{B1} = I_{B2} = 1.0 mAdc)	t_f	—	90	ns

(1) Pulse Test: Pulse Width ≤ 300 μs, Duty Cycle ≤ 2.0%.

Ⓜ **MOTOROLA** *Semiconductor Products Inc.*

GATE SYMBOLS	INPUTS		OUTPUT
	A	*B*	*X*
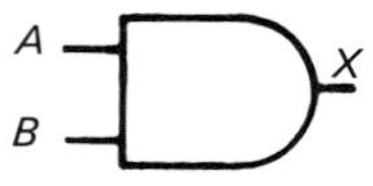 **AND**	0	0	0
	0	1	0
	1	0	0
	1	1	1
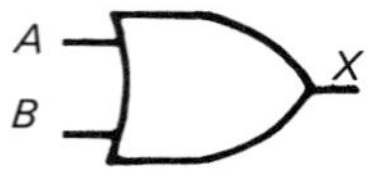 **OR**	0	0	0
	0	1	1
	1	0	1
	1	1	1
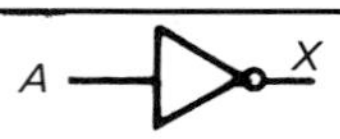 **INVERTER**	1		0
	0		1
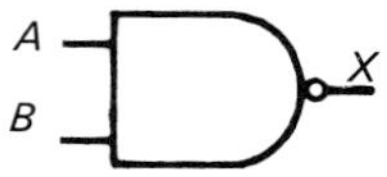 **NAND**	0	0	1
	0	1	1
	1	0	1
	1	1	0
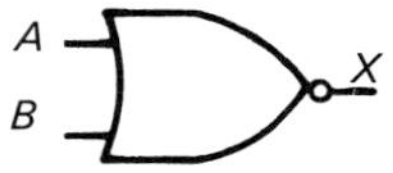 **NOR**	0	0	1
	0	1	0
	1	0	0
	1	1	0
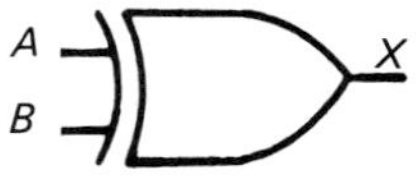 **EXCLUSIVE OR**	0	0	0
	0	1	1
	1	0	1
	1	1	0
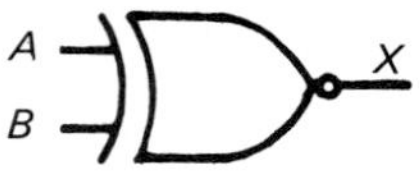 **EXCLUSIVE NOR**	0	0	1
	0	1	0
	1	0	0
	1	1	1

BOOLEAN ALGEGRA

= stands for equal
A is A
$\overline{A}$ is *not* A
A + B stands for A or B, *not* A plus B
A•B is the same as AB and stands for A *and* B

Appendix H   ETCHED CIRCUITS

Many projects in the text suggest the use of etched circuits. You should know how to make them. Copper conductive patterns are etched on thin copper clad plastic board.

There are three general methods of making etched circuits:

- The copper clad board is covered with a thin coating of light-sensitive photo chemical, similar to Kodak Photoresist. Then, the board is exposed to ultraviolet light, through a photographic mask. The board is developed in an etching bath that eats away the undesired parts of the thin copper, leaving only the conductive paths.
- The art of circuit design may be actually printed on the copper clad board with acid resisting ink. A silk screen process is used. After the ink is dry, the board is etched in the usual way.
- The conductive circuits may be painted on the copper clad board. After drying, the board is etched to remove the unwanted copper and leave only the painted circuit.

MAKING AN ETCHED CIRCUIT BOARD

1. Study the schematic diagram of the circuit, and place the components on a sheet of tracing paper. Use a colored pencil to carefully outline each resistor, capacitor, transistor, etc., in its proper place. Remove parts and label each outlined part with its component number such as R_3, C_1, etc. Using a pencil of another color, draw lines representing the wire connections between components. Check and recheck with the schematic drawing.
2. Carefully bend all leads of components downward, and determine the exact spacing between leads. At the point where each lead will be mounted, place a small circle, or pad, about 1/8″ to 1/4″ in diameter. Connect the pads together according to your outline with a heavy black line about 1/8″ wide. Make a neat drawing, using a ruler and appropriate curves.

NOTE: The size of the pads and width of conductive paths will depend on the circuit to be made. Study the photo in Figure H-1. Avoid crossover conductive paths. A rearrangement of components may eliminate crossovers.

3. Cut a piece of copper clad board the size of your circuit design. Using a 4H pencil and carbon paper, transfer the outline of the circuit to the copper clad. With a pointed pencil, mark all centers of holes to be drilled.
4. Drill all holes for component leads with a No. 40 drill. Enlarge with slightly larger drills if necessary. Remove all burrs.
5. Paint the circuit path, very neatly, with an acid-resist paint. Use a No. 2 artist brush or special pen with acid-resist ink. Allow time to dry.
6. Etch the board in a solution of 1.41 specific gravity ferric chloride ($FeCl_3$) for 1-2 hours. Do not leave in solution too long. Stir and turn the board frequently with an old pair of tweezers or tongs.

Figure H-1. A completed circuit board. Note the pads and conductive paths.

NOTE: Some shops warm the solution to hasten the etching.

7. Remove board, inspect, and wash with clear water.
8. Clean off resist paint with rag and lacquer thinner or kerosene.
9. Replace pattern and mark all wire holes. Drill with a No. 54 wire drill. Do not use center punch to mark holes.
10. Mount parts, solder, and clip off excess leads. Do not use excessive heat, but avoid cold solder joints.

Throughout this text, suggested circuit board layouts are supplied. You are encouraged to improve upon them. When mounting components such as a tuning capacitor or a potentiometer on panels, attach wires to board at a pad and run directly to the component.

LESSON IN SAFETY: Keep etchant from your clothes and skin. Etch in a shallow glass dish. Use proper ventilation.

Appendix I

ELECTRONIC PROJECT LAYOUT

In this book, there are many excellent projects for the student or experimenter. The appealing design of the project depends on careful planning and the wise use of many materials. The following suggestions will help you in the fabrication and design of your project.

- Many different chassis designs are available already built through electronic hobby stores and catalogs. However, you may wish to design your own chassis and fabricate it with sheet metal equipment.
- Chassis may be painted or covered with self-adhesive vinyl contact paper.
- Rub-on decals can be purchased from local hobby stores or office supply stores. In the case of a mistake, use transparent tape to lift the decal from the surface.
- Certain areas of the chassis can be highlighted by the use of thin chart tape. These come in solid colors and black, as well as striped designs. See Figure I-1.
- Attractive knobs may be purchased to complement your chassis design.
- Perforated metal or phenolic board may be used to cover a speaker where sound must exit.
- Adhesive felt or rubber feet help keep the chassis from scratching surfaces. Also, screw-on feet are available for larger chassis.
- Always use grommets where wires come through the chassis.
- Good tools are important in the construction and fabrication of electronic projects. Always use the correct tool for the job at hand.

Figure I-1. Some chart and graphic art tape styles.

Dictionary of Terms

abacus: An ancient calculating device still in use today. Made of a frame of rods, representing decimal place-value columns and beads that are used to represent digits.

ac component: The ac portion of a composite signal, which has an axis other than zero. The ac wave rides on the dc component.

ac generator: See alternator.

ac series motor: See universal motor.

acceptor circuit: Series circuit tuned to accept those signals near the resonant frequency of the circuit.

acceptor impurity: Impurity added to intrinsic semiconductor material to create holes for current carriers.

access time: The read time between the instant of calling for data from a storage device and the instant of completion of delivery. The write time between the instant of requesting storage of data and the instant of completion of storage.

active device: An electronic component that can change its response to an external signal. Active devices include transistors, ICs, and vacuum tubes.

active power: See true power.

address: The exact location of a word in a computer memory.

air-core inductor: Inductor wound on insulated form without metallic core. Self-supporting coil without core.

alignment: The adjustment of tuned circuits in amplifier and/or oscillator circuits so that they will produce a specified response at a given frequency.

alkaline cell: A primary cell using carbon and manganese dioxide for the positive electrode and powdered zinc as the negative electrode. Potassium hydroxide, a caustic alkali, is used for the electrolyte.

alnico: Special alloy used to make small permanent magnets.

alpha (α): Current gain of a transistor connected in common-base configuration. Temperature coefficient of resistance.

alternating current (ac): Current of electrons that moves first in one direction and then in the other.

alternator: Generator using slip rings and brushes to connect armature to external circuit. Output is alternating current. Also called ac generator.

American Wire Gauge (AWG): Used in sizing wire by numbers.

ammeter: Meter used to measure current. Meter leads are placed in series with desired circuit.

ampere (A): Unit of measurement of current, representing the flow of 1 coulomb of charge per second past a given point in a circuit.

ampere-hour (A·h): Capacity rating of batteries. A battery rated at 100 A·h will produce, theoretically, 100 amperes for 1 hour.

ampere-turn (A·t): Unit of measurement of magnetomotive force representing the product of amperes times number of turns in a coil of wire.

amplification: The production of an output signal that is greater than the input signal.

amplifier: A device capable of increasing the magnitude of a physical quantity, such as an electric current or mechanical force.

amplitude: Magnitude of maximum instantaneous value (positive or negative) of a periodic quantity.

amplitude distortion: Distortion resulting from nonlinear operation of transistor, when peaks of input signals are reduced or cut off by either excessive input signal or incorrect bias.

amplitude modulation (AM): Transmitter modulation of a signal by varying the strength (amplitude) of the RF carrier at audio rate.

analog: Pertains to the general class of devices or circuits in which the output varies as a continuous function of the input. For example, meters employing a D'Arsonval movement are analog meters.

analog computer: A computer that substitutes for any given physical quantity, a mechanical, electrical, or thermodynamic equivalent quantity that follows in direct proportion the same laws of behavior as the original quantity. In general, the analog computer processes continuous inputs and produces continuous outputs.

AND gate: A logic gate that produces an output only when all of its inputs are energized.

angle of lead (or lag): The angle by which one rotating vector, or ac quantity, leads (or lags) another of the same frequency. Example: In a capacitive circuit, the current *leads* the voltage by a certain angle.

angular velocity (ω): Speed of rotating vector in radians per second. Radian frequency.

anion: Negatively charged ion.

anode: Positive terminal.

antenna: Device for radiating or receiving radio waves.

antiphase: Said of two waveforms that are 180° out of phase.

apparent power (S): In a reactive circuit, the power that at any instant appears to be supplied to a circuit. It is the vector sum of true and reactive powers and is equal to the product of current and voltage. Units of measure are volt-amperes.

Arabic number system: See decimal number system.

arithmetic/logic unit (ALU): Part of a computer's CPU, it contains adders and registers and is where arithmetic and logic functions, such as comparing two numbers, are performed.

armature: Moving or stationary part of a motor or generator that includes the main current-carrying winding wherein an emf is induced. Moving part of a relay.

Armstrong oscillator: An oscillator using tickler coil for feedback.

ASCII: American Standard Code for Information Interchange. A system for encoding alphanumeric characters and other symbols as binary digits.

astable multivibrator: A multivibrator in which each active device alternately conducts and is cut off for intervals of time determined by circuit constants, without use of external triggers. Continuous oscillation. Also called free-running multivibrator.

AT-cut crystal: Crystal cut at a 35° angle with respect to the z-axis of the crystal.

atom: Smallest particle of an element that retains the properties of the element.

atomic number: The number of protons in the nucleus of a given atom.

atomic weight: The relative mass of an atom. For practical purposes, it is equal to the total number of protons and neutrons contained in an atom.

attenuation: Reduction in intensity or amplitude of a signal.

audio frequency (AF): The range of frequencies from 20 Hz to 20 kHz.

audio-frequency amplifier: Amplifier used to amplify audio frequencies of around 15 Hz to 20,000 Hz.

automatic gain control (AGC): Circuit employed to vary gain of amplifier in proportion to input signal strength so output remains at a constant level.

automatic volume control (AVC): A self-acting device that maintains the output of an amplifier or radio receiver constant while the input voltage varies over a wide range.

automation: Employment of automatic self-controlled machinery to replace human labor and control.

auxiliary storage: Device, such as magnetic disk and tape, for storing data in addition to main memory. Also called external storage.

average value: The value of a sine wave equal to the area under the curve of a half cycle divided by the distance along the *x*-axis between 0° and 180°. It is equal to 0.637 times peak value.

band: Group of adjacent frequencies in frequency spectrum.

bandpass filter: Filter circuit designed to pass currents of frequencies within a continuous band and reject or attenuate frequencies above or below the band.

band-reject filter: Filter circuit designed to reject currents in a continuous band of frequencies but pass frequencies above or below the band.

bandswitch: A switch used to select any one of the frequency bands in which an electronic unit may operate.

bandwidth (BW): Band of frequencies allowed for transmitting a modulated signal.

barrier potential: Potential difference across a pn junction due to diffusion of electrons and holes across the junction.

base: The thin section between the emitter and collector of a transistor.

battery: Several voltaic cells connected in series or parallel usually contained in one case.

beat frequency: The resultant frequency obtained by combining two frequencies.

beat-frequency oscillator (BFO): Oscillator whose output is beat with a continuous wave to produce beat frequency in audio range. Used in CW reception.

bel: Dimensionless unit of gain equal to a 10 to 1 ratio of power gain.

beta (β): Current gain of a transistor connected in common-emitter configuration.

bias: A dc voltage applied across the terminals of a transistor to establish the correct operating point.

binary number system: Number system having base of 2. Binary digits are digits 0 and 1.

binary-coded decimal (BCD) number system: Binary system of a sort, based upon a code.

bipolar junction transistor (BJT): Transistor consisting of a thin layer of n- or p-type semiconductor sandwiched between crystals of opposite type. Designated as npn or pnp.

bistable multivibrator: See flip-flop.

bit: A binary digit (0 or 1).

black box: Component containing an unknown and possibly complicated circuit wherein only input and output is of concern.

bleeder: Resistor connected across power supply to discharge filter capacitors.

Boolean algebra: A system for expressing combinations of logic statements—statements that can be either true or false.

bridge circuit: Network of four components connected in series to form a rectangle and a bridge connected across one of the diagonals. Frequently used in electronic measuring instruments.

bridge rectifier: Full-wave rectifier circuit employing four rectifiers in bridge configuration. Also called full-wave bridge rectifier.

brush: A piece of conductive material, usually carbon or graphite, that contacts slip rings or commutators for continuous electrical contact in rotating machines. Brushes provide stationary connection for external circuit.

buffer: A device in a computer for storing information temporarily during data transfers.

bus: An electrical pathway in which current flows.

bypass capacitor: Fixed capacitor that bypasses unwanted ac to ground.

byte: In computers, a string of 8 bits, or 2 nibbles.

cable: A stranded conductor or group of single conductors insulated from each other.

cable TV: A communication system that provides television signals through coaxial cable.

capacitance (C): The ability to store electric charge. Units of measure are farads.

capacitive coupling: Connecting amplifier stages by a capacitor resulting in loss of dc component, preventing dc bias of one stage from affecting dc bias of the next.

capacitive reactance (X_C): Opposition to ac as a result of capacitance. Units of measure are ohms.

capacitor: Device having property of capacitance. It is used to store charge. A simple capacitor consists of two metal plates separated by an insulator.

capacitor-input filter: A power-supply filter in which a capacitor is connected directly across the rectifier output.

capacity: Ability of battery to produce current over a given length of time. Capacity is measured in ampere-hours.

carrier: Usually an RF continuous wave to which modulation is applied. In semiconductors, conducting hole or electron.

carrier frequency: Frequency of transmitting station.

carry bit: A binary digit resulting from an addition that must be transferred to the next higher place value.

cascade: Arrangements of amplifiers where output of one stage becomes input of next, throughout series of stages.

cathode: Negative terminal.

cathode-ray tube (CRT): Electron tube in which electrons emitted from cathode are shaped into a narrow beam and accelerated to high velocity before striking a phosphor-coated viewing screen.

cation: Positively charged ion.

CD ROM: Auxiliary storage medium on compact disc.

center frequency: Frequency of transmitted carrier wave in FM when no modulation is applied.

center tap (CT): Connection made to center of coil.

central computer unit: Part of a computer; comprised of central processing unit and memory.

central processing unit (CPU): The section of a computer that directs, or controls, the major operations in a computer. It is basically comprised of an arithmetic/logic unit and a control unit. CPUs are contained on a microprocessor.

characteristic curve: Graphical representation of characteristics of a circuit or device.

chip: A very small piece of semiconductor material containing transistors, diodes, resistors, capacitors, etc.

choke coil: A high inductance coil used to prevent the passage of pulsating currents but allow steady dc to pass.

circuit: An electrical network consisting of a pathway for current, or conductor, and any other components of the network.

circuit breaker: Safety device that automatically opens circuit if overloaded.

circular mil (cmil): A unit of area used especially for the cross section of wire equal to the area of a circle having a diameter of 1 mil.

circulating current: Inductive and capacitive currents flowing around in a parallel circuit.

citizens band (CB): A band of frequencies allotted to two-way radio communications by private citizens. Operators are not required to pass technical examinations.

class-A amplifier: An amplifier biased so that current flows during entire cycle of input signal.

class-AB amplifier: An amplifier biased somewhere between class A and class B.

class-B amplifier: An amplifier biased so that current flows for approximately one half the cycle of input signal.

class-C amplifier: An amplifier biased so that current flows for much less than one half of each cycle of applied input.

clock: In a computer, an electronic device, usually based on a quartz crystal, that gives off regular pulses used to coordinate operation.

closed circuit: A complete circuit through which current may flow when voltage is applied.

CMOS logic: A transistor logic family using complementary metal oxide semiconductors. Has wide tolerances on supply voltage and low power consumption; prone to damage from electrostatic discharge during handling.

coherent light: Radiant electromagnetic energy of the same, or almost the same, wavelength, and with definite phase relationships between different points in the field.

collector: In a transistor, the semiconductor section that collects the majority carriers.

collector efficiency: The amount of power drawn from the supply that is actually delivered to the load. It is the ratio of ac output power to dc input power.

Colpitts oscillator: An oscillator using a split-stator variable capacitor as feedback circuit.

combination circuit: Circuit that contains elements of both series and parallel circuits. Also called series-parallel circuit.

common base (CB): Transistor circuit in which base is common to input and output circuits.

common collector (CC): Transistor circuit in which collector is common to input and output circuits.

common emitter (CE): Transistor circuit in which emitter is common to input and output circuits.

commutation: The process of changing the alternating current in a generator armature into direct current in the external circuit by a mechanical switch consisting of commutator bars and brushes.

commutator: Group of bars providing connections between armature coils and brushes in dc motors and generators. Mechanical switch to maintain current in one direction in external circuit.

compact disc (CD): A type of optical disk.

complementary-symmetry amplifier: An amplifier consisting of an arrangement of pnp and npn transistors that provides push-pull operation from one input signal.

composite signal: The signal resulting from the total of ac and dc components.

compound generator: Generator that uses both series and shunt windings.

compound motor: A dc motor that uses both series and parallel field coils.

computer: A machine that is capable of accepting data, processing it according to a stored program, and outputting the results.

computer architecture: The design of a total computer system, including its hardware and operating system software. Also called von Neumann architecture.

computer language: Actual set of representations or instructions fed to a computer to make it do certain things.

computer-on-a-chip: See single-chip microcomputer.

conductance (G): A measure of the ability of a circuit to conduct current. Units of measure are siemens.

conduction band: Outermost energy level of atom.

conductivity (σ): The ability of a material to conduct an electric current. It is the reciprocal of resistivity.

conductor: Pathway for carrying electric current. Material that permits free motion of a large number of electrons.

cone: A diaphragm that sets air in motion to create a sound wave in a loudspeaker; usually conical in shape.

continuous wave (CW): Uninterrupted sinusoidal RF wave radiated into space with all wave peaks equal in amplitude and evenly spaced along time axis.

control grid: Grid in electron tube closest to cathode that serves to control the anode current. Grid to which input signal is fed to tube.

control unit: Part of a computer's CPU, it takes instructions from input devices and moves them into memory and generates the signals necessary to make the computer accomplish a given task.

coordinates: Horizontal and vertical positions giving location of a point on a graph.

copper loss: Heat loss in motors, generators, and transformers as a result of wire resistance. Also called I^2R loss.

core iron: Magnetic materials usually in sheet form used to form laminated cores for chokes, transformers, and electromagnets.

core storage: An assembly of magnetic cores, or ferrite toroids, where bits are stored in terms of the direction of magnetization.

coulomb (C): Quantity of charge where 1 C has the equivalent charge of 6.24×10^{18} electrons.

counter emf (cemf): Voltage induced in an inductive circuit by a changing current, the polarity of which is at each instant opposite that of the applied voltage.

coupling: Percentage of mutual inductance between coils.

coupling coefficient (k): Ratio of flux coupling two coils to the total flux produced by the first coil.

covalent bond: Atoms joined together sharing electrons, forming a stable molecule.

crossover network: The network designed to divide audio frequencies into bands for distribution to speakers.

cryogenics: The subject of physical phenomena at temperatures below $-50°C$.

crystal detector: A type of detector that uses the rectification characteristics of a crystal substance such as galena, silicon, germanium, and iron pyrite.

crystal diode: See point-contact diode.

crystal lattice: Structure of material when outer electrons are joined in covalent bond.

crystal oscillator: An oscillator controlled by the piezoelectric effect.

crystal pickup: Phonograph cartridge that produces an electrical output from piezoelectric effect. Also called crystal cartridge.

crystal rectifier: See point-contact diode.

current (I): The movement of electrons through a conductor. Units of measure are amperes.

cutoff: The condition when the EB junction of a transistor has zero bias or is reverse biased, and there is no collector current.

cutoff frequency: The frequency at which the gain of an amplifier drops to 0.707 of maximum gain.

cybernetics: The study of complex electronic computer systems and their relationship to the human brain.

cycle: Set of events occurring in sequence. One 360° rotation of a phasor, or the change of an alternating wave from zero to positive maximum, back through zero to negative maximum, and back to zero.

damped wave: A wave in which successive oscillations decrease in amplitude.

damping: Gradual decrease in amplitude of oscillations in tuned circuit due to energy dissipated in resistance.

D'Arsonval movement: A meter movement used in some analog meters that works by the reaction between current through a moving coil and a fixed, permanent magnet.

data: Computer input and output information; especially, numerical and qualitative values as opposed to programs.

dBm: Abbreviation for decibels above (or below) 1 mW. Loss or gain in reference to arbitrary power level of 1 mW.

dc component: The dc portion of a composite signal equal to the *y*-value of the axis of the ac wave. The ac wave rides on the dc component.

dc generator: Generator with connections to armature through a commutator. Output is direct current.

DE MOSFET: Depletion-enhancement metal-oxide semiconductor field-effect transistor. MOSFET with both depletion and enhancement modes.

debugging: Process of correcting mistakes in a computer program.

decay: Term used to express gradual decrease in values of current and voltage.

decay time: The time taken by a quantity to decay to a stated fraction of its initial value.

decibel: One-tenth of a bel. A unit used to express the relative increase or decrease in power. Unit used to express gain or loss in a circuit.

decimal number system: A number system, having a base of 10. Digits are 0 through 9. Most popular number system in use. Also called Arabic number system.

declination: Angle between true north and magnetic north.

decoder: A combinational circuit that converts binary data into some other number system.

deflection: Deviation from zero of needle in meter. Movement or bending of an electron beam.

deflection angle: Maximum angle of deflection of electron beam in TV picture tube.

degeneration: See negative feedback.

demodulation: Process of removing modulating signal intelligence from carrier wave in radio receiver. Also called detection.

depletion mode: Operation of a field-effect transistor in which current flows when the gate-source voltage is zero, and is increased or decreased by altering the gate-source voltage.

depletion region: Region near pn junction void of current carriers. Region in which mobile carrier charge density is insufficient to neutralize net fixed charge of donors and acceptors. Also called transition layer, or space-charge region.

depolarizer: Chemical agent, rich in oxygen, introduced into a cell to minimize polarization.

detection: See demodulation.

deviation ratio: The ratio between the maximum frequency deviation in FM and the highest modulating frequency.

diac: A two-terminal, three-layer, bidirectional solid-state device. Acts much like two zener diodes that are series connected in opposite directions.

diaphragm: Thin disk used in an earphone for producing sound.

dielectric: Insulation material between plates of a capacitor.

dielectric constant: See relative permittivity.

difference frequency: A signal equal to the difference between the frequencies of two oscillations. In a superheterodyne receiver, all tuned incoming signals are converted to this single frequency. Also called intermediate frequency.

diffusion: Mechanism of current caused by movement of carriers from regions of high concentration to low concentration and not to the presence of an electric field.

digital: Pertains to the general class of devices or circuits in which the output varies in discrete steps. Instruments of this type have a digital readout.

digital computer: A computer that operates with discrete numbers expressed directly as digits, as opposed to the continuously variable quantities of the analog computer.

digital integrated circuit: An IC that operates like a switch to give binary output (0 or 1).

digital readout: An output display, generally thought of as giving information in the form of digits.

diode: Unidirectional, two-element device containing a cathode and anode mainly employed as a rectifier.

diode detector: A detector that uses crystal or electron-tube diodes as the rectifier of an RF signal.

DIP switch: A microminiature switch made in an in-line assembly. A number of switches are contained in one assembly. DIP switches are used, for example, in computers for adopting peripherals such as printers.

direct coupling: Connecting amplifier stages directly without use of coupling component. Amplification is achieved without loss of dc component.

direct current (dc): A steady or pulsating current of one polarity that flows in only one direction.

direct current working voltage (DCWV): The maximum continuously applied dc voltage for which a capacitor is rated.

distortion: The deviations in amplitude, frequency, and phase between input and output signals of amplifier or system.

distributed capacitance: Any capacitance not concentrated within a capacitor, such as that resulting from adjacent conductors or turns on coils. Also called self-capacitance.

domain: In magnetic theory, that region of a magnetic material in which the spontaneous magnetization is all in one direction.

domain theory: Theory of magnetism assuming that atomic magnets produced by movement of planetary electrons around nucleus have a strong tendency to line up together in groups. These groups are called domains.

donor impurity: Impurity added to intrinsic semiconductor material to generate electron carriers.

dopant: Donor or acceptor impurity. Added to intrinsic semiconductor to enhance certain electrical properties.

doping: Adding impurities to semiconductor material.

drift: Mechanism of current caused by motion of a charged particle in an electric field.

drive unit: Mechanism used to carry out reading and writing of auxiliary storage files.

dry cell: A voltaic cell having an immobilized electrolyte. A carbon-zinc cell, composed of a zinc case, carbon positive electrode, and electrolytic paste of ground carbon, manganese dioxide, and ammonium chloride, is a common dry cell.

DTL logic: Diode-transistor logic. A transistor logic family using diodes, resistors, and transistors for its logic circuits.

dynamic characteristic: See load characteristic.

dynamic resistance: Incremental resistance measured over a relatively small portion of the operating characteristic of a device.

dynamic speaker: A speaker comprised basically of a moving coil in a magnetic field and attached cone. Current through the voice coil produces a fluctuationg field that interacts with the magnetic field, causing the cone to move back and forth, which produces sound.

E MOSFET: Enhancement-only metal-oxide semiconductor field-effect transistor. MOSFET with no depletion mode.

eddy currents: Currents induced in the body of a conducting mass by a variation in magnetic flux.

eddy-current loss: Heat loss resulting from eddy currents flowing in a magnetic core.

Edison effect: Effect, first noticed by Thomas Edison, that emitted electrons were attracted to the positive plate in a vacuum tube.

effective value: That value of an ac waveform that has the equivalent heating effect of a direct current. It is equal to 0.707 times peak value. Also called the root-mean-square (rms) value.

efficiency (η): Ratio between output power and input power. In general, ratio of output to input.

electric potential: Electric potential energy per unit charge. Units of measure are volts.

electrically erasable programmable read-only memory (EEPROM): An alterable PROM, which can be quickly erased by discharging voltage and reprogrammed.

electrode: Conductor by which an electric current enters or leaves an electrolyte, electron tube, or semiconductor device.

electrodynamic speaker: Dynamic speaker that uses electromagnetic fixed field.

electrolyte: Acid solution in a cell. A substance in which the conduction of electricity is accompanied by chemical action.

electrolytic capacitor: Capacitor with positive plate of aluminum and negative plate of dry-paste or liquid electrolyte. Dielectric is thin coat of oxide on aluminum plate.

electromagnet: A coil wound on a soft iron core that becomes a temporary magnet whenever current flows through the coil.

electromagnetic radiation: Oscillation of an electric charge, produced by electric and magnetic fields. The resulting wave moves through space at the speed of light.

electromagnetic wave: The radiant energy produced by electromagnetic radiation, including radio waves, light waves, x-rays, etc.

electromotive force (emf): The potential difference that exists across the terminals of a voltage source when no charge flows to an external circuit. Units of measure are volts.

electron: Negatively charged particle of an atom. Majority carrier in an n-type semiconductor material.

electron tube: An evacuated or gas-filled, metal or glass shell, which encloses several elements.

electron volt: A measure of energy. It represents the energy acquired by an electron while passing through a potential of 1 V.

electron-coupled oscillator (ECO): Combination oscillator and power amplifier using electron stream as coupling medium between grid and plate tank circuits.

electronic: Pertaining to vacuum tube or semiconductor devices and circuits. Pertaining to electron flow through a conductor.

electronics: Study, control, and application of electricity.

electrostatic field: Force field around a charged body in which the influence of the body is felt.

element: A substance made up of only like atoms. One of the distinct kinds of substances that either singly, or in combination with other elements, makes up all matter in the universe.

embedded computer: Computer that controls the operation of machines. A type of microcomputer.

emission: Escape of electrons from a surface.

emitter: In a transistor, the semiconductor section from where majority carriers, which become minority carriers in the base section, are emitted.

encoder: A combinational circuit that converts data from some other number system into binary.

energy: That which is capable of producing work.

energy gap: See forbidden band.

enhancement mode: Operation of a field-effect transistor in which no current flows when zero gate voltage is applied, and increasing the gate voltage increases the current.

epitaxy: The physical placement of material on a surface.

erasable programmable read-only memory (EPROM): An alterable PROM, which can be erased by ultraviolet light and reprogrammed.

etched circuit: A type of printed circuit in which conduction paths are formed by coating copper-clad insulation with acid resist, then, placing the board in an acid bath. This eats away the unprotected parts of the copper clad leaving the circuit conductors. Components are mounted and soldered between the conductors to form the completed circuit.

excitation: Application of a signal to an amplifier circuit. Application of energy to an antenna system. Application of current to energize the field windings of a generator.

exciting current: The current that flows through the primary winding of a transformer when no loads are connected to the secondary. This current establishes the magnetic field in the core and furnishes energy for the no-load power losses in the core. Also called no-load current. The current through the field windings of a generator or motor. Also called magnetizing current.

execute: Usually, to run a compiled or assembled program on the computer. To compile or assemble and run a source program.

execute statement: A program statement that indicates the beginning of a job statement in a job control language.

execution cycle: In a computer, the time during which an elementary operation takes place.

external storage: See auxiliary storage.

extremely high frequency (EHF): Frequency band from 30,000 MHz – 300,000 MHz.

extrinsic semiconductor: A semiconductor that depends upon impurities for its electrical properties.

farad (F): Unit of measurement of capacitance. A capacitor has a capacitance of 1 F when a charge of 1 C raises its potential 1 V.

Federal Communications Commission (FCC): A federal board having the power to regulate all electronic communication systems originating in the United States.

feedback: Returning a signal from output back to the circuit input.

fetch: To locate and load into main memory a requested load module, relocating it as necessary and leaving it in a ready-to-execute condition.

fetch cycle: The period during which a machine language instruction is read from memory into the control section of the central processing unit.

field magnet: Electromagnet or permanent magnet that produces a strong magnetic field in a speaker, microphone, motor, generator, or other electrical device.

field-effect transistor (FET): A semiconductor device in which the resistance between source and drain terminals depends on a field produced by a voltage applied to a gate terminal. Types include JFETs and MOSFETs.

figure of merit: See quality factor.

film resistor: Resistor made from depositing resistive film on a ceramic tube with caps that slip over the tube ends to form leads.

filter: Circuit used to separate ac and dc signals, blocking dc while passing ac or preserving dc while removing ac, or used to separate a certain band or certain bands of frequencies.

fixed disk: Built in disk of a disk drive unit, having a magnetic coating for data storage; an auxiliary storage device. Also called hard disk.

flip-flop: A multivibrator circuit having two stable (bistable) states, and the signal is switched back and forth between them. Also called bistable multivibrator.

floppy disk: A flexible disk with magnetic oxide on its surface that is used with computers for auxiliary storage of data or information.

flyback: The shorter of two time intervals comprising a sawtooth wave. The return of the spot of a cathode-ray tube to its starting point after having reached the end of its trace. Also called retrace.

foot-pound: U.S. Customary unit of work.

forbidden band: Energy range between the valence band and conduction band in a semiconductor. Also called energy gap.

format: To prepare a floppy disk by means of computer for storage of information.

forward bias: Connection of potential to produce current across pn junction (i.e., positive of source to p-type semiconductor material, negative to n-type). A forward bias reduces the width of the depletion region and reduces the junction potential.

forward voltage: Value of the voltage drop across a diode when current is in the forward direction. This voltage remains roughly equal to the barrier potential. Also called forward voltage drop.

forward voltage drop: See forward voltage.

Foster-Seeley discriminator: Discriminator circuit used in detection of FM.

free electrons: Electrons that have broken loose from their atomic structures, and which are not bound to a particular atom but circulate among the atoms of a substance.

free-running multivibrator: See astable multivibrator.

frequency band: The frequencies lying between upper and lower limits. Standard frequency bands are: very low, low, medium, high, very high, ultra high, super high, and extremely high frequencies.

frequency departure: Instantaneous change from center frequency in FM as a result of modulation.

frequency deviation: Maximum departure from center frequency at the peak of the modulating signal.

frequency distortion: Distortion resulting from signals of some frequencies being amplified more than others or when some frequencies are excluded.

frequency doubler: An electronic stage having a resonant output circuit tuned to the second harmonic of the input frequency. The output signal will have twice the frequency of the input signal.

frequency (f): Number of times a periodic waveform repeats itself in a unit time. Units of measure are hertz.

frequency modulation (FM): Modulating transmitter by varying frequency of RF carrier wave at an audio rate.

frequency response: Rating of a device indicating its ability to operate over a specified range of frequencies.

frequency swing: The total frequency swing from maximum to minimum. It is equal to twice the deviation.

full adder: An electronic circuit for adding two bits and a third carry bit from a previous stage.

full-wave bridge rectifier: See bridge rectifier.

full-wave rectifier: Rectifier circuit that produces a dc pulse output for each half cycle of applied alternating current.

fundamental frequency: Principal component of a wave. Component with the lowest frequency or greatest amplitude. Component tone of lowest pitch in complex tone.

fuse: Safety protective device that opens an electric circuit if overloaded. Current above rating will melt fusible link.

gain: The amount of amplification in a circuit. The ratios of current, voltage, or power outputs to respective inputs.

gallium arsenide (GaAs): A semiconductor material used in the construction of high-speed transistors and integrated circuits.

galvanometer: Instrument for measuring very small current values.

gamma (γ): Current gain of a transistor connected in common-collector configuration.

gate: A circuit that gives a discrete output only when a predetermined set of discrete input conditions are met. One of the terminals of a FET, an SCR, or a triac.

gauss (G): Cgs unit of flux density.

generator: Rotating electric machine that converts mechanical energy into electrical energy.

germanium: Tetravalent, metallic element used in semiconductors.

gilbert (Gb): Cgs unit of magnetomotive force.

grid: Mesh of fine wire placed between cathode and plate of an electron tube.

grid bias: Voltage between the grid and cathode of an electron tube, usually negative.

grid current: Current flowing in grid circuit of electron tube, when grid is driven positive.

grid-dip meter: A test instrument for measuring resonant frequencies, detecting harmonics, and checking relative field strength of signals.

hacker: A person who uses a computer for nonconstructive purposes or without proper authorization.

half adder: An electronic circuit for adding two bits but not a carry bit.

half-wave rectifier: Rectifier that permits one-half of an alternating current cycle to pass and rejects reverse current of remaining half-cycle. Its output is pulsating dc.

hardware: Physical components of a computer system.

harmonic frequency: Frequency that is a multiple of fundamental frequency. For example, if fundamental frequency is 1000 kHz, then second harmonic is 2 × 1000 kHz, or 2000 kHz; third harmonic is 3 × 1000 kHz, or 3000 kHz; and so on.

Hartley oscillator: An oscillator using inductive coupling of tapped tank coil for feedback.

heat sink: Mass of metal for mounting electronic components. Used for carrying heat away from component.

heater: Resistance heating element used to heat cathode in a vacuum tube.

henry (H): Unit of measurement for inductance. A coil has 1 H of inductance if an emf on 1 V is induced when current through the inductor is changing at a rate of 1 A/s.

hertz (Hz): Unit of measurement for frequency, where 1 Hz equals 1 cps (cycle per second). Named in honor of Heinrich Hertz, who discovered radio waves.

heterodyne: Combine two signals of different frequencies to obtain the difference frequency.

hexadecimal number system: A number system having a base of 16. Convenient for representing 4-bit numbers. Digits are 0 through 9 and A through F.

high frequency (HF): Frequency band from 3 MHz – 30 MHz.

hole: Positive charge carrier. A space left by a removed electron. Majority carrier in a p-type semiconductor material.

hole injection: Creation of holes in n-type semiconductor material by removal of electrons by strong electric field around a point contact.

horsepower (HP): Unit of power where 1 HP is equal to 33,000 ft-lb of work per minute, 550 ft-lb per second, or 746 watts.

hum: Form of distortion introduced in an amplifier as a result of coupling to stray electromagnetic and electrostatic fields or insufficient filtering.

hydrometer: Bulb-type instrument used to measure specific gravity of a liquid.

hysteresis: Property of a magnetic substance that causes magnetization to lag behind force that produces it.

hysteresis loss: Energy loss in a magnetic material as molecules or domains move through cycle of magnetization. Loss due to molecular friction.

I^2R loss: See copper loss.

impedance diagram: Vector diagram represented as a right triangle, where x and y components are resistance and reactance and resultant (hypotenuse) is impedance. Also called impedance triangle.

impedance matching: The matching of two different impedances to obtain maximum transfer of power. Sometimes called Z-matching.

impedance triangle: See impedance diagram.

impedance (Z): The sum of resistance and reactive components in an ac circuit representing the total resistance in that circuit.

impurity: Atoms within a crystalline solid that are foreign to the crystal. See acceptor impurity and donor impurity.

in phase: Said of multiple waveforms of the same frequency when they pass through their maximum and minimum values at the same instant with the same polarity.

induced current: Current that flows as a result of an induced voltage.

induced emf: An emf resulting from the motion of a conductor through a magnetic field, or from a change in the magnetic flux as it moves through a conductor.

induced voltage: Voltage induced in a conductor as it moves through a magnetic field.

inductance (I): Inherent property of an electric circuit that opposes a change in current. Property of a circuit whereby energy may be stored in a magnetic field. Units of measure are henrys.

induction motor: An ac motor operating on principle of a rotating field. Rotor has no electrical connections but receives energy by transformer action from field windings. Motor torque is developed by interaction of rotor current and magnetic field.

inductive circuit: Circuit in which an appreciable emf is induced while current is changing.

inductive reactance (X_L): Opposition to an ac current as a result of inductance. Units of measure are ohms.

inductor: A coil or component with the properties of inductance.

infrasonic frequency: A frequency below the audio-frequency range. Also called subsonic frequency.

initialize: Adjust computer environment to starting configuration by the setting of counters, switches, and addresses to zero.

input: The information fed into a computer. The process of entering information into a computer.

input device: A computer device for entering information into a computer. Examples are keyboards and mice.

input impedance: Impedance offered to an input signal in an input circuit as "seen by" an ac source.

input/output (I/O): Pertaining to all equipment used to transfer information into or out of a computer.

input/output unit: See peripheral equipment.

instantaneous value: The magnitude, at any given instant, of a varying value.

insulator: A substance containing very few free electrons and requiring large amounts of energy to break electrons loose from influence of nucleus. Used, when desired, to prevent current.

integrated circuit (IC): A concentration of transistors, diodes, resistors, and capacitors in a microminiature chip.

interelectrode capacitance: The capacitance between metal elements in an electron tube.

intermediate frequency (IF): See difference frequency.

intermediate-frequency amplifier: Amplifier used to amplify intermediate frequencies.

internal resistance (R_i): The resistance found within a source voltage. A battery or generator has internal resistance. It is represented schematically as a resistor in series with the source.

interrupted continuous wave (ICW): Continuous wave radiated by keying a transmitter into long and short pulses of energy (dashes and dots) conforming to code, such as Morse Code.

interstage: Existing between stages, such as an interstage transformer between two stages of amplifiers.

intrinsic semiconductor: A pure semiconductor without any doping.

inverter: A circuit whose purpose is to produce an output signal that is inverted from the input signal—output positive when input is negative (phase inverter), output 1 when input 0, etc. A NOT gate.

ion: An electrically charged atom with a surplus (negative ion) or deficit (positive ion) of electrons.

ionization: The process by which a neutral atom loses or gains electrons, thereby acquiring a net charge and becoming an ion. Occurs as a result of the dissociation of the atoms of a molecule in solution or of a gas in an electric field.

ionization potential: The energy, expressed in electron volts, needed to remove an electron from a neutral atom or molecule.

ionosphere: Atmospheric layer from 40 to 350 miles above Earth, containing a high number of positive and negative ions.

isolation transformer: Transformer with a one-to-one turns ratio.

JFET: Junction field-effect transistor. Semiconductor device in which output current is controlled by voltage on the input.

joule (J): SI unit of work, where 1 joule is equal to 1 newton-meter.

junction diode: A diode having a pn junction.

junction potential: Potential across a pn junction equal to the difference between the applied voltage and the barrier potential.

key: Manually operated switch used to interrupt RF radiation of transmitter.

keying: Process of causing CW transmitter to radiate an RF signal when key contacts are closed.

kilo (k): Prefix meaning one thousand times.

kilowatt-hour (kWh): Unit of energy equal to 1000 watts of power supplied for 1 hour. Power for home and industrial usage is priced by the kilowatt-hour.

Kirchhoff's current law: At any junction of conductors in a circuit, the algebraic sum of currents is zero.

Kirchhoff's voltage law: The algebraic sum of voltages around a circuit is zero.

laminations: Thin sheets of steel used in cores of transformers, motors, and generators.

large scale integration (LSI): An integrated circuit with 100 to 999 gates.

large-signal amplifier: See power amplifier.

laser: A device that emits a narrow, intense beam of coherent light achieved through light amplification by stimulated emission of radiation.

laser beam: Beam of light emitted from a laser.

law of magnetism: Law stating that unlike poles attract and like poles repel.

LC filter: Filter employing a capacitor and an inductor, where capacitor is in parallel with and inductor is in series with the load. It is used for reducing ripple.

lead-acid cell: Secondary cell that uses lead peroxide and spongy lead for plates and sulfuric acid and water for electrolyte. Part of a lead-acid storage battery.

lead-acid storage battery: Battery commonly used in automobiles.

left-hand rule: Gives correct directional relationships for straight conductors, coils, and conductors in generator armatures based on electron flow.

Lenz's law: Induced emf in any circuit is always in such a direction as to oppose the effect that produces it.

light-emitting diode (LED): A two-element junction diode that gives off light when biased in the forward direction.

linear: Having an output that rises or falls in direct proportion to the input.

linear amplifier: An amplifier whose output is an amplified replica of the input.

linear detector: Detector using linear portions of characteristic curve on both sides of knee. Output is proportional to input signal.

linear device: Electronic device having a current-voltage relationship that is given by a straight line. A device in which a change in input produces a proportional change in output.

linear integrated circuit: Amplifier-type IC that controls varying voltages.

lines of force: Graphic representation of electrostatic and magnetic fields showing direction and magnetic flux.

lithium battery: A primary cell using lithium metal for one electrode. Has a relatively long life.

load: Resistance connected across a circuit that determines current and energy used.

load characteristic: Relation between the instantaneous values of a pair of variables such as electrode voltage and electrode current, when all dc supply voltages are constant. Also called dynamic characteristic.

load module: A program in a form suitable for loading into memory and executing.

loading a circuit: Effect of connecting a voltmeter across a circuit. Meter will draw current and effective resistance of circuit is lowered.

local oscillator: Oscillator in superheterodyne receiver, the output of which is mixed with an incoming signal to produce the intermediate frequency.

lodestone: See natural magnet.

logic function: One of the Boolean algebraic functions, such as AND, OR, and NOT.

loudspeaker: Device used to convert electrical energy into sound energy.

low frequency (LF): Frequency band from 30 kHz—300 kHz.

lower sideband: Frequency equal to difference between carrier and modulating frequencies.

L-pad: A combination of two variable resistors, one in series and the other across a load, used to vary the output of an audio system and match impedances.

magnet: Substance that has the property of magnetism.

magnetic amplifier: Transformer-type device employing a dc control winding. Control current produces more or less magnetic core saturation, thus varying output voltage of amplifier. It is used to obtain power gain.

magnetic circuit: Complete path through which magnetic lines of force may be established under influence of magnetizing force.

magnetic disk: A flat circular plate with a magnetizable surface on which data can be stored by selective magnetization of portions of the flat surface; an auxiliary storage device, such as a fixed or floppy disk.

magnetic field: Space surrounding a magnet or electromagnet where magnetic forces can be detected.

magnetic field intensity (H): Magnetomotive force per unit length of a magnetic circuit. Cgs units are oersteds. Also called magnetic field strength, or magnetizing force.

magnetic field strength: See magnetic field intensity.

magnetic flux density (B): Magnetic flux per unit cross-sectional area.

magnetic flux (Φ): Total lines of force in a magnetic field.

magnetic lines of force: Invisible lines used to represent a magnetic field. Lines run in a continuous loop through space, directed from north to south pole, and returning through the magnet. The magnitude of the magnetic flux is indicated by the number of lines.

magnetic materials: Materials, such as iron, steel, nickel, and cobalt, that are attracted to a magnet.

magnetic pickup: Phonograph cartridge that produces an electrical output from armature in magnetic field. Armature is mechanically connected to reproducing stylus. Also called magnetic cartridge.

magnetic poles: Points of maximum attraction on a magnet, designated as north and south poles.

magnetizing current: See exciting current.

magnetizing force: See magnetic field intensity.

magnetomotive force (mmf): Force that produces flux in a magnetic circuit.

magnetostriction: The effect of a change in dimension of certain elements when placed in a magnetic field.

main memory: The internal, high-speed, working storage of a computer, as opposed to auxiliary storage.

mainframe computer: Large computer with capability of serving hundreds of users at one time.

majority carrier: The predominate charge carrier in a semiconductor. Electrons are the majority carriers in n-type semiconductors. Holes are the majority carriers in p-type semiconductors.

matter: Anything that takes up space and has mass.

maximum power transfer: Condition that exists when resistance of load equals internal resistance of source.

maximum value: See peak value.

maxwell (Mx): Cgs unit of magnetic flux, where 1 maxwell equals 10^{-8} weber.

medium frequency (MF): Frequency band from 300 kHz—3000 kHz.

medium scale integration (MSI): An integrated circuit with 12 to 99 gates.

mega (M): Prefix meaning one million times.

memory: Part of computer where data is stored; especially, the internal, high-speed working storage of a computer, as opposed to auxiliary storage.

memory capacity: The total number of discrete units of information that can be stored in an electronic memory device; units given bytes.

metal-oxide semiconductor (MOS): A semiconductor with an insulating layer of oxide material. Material used in ICs—allows for high component density.

mho: Former unit of conductance now called siemen.

mica capacitor: Capacitor made of metal foil plates separated by sheets of mica.

micro (μ): Prefix meaning one millionth of.

microcomputer: Smaller, stand-alone as opposed to networked, computer. Personal computer or embedded computer. Uses a microprocessor for CPU.

microphone: Energy transducer that changes sound energy into electrical energy.

microprocessor: A central processing unit on one integrated circuit chip.

midrange: A speaker designed to handle the middle range of audio frequencies.

mil: One-thousandth of an inch.

mil-foot: Unit length of wire measuring 1 mil in diameter and 1 foot long.

milli (m): Prefix meaning one-thousandth of.

milliammeter: Meter that measures currents in milliampere range.

minicomputer: Midsize computer used where networking capability is desired.

minority carrier: The less predominant charge carrier in a semiconductor. Holes are minority carriers in n-type semiconductors and electrons are in p-type semiconductors.

mismatch: Incorrect impedance matching of load to source.

mobility: The ease with which current carriers move through a semiconductor. Units are cm²/V·s.

modem: A device that modifies computer data in a way that allows transmission and reception over telephone lines.

modulation: Process by which amplitude or frequency of sine wave signal is made to vary according to variations of another signal called a modulation signal.

modulation product: Sideband frequencies resulting from modulation of a radio wave.

module: A functional assembly of wired electronic components—a discrete electronic building block.

molecule: A chemically bonded group of two or more of the same or different atoms. A neutral group of atoms that act as a unit.

monostable multivibrator: A multivibrator with one stable state and one unstable state; a trigger signal is required to drive the unit into the unstable state, where it remains for a predetermined time before returning to stable state. Also called one-shot multivibrator.

MOSFET: Metal-oxide semiconductor field-effect transistor. A field-effect transistor having a gate that is insulated from the semiconductor substrate by a thin layer of silicon dioxide. Has no pn junction structure. Types include E MOSFETs and DE MOSFETs.

motor: Rotating electric machine that converts electrical energy into mechanical energy.

movement: The moving coil of an electromechanical meter.

moving coil: Coil suspended or pivoted in a magnetic field.

multimeter: A combination voltmeter, milliammeter, and ohmmeter. Also called a VOM (volt-ohm-milliammeter).

multiplex: Transmit several messages in one or both directions over a single transmission path. Used in radio and telephone communications, for example.

multiplier: Resistance connected in series with a meter movement to increase its voltage range.

mutual inductance (*M*): The property that exists between two conductors when a changing magnetic flux in one links with the other and causes an emf in it. Units of measure are henrys. Two coils have mutual inductance of 1 henry when a current change of 1 ampere per second in one coil induces 1 volt in the other.

NAND gate: A NOT AND gate. Provides a logic high output (1) if any input is logic low.

natural magnet: Magnets found in natural state in form of mineral called magnetite. Also called lodestone.

negative feedback: Feedback that is 180° out of phase with input signal so it subtracts from input. Also called degeneration.

negative resistance: Property exhibited when an increase in voltage produces a decrease, rather than an increase, in current.

network: Two or more interrelated circuits. An interconnected system of transmission lines used for joining together computers or computer peripherals.

neutron: Electrically neutral particle contained in the nucleus of an atom.

nibble: In computers, a string of 4 bits, or 1/2 byte.

NICAD cell: See nickel-cadmium cell.

nickel-cadmium cell: Rechargeable, alkaline cell with paste electrolyte. Hermetically sealed. Used in aircraft. Also called NICAD cell.

no-load current: See exciting current.

no-load voltage: Terminal voltage of battery or power supply when not connected to a load.

nonlinear: Pertaining to a response that is other than directly or inversely proportional to a given variable.

nonlinear device: Electronic device having a current-voltage relationship that is not given by a straight line. A device in which a change in input does not produce a proportional change in output.

nonvolatile memory: Computer memory, the contents of which are not lost when power is removed. ROM is nonvolatile memory.

NOR gate: A NOT OR gate. Provides a logic high output (1) only if all inputs are logic low.

NOT gate: A logic gate that inverts the input signal so that the output is always opposite of the input. Also called an inverter.

n-type conductivity: Conduction by electrons in n-type semiconductor.

n-type semiconductor: Semiconductor that uses holes as majority carrier.

nucleonics: The branch of physical science dealing with all phenomena of the atomic nucleus, including the release of energy from it.

nucleus: Core of the atom. Contains protons and neutrons.

null indicator: A meter designed to indicate the balance of a circuit; indicator for no current or no voltage in a circuit.

number crunching: The quick processing of large quantities of numbers.

octal number system: A number system, having a base of 8. Digits are 0 through 7.

oersted (Oe): Cgs unit of magnetic field intensity.

ohm (Ω): Unit of measurement of resistance.

ohmmeter: Meter used to measure resistance. Meter leads are placed in parallel with resistance to be measured.

Ohm's law: Mathematical relationship between current, voltage, and resistance stating that when a voltage is applied to a metal conductor, the current moving through the conductor is proportional to the applied voltage. Discovered by Georg Simon Ohm.

Ohm's law for magnetic circuits: Mathematical relationship between magnetic flux, magnetomotive force, and reluctance stating that when an mmf is applied to a magnetic material, the magnetic flux through the material is proportional to the mmf.

ohms per volt (Ω/V): Units of measurement of sensitivity.

one-shot multivibrator: See monostable multivibrator.

open circuit: A circuit with an incomplete path for current.

operating system software: A set of programs and routines that guides a computer in the performance of its tasks, assists the programs with certain supporting functions, and increases the usefulness of the computer's hardware.

operational amplifier: High-gain amplifier with negative feedback.

optical disk: A type of disk storage device containing submicroscopic pits that are sensed by a laser beam.

OR gate: A logic gate that produces an output whenever any one (or more) of its inputs is energized.

orbital electron: See planetary electron.

oscillator: An electronic device that converts direct current to alternating current of specified frequencies and amplitudes.

oscilloscope: Test instrument using cathode-ray tube, permitting observation of signals.

output: The current, voltage, power, or driving force delivered by a circuit or device. Information transferred from a central computer unit to output device or external storage.

output: The information sent out of a computer. The process of extracting information from a computer.

output device: Peripheral computer unit. The target for computer output. Examples are CRTs and printers.

overmodulation: Condition when modulating wave exceeds amplitude of continuous carrier wave, resulting in distortion.

parallel: Said of two or more circuit elements so connected that the total current flow is divided between them. Also called shunt.

parallel circuit: Circuit that contains two or more paths for electron flow supplied by a common voltage source.

parallel processing: A technique that involves the execution of more than one instruction or part of an instruction at the same time.

parallel resonant circuit: Resonant circuit of an inductor and capacitor in parallel. Current in capacitive branch is 180° out of phase with inductive current. Current is minimum; impedance is maximum. Also called a tank circuit.

parameter: A constant that is constant under a given set of conditions but may be different under other conditions.

parasitic oscillation: Oscillation in a circuit resulting from circuit components or condition and occurring at frequencies other than that desired.

passive device: Electronic component that does not have the ability to change its output response based on the external input signal. Some common passive devices are resistors, capacitors, inductors, and diodes.

peak inverse voltage (PIV): The maximum voltage that can be applied to a diode in a reverse direction without destruction.

peak value: The maximum instantaneous value of a varying current, voltage, or power. For a sine wave, it is 1.414 times the effective value. Also called maximum value.

peak-to-peak value: Measured value of periodic waveform from positive to negative maximum, equal to twice peak value.

pentavalent: Semiconductor donor impurity, having five valence electrons.

percent modulation: Maximum deviation from normal carrier value as result of modulation expressed as a percentage.

percent ripple: Ratio of rms value of ripple voltage to average value of total output voltage expressed as a percentage.

period: Time for one complete cycle, equal to the reciprocal of frequency.

periodic wave: Wave repeating itself regularly in time and in form.

peripheral equipment: Any computer hardware aside from the central computer unit. Includes input, output, and auxiliary storage devices. Also called input/output, or I/O, units.

permanent magnet (PM): Bars of steel and other substances that have been permanently magnetized.

permeability (μ): The measure of the ease with which magnetic flux may be established in a substance, given by the ratio of magnetic flux density to magnetic field intensity.

permeability of free space (μ_0): Permeability of air or vacuum, equal to $4\pi \times 10^{-7}$ W/A·m.

permittivity (ϵ): A measure of the ability of a dielectric to permit the establishment of an electrostatic field between two plates.

permittivity of free space (ϵ_0): Permittivity of air or vacuum, equal to 8.85×10^{-12} F/m.

personal computer: Desktop or laptop computer. A type of microcomputer.

phase: Relationship between two phasors in respect to angular displacement.

phase angle: The angle between two vectors representing periodic functions that have the same frequency.

phase distortion: Distortion resulting from shift of phase of some signal frequencies.

phase inverter: See inverter.

phase splitter: Amplifier that produces two waves that have opposite polarities of each other.

phasor: Any quantity that has both magnitude and direction and that varies in time. Also called rotating vector.

phasor diagram: An arrangement of phasors showing the relationships between alternating quantities having the same frequency. Similar to a vector diagram.

phonon: Quantum of an acoustic mode of thermal vibration in a crystal lattice.

photocell: See photoelectric cell.

photoconductive cell: Semiconductor device whose resistance varies inversely as the light intensity. Also called photoresistive cell, or photoresistor.

photodiode: A photoconductive cell comprised of a pn junction diode that conducts upon exposure to light energy.

photoelectric cell: A device that converts light energy to electrical energy. A photovoltaic, photoconductive, or photoemissive cell. Also called photocell.

photoelectric effect: The property of certain substances to emit electrons when subjected to light.

photoelectric emission: Emission of electrons as a result of light striking the surface of certain materials.

photoelectrons: Electrons emitted as a result of light.

photon: Quantum of electromagnetic wave energy. LEDs emit photons, which give off light.

photoresistive cell: See photoconductive cell.

photoresistor: See photoconductive cell.

photosensitive: Characteristic of a material that emits electrons from its surface when energized by light.

phototube: Vacuum tube employing photosensitive material as its emitter or cathode.

photovoltaic cell: A device that generates a potential difference when illuminated with a strong light.

pi filter: Filter consisting of two capacitors and an inductor connected in a π configuration.

piezoelectric effect: Property of certain crystals of changing shape when impressed with an emf, or vice versa.

pitch: Property of sound determined by its frequency.

pixels: The smallest part of an electronically coded image; the dots on a video screen.

planetary electron: An electron moving in an orbit around the nucleus of an atom, as opposed to a free electron. Also called orbital electron.

plate: Anode of a vacuum tube. Element in a tube that attracts electrons.

plug-in circuit: A total or partial circuit, usually a printed circuit board, that can be plugged into a piece of equipment. It may be rapidly removed or replaced during service.

PM speaker: A speaker employing a permanent magnet as its field.

pn junction: The line of separation between n- and p-type semiconductor materials.

point-contact diode: Diode formed by small semiconductor crystal and catwhisker. Also called crystal rectifier, or crystal diode.

polarity: Property of a system that has two points of opposite characteristics, such as two opposite charges or two opposite magnetic poles.

polarization: Defect in voltaic cell caused by hydrogen bubbles surrounding positive electrode and effectively insulating it from chemical reaction.

pole: North or south end of a magnet or electromagnet. Output terminal of a switch.

positive feedback: Feedback that is in phase with the input signal so it adds to the input. Also called regeneration.

potential difference: Difference in potential between two points. The force moving electrons along in a conductor. Potential difference is physically measurable; electric potential is not. Units of measure are volts. Also called voltage.

potentiometer: A three-terminal adjustable resistance element used as a voltage divider. Called a pot, for short.

power amplifier: Final stage of an amplifier intended for driving speakers or other transducers. Designed to give substantial power output as distinct from providing voltage gain. Also called a large-signal amplifier.

power factor (cos θ): In a reactive circuit, the ratio of power used to power supplied, or true power to apparent power. Tells how much of power supplied is actually used.

power (P): Rate of doing work. Types include true power, apparent power, and reactive power. Units of measure are watts, volt-amperes, volt-amperes reactive.

power supply: Electronic circuit designed to provide various ac and dc voltages for equipment operation. Circuit may include transformers, rectifiers, filters, and regulators.

power transistor: Transistors designed to deliver a specified output power level.

preamplifier: Sensitive, low-level amplifier with sufficient output to drive a standard amplifier.

prefix: A multiplier preceding a unit of measurement, making the unit larger or smaller.

primary cell: Cell that cannot be recharged.

primary winding: Coil of a transformer that receives energy from an ac source.

printed circuit: A circuit made from very thin layers of conductive material, such as copper, that is stenciled, printed, or otherwise deposited on an insulated base. Component parts may also be printed or actual components soldered in place.

probe: A pointed, metal end of a test lead, designed to contact specific points to be measured in a circuit.

program: Step-by-step sequence of instructions given to a computer.

programmable read-only memory (PROM): ROM that is irreversibly programmed by the user after the IC is made.

proton: Positively charged particle contained in the nucleus of an atom.

p-type conductivity: Conduction by holes in p-type semiconductor.

p-type semiconductor: Semiconductor that uses electrons as majority carrier.

pulse: Sudden rise and fall of an electrical signal.

pulse code: A train of pulses used to transmit information.

pulse code modulation (PCM): Pulse modulation in which the signal is sampled periodically, and each sample is quantized and transmitted as a digital binary code.

pulse decay time: The time required for a pulse to drop from 90% to 10% of its peak value.

pulse droop: Distortion characterized by a slight decrease in amplitude of an otherwise flat-topped rectangular pulse.

pulse duration: The time interval between the first and last instants at which the instantaneous value reaches a stated fraction of the peak pulse value.

pulse modulation: Use of pulse code to modulate a transmitter. Modulation may involve changes of pulse amplitude, position, number, or duration.

pulse repetition frequency (PRF): See pulse repetition rate.

pulse repetition period: The reciprocal of pulse repetition frequency.

pulse repetition rate (PRR): The average number of times per second that a pulse is transmitted. Also called pulse repetition frequency.

pulse rise time: The time required for a pulse to rise from 10% to 90% of its peak value.

pulse spike: An unwanted pulse of relatively short duration superimposed on a main pulse.

pulse sync: A pulse sent by a TV transmitter to synchronize the scanning of the receiver with the transmitter. A pulse used to maintain a predetermined speed and/or phase relation.

pulse train: A group or sequence of pulses of similar characteristics.

pure power: See true power.

push-pull amplifier: An amplifier consisting of two similar transistors, connected for push-pull operation.

push-pull operation: Operation of a class-B amplifier, in which two transistors are connected in parallel to allow an output current for both positive and negative halves of the input signal.

push-pull oscillator: A balanced oscillator employing two similar transistors in phase opposition.

quality factor: A measure of the relationship between stored energy and rate of dissipation in certain electric elements, structures, or materials. In an inductor or capacitor, the ratio of reactance to effective series resistance at a given frequency. A measure of the sharpness of resonance or frequency selectivity of a mechanical or electrical system. Also called Q-factor, or figure of merit.

quantum: Discrete quantities of any physical property, such as momentum, energy, mass, etc. For example, photons and phonons are discrete packages (quantums) of energy. A definite amount of energy acquired or given off that results in an electron moving to a higher or a lower energy level.

quiescent: At rest. In particular, the condition of a circuit with no ac signal applied, strictly the dc level.

radian: An angle, the measure of which is defined by two radii of a circle and an arc joining them, all of the same length. The number of radians in any circle is always 2π (6.28).

radian frequency: See angular velocity.

radiation detector: A device used to detect the presence and level of radiation.

radio detector: Type of FM detector.

radio frequency choke (RFC): A choke coil having high impedance to RF currents.

radio frequency (RF): The range of frequencies from 20 kHz to 100 GHz.

radio spectrum: Division of electromagnetic spectrum used for radio.

radio wave: A complex electrostatic and electromagnetic field radiated from a transmitter antenna.

radio-frequency amplifier: Amplifier used to amplify radio frequencies.

random access: Pertaining to storage media where data is accessed at random as opposed to sequentially. Computer memory is random access. Magnetic disks are random access.

random-access memory (RAM): Active, read/write memory of a computer, where data can be both read and readily overwritten. Volatile memory; provides temporary storage of data while the computer is operating.

ratio detector: A detector for FM signals. It is based upon the ratio of output voltages of two diodes, which is detected as the intelligence in the signal.

RC coupling: Connecting amplifier stages by a combination of resistive and capacitive elements.

RC oscillator: An oscillator depending on charge and discharge of capacitor in series with resistance. The frequency is determined by RC elements.

reactance: Opposition to ac as a result of inductance or capacitance.

reactive power (Q): In a reactive circuit, power absorbed by a reactive element. It is apparent power times the power factor. Units of measure are volt-amperes reactive.

read time: Complete time interval in reading from a storage device.

reading: In a computer, the extracting of information from storage by the CPU.

read-only memory (ROM): An unalterable computer memory from which data may be extracted but not stored. Nonvolatile memory.

read/write head: Mechanism used to sense magnetic fields on magnetic storage media and record information by altering the magnetic fields.

rectifier: A device that converts ac into pulsating dc. A diode is a rectifier.

reed relay: Relay that has switch contacts consisting of reeds, which are hermetically sealed in a glass capsule filled with an inert gas.

regeneration: See positive feedback.

register: In a computer, short-term storage usually having a capacity of one computer word.

regulation: Voltage change that takes place in output of generator or power supply when load is changed. Also called voltage regulation.

reject circuit: Parallel tuned circuit at resonance. Rejects signals at resonant frequency.

relative permeability (μ_r): Permeability of a substance relative to air, given by the ratio of the permeability of the substance to that of free space.

relative permittivity (ϵ_r): Permittivity of a substance relative to free space, given by the ratio of the capacitance of a capacitor with a given dielectric to the capacitance of an otherwise identical capacitor except having air for its dielectric. Also called dielectric constant.

relaxation oscillator: Nonsinusoidal oscillator whose frequency depends upon the time required to charge and discharge a capacitor or an inductor through a resistor.

relay: An electromagnetic switch.

reluctance: Resistance to flow of magnetic lines of force.

repulsion motor: A single-phase motor in which the stator winding is connected to the source of power and the rotor winding to the commutator. Brushes on the commutator are short-circuited and are placed so that the magnetic axis of the rotor winding is inclined to that of the stator winding. This type of motor has a varying speed characteristic.

repulsion-induction motor: A constant or variable-speed repulsion motor with a squirrel-cage winding in the rotor, in addition to the regular winding.

repulsion-start induction motor: See repulsion-start induction-run motor.

repulsion-start induction-run motor: A single-phase motor that has the same windings as a repulsion motor but operates at a constant speed. The rotor winding is short-circuited to give the equivalent of a squirrel-cage winding. It starts as a repulsion motor, but operates as an induction motor with constant-speed characteristics. Also called repulsion-start induction motor.

residual magnetism: Magnetism remaining in a magnetic material after the magnetizing force is removed.

resistance (R): Quality of an electric circuit that opposes flow of current through it. Units of measure are ohms.

resonance: Condition in a circuit when, for an applied alternating voltage of a given frequency, the inductive and capacitive reactances are equal.

resonant circuit: A circuit containing both inductance and capacitance in series or in parallel, and which exhibits the condition of resonance. Also called tuned circuit.

resonant frequency (f_r): Frequency at which a tuned circuit is in resonance.

resultant: A vector representing the total effect produced by two or more vectors. In a phasor or impedance diagram, it is the hypotenuse.

retentivity: Ability of a material to retain magnetism after magnetizing force is removed.

retrace: See flyback.

reverse bias: A bias voltage applied to a semiconductor junction with polarity such that little or no current flows. It is the opposite of forward bias. Reverse bias across a pn junction would be negative source connected to p-type semiconductor and positive source to n-type semiconductor material. A reverse bias increases the width of the depletion region and increases the junction potential.

reverse voltage: Value of the reverse bias voltage applied to a diode.

rheostat: A two-terminal adjustable resistance element used to vary load current.

right-hand rule: Gives correct directional relationship for conductors in motor armatures based on electron flow.

ripple voltage: The ac component of dc output of power supply due to insufficient filtering.

rms value: Abbreviation for root-mean-square value.

root-mean-square value: See effective value.

rotating field: The magnetic field in the stator of induction motors, which appears to rotate around the stator from pole to pole.

rotating vector: See phasor.

rotor: The rotating member of a motor or generator. The movable plates of a variable capacitor.

RTL logic: Resistor-transistor logic. A transistor logic family using resistors and transistors for its logic circuits.

salient pole: A pole on which is mounted a field coil of a generator or motor.

satellite: Body moving in orbit around a planet. An electronic device placed in orbit around the earth to receive and transmit information via radiation of electromagnetic waves.

saturation: The operating condition of a transistor when an increase in base current produces no further increase in collector current. The state of magnetism beyond which a further increase in magnetizing force produces no further increase in flux density of the material. A circuit condition whereby an increase in the input signal no longer produces a change in the output.

sawtooth generator: A generator whose output voltage has a sawtooth waveform. A sweep oscillator in an oscilloscope.

sawtooth wave: A periodic wave shaped much like the teeth of a saw.

schematic: Diagram of electronic circuit showing electrical connections and identification of various components.

screen grid: Second grid in electron tube between grid and plate, used to reduce interelectrode capacitance.

second harmonic distortion: Distortion of wave by addition of its second harmonic.

secondary cell: Cell that can be recharged by reversing chemical action with electric current. Also called a storage cell.

secondary emission: Emission caused by impact of other electrons striking surface.

secondary emission: Emission of electrons as a result of electrons striking plate of electron tube.

secondary winding: Coil that receives energy from primary winding by mutual induction and delivers energy to the load.

selectivity: Relative ability of a receiver (or circuit) to select a desired signal while rejecting all others.

self-excited generator: Generator whose field windings are excited by current from its output.

self-inductance: The property that determines the amount of counter emf produced in a circuit by a changing magnetic field produced by a changing current.

semiconductor: Conductor with resistivity somewhere in range between conductors and insulators.

semiconductor diode: Solid-state device usually designed to permit electron flow in one direction and block flow in the other direction.

sensitivity: Ability of a circuit to respond to small signal voltages. Indication of the loading effect of a meter, expressed in units of ohms per volt.

separately excited generator: Generator whose field windings are excited by a separate dc source.

sequential access: See serial access.

serial access: Pertaining to storage media where data is accessed sequentially as opposed to randomly. Magnetic tapes are serial access. Also called sequential access.

series: Said of circuit elements so connected that the same current passes through each.

series circuit: Circuit that contains only one possible path for electron flow supplied by a common voltage source.

series generator: Generator having field windings in series with the armature and load

series motor: A dc motor that has field coils connected in series with the armature circuit.

series resonant circuit: Resonant circuit of an inductor and capacitor in series. Net reactance is zero: impedance is minimum; current is maximum.

series-parallel circuit: See combination circuit.

shaded-pole motor: Single-phase induction motor using a salient-pole stator, each of which is split to accommodate a short-circuiting ring called a shading coil. This coil produces a sweeping movement of field across pole face for starting.

shading coil: A copper ring set into part of the pole piece of a small ac motor to produce a split-phase effect for starting purposes.

shielding: Enclosure around components in a circuit to minimize effects of stray magnetic and radio frequency fields.

shift register: A special type of register in which data can be moved (shifted) to the left or right.

short: See short circuit.

short circuit: Faulty connection of low resistance between two points of a circuit, resulting in excessive current. Also called a short.

shunt: Parallel resistor used to conduct excess current around a meter moving coil to increase range of meter. See parallel.

shunt generator: Generator having field windings in parallel with armature and load.

shunt motor: A dc motor that has field coils connected in parallel with the armature circuit.

sideband: Frequency above or below carrier frequency as a result of modulation.

siemen: Unit of measurement of conductance.

signal: A time-varying electric or electromagnetic waveform.

silicon: Tetravalent, nonmetallic element used in semiconductors.

silicon controlled rectifier (SCR): A three-terminal semiconductor rectifier that can be controlled. It acts as an open circuit until switched into a conducting state by a gate signal.

silver oxide cell: Compact primary cell with a button-type construction; used in watches.

sine wave: An ac waveform of a single frequency whose displacement is the sine of an angle proportional to time or distance.

single-chip microcomputer: Circuitry for an entire computer on a single chip. Holds CPU, memory, and input/output interface circuits. Also called a computer-on-a-chip.

single-ended amplifier: An amplifier whose final power stage is a single transistor.

single-phase generator: Generator that produces a single phase of alternating current.

single-phase motor: Motor that operates on single-phase alternating current.

sinusoidal: Said of an alternating quantity, the value of which plotted against time produces a sine wave.

slip ring: Metal ring for making continuous electrical contact between stationary and rotating parts of an alternator. Stationary brushes sliding on ring provide connection for external circuit.

small-scale integration (SSI): An integrated circuit with up to 12 gates per chip.

software: Stored program of a computer.

solenoid: Coil of wire, without a core, which possesses the characteristics of a magnet when carrying electric current. An electromagnet with a movable core.

solid state: Pertaining to circuits and components using semiconductors.

space-charge region: See depletion region.

spacistor: A four-element semiconductor, similar to a transistor. It uses a space-charge region. Its main advantage is its adaptability to ultra-high frequencies.

speaker voice coil: See voice coil.

specific gravity: Weight of a liquid relative to an equal volume of water. The specific gravity of water is 1.

split-phase motor: Single-phase induction motor that develops starting torque by phase displacement between field windings.

square mil: Actual cross-sectional area in square mils of a wire whose diameter is given in mils. There is 0.7854 square mils in 1 circular mil.

squirrel-cage rotor: Rotor used in an induction motor made of bars placed in slots of rotor core and all joined together at ends.

stability (S): The ability to stay on a given frequency or in a given state without undesired variation.

stage: Each amplifier in a cascaded arrangement.

static characteristics: The relationship, usually represented by a graph, between a pair of variables with all other conditions held constant.

static charge: Negative or positive charge on a body.

static electricity: Electricity at rest as opposed to electric current.

stator: The stationary member of a motor or generator that contains the parts for the magnetic circuit and associated windings. The stationary plates of a variable capacitor.

steady state: Fixed, nonvarying condition.

step-down transformer: Transformer with turns ratio greater than one. The output voltage is less than the input voltage.

step-up transformer: Transformer with turns ratio less than one. The output voltage is greater than the input voltage.

storage cell: See secondary cell.

stylus: Phonograph needle made of diamond or steel, for example, which follows grooves in a record.

subharmonic: A frequency below harmonic, usually a fractional part of the fundamental frequency.

subsonic frequency: See infrasonic frequency.

super high frequency (SHF): Frequency band from 3000 MHz−30,000 MHz.

superconductor: A conductor made of a material having zero resistivity at absolute zero.

superheterodyne receiver: Radio receiver in which incoming signal is converted to fixed intermediate frequency before detecting audio signal component.

supersonic frequency: See ultrasonic frequency.

suppressor grid: Third grid in electron tube, between screen grid and plate, used to repel or suppress secondary electrons from plate.

surface-alloy transistor: A silicon junction transistor, in which aluminum electrodes are deposited in shallow pits etched on both sides of a thin silicon crystal, forming p regions.

surface-barrier transistor: A transistor so constructed that the interfaces performing the collection and emission of carriers are located at the surface of the semiconductor crystal.

sweep circuit: A sweep oscillator, sweep amplifier, or any other stage used to produce the deflection voltage or current for a cathode-ray tube.

sweep generator: See sweep oscillator.

sweep oscillator: An oscillator used to develop a sawtooth voltage that can be amplified to deflect the electron beam of a cathode-ray tube; used in oscilloscopes. Also called sweep generator.

switch: Device for directing or controlling current flow in a circuit.

sync pulse: Abbreviation for synchronization pulse, used for triggering an oscillator or circuit.

synchronous: Having the same period or frequency.

synchronous motor: An induction motor that runs at synchronous speed using field magnets excited with direct current.

synchronous speed: The speed of rotation of the magnetic field in an induction motor.

system: An assembly of component parts linked together into an organized whole to perform some regulated action.

tank circuit: See parallel resonant circuit.

tap: A connection made to a coil at a point other than at the ends of the coil.

tape drive: The mechanism that physically moves a tape past a stationary head. Also called tape transport.

tape transport: See tape drive.

telemetering: See telemetry.

telemetry: The transmission of measurements recorded on instruments by means of wires, radio waves, or other means to a remote point. Also called telemetering.

television: A system for transmitting and receiving, by radio or over wires, visual images.

temperature coefficient of resistance (α): The rate of change of resistance of a conductor with respect to temperature. If alpha is positive, resistance varies directly with temperature. If alpha is negative, resistance varies inversely with temperature.

tetravalent: An element with four valence electrons.

tetrode transistor: A transistor with four elements, usually using either two emitters or two bases.

thermal runaway: In a transistor, regenerative increase in collector current and junction temperature.

thermionic emission: Process where heat produces energy for release of electrons from surface of emitter.

thermistor: A nonlinear semiconductor resistor, having a high negative temperature coefficient of resistance.

three-phase current: A current delivered through three wires. Each wire serves as the return for the other two. A phase difference of 120° exists between each of the three current components.

thyristor: A semiconductor device of three or more junctions that acts as an open or closed switch. Diacs, triacs, and SCRs are thyristors.

tickler coil: Coil used to feed energy back from output to input circuit.

time constant: In the RL series circuit, the time required for the current to change by 63.2%. In the RC series circuit, the time required for the voltage to change by 63.2%.

tone control: Adjustable filter network to emphasize either high or low frequencies in output of audio amplifier.

toroid: A doughnut-shaped solid.

toroidal coil: A coil wound in the shape of a toroid.

transceiver: A combined transmitter and receiver.

transducer: Device by which one form of energy may be converted to another form. Examples include generators, motors, microphones, and loudspeakers.

transfer characteristic: Relation between input and output characteristics of a device.

transformer: Device that transfers energy from one circuit to another by mutual induction.

transformer coupling: Connecting amplifier stages by a transformer resulting in loss of dc component. Used in high frequency applications.

transient response: Response to momentary signal or force.

transistor: An active semiconductor device used as an amplifier, detector, or switch.

transistor socket: A small device in which the leads of a transistor are placed to provide ease in connection to circuit and to permit replacement of transistor.

transition layer: See depletion region.

transmission line: Wire or wires used to conduct electrical energy.

transmitter: Device for converting intelligence into electrical impulses for transmission through lines or through space from radiating antenna.

triac: A three-terminal solid-state device that acts, in effect, like a bidirectional silicon controlled rectifier.

trivalent: Semiconductor acceptor impurity, having three valence electrons.

true power (*P*): The power that is dissipated by a circuit, equal to the product of current and voltage. It is resistive power. Units of measure are watts. Also called pure power, or active power.

truth table: A representation of Boolean algebra, given by a table of input and resulting output conditions; used in digital electronics.

TTL logic: Transistor-transistor logic. A transistor logic family using transistors for its logic circuits.

tune: The process of bringing a circuit into resonance by adjusting one or more variable components.

tuned amplifier: Amplifier employing tuned circuit for input and/or output coupling.

tuned circuit: See resonant circuit.

turns ratio: Ratio of number of turns of primary winding to number of turns of secondary winding.

tweeter: A speaker designed to handle the higher audio frequencies.

ultra high frequency (UHF): Frequency band from 300 MHz–3000 MHz.

ultrasonic frequency: A frequency above the audio-frequency range. Also called supersonic frequency.

unity coupling: Perfect magnetic coupling between two coils, such that all the magnetic flux of the primary passes through the entire secondary. Coupling coefficient is 1.

universal motor: A series motor that also operates on ac. Fractional horsepower ac/dc motor. Also called ac series motor.

universal time constant chart: A graph with curves representing growth and decay of voltages and currents in RL and RC circuits.

upper sideband: Frequency equal to sum of carrier and modulating frequencies.

valence electron: Electron in the *s* and *p* subshells in the valence shell of an atom.

valence shell: The outer orbital shell of an atom.

valid logic high: The operating range for an IC gate to be turned on, representing a 1.

valid logic low: The operating range for an IC gate to be turned off, representing a 0.

varistor: A nonlinear semiconductor resistor, the resistance of which markedly varies as the inverse of voltage.

vector: A quantity that has both magnitude and direction; represented graphically as an arrow.

vector diagram: Diagram comprised of individual vector components and their combined resultant.

very high frequency (VHF): Frequency band from 30 MHz–300 MHz.

very low frequency (VLF): Frequency band from 10 kHz–30 kHz.

video cassette recorder (VCR): An electronic machine used to record and play video media, such as television programs or movies.

voice coil: Small coil attached to the cone of a dynamic speaker. Coil and cone move back and forth by electric impulses of an applied signal. Also called speaker voice coil.

volatile memory: Computer memory, the contents of which are lost when power is removed. RAM is volatile memory.

volt (V): Unit of measure of electric potential and potential difference, where 1 volt equals 1 joule per coulomb.

voltage: See potential difference.

voltage amplifier: Amplifier designed to increase voltage capable of delivering only small amounts of current.

voltage divider: Tapped resistor, series resistors, or potentiometer across source voltage to produce multiple voltages.

voltage doubler: Rectifier circuit that produces a dc output that is about twice the peak value of the applied ac voltage.

voltage drop: Voltage measured across two leads, or terminals, of a device, such as a resistor.

voltage multiplier: Rectifier circuit that produces output voltage at a multiple greater than input voltage–possibly double, triple, or quadruple.

voltage regulation: See regulation.

voltaic cell: Cell produced by suspending two dissimilar elements in acid solution. Electromotive force is developed by chemical action.

volt-ampere reactive (VAR): Unit of measurement of reactive power.

volt-ampere (VA): Unit of measurement of apparent power.

voltmeter: Meter used to measure voltage. Meter leads are placed in parallel with voltage to be measured.

volt-ohm-milliammeter (VOM): See multimeter.

von Neumann architecture: See computer architecture.

watt (W): Unit of measurement of true power.

watt-hour meter: Meter used by electric utility companies to measure industrial or residential energy consumed. Meter reads in kilowatt-hours.

wattmeter: Meter used to measure power in watts.

wave: A disturbance in a medium that is a function of time and space or both, such as audio or radio wave. Energy may be transmitted by waves.

waveform: The shape of a wave derived from plotting its instantaneous values during a cycle against time.

wavelength (λ): The length in space occupied by one cycle of a periodic wave.

wheatstone bridge: A bridge circuit for determining the resistance value of an unknown component by comparison to one of known value.

woofer: A speaker designed to handle the lower audio frequencies.

word: The fundamental unit of storage capacity for a digital computer, almost always considered to be more than 8 bits in length. The number of bits that a computer can store at a single memory location, treated by the computer as a unit.

work: The magnitude of force times the distance the force is applied. Units of measure are foot-pounds and joules.

write once, read many (WORM): Type of optical storage device.

write protect: Disk feature used to prohibit writing to it. Guards against the mistaken erasure of important data.

write time: Complete time interval in writing to a storage device.

writing: In a computer, the sending of information to storage by the CPU.

x-axis: The horizontal axis of a graph.

XNOR gate: An Exclusive NOR gate. Provides a logic high output (1) only if all inputs are logic high or all inputs are logic low.

XOR gate: An Exclusive OR gate. Provides a logic high output (1) whenever any, but not all, inputs are logic high.

y-axis: The vertical axis of a graph. Axis drawn perpendicular to two parallel faces of a quartz crystal.

Y-cut crystal: A crystal cut in such a way that its major flat surfaces are perpendicular to the y-axis of the original quartz crystal.

z-axis: In a three-dimensional rectangular coordinate system, the axis perpendicular to the x- and y-axes. Optical axis of a crystal.

zener diode: A pn junction diode that makes use of the breakdown properties of a pn junction. The diode is designed to conduct in the reverse direction when its value of breakdown voltage is reached. Beyond this point, the diode will maintain a relatively constant voltage despite variations in current.

zener voltage: The reverse voltage at which the breakdown occurs in a zener diode.

INDEX